W0261897

FORTSCHRITTE
DER CHEMIE ORGANISCHER
NATURSTOFFE

PROGRESS IN THE CHEMISTRY
OF ORGANIC NATURAL PRODUCTS

PROGRÈS DANS LA CHIMIE
DES SUBSTANCES ORGANIQUES
NATURELLES

HERAUSGEGEBEN VON   EDITED BY   RÉDIGÉ PAR

L. ZECHMEISTER
CALIFORNIA INSTITUTE OF TECHNOLOGY, PASADENA

EINUNDZWANZIGSTER BAND
TWENTY-FIRST VOLUME   VINGT-ET-UNIÈME VOLUME

VERFASSER   AUTHORS   AUTEURS

R. BANGERT · J. BONNER · H. BROCKMANN · L. CROMBIE
L. JAENICKE · C. KUTZBACH · A. D. MEBANE · H. MUXFELDT
W. OROSHNIK

MIT 14 ABBILDUNGEN   WITH 14 FIGURES   AVEC 14 ILLUSTRATIONS

WIEN · SPRINGER-VERLAG · 1963

LIBRARY OF CONGRESS CATALOG CARD NUMBER AC 39-1015

ISBN-13: 978-3-7091-7150-9   e-ISBN-13: 978-3-7091-7149-3
DOI: 10.1007/978-3-7091-7149-3

Inhaltsverzeichnis.
Contents. — Table des matières.

The Biosynthesis of Rubber. By JAMES BONNER, California Institute of Technologie, Pasadena, California 1

   I. Distribution of Rubber 1

   II. Latex 2
      Structure and Configuration 2

   III. Biogenesis of the Monomer 4

   IV. Polymerization 10

   V. Further Problems 11

   References 13

The Polyene Antifungal Antibiotics. By W. OROSHNIK, Central Research Laboratory *Shulton*, Clifton, New Jersey, and A. D. MEBANE, Ortho Research Foundation, Raritan, New Jersey 17

   I. Introduction 18

   II. Ultraviolet Spectra 19
      1. General Observations 19
      2. The Tetraenes 23
      3. The Pentaenes 24
      4. The Methylpentaenes 24
      5. The Hexaenes 24
      6. The Heptaenes 24

   III. Structural Elucidation 26
      1. General Features 26
      2. Mycosamine 28
      3. Retro-Aldol Cleavage 30
      4. Fungichromin (Lagosin) 32
      5. Filipin 37
      6. Other Methylpentaenes 39
      7. Pimaricin 40
      8. Nystatin and Other Tetraenes 43
      9. Pentaenes and Hexaenes 45
      10. Trichomycin and Other Heptaenes 46

   IV. Biogenetic Relationships 51

   V. Tables 56
      1. Typical Tetraenes: Spectral Data 56
      2. Typical Pentaenes: Spectral Data 56
      3. Typical Methylpentaenes: Spectral Data 57
      4. Typical Hexaenes: Spectral Data 57
      5. Typical Heptaenes: Spectral Data 57

6. Tetraenes: Physical and Chemical Properties 58
7. Pentaenes: Physical and Chemical Properties 62
8. Methylpentaenes: Physical and Chemical Properties 64
9. Hexaenes: Physical and Chemical Properties 66
10. Heptaenes: Physical and Chemical Properties 66

References 72

Die Chemie der Tetracycline. Von H. MUXFELDT und R. BANGERT, Department of Chemistry, The University of Wisconsin, Madison, Wisconsin 80

I. Einleitung 80

II. Konstitutionsaufklärung 82

1. Terramycin 82
Alkalischer Abbau 83
Saurer Abbau 87
Reduktiver Abbau 90

2. Aureomycin 91

3. 6-Desmethyl-tetracycline 96

4. 5a,11a-Dehydro-7-chlor-tetracyclin 96

5. 2-Acetyl-2-descarboxamido-tetracycline 97

III. Weitere chemische Eigenschaften 98

1. Reaktionen am $C_{(2)}$ 99

2. Reaktionen am $C_{(4)}$ 100

3. Reaktionen am $C_{(6)}$ 102

4. Reaktionen am $C_{(11a)}$ und $C_{(12a)}$ 108

IV. Biogenese der Tetracycline 113

V. Versuche zur Synthese von Tetracyclinen 115

Literaturverzeichnis 116

Anthracyclinone und Anthracycline (Rhodomycinone, Pyrromycinone und ihre Glykoside). Von HANS BROCKMANN, Organisch-chemisches Institut der Universität Göttingen 121

I. Einleitung 122

II. Isolierung der Anthracyclinone und Anthracycline 123

1. Gewinnung der ε-Pyrromycinon-glykoside und Pyrromycinone 124
Cinerubin A und B 124
Pyrromycin und Pyrromycinone 124

2. Gewinnung der Rhodomycinone, Iso-rhodomycinone und ihrer Glykoside 125
Trennung von Rhodomycinon/Iso-rhodomycinon-Gemischen 126
Trennung von Rhodomycinen und Iso-rhodomycinen 127

III. Die Anthracyclinone 127

1. Vorbemerkungen zur Struktur der Anthracyclinone 127

2. Zur Konstitutionsermittlung der Anthracyclinone 130
Die Aufklärung des Chromophors 131
Die Anellierung des alicyclischen Ringes 133

Die Substituenten an Ring *A* 134
Schreibweise und Bezifferung der Anthracyclinon-Formeln 137

3. Konstitution der Anthracyclinone 137
   A. Iso-rhodomycinone 137
      ε-Iso-rhodomycinon 137
      ζ-Iso-rhodomycinon 139
      β-Iso-rhodomycinon 140
   B. Rhodomycinone 140
      ε-Rhodomycinon 140
      ζ-Rhodomycinon 141
      β-Rhodomycinon 141
      γ-Rhodomycinon 145
      δ-Rhodomycinon 145
   C. Pyrromycinone 146
      η-Pyrromycinon 146
      ε-Pyrromycinon 151
      ζ-Pyrromycinon 151
   D. Aklavinone 154
      Aklavinon 154
      7-Desoxy-aklavinon 154

4. Die KMR-Spektren der Anthracyclinone 155

5. Zur Stereochemie der Anthracyclinone 160

6. Zur Biogenese der Anthracyclinone 163

IV. Die Anthracycline 170
  1. Die Zucker der Anthracycline 171
     Rhodosamin 171
     2-Desoxy-*L*-fucose 173
     Rhodinose 173
  2. Anthracycline des ε-Pyrromycinons 174
     Pyrromycin 174
     Cinerubine 175
     Rutilantine 176
  3. Anthracycline der Rhodomycinone 176
     Rhodomycin A 176
     Rhodomycin B 177
     γ-Rhodomycine 177
     Iso-rhodomycin A 178
     Antibiotica der Mycetin-Violarin-Gruppe 179
  4. Anthracycline des Aklavinons 179
     Aklavin 179

Literaturverzeichnis 179

Folsäure und Folat-Enzyme. Von L. JAENICKE und C. KUTZBACH,
  Physiologisch-chemisches Instiut der Universität Köln 183

I. Einleitung 184

II. Das Vitamin Folsäure 187
  1. Entdeckung der Folsäure und ihrer Konjugate 187

2. Konjugat-spaltende Enzyme 190

3. Vorkommen, Bedarf und Ausscheidung 190

III. Auf- und Abbau der Folsäure-Cofaktoren 192

1. Biogenese der Folsäure 192

2. Biologischer Abbau der Folsäure 195

3. Enzymatische Reduktion der Folsäure zum Cofaktor 197

IV. Chemie der Folat-Verbindungen 199

1. Folsäure 199

   a. Isolierung 199

   b. Konstitution und physikalische Eigenschaften 200

   c. Chemische Eigenschaften 201

   d. Folsäure-Synthesen 203

2. Reduktion von Folsäure 204

   a. Dihydrofolsäure und das Problem der Dihydrofolat-Isomerie 204

   b. 5,6,7,8-Tetrahydro-folsäure 207

3. Mit Einkohlenstoff-Körpern substituierte Folsäuren 209

   a. 10-Formyl-folsäure 209

   b. 10-Formyl-tetrahydrofolsäure 209

   c. 5-Formyl-tetrahydrofolsäure 211

   d. 5,10-Methinyl-tetrahydrofolsäure 212

   e. 5-Formimino-tetrahydrofolsäure 215

   f. 5,10-Methylen-tetrahydrofolsäure 216

   g. 5-Methyl-tetrahydrofolsäure 219

4. Folsäure-Analoge 220

5. Spektren von Folat-Verbindungen 222

6. Analyse und Trennung von Folsäure-Verbindungen 225

   a. Chemische Verfahren 225

   b. Polarographie 225

   c. Mikrobiologische Methoden 226

   d. Chromatographische Trennung 227

V. Das Einkohlenstoff-Reservoir 229

1. Herkunft der Ameisensäure 229

2. Glycin als Quelle von Einkohlenstoffkörpern 231

3. Der Einkohlenstoff-Donator Serin 232

4. Herkunft der Methylgruppe 233

VI. Folat-katalysierte Enzym-Reaktionen 236

1. Der Transhydroxymethylierungs-Cyclus 236

   a. Serin-Aldolase 236

   b. Transhydroxymethylierungs-Reaktionen 239

   c. Methylentetrahydrofolat-Dehydrogenase 239

2. Methylengruppen-Genese 240

   a. Thymidylat-Bildung 240

   b. Methylentetrahydrofolat-Reduktase: 5-Methyl-tetrahydrofolsäure .. 242

   c. Methionin-Bildung 243

      α. Die Gesamt-Reaktion 243

      β. Zusammenhänge zwischen Folsäure und Vitamin B_{12} 244

      γ. Der Acceptor der Methylgruppe 245

3. Transformylierungs-Cyclen 246
    a. Abbau von Histidin 246
    b. Deacylase und Glutamyl-Transferase 247
    c. Aktivierte Ameisensäure im Purin-Stoffwechsel.......... 248
      α. Vergärung von Purinen 248
      β. Tetrahydrofolat-Formylase......................... 249
      γ. Transformylierungen 251

VII. Zusammenfassung 253

Literaturverzeichnis 254

Chemistry of the Natural Rotenoids. By L. CROMBIE, Department of Chemistry, University of London King's College, London 275

I. Introduction 275

II. General Remarks on Rotenone and the Rotenoids.................. 276
    1. Isolation 278
    2. Colour Tests....................................... 278
    3. Nomenclature..................................... 279

III. Stereochemistry of Rotenone 279

IV. Chemistry of Rotenone.................................... 284

V. The Rotenolones and Isorotenolones.......................... 290
    1. The A and B Series................................. 290
    2. The C and D Series................................. 294

VI. The Rotenoids 295
    1. Stereochemistry................................... 295
    2. Deguelin 296
    3. Elliptone... 297
    4. Munduserone 298
    5. α-Toxicarol....................................... 299
    6. Sumatrol... 301
    7. Malaccol 302
    8. Pachyrrhizone 303
    9. Erosone 304
   10. Dolineone 304

VII. Biogenesis and Biogenetic Connections of the Rotenoids 305

VIII. Synthesis in the Rotenoid Group 309

Addendum 316

References... 316

Namenverzeichnis. Index of Names. Index des Auteurs 326

Sachverzeichnis. Index of Subjects. Index des Matières 340

The Biosynthesis of Rubber.

By JAMES BONNER, Pasadena, California.

With 2 Figures.

Contents.

		Page
I.	Distribution of Rubber	1
II.	Latex	2
	Structure and Configuration	2
III.	Biogenesis of the Monomer	4
IV.	Polymerization	10
V.	Further Problems	11
	References	13

Of the major biosynthetic pathways of the plant, that associated with the synthesis of rubber has been among the last to be elucidated. All that we know today concerning isoprenoid biogenesis is information acquired since 1949, much of it since 1956. Today, however, rubber biosynthesis may be considered as a problem solved. We can plot out in intimate detail the complete pathway by which carbon atoms present in common plant metabolites such as carbohydrates are converted to the polyisoprene molecule.

I. Distribution of Rubber.

All species of higher plants possess, it is believed, the capability of synthesizing varied species of isoprenoids. Amongst the widely distributed isoprenoids are the carotenoids, the steroids, the long-chain isoprenoid alcohols, such as phytol, and the isoprenoid tails of the ubiquinones. Ubiquitous distribution is not, however, a property of rubber. Rubber is synthesized by but some 2000 of the 400000 species of higher plants. The distribution of the ability to synthesize rubber appears to be random through the higher plant family tree, with the exception that such ability does not occur in any monocotyledonous plant such as, for example, the grasses, nor in any of the gymnosperms, as, for example, the pines and their allies (*13*). Of the 2000 species which produce rubber, only a few produce it in large quantity. Amongst the principal rubber accumulating plants are, *Ficus elastica* ROXB. (the rubber plant), *Parthenium argentatum*

Gray (the guayule), and *Taraxicum kok saghyz* Rodin (the Russian dandelion). The rubber tree of commerce, *Hevea braziliensis* Muell. supplies, however, all but an infinitesimal amount of the world's cultivated natural rubber. A small number of tropical species, including *Palaquium gutta* Burck and *Mimusops balata* Gaertn., produce an isomer of rubber, gutta.

II. Latex.

In the rubber producing species, rubber is contained in general in specialized cells, the latex vessels. Latex is the protoplasm of these cells and contains therefore, in addition to rubber particles stabilized by protein, all of the usual intracellular components, enzyme molecules, mitochondria, nuclei, and in all probability, ribosomes, as well as low-molecular weight metabolites. The enzymes of latex include those which participate in the biosynthesis of rubber as will be discussed below.

The concentration of rubber in latex may be very high indeed, 30% in the case of *Hevea braziliensis*.

The latex vessels of Hevea and of the other principal rubber-producing species are located in the bark, and are interconnected to form a vast continuous network *(Figs. 1, 2)*. Latex within this network is under positive hydrostatic pressure *(11)*.

If a cut into the lactiferous system is made, latex pours out through the opened vessels, the flow continuing for minutes or hours depending upon the species involved. This flow ultimately ceases, probably often due to coagulation of the latex by the action of microorganisms. The vessel again takes up water to restore its hydrostatic pressure to normal, and in Hevea in particular, rubber is again synthesized. It is this feature of Hevea, the feature of rapid regeneration of the rubber content of the latex vessels, which makes it the rubber plant of choice. A rubber tree may be tapped on alternate days throughout the bulk of its life, and yield a constant or increasing amount of rubber, over the course of 30 years or more.

Structure and Configuration.

Rubber and gutta are both polyisoprenoids, in which isoprene residues are linked together through 1,4-linkages. The polymer molecule is of the order of 500 to 5000 residues long, in the case of rubber; of the order of 100 residues long in the case of gutta (Guttapercha, Balata). In rubber (VI, p. 8) the double bonds of each isoprene residue possess *cis* configuration, and rubber is therefore an all-*cis*-polyterpene. In gutta, on the other hand, all isoprene double bonds possess *trans* configuration. Hybrid molecules containing both *cis* and *trans* double bonds do not appear to occur, either in gutta or in rubber *(20)*.

Fig. 1. High yielding tree of *Hevea brasiliensis* under tapping and with latex dripping from tapping cut. The sloping tapping cut intersects the latex vessels at approximately right angles. At each tap a fresh ca. 1 mm. of bark is removed, so the tapping cut moves downward. [Courtesy, Rubber Research Institute of Malaya, Kuala Lumpur.]

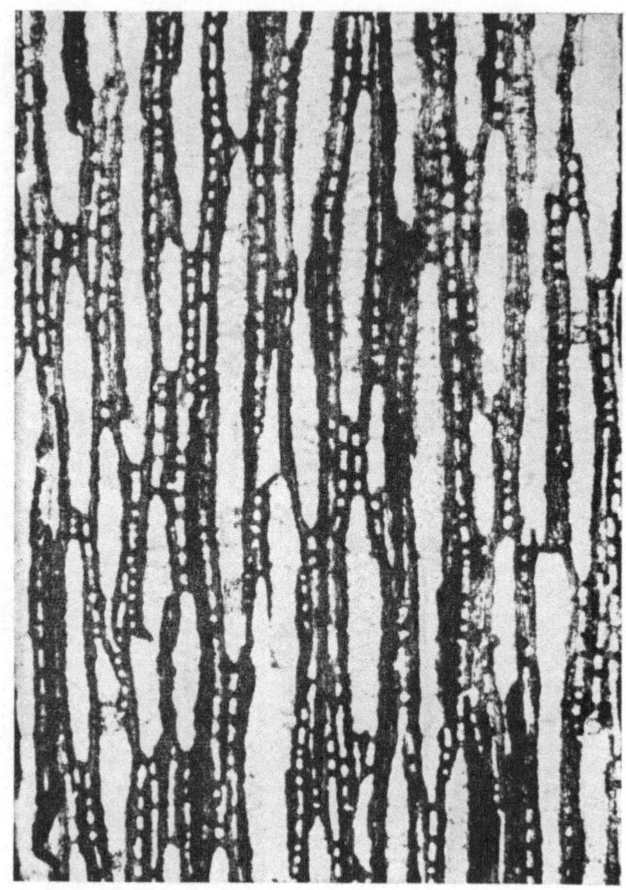

Fig. 2. Tangential section through phloem in bark of high yielding tree of *Hevea brasiliensis*. The dark cells are the anastomosing latex vessels. Magnification ca. 100×. [Courtesy, Rubber Research Institute of Malaya, Kuala Lumpur.]

III. Biogenesis of the Monomer.

The basic precursor of rubber, and in fact of all isoprenoids, is the acetate molecule. All carbon atoms of rubber are derived from acetate (*10, 12*). The evidence which supports this view is the fact that when rubber-producing plants are supplied with large amounts of uniformly C^{14}-labeled acetate, the acetate carbon is not only preferentially transformed to rubber, but in addition, that the carbon atoms of the rubber formed possess the specific radioactivity of the acetate supplied (*4, 10*). Carbon atoms of acetate are therefore transformed without any great dilution into carbon atoms of rubber. Acetate forms also the source of carbon for other isoprenoids, as, for example, sterols. This was first demonstrated by BLOCH (9) for cholesterol, and later by OTTKE et al. (35) for the instance of ergosterol synthesis by Neurospora, a fungus. It is now clear that acetate serves as the principal source of carbon in the biosynthesis of all isoprenoids.

Although early work on rubber biosynthesis and on acetate as a precursor of such synthesis was carried on with the guayule, *Parthenium argentatum*, it is now clear that acetate is the basic precursor for rubber biosynthesis in all rubber plants that have been investigated. These include the Russian dandelion, Kok saghyz (36), Hevea (48, 36), and seventeen further, less fashionable rubber producers (47).

One principal problem in the study of rubber biogenesis has been determination of the isoprenoid monomer from which rubber is polymerized, and of the biosynthetic pathway by which this monomer is synthesized from the carbon atoms of acetate.

BLOCH (9) early established the positions in the isoprenoid carbon skeleton occupied by the carboxyl and methyl carbon atoms of acetate, work which he did in connection with the biogenesis of cholesterol (9). The positions occupied by the two different carbon atoms of acetate, in the isoprene skeleton are illustrated in *Chart 1*. It may be seen from

$$\cdots-\left[\begin{array}{c} C^* \\ \diagdown \\ C^*\diagup \end{array}C^\circ{=}C^*{-}C^\circ\right]-\cdots$$

Chart 1. Contribution of Carboxyl and Methyl Carbon Atoms of Acetate to the Isoprenoid Carbon Skeleton.

C* atoms derived from methyl carbons of acetate.
C° atoms derived from carboxyl carbons of acetate.

the Chart that the 5-carbon isoprenoid monomer contains three carbon atoms derived from the methyl carbon of acetate, and two carbon atoms derived from the carboxyl carbon. That the distribution of carbon atoms found by BLOCH for cholesterol is true also for rubber has been demonstrated by TEAS et al. (48) and by PARK and BONNER (36).

Clearly, however, the formation of the 5-carbon isoprene molecule must involve condensation of three acetate molecules in some manner, together with the elimination of one carboxyl carbon atom. The steps in the condensation of three acetate molecules involve the acetyl-transferring coenzyme A (CoA). The basis of our knowledge of CoA chemistry was laid by LIPMANN (24, 27), and the general function of CoA as a carrier of acetyl groups by LYNEN et al. (32). Plant systems contain enzymes for the formation of acetyl CoA (*34*), and also for catalysis of two successive acetyl condensing reactions, which as shown in *Chart 2*,

$$3\ CH_3COOH + 3\ CoA + 3\ ATP \rightleftharpoons 3\ CH_3COCoA + 3\ AMP + 3\ P_1\!-\!P_1$$

Acetate.

$$CH_3COCoA + CH_3COCoA \rightleftharpoons CH_3COCH_2COCoA + CoA$$

Acetyl CoA.   Acetyl CoA.   Acetoacetyl CoA.

$$CH_3COCH_2COCoA + CH_3COCoA \rightleftharpoons CH_3\underset{\underset{CH_2COOH}{|}}{\overset{\overset{OH}{|}}{C}}CH_2COCoA$$

Acetoacetyl CoA.   Acetyl CoA.

(I.) β-Hydroxy-β-methyl-glutaryl CoA.

Chart 2. Formation of β-Hydroxy-β-methyl-glutaryl CoA from Acetate. ($P_1\!-\!P_1$ = pyrophosphate.)

result in the successive formation of acetoacetyl CoA (*34*), and of the 6-carbon branched chain of β-hydroxy-β-methyl-glutaryl CoA (I). The first two of these steps, formation of acetyl CoA and of acetoacetyl CoA, are steps which are general to living creatures, plant, animal and microorganism alike. The acetate-activating step was discovered and described by Lipmann (*28*) and Beinert et al. (*8*). The acetoacetyl synthesizing step was studied in particular by Chou and Lipmann (*15*), Stadtman et al. (*42*), and Lynen (*33*). The following steps result in the formation of β-hydroxy-β-methyl-glutaryl CoA. This compound is a key intermediate in isoprenoid metabolism, and although it participates in other metabolic pathways also, was discovered during the course of studies on isoprenoid biosynthesis (*7*). The enzyme responsible for condensation of acetyl CoA with acetoacetyl CoA to yield β-hydroxy-β-methyl-glutaryl CoA has been characterized for plants by Johnston et al. (*23*), and for animals by Rabinowitz et al. (*40*) and by Bachhawat et al. (*6*). β-Hydroxy-β-methyl-glutarate has also been isolated from plants (*26*). That the reaction leading to β-hydroxy-β-methyl-glutaryl CoA formation should be formulated as shown in *Chart 2*, has been shown by Rudney (*41*), in a series of labeling experiments.

β-Hydroxy-β-methyl-glutaryl CoA is involved in two principal pathways. One of these leads to the formation of isoprenoids, the other, by removal of a carboxyl group, to β-methylcrotonyl CoA. β-Methylcrotonic acid, $CH_3C(CH_3)=CHCOOH$, which has been known as a natural product for many years (*5*), possesses a carbon skeleton identical to that of the isoprenoid unit. C^{14}-labeled β-methylcrotonic acid is incorporated by rubber-forming plants into rubber, although in small yield (*4*). Coon et al. (*16*) have studied the reactions leading from β-hydroxy-β-methyl-glutaryl CoA to β-methylcrotonyl CoA and have in the course of their work identified adenylphosphoryl-CO_2 as the form by which CO_2 is released or added in the transformation. COON et al. (*16*) have further shown that β-methylcrotonic acid participates in the oxidation of 5-carbon compounds which contain the isopentenyl structure, and is an intermediate in the degradation of leucine and valine; these are converted first to β-methylcrotonyl CoA, thence to β-hydroxy-β-methyl-glutaryl CoA (I), and thence backward to acetyl CoA.

In the pathway leading to isoprenoid formation, β-hydroxy-β-methyl-glutaryl CoA becomes reduced. That this is so was first indicated by the work of TAVORMINA et al. (*45*) and of TAMURA (*44*). These two groups of workers, of the Merck Laboratories and Tokyo University, respectively, had been engaged in the isolation of growth factors for lactobacilli and simultaneously, or approximately so, isolated a mono-carboxylic acid required as a growth factor for a particular strain of this

$$
\begin{array}{ccc}
\overset{\displaystyle OH}{\underset{\displaystyle CH_2COOH}{CH_3\!-\!\overset{|}{\underset{|}{C}}\!-\!CH_2COCoA}} + 2\ TPNH^* & \rightarrow & \overset{\displaystyle OH}{\underset{\displaystyle CH_2COOH}{CH_3\!-\!\overset{|}{\underset{|}{C}}\!-\!CH_2CH_2OH}} + 2\ TPN + CoA
\end{array}
$$

(I.) β-Hydroxy-β-methyl-glutaryl CoA.   (II.) Mevalonic acid.

Chart 3. Formation of Mevalonic Acid from β-Hydroxy-β-methyl-glutaryl CoA.

organism (*41 a*). This compound, which has the structure shown in *Chart 3*, has been given the name mevalonic acid. It possesses a carbon skeleton identical with that of β-hydroxy-β-methyl-glutaryl CoA (*22*). Since mevalonic acid (II) is obviously related in structure to β-hydroxy-β-methylglutarate, TAVORMINA et al. *(45)* tested it for its ability to be incorporated into cholesterol by a crude liver enzyme system. They demonstrated that it is incorporated into the steroid in high yield, and with elimination of the carboxyl carbon atom $C_{(1)}$. Subsequent investigations have shown that mevalonic acid is similarly metabolized in high yield to simple terpenes (*43*), to carotenes (*39*), to squalene (*2, 18*) and to rubber (*36*), the product formed depending on the nature of the tissue or organism used.

The reduction of β-methylglutaryl CoA to mevalonic acid (II) is simply a two-step reduction. In the first step, the CoA-bearing carboxyl group of β-methyl-β-glutaryl is reduced by TPNH, with the elimination of CoA. Reaction with a further molecule of TPNH reduces the aldehyde to the level of the primary alcohol. The reduction is essentially irreversible; and a molecule, once it has been converted to mevalonic acid, as is shown in Chart 3, is essentially committed to the isoprenoid pathway (*19*).

---

* TPNH = reduced triphosphopyridine nucleotide.

The labeled rubber is degraded by ozonolysis to yield labeled levulinic acid. The levulinic acid is degraded to iodoform (unlabeled) and succinate (labeled). The succinate is degraded by pyrolysis to yield its carboxyl groups as barium carbonate (unlabeled). The label of 2-C^{14}-mevalonic acid is therefore transformed into rubber as expected on the basis of the following formulation.

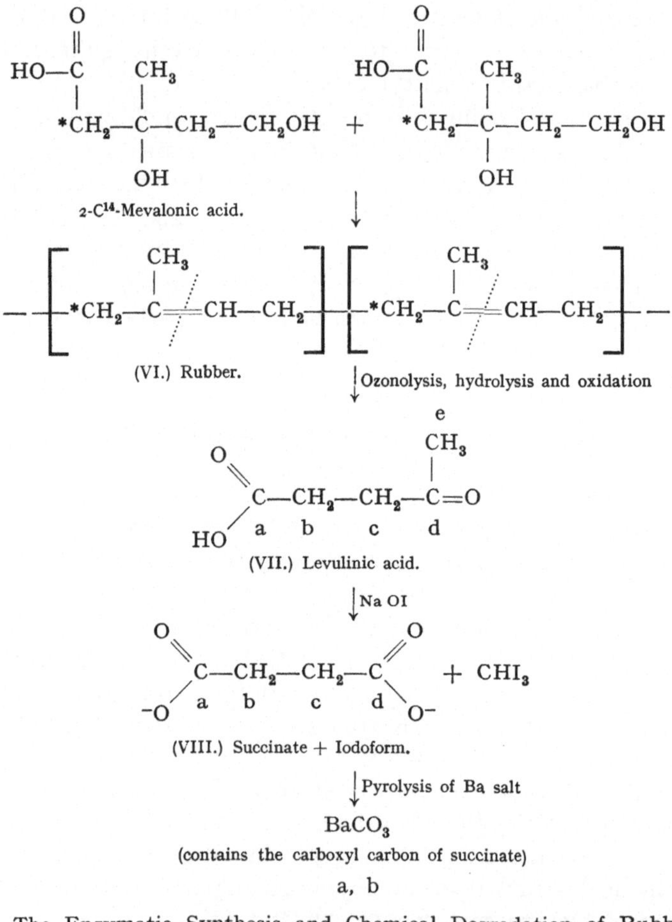

$$\underset{\text{2-}C^{14}\text{-Mevalonic acid.}}{\overset{\displaystyle HO-\overset{\overset{\textstyle O}{\|}}{C}\quad CH_3}{*CH_2-\underset{\underset{\textstyle OH}{|}}{\overset{|}{C}}-CH_2-CH_2OH}} \;+\; \overset{\displaystyle HO-\overset{\overset{\textstyle O}{\|}}{C}\quad CH_3}{*CH_2-\underset{\underset{\textstyle OH}{|}}{\overset{|}{C}}-CH_2-CH_2OH}$$

$$-\left[*CH_2-\overset{\overset{\textstyle CH_3}{|}}{C}=CH-CH_2 \right]\left[*CH_2-\overset{\overset{\textstyle CH_3}{|}}{C}=CH-CH_2 \right]-$$

(VI.) Rubber.   Ozonolysis, hydrolysis and oxidation

$$\underset{\text{(VII.) Levulinic acid.}}{\overset{\displaystyle e}{\underset{HO}{\overset{O}{\diagup}}C\underset{a}{-}CH_2\underset{b}{-}CH_2\underset{c}{-}\overset{\overset{\textstyle CH_3}{|}}{C}\underset{d}{=}O}}$$

Na OI

$$\underset{\text{(VIII.) Succinate + Iodoform.}}{\overset{O}{\underset{{}^-O}{\diagup}}C\underset{a}{-}CH_2\underset{b}{-}CH_2\underset{c}{-}\overset{O}{\underset{O^-}{\diagdown}}C}_{d} \;+\; CHI_3$$

Pyrolysis of Ba salt

$$BaCO_3$$

(contains the carboxyl carbon of succinate)

a, b

Chart 4. The Enzymatic Synthesis and Chemical Degradation of Rubber from Latex Incubated with 2-C^{14}-Mevalonic Acid.

That rubber formation proceeds from mevalonic acid without randomization or rearrangement of carbon atoms has been shown for the enzymatic system of Hevea latex by PARK and BONNER (36), and by KEKWICK et al. (25). In these experiments 2-C^{14}-mevalonic acid was incubated with latex, together with ATP. The rubber (VI) enzymatically produced in the reaction mixture was isolated and degraded by reactions appropriate for the separate collection of the individual carbon atoms of the isoprene moiety. The results *(Chart 4)* indicate clearly that substantially all of the label of the carbon atom-2 of mevalonic acid is contained specifically in the 4-positions of the isoprene moieties of rubber. Mevalonic acid is therefore incorporated without randomization, and in a way consistent with the information already available for the incorporation of the label of acetate into rubber and into the carbon skeletons of β-hydroxy-β-methyl-glutaryl CoA (I, p. 7) and of mevalonic acid.

The reactions involved in the transformation of mevalonic acid to the monomer for isoprenoid polymerization have been studied in yeast by TCHEN (46), BLOCH (9) and their colleagues, and by LYNEN and his associates (1, 21). That the transformations found for yeast occur too in higher plant systems has been demonstrated by POLLARD et al. (37). These transformations summarized in *Chart 5* involve first the

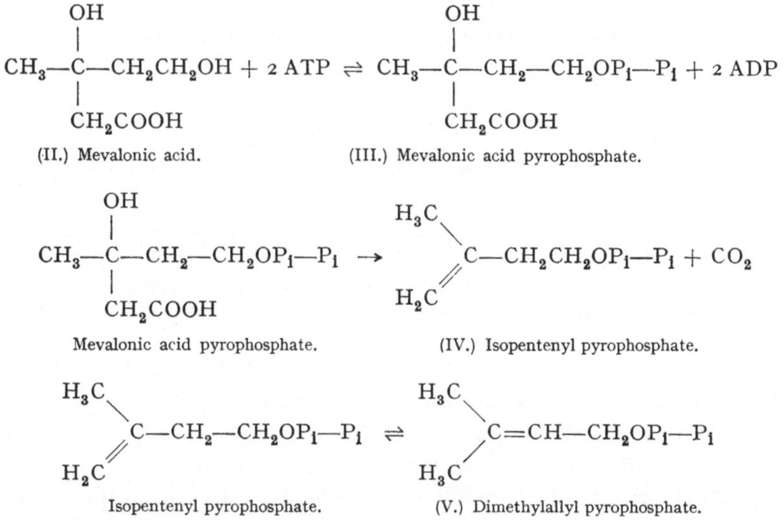

$$
\underset{\substack{| \\ CH_2COOH}}{\overset{\substack{OH \\ |}}{CH_3\!-\!C\!-\!CH_2CH_2OH}} + 2\,ATP \;\rightleftharpoons\; \underset{\substack{| \\ CH_2COOH}}{\overset{\substack{OH \\ |}}{CH_3\!-\!C\!-\!CH_2\!-\!CH_2OP_1\!-\!P_1}} + 2\,ADP
$$

(II.) Mevalonic acid.   (III.) Mevalonic acid pyrophosphate.

$$
\underset{\substack{| \\ CH_2COOH}}{\overset{\substack{OH \\ |}}{CH_3\!-\!C\!-\!CH_2\!-\!CH_2OP_1\!-\!P_1}} \;\rightarrow\; \underset{H_2C}{\overset{H_3C}{\diagdown}}C\!-\!CH_2CH_2OP_1\!-\!P_1 + CO_2
$$

Mevalonic acid pyrophosphate.   (IV.) Isopentenyl pyrophosphate.

$$
\underset{H_2C}{\overset{H_3C}{\diagdown}}C\!-\!CH_2\!-\!CH_2OP_1\!-\!P_1 \;\rightleftharpoons\; \underset{H_3C}{\overset{H_3C}{\diagdown}}C\!=\!CH\!-\!CH_2OP_1\!-\!P_1
$$

Isopentenyl pyrophosphate.   (V.) Dimethylallyl pyrophosphate.

Chart 5. Transformation of Mevalonic Acid to Isopentenyl Pyrophosphate and Dimethylallyl Pyrophosphate.

phosphorylation of mevalonic acid by ATP, with formation of mevalonic acid-5-phosphate. The enzyme involved in this reaction has been designated as mevalonic acid kinase (46). A second phosphate is then transferred to mevalonic acid phosphate by phosphomevalonic acid kinase (14, 21). The resulting mevalonic acid pyrophosphate (III) is then attacked by mevalonic acid pyrophosphate decarboxylating enzyme. In this reaction which is highly characteristic of the isoprenoid pathway, the carboxyl group of mevalonic acid is eliminated as CO_2, simultaneously with the elimination of one molecule of water between carbon atoms 2 and 3 to yield Δ-3-isopentenyl pyrophosphate (IpPP) (IV) (14, 30).

Isopentenyl pyrophosphate is an extremely effective substrate for rubber formation by Hevea latex. It is utilized some ten times more rapidly than is mevalonic acid, and is converted with essentially 100% efficiency so that all of the substrate supplied to latex is transformed to rubber (*3, 31*).

IV. Polymerization.

The mechanism of the polymerization of isopentenyl pyrophosphate has been studied in detail, particularly by the group of LYNEN (*1*). In the first step, isopentenyl pyrophosphate (IV) is isomerized by double bond shift to form dimethylallyl pyrophosphate (V) (*29*). This step is catalyzed

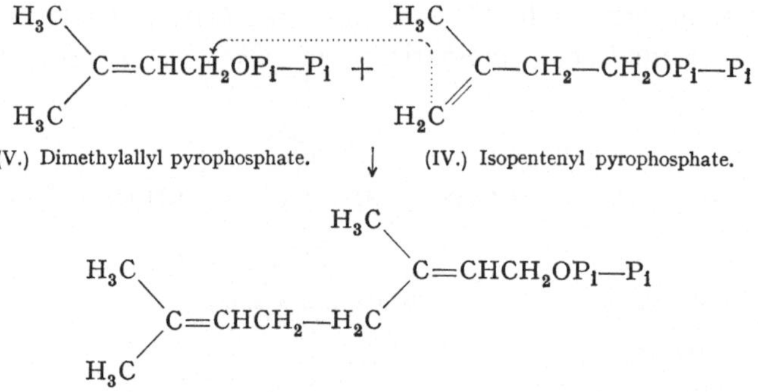

$$H_3C{\diagdown}\atop{H_3C\diagup}C=CHCH_2OP_1-P_1 \quad + \quad H_3C{\diagdown}\atop{H_2C\diagup}C-CH_2-CH_2OP_1-P_1$$

(V.) Dimethylallyl pyrophosphate.   ↓   (IV.) Isopentenyl pyrophosphate.

$$\begin{array}{c}H_3C\diagdown\\ \\ H_3C\diagup\end{array}C=CHCH_2-H_2C\diagup\begin{array}{c}H_3C\diagdown\\ C=CHCH_2OP_1-P_1\end{array}$$

(IX.) Geraniol pyrophosphate *(cis)* or nerol pyrophosphate *(trans)*.

Chart 6. Polymerization of Isopentenyl Pyrophosphate.
(One molecule of dimethylallyl pyrophosphate serves as the chain initiator.)

by the enzyme isopentenyl pyrophosphate isomerase (*1*). One molecule of dimethylallyl pyrophosphate then serves as an acceptor for one molecule of isopentenyl pyrophosphate, with the elimination of pyrophosphate and the formation of one molecule of C_{10}-di-isoprenoid alcohol pyrophosphate *(Chart 6)* (*50*). It has been suggested (*17*) that the polymerization proceeds by the loss by the dimethylallyl pyrophosphate of its pyrophosphate group and temporary formation of electron-deficient species of dimethylallyl molecule which then attacks isopentenyl pyrophosphate with the resultant shift of the double bond of the latter (*38*).

Further extension of the polyisoprene chain probably involves merely the addition of further isopentenyl-pyrophosphate units, that is, dimethyl-allyl-pyrophosphate serves only as the chain initiator. In yeast and in liver, polymerization proceeds to the sesquiterpene alcohol pyrophosphate, farnesol pyrophosphate, which is in turn produced from the C_{10}-geraniol pyrophosphate (IX). The farnesol pyrophosphate thus produced forms the substrate for the synthesis of squalene, and hence of sterols. In the plant systems studied by POLLARD et al. (*37*), geraniol pyrophosphate once formed is the substrate for the formation of farnesol pyrophosphate,

References, pp. 13—16.

which in turn serves as a substrate for the formation of the di-terpene alcohol pyrophosphate containing four isoprenoid units (*49*). The further polymerization steps may be thought of as precisely analogous to the first. Thus, the C_{10}-unit geraniol pyrophosphate may attack a further molecule of isopentenyl-pyrophosphate, etc.

The much studied polymerization of isopentenyl-pyrophosphate by yeast and liver enzyme systems results in the formation of the *trans* isomer of geraniol pyrophosphate, as well as of the *trans-trans* isomer of farnesol pyrophosphate. The plant systems studied by POLLARD et al. (*37*) which are concerned with carotene biosynthesis also involve exclusively the production of *trans* isomers. Of the isopentenyl-pyrophosphate polymerization systems thus far studied, only that concerned with rubber results in the formation of the *cis* isomer. The determination of the configuration of each C_5-unit as it is added to the growing terpene chain is quite evidently enzymatically operated. This determination is in turn made by the geometry of the proton shift which takes place in the new allyl double bond as it is produced. It depends solely upon which of the two protons is transferred in the formation of the new double bond.

It will be of great and general interest to rigorously separate and characterize the enzymes responsible for *cis* and *trans* isoprenoid polymers, and to discover the mechanism used by nature for the determination of configuration.

V. Further Problems.

Further interesting unsolved questions concerning rubber biogenesis are still with us, even though we now know the path of carbon in detail. It would be of interest, for example, to know how many polymerization enzymes are involved in the formation of a polyisoprenoid. Is, for example, one enzyme sufficient for the production of a rubber molecule, or are there perhaps different isopentenyl-pyrophosphate polymerases, one appropriate for dealing with short oligo-isoprenoids, a second appropriate for dealing with intermediate chain length isoprenoids, and a third appropriate for dealing with very long chain lengths poly-isoprenoids. It would be of interest too to know what mechanism is involved in the quenching of chain growth. What, for example, tells the yeast system to stop with the polymerization of three isopentenyl-pyrophosphate moieties, and yet permits Hevea latex to proceed to much higher level of polymerization? Chain growth might cease owing to the absence of a suitable polymerase for the making of higher level polymers, or it might conceivably be due to the presence of an enzyme which specifically de-activates the growing chain after it has reached an appropriate length.

The problem of the regulation of the concentration of rubber in latex also poses interesting questions for the rubber biochemist of the future. This problem may be stated as follows. After the coagulation of latex at a tapping cut the latex vessels fill with water as outlined above. The rubber concentration in the latex is now low, due to the draining of the latex vessels by tapping. As the hydrostatic pressure in the latex vessel is restored, rubber synthesis takes place, and the rubber concentration in the latex rises. Rubber continues to be synthesized until it reaches the concentration in the latex characteristic of the particular strain of rubber plants concerned. In Hevea the final concentration of latex may attain, as pointed out above, a value as high as 30% of the latex weight. When this final and appropriate value is attained, rubber synthesis ceases. Rubber concentration in the latex remains constant until a further tapping cut has been made, the latex vessel again emptied, and latex again coagulated at the tapping cut. The synthesis of rubber therefore takes place in the latex vessel in response to low rubber concentration, and is stopped by high rubber concentration. There is evidently a negative feedback between concentration of rubber in the latex vessel and the activity of one or more of the enzymes in the sequence responsible for rubber synthesis.

The inhibition of rubber synthesis by high rubber concentrations cannot be exerted at the level of the final polymerase step, because if it were at this step, IpPP would accumulate in latex vessels, under conditions of high latex concentration. That IpPP does not accumulate in latex vessels is indicated, for example, by the fact that latex of high rubber concentration does not synthesize rubber in response to mere dilution. The inhibition must therefore occur at an earlier step or steps, perhaps as far back as the biogenesis of mevalonate from acetate, or even the biogenesis of acetate itself. That the block may occur before the production of acetate is indicated by that fact that latex can metabolize acetate to rubber, as well as by the fact that high levels of mevalonate are not characteristic of latex.

The inhibition of enzymatic synthesis of rubber by high rubber concentration in latex is thus reminiscent of the negative feedback loop inhibition of whole enzymatic pathways by high concentrations of ultimate product in bacteria. Study of the nature of the feedback loop inhibition involved in rubber biosynthesis might make it possible to circumvent the inhibition and to attain even higher concentrations of rubber in latex.

A final, and important and interesting, problem of rubber biogenesis is that of the origin of the enzymes of the rubber biosynthetic pathway. At each tapping large amounts of the responsible enzymes leave the latex vessel in the company of the rubber. That the well-known serum

References, pp. 13—16.

proteins of latex include the rubber biosynthetic enzymes is evidenced by the fact that latex contains, as shown above, all of the enzymes required for the transformation of acetate to mevalonate to IpPP to rubber. Clearly, therefore, after a tapping, and during regeneration of the turgor pressure of the latex vessel, a new supply of enzymes for the biosynthesis of rubber must be manufactured within the latex vessel.

A great deal is now known about the biogenesis of enzyme molecules. It is known that such biogenesis occurs upon the surface of characteristic cytoplasmic particles, the ribosomes, whose duty it is to synthesize individual enzyme molecules. It is known, too, that the informational content of the ribosome, the information required to properly assemble an appropriate sequence and abundance of the amino acid residues characteristic of a particular enzyme, is contained in the ribosome in the form of a molecule of messenger RNA. Ribosomes and their accompanying messenger RNA molecules are synthesized within the nucleus of the cell. Although a few nuclei are swept out of the latex vessel in the gush of latex which flows out as a result of a tapping cut, still the bulk of the nuclei of the latex vessel appear to remain within the vessel. They are ordinarily disposed close to the latex vessel wall, and in a peripheral layer of cytoplasm which does not flow out in response to the tapping cut. It is no doubt the function of these nuclei to provide the new ribosomes and messenger RNA molecules required for the synthesis of the new rubber synthesizing enzymes. The synthesis of new ribosomes and of new messenger RNA molecules by the nuclei of the latex vessel would require, of course, a bountiful supply of energy in the form of ATP, of the nucleoside triphosphates required for RNA synthesis, and of the amino acid molecules from which enzymes are constructed. The source of the ATP, and of the metabolites required for ribosome, RNA and enzyme production within the latex vessel, is however unknown.

The very concept that the enzyme molecules of latex are generated anew after each tapping is a logical deduction rather than an accomplished experimental fact. The study, therefore, of the way in which the rubber-synthesizing machinery is reconstituted within the vessel after each tapping may well prove to be one of the most fruitful and most interesting of the remaining problems of the biogenesis of rubber.

References.

1. AGRANOFF, B. W., H. EGGERER, U. HENNING and F. LYNEN: Biosynthesis of Terpenes. VII. Isopentenyl Pyrophosphate Isomerase. J. Biol. Chem. **235**, 326 (1960).
2. AMDUR, B. H., H. RILLING and K. BLOCH: Enzymatic Conversion of Mevalonic Acid to Squalene. J. Amer. Chem. Soc. **79**, 2646 (1957).

3. ARCHER, B. L., G. AYREY, E. G. COCKBAIN and G. P. McSWEENEY: Incorporation of [1-C^{14}]-Isopentenyl Pyrophosphate into Polyisoprene. Nature **189**, 663 (1961).

4. ARREGUIN, B., J. BONNER and B. J. WOOD: Studies on the Mechanism of Rubber Formation in the Guayule. III. Experiments with Isotopic Carbon. Arch. Biochem. Biophys. **31**, 234 (1951).

5. ASAHINA, Y.: Notiz über Seneciosäure. Arch. Pharm. **251**, 355 (1913).

6. BACHHAWAT, B. K., W. G. ROBINSON and M. J. COON: CO_2 Fixation in Heart Extracts by β-Hydroxy-Isovaleryl Coenzyme A. J. Amer. Chem. Soc. **76**, 3098 (1954).

7. BANDURSKI, R., T. COYLE and J. BONNER: Studies on the Mechanism of Rubber Biosynthesis. Biology 1953, Calif. Inst. of Tech., pp. 118/119.

8. BEINERT, H., D. E. GREEN, P. HELE, H. HIFT, R. W. VON KORFF and C. V. RAMAKRISHNAN: The Acetate Activating Enzyme System of Heart Muscle. J. Biol. Chem. **203**, 35 (1953).

9. BLOCH, K.: The Biological Synthesis of Cholesterol. Recent Progr. in Hormone Res. **6**, 111 (1951).

10. BONNER, J.: Synthesis of Isoprenoid Compounds in Plants. J. Chem. Educ. **26**, 628 (1949).

11. — Plant Biochemistry. New York: Academic Press. 1950, p. 27.

12. BONNER, J. and B. ARREGUIN: The Biochemistry of Rubber Formation in the Guayule. I. Rubber Formation in Seedlings. Arch. Biochem. **21**, 109 (1949).

13. BONNER, J. and A. W. GALSTON: The Physiology and Biochemistry of Rubber Formation in Plants. Bot. Rev. **13**, 543 (1947).

14. CHAYKIN, S., J. LAW, A. H. PHILLIPS, T. T. TCHEN and K. BLOCH: Phosphorylated Intermediates in the Synthesis of Squalene. Proc. Nat. Acad. Sci. (USA) **44**, 998 (1958).

15. CHOU, T. C. and F. LIPMANN: Separation of Acetyl Transfer Enzymes in Pigeon Liver Extract. J. Biol. Chem. **196**, 89 (1952).

16. COON, M. J., W. G. ROBINSON and B. K. BACHHAWAT: Enzymatic Studies on the Biological Degradation of the Branched Chain Amino Acids. In: Amino Acid Metabolism, p. 431. Baltimore: Johns Hopkins Press. 1955.

17. CORNFORTH, J. W. and G. POPJÁK: Mechanism of Biosynthesis of Squalene from Sesquiterpenoids. Tetrahedron Letters **19**, 29 (1959).

18. DITURI, F., J. L. RABINOWITZ, R. P. HULLIN and S. GURIN: Precursors of Squalene and Cholesterol. J. Biol. Chem. **229**, 825 (1957).

19. FERGUSON, J. J., Jr., I. F. DURR and H. RUDNEY: The Biosynthesis of Mevalonic Acid. Proc. Nat. Acad. Sci. (USA) **45**, 499 (1959).

20. HENDRICKS, S. B., S. G. WILDMAN and E. J. JONES: Differentiation of Rubber and Gutta Hydrocarbons in Plant Materials. Arch. Biochem. **7**, 427 (1945).

21. HENNING, U., E. M. MÖSLEIN and F. LYNEN: Biosynthesis of Terpenes. V. Formation of 5-Pyrophospho Mevalonic Acid by Phospho Mevalonic Kinase. Arch. Biochem. Biophys. **83**, 259 (1959).

22. HOFFMAN, C. H., A. F. WAGNER, A. N. WILSON, E. WALTON, C. H. SHUNK, D. E. WOLF, F. W. HOLLY and K. FOLKERS: Synthesis of *DL*-3,5-Dihydroxy-3-Methylpentanoic Acid (Mevalonic Acid). J. Amer. Chem. Soc. **79**, 2316 (1957).

23. JOHNSTON, J. A., D. W. RACUSEN and J. BONNER: Metabolism of Isoprenoid Precursors in a Plant System. Proc. Nat. Acad. Sci. (USA) **40**, 1031 (1954).

24. KAPLAN, N. O. and F. LIPMANN: The Assay and Distribution of Coenzyme A. J. Biol. Chem. **174**, 37 (1948).

25. KEKWICK, R. G. O., B. L. ARCHER, D. BARNARD, G. M. C. HIGGINS, G. P. McSWEENEY and C. G. MOORE: Incorporation of *DL*-(2 C^{14}) Mevalonic Acid Lactone into Polyisoprene. Nature **184**, 268 (1959).

26. KLOSTERMAN, H. J. and F. SMITH: Isolation of β-Hydroxy-β-Methylglutaric Acid from the Seed of Flax *(Linum usitatissimum)*. J. Amer. Chem. Soc. **76**, 1229 (1954).

27. LIPMANN, F.: Acetylation of Sulfanilimide by Liver Homogenates and Extracts. J. Biol. Chem. **160**, 173 (1945).

28. LIPMANN, F., M. E. JONES, S. BLACK and R. M. FLYNN: Enzymatic Pyrophosphorylation of Coenzyme A by Adenosine Triphosphate. J. Amer. Chem. Soc. **74**, 2384 (1952).

29. LYNEN, F., B. W. AGRANOFF, H. EGGERER, U. HENNING und E. M. MÖSLEIN: γ,γ-Dimethyl-Allyl-Pyrophosphat und Geranyl-pyrophosphat, biologische Vorstufen des Squalens (Zur Biosynthese der Terpene, VI). Angew. Chem. **71**, 657 (1959).

30. LYNEN, F., H. EGGERER, U. HENNING und I. KESSEL: Farnesyl-Pyrophosphat und 3-Methyl-Δ^3-Butenyl-1-Pyrophosphat, die biologischen Vorstufen des Squalens (Zur Biosynthese der Terpene, III). Angew. Chem. **70**, 738 (1958).

31. LYNEN, F. und U. HENNING: Über den biologischen Weg zum Naturkautschuk. Angew. Chem. **72**, 820 (1960).

32. LYNEN, F., E. REICHERT und L. RUEFF: Zum biologischen Abbau der Essigsäure VI („Aktivierte Essigsäure", ihre Isolierung aus Hefe und ihre chemische Natur). Liebigs Ann. Chem. **574**, 1 (1951).

33. LYNEN, F., L. WESSELY, O. WIELAND und L. RUEFF: β-Oxydation der Fettsäuren. Angew. Chem. **64**, 687 (1952).

34. MILLERD, A. and J. BONNER: Acetate Activation and Acetoacetate Formation in Plant Systems. Arch. Biochem. Biophys. **49**, 343 (1954).

35. OTTKE, R. C., E. L. TATUM, I. ZABIN and K. BLOCH: Ergosterol Synthesis in Neurospora. Federat. Proc. (Amer. Soc. exp. Biol.) **9**, 212 (1950).

36. PARK, R. B. and J. BONNER: Enzymatic Synthesis of Rubber from Mevalonic Acid. J. Biol. Chem. **233**, 340 (1958).

37. POLLARD, C., J. BONNER, A. J. HAAGEN-SMIT and C. C. NIMMO: Transformations of Mevalonic Acid by an Enzyme System from Peas. Plant Physiol. (In press.)

38. POPJÁK, G.: Biosynthesis of Derivatives of Allylic Alcohols from 2 C^{14} Mevalonate in Liver Enzyme Preparations and Their Relation to Synthesis of Squalene. Tetrahedron Letters **19**, 19 (1959).

39. PURCELL, A. E., G. A. THOMPSON, Jr. and J. BONNER: The Incorporation of Mevalonic Acid into Tomato Carotenoids. J. Biol. Chem. **234**, 1081 (1959).

40. RABINOWITZ, J. L. and S. GURIN: Biosynthesis of Cholesterol and β-Hydroxy-β-Methylglutaric Acid by Extracts of Liver. J. Biol. Chem. **208**, 307 (1954).

41. RUDNEY, H.: The Synthesis of β-Hydroxy-β-Methylglutaric Acid in Rat Liver Homogenates. J. Amer. Chem. Soc. **76**, 2595 (1954).

41a. SKEGGS, H. R., L. D. WRIGHT, E. L. CRESSON, G. D. E. MacRae, C. H. HOFFMAN, D. E. WOLF and K. FOLKERS: Discovery of a New Acetate-replacing Factor. J. Bacteriol. **72**, 519 (1956).

42. STADTMAN, E. R., M. DOUDOROFF and F. LIPMANN: The Mechanism of Acetoacetate Synthesis. J. Biol. Chem. **191**, 377 (1951).

43. STANLEY, R. G.: Terpene Formation in Pine from Mevalonic Acid. Nature **182**, 738 (1958).

44. TAMURA, G.: Hiochic Acid or New Growth Factor for *Lactobacillus homohiochi* and *L. heterohiochi*. J. Gen. Appl. Microbiol. **2**, 431 (1956).

45. TAVORMINA, P. A., M. H. GIBBS and J. W. HUFF: The Utilization of β-Hydroxy-β-Methyl-δ-Valerolactone in Cholesterol Biosynthesis. J. Amer. Chem. Soc. **78**, 4498 (1956).

46. Tchen, T. T.: Mevalonic Kinase: Purification and Properties. J. Biol. Chem. **233**, 1100 (1958).
47. Teas, H. J. and J. Bonner: Rubber Biosynthesis. Revue Generale du Caoutchouc **37**, 1143 (1960).
48. Teas, H. J., L. Polhemus and J. C. Montermoso: Abstr., Conference on Radioactive Isotopes in Agriculture, (U. S. Dept. of Agric.), January, 1956.
49. Thompson, G. A., Jr., A. E. Purcell and J. Bonner: A Carotene Precursor; its Proposed Structure and Place in Biosynthetic Sequence. Plant Physiol. **35**, 678 (1960).
50. Witting, L. A. and J. W. Porter: A Geraniol Derivative—An Intermediate in the Biosynthesis of Squalene by a Rat Liver Enzyme System. Biochem. Biophys. Res. Comm. **1**, 341 (1959).

(Received, November 14, 1962.)

The Polyene Antifungal Antibiotics.

By **W. OROSHNIK**, Clifton, New Jersey and
A. D. MEBANE, Raritan, New Jersey.

With 2 Figures.

Contents.

		Page
I.	Introduction	18
II.	Ultraviolet Spectra	19
	1. General Observations	19
	2. The Tetraenes	23
	3. The Pentaenes	24
	4. The Methylpentaenes	24
	5. The Hexaenes	24
	6. The Heptaenes	24
III.	Structural Elucidation	26
	1. General Features	26
	2. Mycosamine	28
	3. Retro-Aldol Cleavage	30
	4. Fungichromin (Lagosin)	32
	5. Filipin	37
	6. Other Methylpentaenes	39
	7. Pimaricin	40
	8. Nystatin and Other Tetraenes	43
	9. Pentaenes and Hexaenes	45
	10. Trichomycin and Other Heptaenes	46
IV.	Biogenetic Relationships	51
V.	Tables	56
	1. Typical Tetraenes: Spectral Data	56
	2. Typical Pentaenes: Spectral Data	56
	3. Typical Methylpentaenes: Spectral Data	57
	4. Typical Hexaenes: Spectral Data	57
	5. Typical Heptaenes: Spectral Data	57
	6. Tetraenes: Physical and Chemical Properties	58
	7. Pentaenes: Physical and Chemical Properties	62
	8. Methylpentaenes: Physical and Chemical Properties	64
	9. Hexaenes: Physical and Chemical Properties	66
	10. Heptaenes: Physical and Chemical Properties	66
References		72

I. Introduction.

During the early 1950's there appeared reports, from time to time of the presence among the products elaborated by actinomycetes of antifungal antibiotics which exhibited very similar and very characteristic multipeaked ultraviolet absorption spectra. In 1954, with a good number of examples on record, these spectra were analyzed and identified as those of straight-chain conjugated polyenes, comprising tetraenes, pentaenes, hexaenes and heptaenes (*85*, *130*). These antibiotics have since been commonly referred to as the polyene antifungal antibiotics to distinguish them from a host of other miscellaneous antibiotics which also have antifungal properties. Within the next few years, reports of discoveries of new members of this class multiplied rapidly, and almost sixty are now known.

Unquestionably, a number of these will eventually be found to be identical with others, as has already happened in several instances: for example, the tetraene "tennecetin" proved to be a rediscovery of pimaricin (*34*), and in the methylpentaenes "moldcidin B" has been identified with pentamycin (*83*), and "lagosin" appears to be indistinguishable from fungichromin (*22*).

Those that have been purified have turned out to be of fairly high molecular weight (ca. 700–1300) and all appear to be substances of rather similar molecular structure. So far only three, pimaricin, fungichromin (lagosin) and filipin, have been structurally elucidated.

As a group they show very poor solubility in the common organic solvents and in water but can be dissolved to a reasonable extent in very polar solvents like pyridine, dimethylformamide and dimethylsulfoxide. Some, containing basic nitrogen functions or acidic groups, can be dissolved in aqueous acidic or alkaline media, respectively; or as in the majority of cases, where both acidic and basic groups are present, solution can be attained in either medium. However, in such solutions these antibiotics are quite unstable. Heat and light also cause rapid deterioration, and, as would be expected, this sensitivity increases with increasing length of the conjugated polyene chain. In the dry state, however, and in the absence of heat and light, the polyene antibiotics can be kept intact for indefinite periods.

Biologically, all of the polyene antibiotics have in common a very pronounced activity against yeasts and fungi, but no significant antibacterial properties. Three of them are in general medical use in the treatment of certain fungal infections: nystatin and trichomycin for infections of *Candida albicans* and amphotericin B for cryptococcal meningitis and other systemic mycoses. All have been produced by actinomycetes of the immense genus *Streptomyces*. Strains of *Streptomyces* producing these polyenes are quite common: screening programs in the

United States, for example, have revealed them in 28 out of 93 strains tested (*90*), and in Italy in 130 out of 1000 (*25*; see also *11, 116, 126*). If a Streptomycete produces an antifungal agent at all, the chances are about 95 out of 100 that it will prove to be a polyene (*11*). The polyene antibiotic may be accompanied by an antibiotic of non-polyene nature; for example, *St. noursei*, the source of nystatin, simultaneously produces the chemically unrelated cycloheximide (*54*), while rimocidin and oxytetracycline are co-products of *St. rimosus* (*29*). The simultaneous production of three different types of antibiotic is not rare (*5, 26, 30, 92, 113*).

The isolation procedure normally begins by extracting the mycelial filter cake with an aqueous alcohol or aqueous acetone. In some cases it is preferable to extract the whole culture broth or even the filtrate, since the polyene antibiotics may appear "dissolved" owing to the presence of concurrently produced dispersing agents (*11*). The solvent is removed under vacuum until only water remains, the precipitated antibiotic taken up with butanol, and the butanol extract vacuum-dried or precipitated with ether.

Purification is apt to be difficult because of low solubility and instability to manipulation; most of the members of the group have not yet been obtained in the pure crystalline state. This difficulty is aggravated by the fact that the parent organism sometimes produces a mixture of closely-related polyene antibiotics, a situation which often prevails with the hexaenes and heptaenes. Sometimes these congeners can be separated by special expedients, as in the case of the amphotericins, where the tetraene component, amphotericin A, can be separated from the heptaene component, amphotericin B, by its greater solubility in methanolic calcium chloride (*125*); more often, countercurrent distribution must be resorted to. A solvent mixture (52 water: 40 dioxane: 30 pyridine: 18 *n*-butyl acetate), recommended by HOSOYA and HAMAMURA (*58*) for the recrystallization of the heptaene trichomycin has been found useful for other similar compounds (*51*). Perhaps the most tractable as regards purification, probably because of their lack of carboxyl and amine functions, are the methylpentaenes, all of which can be recrystallized, e. g. from methanol (*122, 69, 118*).

II. Ultraviolet Spectra.

1. General Observations.

In spite of the paucity of accurate physical and chemical data for most of these antibiotics, due in large part to the difficulties encountered in purification, considerable information regarding the structures of their chromophores can be obtained from their ultraviolet spectra. All show spectra of virtually the same shape, differing only in wavelength (*Figs. 1* and *2*). The main absorption band is resolved into a regular series of sharp, narrow peaks separated by deep valleys, four or five of these

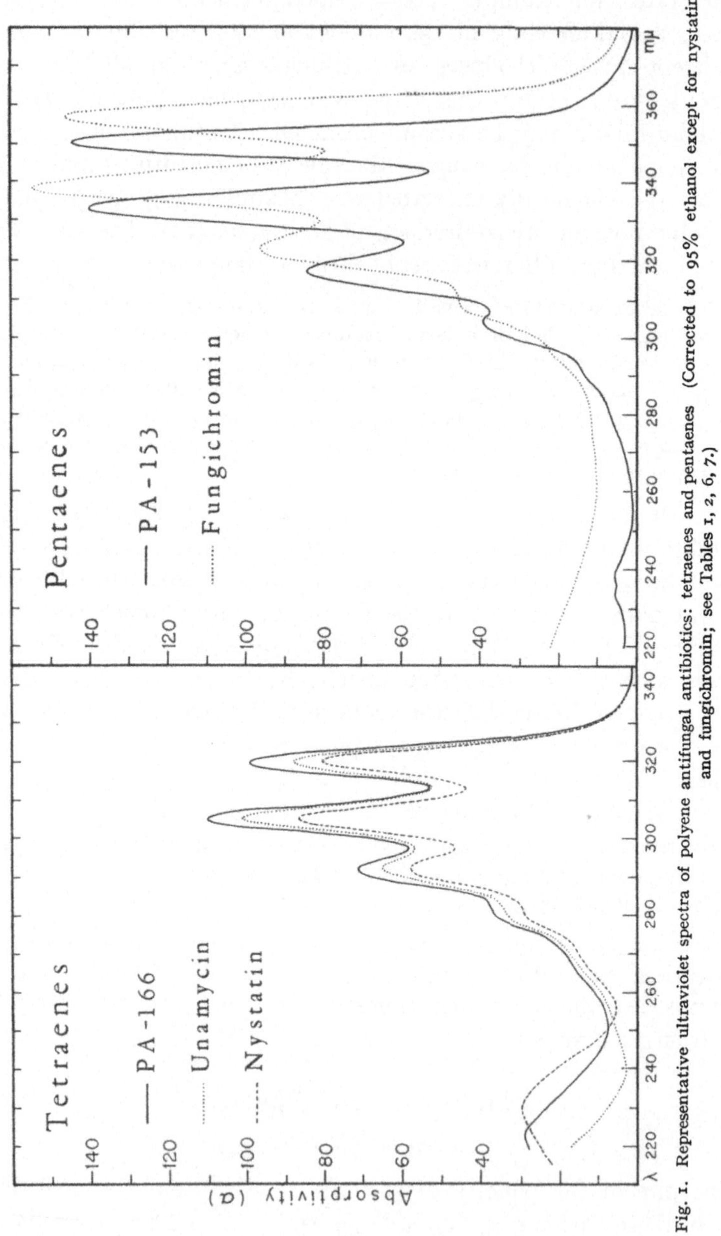

Fig. 1. Representative ultraviolet spectra of polyene antifungal antibiotics: tetraenes and pentaenes (Corrected to 95% ethanol except for nystatin and fungichromin; see Tables 1, 2, 6, 7.)

being discernible in the curve. As the illustrations show, the peak of longest wavelength, which is also the narrowest (the "vibrationless transition"), is either the strongest in the series or but slightly weaker than the second peak.

References, pp. 72—79.

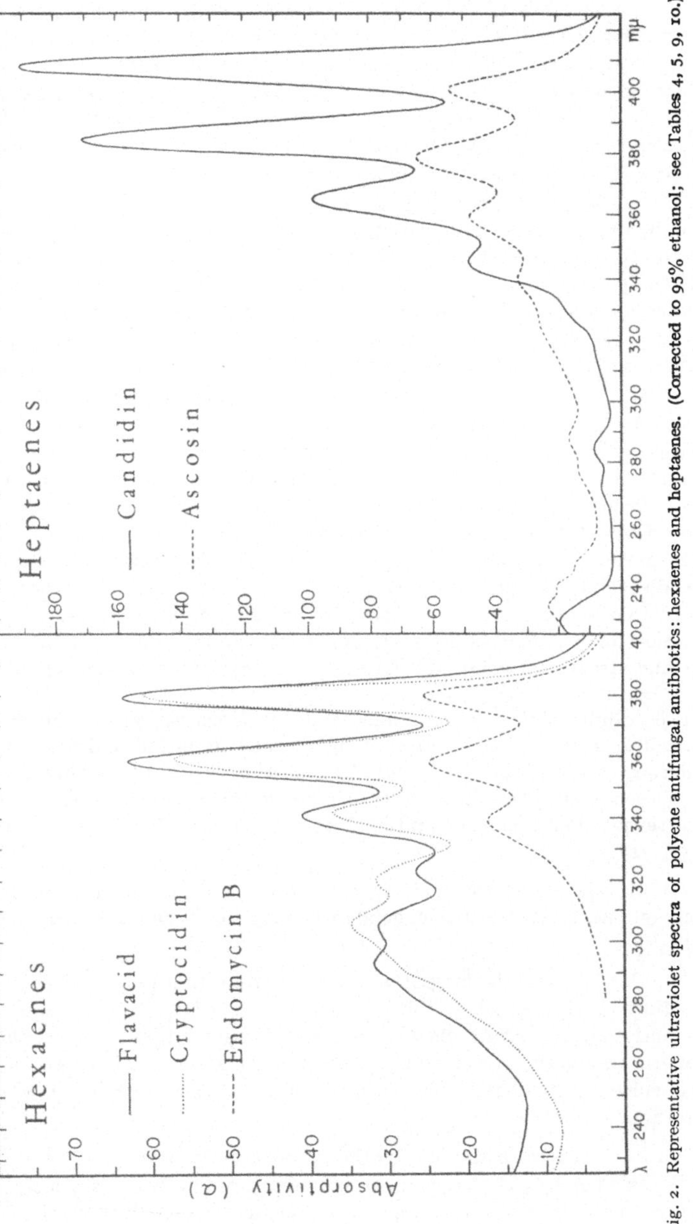

Fig. 2. Representative ultraviolet spectra of polyene antifungal antibiotics: hexaenes and heptaenes. (Corrected to 95% ethanol; see Tables 4, 5, 9, 10.)

Pronounced "vibrational fine structure" of this type, showing an approach to a regular series of discrete spectral "lines", is characteristic of homogeneous, rigid, coplanar systems of conjugation in which oscillation of the photo-excited electrons can occur with little damping. Under ordinary conditions* it is exhibited only by linear polynuclear aromatics, unhindered linear polyenes, and polyacetylenes. Polyacetylenes, whose spectra display the most spectacular "fine structure" of all (62a), can be immediately differentiated from polyenes by the vibrational peak-spacing, which is 200–210/mm (2000–2100 cm^{-1}, 60–63 fresnels) for poly-ynes, and 140–160/mm (1400–1600 cm^{-1}, 42–48 fresnels) for polyenes**. Mixed chromophores, containing both ethylenic and acetylenic bonds in conjugation, show spectra either blurred by the concurrent presence of two incommensurable peak-spacings, or predominantly of poly-yne type.

A carbonyl group in conjugation with a polyene system always "degrades" (blurs) the spectral fine structure, which in polyene ketones or short-chain polyene acids is completely obliterated. In polyene acids or esters containing five or more conjugated double bonds, however, the degradation of shape is less drastic. A second carbonyl group at the other end of the polyene system is spectrally about equivalent to an alkyl group, since the polarity of oxygen does not allow simultaneous conjugation of two oppositely-oriented carbonyl bonds (103a).

The fine-structure degradation produced by *alkyl substitution* of polyenes, unlike that produced by carboxylic conjugation, is characteristically accompanied by the *selective depression of the longest-wavelength peak*, with the result that most polyenes, from dienes to carotenoids, exhibit an absorption curve dominated by the second peak***. Alkyl substitution also produces a bathochromic effect whose magnitude varies according to the location of the substituent, but this does not run parallel with the fine-structure-degrading effect†, which is evidently caused by interaction between the substituent and neighboring hydrogen atoms. Terminal substitution, for example, has the greatest bathochromic effect but the least effect on the fine structure; while internal methyl groups, though they produce batho-

---

* I. e., dilute solution at room temperature. Of course, spectra observed under such conditions are seriously affected by thermal agitation and solvent-molecule participation. Under more nearly "ideal" conditions (such as very low temperature or the gaseous state) in which these disturbing effects are minimized, absorption spectra reveal much more vibrational fine structure than is ordinarily to be seen (53, 71).

** It may be noted that this spacing corresponds to the Raman or infrared "stretching" peaks of the triple bond (ca. 2200 cm^{-1}) and the double bond (ca. 1640 cm^{-1}).

*** In this Review, the longest-wavelength absorption peak will be referred to as the "first" peak, and $\lambda\lambda_{max}$ will be cited in that order. This deviation from the usual practice among organic chemists is desirable here not only because the longest-wavelength peak is the one of primary theoretical significance but also for practical reasons, since it dominates the spectra of these chromophores even at room temperature.

† This is in contradistinction to the bathochromic shifts produced by *solvents*, where elevation of λ_{max} is accompanied by blurring of fine structure; there is a closer analogy to the effects of *heat*, as is reasonable, since in both cases the effects are primarily mechanical. Low temperatures, with their quieting of molecular oscillations, in fact "relieve" these mechanical effects of substituents to a quite surprising extent: even a compound whose substituents overtly obstruct coplanarity, resulting at room temperature in a grossly degraded absorption spectrum with both λ_1 and λ_2 depressed below λ_3, may show at − 185° a spectrum much like that of a wholly unsubstituted polyene (71).

References, pp. 72—79.

chromic shifts of only 3 to 5 mμ, have a degrading effect on fine structure comparable to that of an internal *cis*-double bond*.

In the light of these generalizations, the chromophores responsible for the highly-structured spectra of the fungicidal antibiotics can be unambiguously identified; only in the case of the longest-wavelength group does the spectrum leave room for more than one possible interpretation.

2. The Tetraenes.

As *Table 1* (p. 56) illustrates, even today all of the antibiotics of the shortest-wavelength group (*124*) show $\lambda\lambda_{max}$ all of which fall within 1 mμ of the average values, 319, 304.5, and 291 mμ. Variations of this magnitude between spectra from different sources are small enough to be attributed to instrumental differences. To all appearances, the chromophore in all of these compounds is identical.

Table 1 (p. 56) demonstrates that the peak-spacing observed in this spectrum (and in all the others of this series) is that characteristic of polyenes.

The sharp development of the fine structure, even in this group where it is least pronounced, excluded the presence of a mixed type of chromophore (a polyenyne, polyenone, or polyenoic-acid derivative) and reduced the possibilities to one: a not-too-heavily-alkylated polyene, which by comparison with known examples could only be a tetraene. One of these reference compounds, the α,ω-disubstituted all-*trans* tetraene β-parinaric acid, proved to match the antibiotic spectrum precisely in shape, with a 3.5 mμ difference in λ_{max} (see Table 1) which is attributable to the difference in size of the terminal substituent. (This effect of the size of the substituents can also be seen in the difference between the $\lambda\lambda_{max}$ of β-parinaric acid and those of all-*trans*-decatetraene.)

In view of the fact that a terminal *cis* double bond in a polyene chain causes a bathochromic shift in the ultraviolet spectrum and only slight fine structure degradation, the possibility of such a configuration in the tetraene antibiotics could not be ignored on the basis of ultraviolet evidence alone. To rule out this possibility in the case of the tetraene-antibiotic chromophore, it is necessary to invoke further evidence such as the stability of the spectrum to stereoisomerizing reagents and the absence of *trans* —CH=CH— peaks in the 10.1–10.6 μ region in the infrared. The latter fact also disposes of the possibility that an internal methyl group might be present, although ultraviolet fine structure considerations alone would tend to rule out that hypothesis.

---

* I. e., not one of the two double bonds that terminate the polyene chain. The spectral effect of a *cis* double bond is quite different in the two cases: whereas an internally *cis* polyene shows $\lambda\lambda_{max}$ 3–5 mμ lower than those of its all-*trans* isomer, in the case of a terminal *cis* configuration the reverse relationship is observed. In both cases, however, the *cis* isomer shows a slight degradation of fine structure as compared to the *trans*.

On the basis of the ultraviolet and infrared evidence, the chromophore of this group of antifungal antibiotics can therefore be written, with considerable confidence, as all-*trans* R—$(CH=CH)_4$—R'.

3. The Pentaenes.

The antibiotics listed in *Table 2* (p. 56) again share a single chromophore, which shows a $\Delta\lambda_1$ from the unsubstituted decapentaene of only $15.5 \pm 1 \, m\mu$. The first peak is now equal to the second, or even (as in Fig. 1) slightly higher. The parallel to the tetraenes leaves no doubt that all-*trans* R—$(CH=CH)_5$—R' represents the structure of this chromophore.

4. The Methylpentaenes.

Differing from the pentaenes described in Table 2 are a group whose spectra are bathochromically displaced by 5–$6 \, m\mu$ (*Table 3*, p. 57). The first such antibiotic, fungichromin, was reported in 1954 (*118*). In early 1955 a group of Upjohn workers reported a second example, filipin (from a *Streptomyces* obtained from Philippine Islands soil) and stated that "filipin and fungichromin appear to be the first reported members of a new family" (*134*).

The same chromophore has since recurred in six other antibiotics (cf. p. 64). The 6-$m\mu$ bathochromic effect points either to a terminal *cis* double bond or to an additional substituent which must likewise be located on a terminal double bond, since filipin readily takes up atmospheric oxygen with conversion to the typical tetraene chromophore (*134*). Here the infrared curve is decisive: it shows no *cis* CH=CH—, but displays a conspicuous trisubstituted double bond peak at $11.9 \, \mu$, showing that the latter alternative is the correct one. By 1958 it had been established by chemical methods that the chromophore is in fact all-*trans* R—$C(CH_3)=CH$—$(CH=CH)_4$—R'; this group of compounds may therefore be referred to as the "methylpentaenes".

5. The Hexaenes.

So far, only four antibiotics have been reported whose spectra fall into this class. *Table 4* (p. 57) gives the available data. It will be noted (cf. *Fig. 2*, p. 21) that as the polyene chain is lengthened, the fine structure continues to increase: the first peak is now the highest in the spectrum. The λ_1 difference from dodecahexaene is only $13.7 \, m\mu$. These spectral data are unambiguous, and it can be concluded without reservation that these antibiotics are α,ω-disubstituted all-*trans*-hexaenes.

6. The Heptaenes.

Interpretation of the heptaene spectra (cf. *Table 5*, p. 57) was complicated by the presence of stereoisomers. In contrast to the shorter chromophores, many of the heptaenes are not found in their all-*trans* form. Indeed, some of them present spectra so conspicuously degraded in both λ_{max} and fine structure that the presence of more than one *cis*-bond seems possible: e. g. ascosin (Fig. 2), λ_1 399 mμ; candicidin and trichomycin (52), λ_1 400 mμ (in methanol). Since the *cis*-peak* is not very prominent in the spectra of these *cis* heptaenes**, it may be tentatively inferred that the *cis* double bond present in them is adjacent to the terminal ones—i. e., the second and/or the sixth.

These originally *cis* heptaenes easily undergo light-induced stereo-isomerization to products that show spectra higher in λ_{max} by 5 mμ and with a much more prominent first peak. This isomerization is likely to occur to a greater or lesser extent during isolation: the Rutgers workers (90) noted, for example, that the heptaenes they encountered varied in λ_1 from 401 to 405 mμ, and the change in spectrum of "trichomycin A" between its countercurrent-purified state and its later crystalline state (Table 5) is probably of the same kind. The spectrum at this stage (λ_1 404 + mμ in methanol, 406 — mμ in ethanol) fits neatly into the tetraene-pentaene-hexaene sequence; and if it could be presumed to represent the all-*trans* isomer, this would imply that the heptaenes too are α,ω-disubstituted polyenes. However, the spectrum of this photo-stable isomer does deviate in respect to *shape*, for the first peak is still somewhat weaker than the second; some sort of hindrance is therefore present here that is absent in the pentaenes and hexaenes.

NAKANO (77) found that the 404-mμ form of trichomycin could be further isomerized, by acetic acid in aqueous acetone, to an isomer showing a spectrum of clearly "all-*trans*" (unhindered) form, with the first peak (λ_{max} 409.5 mμ in methanol) the strongest. This λ_1 value points unequivocally to the presence of *three* substituents, not two. As in the methylpentaenes, the infrared spectrum locates the third substituent: the single *trans* CH=CH peak at 9.98 μ rules out internal alkylation, while a strong 11.9 μ peak places the third substituent (unquestionably a methyl group) at the end of the chromophore.

However, although NAKANO's "all-*trans*-trichomycin" is clearly trisubstituted, it seems more likely that it was formed by allylic re-arrangement than by simple stereoisomerization: the inference to be drawn

---

* The *cis*-peak, whose $\lambda\lambda_{max}$ occur at wavelengths ca. $1/\sqrt{2}$ those of the major $\lambda\lambda_{max}$, has been interpreted as resulting from the photo-excitation of electronic vibration in the first-overtone mode (*140*). Its prominence is a measure of the bending of the chromophore by the *cis* configuration (greatest when the *cis* double bond is centrally located) and hence can be used as a clue to the position of the *cis* double bond [cf. ZECHMEISTER (*140*)].

** Although some of the published heptaene spectra do show conspicuous peaks around 300 mμ (*cis*-peak, ca. 289 mμ), this seems likely to be due to contamination with tetraenic congeners.

is then that the original chromophore was —C(CH$_3$)(OH)—(CH=CH)$_6$— —CH=C<, leaving it still equivocal as to whether this was di- or tri-substituted, though the former is certainly more probable.

Another group of heptaenes evidently occurs in nature as all-*trans* isomers, and no definite evidence for any change in their λ_{max} is on record: this is the group typified by candidin (Fig. 2, p. 21) and amphotericin B, with hindrance-free spectra and λ_1 ca. 406 mμ in methanol. The 2-mμ difference between the observed λ_{max} and that expected for the disubstituted heptaene is probably significant, and it is not possible to say with confidence whether this chromophore carries an additional methyl group or not. The IR spectra, which show a double *trans*-CH=CH peak in this group [additional peak near 10.25 μ (*128*, *40*)] might be taken to support a *non*-terminally-methylated structure; another possibility would be a difficultly isomerizable terminal *cis* double bond.

It may be of interest to refer here to a peculiar type of spectral "degradation" caused by physical conditions, hitherto observed only in the heptaenes, which has caused more than one investigator some perplexity. A dramatic alteration in the spectrum reminiscent of the effects of severe steric hindrance may occur when an aqueous alkaline solution of the compound is neutralized; on adding either alkali or alcohol, the distorted spectrum returns to its original shape. This effect has been encountered with trichomycin (*76*), candicidin (*75*), F-17-C (*26*), amphotericin B (*12*), and candidin (*110*), where the situation was made still more confusing because some less-pure fractions failed to show it (*130*). What is happening under these conditions is not a change in the chromophore, but in its physical environment: the polyene molecules, inadequately solvated, are associating with each other in clusters of two or more, tending toward micelle formation (*76*) or colloidal dispersion (*12*, *128*). Addition of a more powerful solvent, such as alcohol, restores the normal spectrum. This is an effect more familiar to the dye chemist than to the investigator of natural products: it has been encountered repeatedly in poorly soluble dyes of various structures, a recent example being sulfonated phthalocyanine, which shows in aqueous alcohol a reversible spectral degradation precisely like that of trichomycin when the concentration of either dye or water is increased. This was shown to be due to association of the molecules to "dimers" and "tetramers" (*101*); see also (*91*).

III. Structural Elucidation.

1. General Features.

Certain generalizations about the structure of these antibiotics, aside from their chromophores, are apparent from the data in *Tables 6–10*, pp. 58–66. They contain only C, H, O, and N*. One subclass, the methylpentaenes, stands out in being devoid of nitrogen, as well as of the carboxyl function which is found in almost all of the others. The one exception within the methylpentaenes is moldcidin A, which contains both amino and carboxyl groups.

The nitrogen present in the other polyenes is normally basic, the one exception being the tetraene unamycin, where the nitrogen appears to be neutral. Usually only one nitrogen atom per molecule has been found; however, one pentaene (capacidin) and several heptaenes contain two. The status of nitrogen (if any) in the hexaenes is not clear. N-methyl groups have not been found in any case examined.

The carboxyl group is present as a zwitterion, indicated in the infrared spectra by a strong peak at 6.35–6.40-μ. Only in unamycin, which has a non-basic nitrogen, is the 5.9-μ peak of an un-ionized carboxylic acid observed. Perimycin is the only example outside of the methylpentaenes that has no carboxyl function.

Invariably present in the infrared spectrum is a sharp peak at ca. $5.83\ \mu$, indicating the presence of an ester or unstrained lactone**. That this function is in fact a lactone was first stated by DUTCHER (*38*) in the case of nystatin on the basis of the fact that after saponification no fragment smaller than the entire molecule could be obtained. To judge by the infrared curves, a similar lactone linkage must be present in every member of the group examined. In 1958 it was recognized by DHAR, THALLER and WHITING (*31*) that the methylpentaene, "lagosin" was in fact a macrocyclic lactone belonging to the then recently recognized (*136*) class of "macrolide" antibiotics; a similar structure was subsequently found in pimaricin and filipin, and there is every probability that all of the polyene antifungal antibiotics are macrolidic lactones.

Another feature common to all of these antibiotics, and sharing responsibility with the zwitterion for their comparative insolubility in organic solvents, is the presence of a large number of hydroxyl groups: this number is now known to be six in pimaricin, ten in fungichromin and filipin, probably thirteen in nystatin (*36*), and fourteen in amphotericin B (*42*). Inspection of the molecular formulas indicates that there are generally about twice as many hydroxyls as double bonds. When so many hydroxyl groups are present, it is difficult to

---

* Aliomycin is reported to contain sulfur, but its purity is not above suspicion.

** The λ_{max} of this peak (5.80–$5.87\ \mu$ in salt wafers or mulls) is higher than the literature values for unconjugated lactones, and in the case of pimaricin (*89*) a smilar peak has been attributed to a C=C—COO system. However, it may be noted that the lactone function of the saturated macrolide dihydroerythromycin gives rise to a peak at the same position.

determine their number with certainty: DJERASSI et al. (35) were able to satisfy themselves that filipin is $C_{37}H_{62}O_{12}$, containing ten hydroxyls, rather than $C_{35}H_{58}O_{11}$ with nine as suggested by the Oxford workers (32), or $C_{33}H_{54}O_{10}$ with eight as previously thought (13), only by close examination of the NMR spectrum of the peracetate. The olefinic protons of the methylpentaene chromophore, whose number is known to be nine, produce a group of overlapping peaks between $\delta\,5.46$ and $\delta\,6.81$, and the —COO—CH< protons another cluster between 4.23 and 5.46. Since the measured total areas of these peak clusters gave a ratio convincingly close to 9/11, it could be concluded that the molecule contained eleven acylated secondary hydroxyls, viz., the lactone and ten acetate residues.

2. Mycosamine.

A major advance in the structural elucidation of the polyene antibiotics was made in 1956 when DUTCHER's group at Squibb succeeded, after many deceptively negative results (41), in demonstrating the presence of a carbohydrate moiety in amphotericin B (42). By vigorous acetolysis it was possible to cleave from the molecule a sugar which proved to be the site of the basic nitrogen atom. There was thus obtained a mixture of crystalline tri- and tetraacetates of a new deoxyhexosamine, $C_6H_{13}O_4N$, which was named "mycosamine"; by periodate degradation of its N-acetyl derivative and the corresponding methyl glycoside (133), it was subsequently shown to have the structure (I). The configuration illustrated, which represents the 3-amino analogue of D-rhamnose (6-deoxy-D-mannose), was established in 1961 by synthesis ($130\,a$). (The methylglycoside preferentially forms the pyranoside α-anomer (I p) as shown, but it is not certain whether this applies also to the glycoside linkage existing in the antibiotic.) It is interesting to note that mycaminose (II), which occurs in carbomycin and several other macrolides, is now known ($93\,a$) to be the 3-dimethylamino analogue of 6-deoxy-D-glucose, epimeric with mycosamine.

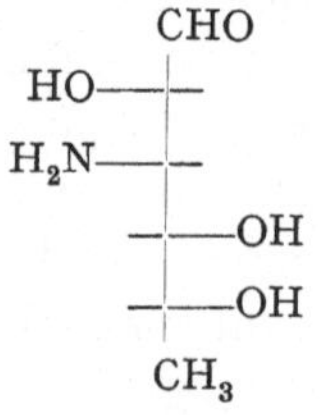

(I.) Mycosamine. 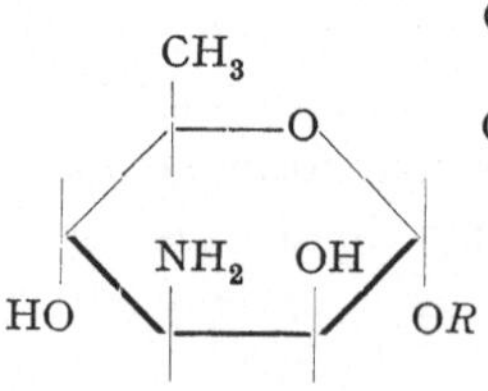   (I p.) (α-pyranoside) 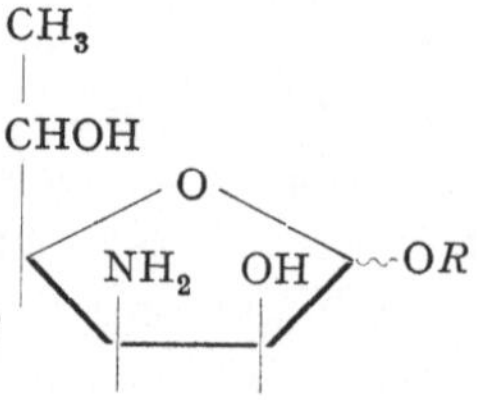   (I f.) (furanoside) 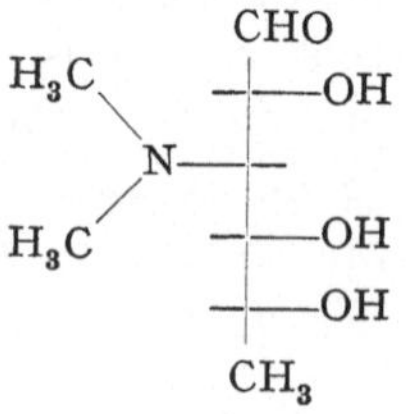   (II.) Mycaminose.

Mycosamine has been found to be present in nearly all of the polyene antibiotics that have been examined. It has been isolated from nystatin (38, 42), pimaricin (89), candidin (99), and trichomycin (77), and is known to be demonstrable in several other members of the group as well (*98*). However, amino sugars differing from mycosamine have already been encountered. Sulfuric-acid-catalyzed hydrolysis of the unique non-carboxylic heptaene, perimycin, liberated a new amino sugar which was given the name "perosamine" (*18*).

All that has yet been published on perosamine is the fact that, unlike mycosamine and mycaminose, it is not a 6-deoxy sugar (since it yields no acetaldehyde on periodate fission). It is a primary amine: no N-methylated amines have yet been observed in this series. The only other antibiotic in which perosamine has been detected is the tetraene PA-166 (*98*).

The ease of scission of the glycoside link apparently is not the same in all of the antibiotics, a circumstance which must reflect structural differences. In pimaricin and candidin (*98*) the sugar is rather readily freed by normal acid-catalyzed hydrolysis or alcoholysis, whereas in amphotericin B and nystatin these procedures were not effective. In the case of trichomycin A, NAKANO (*77*) found that mycosamine could be removed readily by alcoholysis of the perhydro polyene, though it was isolable in a pure state only if the amine had first been protected by dinitrophenylation. In this case it was noted that perhydrogenation of trichomycin A also liberated some mycosamine—a hydrogenolysis which suggested that the glycosidic ether was allylic (i. e., adjacent to the heptaene chromophore).

A point that has not been unequivocally cleared up is the curious circumstance that many mycosamine-containing antibiotics liberate ammonia on alkaline treatment [e. g. pimaricin (*89*)]. Although mycosamine itself might be expected to lose ammonia under alkaline conditions, its glycosides would not, indicating that mycosamine is freed under these seemingly inappropriate conditions. To account for the fission of the glycoside bond with alkali, it seems necessary to suppose that the glycosidic oxygen is aldolic (β- to a carbonyl group)* or, more subtly, *potentially* aldolic. In PATRICK's formula for pimaricin (*89*) (see p. 42) this C—O— bond is both allylic and potentially aldolic.

The isolation of the mycosamine fragment did more than identify the N atom: it enabled DUTCHER to draw a sweeping structural deduction: "Inspection of the empirical formula of perhydroampho-

---

* Similar alkali-induced eliminations of sugar moieties are frequently observed in the previously known macrolides, where they are accounted for by structures such as —CH(O—Glyc.)—CH(CH$_3$)—COO— (methymycin); —CH(O—Glyc.)—CH(CH$_3$)—CH(O—Glyc.)—CH(CH$_3$)—COO— (erythromycin); and —CH(O—Glyc.)—CH(OCH$_3$)—CH(OOCCH$_3$)—CH$_2$—COO— (carbomycin), in all of which the carbonyl group activating the β-elimination(s) is that of the *lactone*. Whether there is an exact structural analogy here remains to be seen.

tericin B $(C_{46}H_{87}O_{20}N)$ reveals that after subtraction of the known functional groupings (carboxyl, lactone, and glycoside), the remainder of the molecule is strictly saturated in composition, implying that the remainder of the oxygen atoms must be of hydroxyl nature." Moreover, "Titration studies with periodic acid do not suggest a contiguous arrangement for more than two or three" of these twelve remaining hydroxyls (*42*). The same conclusions could be drawn in the case of nystatin (*38*). As will be seen below, *a long chain of non-contiguous hydroxyls*, such as first deduced by Dutcher, is indeed a characteristic feature of these molecules; and the conclusion that *the lactone ring is the only ring present* is also one of general validity.

3. Retro-Aldol Cleavage.

Save for the scission of the glycosidic ether linkage, all degradative reactions which break down these molecules into smaller fragments require the cleavage of carbon-carbon bonds. One of these, the production of an aldehyde, presumably by the "retro-aldol fission" of an original aldol system *(Chart 1)*, has been found to occur spontaneously under

$$R\text{—}\underset{\underset{\text{OH}}{|}}{\text{CH}}\text{—}\underset{\underset{\text{O}}{\|}}{\overset{|}{\text{C}}}\text{—}\text{C}\text{—}R' \xrightleftharpoons{\text{HO}^-} R\text{—CHO} + \underset{\underset{\text{O}}{\|}}{\overset{|}{\text{HC}}}\text{—C}\text{—}R'$$

Chart 1. The Retro-aldol Reaction.

fairly gentle alkaline conditions, e. g. 3 days at room temperature in $1\,N$ sodium hydroxide (*77*). The formation of *hexanaldehyde* ($R = n$-amyl) in this way was first observed in 1959 by Cope's (*23*) and Whiting's groups (*33*) working independently on fungichromin and lagosin (as a matter of fact, this is the only small fragment that can be successfully cleaved from the fungichromin molecule). Almost immediately thereafter, the same product was reported from the similar methylpentaene, filipin (*13*). From the heptaene trichomycin A, there is obtained by retro-aldol cleavage in alkali an entirely different carbonyl compound, viz. *p-amino-acetophenone* (*77*), which proves to account for the second nitrogen atom known by analysis to be present in this molecule; but here the sense of the cleavage cannot be the same as that of hexanal, for ultraviolet evidence demonstrates that trichomycin contains a *p*-aminobenzoyl chromophore. The retro-aldolization in this case must therefore be formulated as shown in *Chart 1* with $R' = H_2N\text{—}C_6H_4\text{—}$ and R corresponding to the remainder of the trichomycin molecule (*77*).

From perimycin, probably by the same mechanism, still another product is obtained, *p-aminophenylacetone* (*18*) ($R' = p\text{-}H_2N\text{—}C_6H_4\text{—}CH_2\text{—}$), but from alkali treatment of nystatin, the amphotericins, or candidin, it seems that no side-chain fragment of this sort is isolable.

Alkali treatment can also cause more deep-seated rupture, cleaving the fundamental carbon chain, presumably also by retro-aldolization. In this way chain fragments have been isolated which contain the chromophore in a much altered form, evidently having undergone two successive stages of alkali-catalyzed reaction. Thus, the heptaene trichomycin A, after overnight treatment with 2 N-sodium hydroxide, yielded an *octa*enal of undetermined structure (*77*); in a similar reaction, the tetraene pimaricin gave orange crystals, λ_{max} 375 mμ (methanol), proved by degradation to be 13-hydroxytetradecapentaenal (III) (*89*). (This product, incidentally, furnishes unambiguous confirmation of the structure of the tetraene chromophore.) PATRICK (*89*) postulated the following mechanism as the most plausible reconstruction of this rather remarkable transformation (*Chart 2*).

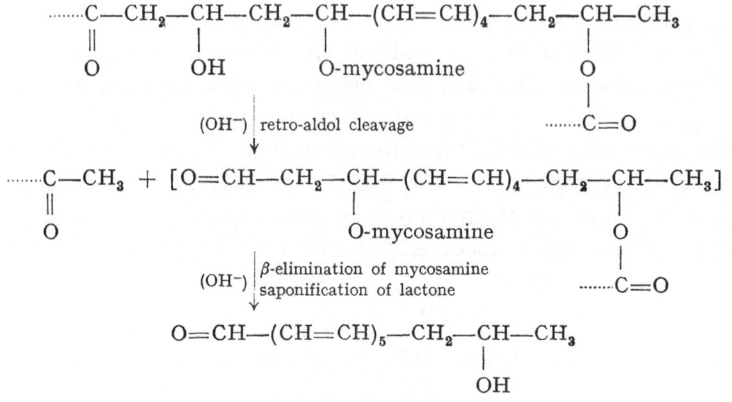

$$\cdots\cdots C-CH_2-CH-CH_2-CH-(CH=CH)_4-CH_2-CH-CH_3$$

with pendant groups: $\parallel$ under first C gives O; OH under second CH; O-mycosamine under the third CH; O under the CH near the end.

$$(OH^-)\ |\ \text{retro-aldol cleavage}$$

$$\cdots\cdots C-CH_3 + [O=CH-CH_2-CH-(CH=CH)_4-CH_2-CH-CH_3]$$

with $\parallel$ O under the first C; O-mycosamine under the CH; O and $\cdots C=O$ under the final CH.

$$(OH^-)\ |\ \beta\text{-elimination of mycosamine}$$
$$\text{saponification of lactone}$$

$$O=CH-(CH=CH)_5-CH_2-CH-CH_3$$
$$|$$
$$OH$$

(III.) 13-Hydroxytetradecapentaenal.

Chart 2. Alkaline Cleavage of Pimaricin.

It will be observed that this single reaction, as formulated in Chart 2, involves no less than seventeen carbon atoms—half of the total number present in pimaricin.

That the lactone ring does actually extend to the position shown, enclosing the entire polyene chain, would of course not be established by this reaction alone; in fact, no coercive evidence on this point was presented by the Lederle workers (*89*).

However, there can be no doubt that PATRICK's placement of the chromophore inside the lactone ring is correct, in the light of the complete structural solutions independently arrived at by the research teams of COPE (M. I. T.) (*22*), WHITING (Oxford) (*31–33*), and DJERASSI (Stanford)

(*35*) for three* of the methylpentaenes, viz., *lagosin, fungichromin* and *filipin*.

These molecules, though large, and in practice most awkward to deal with, actually proved to be very simple in construction. As has already been mentioned, they lack such "complications" as an amino sugar or a carboxyl group; they contain a pentaene which appears from its spectrum to be terminally trisubstituted, a carbonyl group absorbing at $5.83\,\mu$ which by analogy is probably a lactone, and a liberal amount of oxygen. It turned out that they contain *nothing else*: the only ring present is that of the lactone, the only unsaturation is that of the pentaene, and all of the remaining oxygen atoms are hydroxylic.

4. Fungichromin (Lagosin).

The Oxford workers were the first to publish their solution, completed in 1960. "Lagosin" or Glaxo-A-246, a crystalline methylpentaene from a soil-dwelling *Streptomyces* (*11, 13a*), was originally (*31*) formulated as $C_{41}H_{66-70}O_{14}$, but eventually revised to $C_{35}H_{58}O_{12}$ (*32*). The whole structure was deduced from a careful study of the products of a single reaction, periodic-acid fission. The most obvious product was that derived from the chromophore, which now turned up conjugated to a carbonyl group. A hydroxyl and an ester group were also present. Saponification removed a large polyhydroxyacid moiety and freed a dihydroxyaldehyde. The two hydroxyl groups of this product were shown to be adjacent, by periodic acid cleavage to acetaldehyde and the identifiable dialdehyde (IV)

$$
\begin{array}{c}
\mathrm{CH_3} \\
| \\
\mathrm{O{=}CH{-}C{=}CH{-}(CH{=}CH)_4{-}CHO}
\end{array}
$$

(IV.) 2-Methyldodecapentaenedial.

$$
\begin{array}{c}
\mathrm{CH_3} \\
| \\
\mathrm{CH_3{-}CH{-}CH{-}C{=}CH{-}(CH{=}CH)_4{-}CH{=}O} \\
\;\;\;\;\;| \;\;\; | \\
\;\;\;\;\mathrm{OH}\;\;\mathrm{OH}
\end{array}
$$

(V.)

$$
\begin{array}{c}
\mathrm{CH_3} \\
| \\
\mathrm{O{=}CH{-}C{=}CH{-}(CH{=}CH)_4{-}CH{-}CH{-}CH_3} \\
\;| \;\;\; | \\
\;\mathrm{OH}\;\;\mathrm{OH}
\end{array}
$$

(VI.)

---

* Or, as it now appears, *two*, since the structures independently found for lagosin and fungichromin were the same, and it seems nearly certain that the compounds are identical (*22*). However, we will here retain the name "lagosin" for the antibiotic studied by the Oxford group, although in future this name may be dropped as a synonym.

References, pp. 72—79.

(structure proved by conversion to isotridecanoic acid). This degradation verified the methylpentaene chromophore, and showed that the dihydroxy-aldehyde mentioned above must have been either (V) or (VI).

That (VI) was correct was shown by derivatizing the aldehyde group (through condensation with acetone) before the second periodic acid fission: the keto-aldehyde thus obtained could be identified as H_3C—CO—CH=CH—$C(CH_3)$=CH—$(CH=CH)_4$—CHO rather than the isomeric product deriving from (V). The position of the ester group in the precursor, the glycol monoester, could be settled by the observation that the free hydroxyl was allylic, being oxidizable by manganese dioxide or chloranil to a conjugated ketone. These reactions therefore established (VII) as the structure of nearly half of the molecule (31).

$$\begin{array}{c}
\qquad\qquad CH_3 \\
\qquad\qquad | \\
C\!-\!-CH\!-\!C\!=\!CH\!-\!(CH\!=\!CH)_4\!-\!CH\!-\!CH\!-\!CH_3 \\
|\quad\ | \qquad\qquad\qquad\qquad |\quad\ | \\
OH\ OH \qquad\qquad\qquad\qquad OH\ O \\
\qquad\qquad\qquad\qquad\qquad\qquad\qquad | \\
\qquad\qquad\qquad\qquad\qquad\qquad\ C\!=\!O
\end{array}$$

(VII.)

At this point (early 1958) it was clear that the antibiotic was a "macrolide" whose macrocyclic lactone ring enclosed the entire pentaene chromophore. This finding sets a minimum size for the ring. WHITING (31) pointed out that the ultraviolet spectrum, with its sharp vibrational structure and absence of *cis*-peak, was consistent only with a *trans* configuration of (at least) the three internal double bonds and normal unstrained "transoid" conformation of the connecting single bonds:

But to close an unstrained ring from end to end of this rigid ten-carbon system requires, as models show, a bridge of at least thirteen more atoms if one terminal bond is *cis;* fifteen, if both are *trans.* Hence, "the lactone ring must be of at least twenty-three atoms".

By 1959 (33), the nature of this bridge around the chromophore had been almost completely elucidated. In confirmation of the findings of the M. I. T. group (23) working on fungichromin, the formation of hexanal by retro-aldol fission had been established; and the location of this aldehydogenic side-chain next to the lactone carbonyl had been confirmed by success in the arduous enterprise of identifying the carbon skeleton of the remainder of the molecule. To this end, the polyhydroxylated acid formed on saponification of the (VI) ester was isolated, exhaustively deoxygenated (sodium borohydride, phosphorus-hydriodic acid, Raney nickel-alkali, platinum-hydrogen) and converted to its methyl ester. This gave "a very complex mixture" which could, however, be separated by preparative-scale gas chromatography on a 5000-plate column. The major component (ca. 30%), after rechromatography, was identifiable by its mass spectrum as the methyl ester of the C_{19}-acid (VIII),   thus

$$C_6H_{13}\text{—}CH\text{—}(CH_2)_{10}\text{—}CH_3$$
$$|$$
$$COOH$$

(VIII.) 2-*n*-Hexyltridecanoic acid.

accounting for a total of thirty-four carbon atoms. It was thought that seven still remained undetected (presumably lost as a third fragment from the original periodate oxidation, which was known to have broken *two* bonds, since it consumed two equivalents), but on subsequent search, no such seven-carbon fragment could be found. Eventually, re-examination of the molecular-weight evidence indicated that the empirical formula should be reduced to $C_{35}H_{58}O_{12}$ (*32*), which implied that the third unobserved fragment had actually contained only *one* carbon atom; and when the periodate fission was repeated in aqueous suspension, it was in fact possible to isolate formic acid.

The carbon skeleton was therefore completely accounted for, and it remained only to determine the position of the remaining seven oxygen atoms (none of which was carbonylic; all, consequently, in the form of hydroxyl). As it happened, this task could be achieved at a single stroke, for there is but one way in which seven hydroxyls can be distributed on the C_{19}-skeleton so as to fulfil the conditions that no two shall be adjacent (since the C_{19}-moiety was stable to periodate) and that hexan-aldehyde be obtainable by retro-aldolization. The resulting complete structure (IX) for lagosin (*32*) is shown below.

In the meantime, the M. I. T. group under Cope was establishing the structure of fungichromin by the application of a different sequence of degradations, with results which were published in full in early 1962 (*22*). The structure thus found for fungichromin was also (IX), and direct

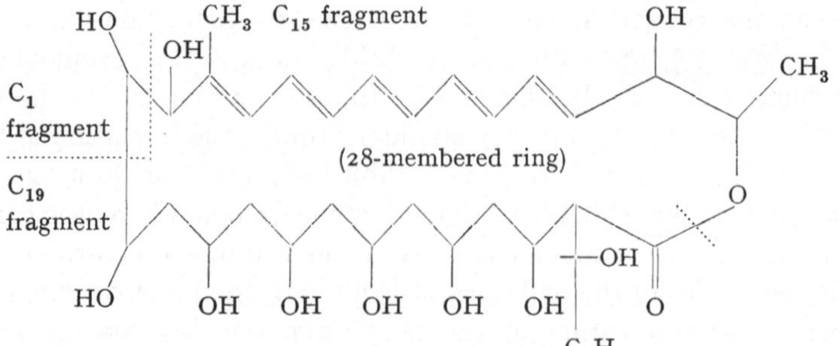

HO   CH₃ C₁₅ fragment   OH

OH

C_1 fragment   CH₃

(28-membered ring)

C_{19} fragment   O

—OH

HO   OH OH OH OH OH   O

C_5H_{11}

(IX.) Fungichromin (Lagosin).

comparison showed that the only detectable (slight) differences between fungichromin and lagosin were in the X-ray pattern and the $[\alpha]_D$ value (10% higher for fungichromin). Almost certainly the compounds are in fact identical.

Here too the originally-determined empirical formula, $C_{35}H_{60}O_{13}$ (*118*), was found to be incorrect, since after absorption of five equivalents of hydrogen the crystalline perhydro derivative analyzed as $C_{35}H_{68-70}O_{12}$; the apparent loss of an oxygen was explained by the finding that the original fungichromin crystals were in fact a stable monohydrate. The 5.83-μ lactone peak in the infrared was observed to be unchanged after perhydrogenation, showing that it could *not* in fact originate from an α,β-unsaturated ester, as its wavelength might suggest.

The degradative approach adopted at M. I. T. was to "obtain a derivative retaining all of the carbon atoms of fungichromin in their original arrangement and yet simple enough for direct proof of structure" by exhaustive deoxygenation of perhydro-fungichromin (similar to the Oxford group's treatment of their C_{19}-acid). Lithium aluminum hydride followed by phosphorus-hydriodic acid, lithium aluminum hydride again, and finally catalytic hydrogenation gave in 13% overall yield a *homogeneous* saturated hydrocarbon, whose molecular weight by mass spectrometry was 492, corresponding to $C_{35}H_{72}$. The mass spectrum exhibited prominent peaks at 323 (C_{23}) and 407 (C_{29}) which showed easy loss of C_6- or C_{12}-fragments, and suggested the structure 7,21-dimethyltritriacontane (X); the identification was proved by direct comparison with a synthetic specimen of this hydrocarbon. Thus, the entire carbon skeleton was conclusively verified from this single derivative.

$$H_3C\text{---}(CH_2)_5\text{---}\underset{\underset{CH_3}{|}}{CH}\text{---}(CH_2)_{13}\text{---}\underset{\underset{CH_3}{|}}{CH}\text{---}(CH_2)_{11}\text{---}CH_3$$

(X.) 7,21-Dimethyltritriacontane.

$$HOCH_2\text{---}\underset{\underset{CH_3}{|}}{CH}\text{---}(CH_2)_9\text{---}\underset{\underset{OH}{|}}{CH}\text{---}\underset{\underset{OH}{|}}{CH}\text{---}CH_3$$

(XI.)

To locate the oxygen atoms, periodic acid fission (in this case, of perhydrofungichromin) was again resorted to. Lithium aluminum hydride reduction then gave two crystalline polyols, $C_{19}H_{40}O_8$ and $C_{15}H_{32}O_3$, which could be deoxygenated to hydrocarbons identifiable by transpose mass spectrum as $C_6\text{---}C(CH_3)\text{---}C_{11}$ and $C\text{---}C(CH_3)\text{---}C_{12}$. The eight hydroxyls of the C_{19}-polyol could be located unambiguously on its carbon skeleton by the same two criteria previously mentioned, i. e.

presence of $n\text{-}C_5H_{11}$—CH(OH)—, absence of any vicinal glycol. As for the C_{15}-triol, two of its three hydroxyls were present as H_3C—CH(OH)— —C(OH)< (with periodic acid → acetaldehyde in 76% yield), and origin by scission of the C_{35}-carbon skeleton requires the third to be located at the other end of the chain; hence it can only have structure (XI). The presence of a periodate-sensitive α-glycol in this fragment must mean that, before lithium aluminum hydride treatment, one of the two glycol hydroxyls had been protected by esterification (the lactone). This was verified by demonstrating that alkaline saponification of perhydrofungichromin does in fact "unmask" a glycol, enabling it to consume a third equivalent of periodic acid. That the esterified hydroxyl is the "outer" one *(Chart 3)* was deducible from the conversion of fungichromin ozonide (via hydrogenation, periodic-acid fission, and treatment with lithium aluminum hydride) to propylene glycol, in 39% yield. This must have arisen by glycol fission of (XII), a reaction which would not have occurred if the monoester had been (XIII).

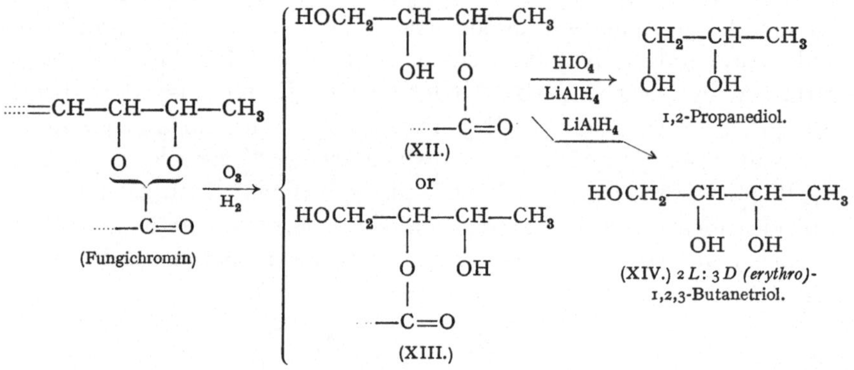

$$\cdots\text{=CH—CH—CH—CH}_3 \;(\text{with } O, O, \cdots\text{—C=O}) \;(\text{Fungichromin}) \xrightarrow[\text{H}_2]{\text{O}_3}$$

$$\left\{\begin{array}{l}\text{HOCH}_2\text{—CH—CH—CH}_3 \;(\text{OH, O, } \cdots\text{—C=O}) \quad (\text{XII.})\\[1ex]\text{or}\\[1ex]\text{HOCH}_2\text{—CH—CH—CH}_3 \;(\text{O, OH, } \cdots\text{—C=O}) \quad (\text{XIII.})\end{array}\right.$$

$$\xrightarrow[\text{LiAlH}_4]{\text{HIO}_4} \quad \text{CH}_2\text{—CH—CH}_3 \;(\text{OH, OH}) \quad \text{1,2-Propanediol.}$$

$$\xrightarrow{\text{LiAlH}_4} \quad \text{HOCH}_2\text{—CH—CH—CH}_3 \;(\text{OH, OH}) \quad (\text{XIV.}) \; 2\,L\!:\!3\,D\;(erythro)\text{-1,2,3-Butanetriol.}$$

Chart 3. Identification of the Esterified Hydroxyl Group in Fungichromin.

An interesting further point is that the propylene glycol was found to be optically active (*D*-enantiomer). By repeating the reactions with omission of the periodic acid treatment, compound (XIV) was obtained in 45% yield, thus establishing the configuration of these two hydroxyls.

Although all of this work was done on perhydrofungichromin, there is no ambiguity about the nature and location of the chromophore, which the M. I. T. workers had already verified as early as 1958 by the isolation in good yield of the crystalline dialdehyde (IV, p. 32) from periodate cleavage of saponified fungichromin (*23*).

With this information available, it was only necessary to "assemble" the C_{15}- and C_{19}-fragments [interpolating a —CHOH—, as dictated by the total carbon skeleton, (X)] and "insert" the five double bonds of the chromophore, to arrive at structure (IX) in a direct and elegant fashion.

References, pp. 72—79.

5. Filipin.

Filipin is essentially identical to fungichromin in its ultraviolet spectrum and resembles it also in lacking amine and carboxyl functions. However, the two can be clearly separated by paper chromatography, filipin being more rapidly eluted. The saponification equivalent suggested an empirical formula of $C_{30}H_{50}O_{10}$ (*134*), on which basis its hydrogen uptake on catalytic hydrogenation was four equivalents. As noted by DUTCHER in 1957 (*38*), this could not be correct, in view of the obvious presence of *five* double bonds, so that the molecular weight must actually be about 5/4 that originally assumed.

The required 25% adjustment was ultimately made, by way of "$C_{32-33}H_{54}O_{10}$" in 1959 (*13*) and "$C_{35}H_{58}O_{11}$" in 1960, as proposed by WHITING et al. (*32*) on the hypothesis, which was nearly correct, that filipin would prove to be deoxy-"lagosin". Eventually, as mentioned above, the formula was established as $C_{37}H_{62}O_{12}$ (hydrogen uptake exactly five equivalents), by NMR-spectral assay of sec.-acyloxy groups in the fully acetylated derivative (*35*).

As with "lagosin", alkali treatment freed hexanaldehyde (*13*), indicating an aldolic side-chain which (for lack of any other carbonyl group) must necessarily be α- to the lactone.

Although the filipin chromophore is clearly the same as that of fungichromin, a difference in its environment is indicated by the facile selective removal of the terminal trisubstituted double bond with uptake of one oxygen atom [e. g. by spontaneous air oxidation of a concentrated alcoholic solution (*134*)], a reaction which was not noted for fungichromin or "lagosin"*. This difference was confirmed when filipin was found to contain no vicinal glycol group (no reaction with periodic acid) until after saponification of the lactone, which unmasked a glycol that consumed periodic acid with production of acetaldehyde and a pentaenal of λ_{max} 386 mμ (*13*).

Since the filipin carbon chain could not be broken beyond the chromophore by periodic acid, a less selective reaction, nitric-acid oxidation, had to be resorted to: by this degradation, perhydrofilipin gave a mixture of acids whose largest identifiable member (by gas chromatography of the methyl esters) was α-methyldodecanedioic acid (XV). More abundant was the next lower homologue, α-methyl-undecanedioic acid, but this could be considered an artifact of over-oxidation. In fact, it was later found (*35*) that by starting not with

---

* The nature of this reaction, which evidently requires the presence of the homoallylic methylene group, has not been elucidated, but the 318, 303, 290 mμ tetraene system produced is apparently stable to concentrated sulfuric acid, with which it gives the typical wine-red tetraene color instead of the blue-violet of the original filipin (*134*).

perhydro-filipin but with the aldehyde formed by periodate cleavage of its saponification product, (XV) could be made the principal oxidation product as would be expected; an effect which incidentally showed that the masked glycol must be located at the *un*methylated

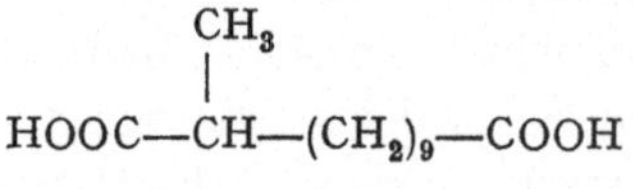

$$\overset{\displaystyle CH_3}{\underset{}{|}}$$
HOOC—CH—(CH_2)_9—COOH

(XV.) α-Methyldodecanedioic acid.

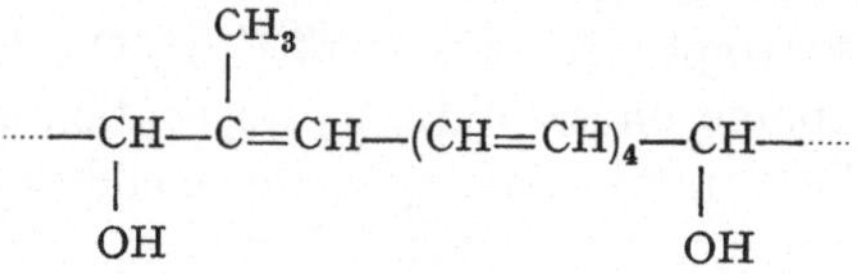

$$\cdots\cdots—CH—\overset{CH_3}{\overset{|}{C}}=CH—(CH=CH)_4—CH—\cdots$$
$$\quad\;\; | \qquad\qquad\qquad\qquad\qquad\quad |$$
$$\quad\;\; OH \qquad\qquad\qquad\qquad\qquad OH$$

(XVI.)

(no OH here)

$$\cdots—C—CH—\overset{CH_3}{\overset{|}{C}}=CH—(CH=CH)_4—CH—CH—CH_3$$

with OH below the second C, and O—O below the —CH—CH—CH_3 group connecting to:

C=O
C—
H—C—OH
C_5H_{11}

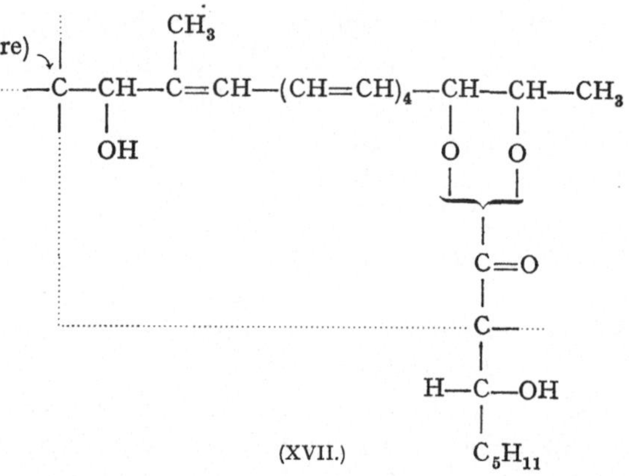

(XVII.)

end of the chromophore. As these α-methylated dibasic acids were not obtainable from filipin itself with nitric acid, they evidently represent the perhydrogenated chromophore, which must accordingly have been (XVI) as in fungichromin (*23*). In view of the periodate and retro-aldol results, this could be expanded to (XVII), or—not yet excluded in 1959—the isomer having the methyl group at the other end of the chromophore (*13*).

Nothing was known about the remainder of the ring except that it contained a number of non-adjacent (periodic acid-resistant) hydroxyl groups and no other functional groups. In view of the extensive resemblance to fungichromin-"lagosin" disclosed by this publication, it was natural for the Oxford workers to venture the suggestion (*32*) that filipin differed from "lagosin" *only* in lacking the central homoallylic hydroxyl which permits the periodate cleavage of that molecule. This proved to be correct except for the assumption that both molecules had a C_{35}-skeleton: the establishment of the C_{37}-formula showed that the chain of —CH_2—CH(OH)-units in filipin must contain seven "links" instead of six.

As in the previous cases, the situation is one in which the complete structure of this chain, including its carbon skeleton, may actually be deduced *without* any direct degradative evidence. It is only necessary to know that a $C_{15}H_{30}O_8$ fragment of some sort connects the chromophore to the α-carbon atom of the lactone, and that this fragment contains no ethers or primary hydroxyl groups and is inert to periodic acid: the formula (XVIII) then follows immediately as the only possible structure that meets these specifications. The absence of primary hydroxyl must however be established, in order to exclude branched-chain solutions such as (XIX). In the case of filipin the absence of primary alcohol groups was verified by consistent inability to prepare a trityl ether; formula (XX)* for the antibiotic then followed directly (*35*).

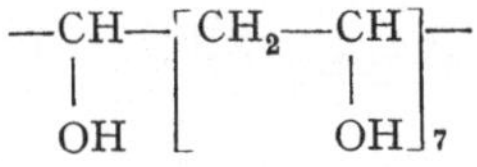

$$-\text{CH}-\left[\text{CH}_2-\text{CH}\right]_7-$$
$$\quad\ \ |\qquad\qquad |$$
$$\ \ \text{OH}\qquad\quad \text{OH}$$

(XVIII.)

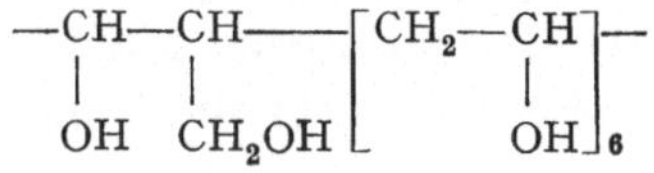

$$-\text{CH}-\text{CH}-\left[\text{CH}_2-\text{CH}\right]_6-$$
$$\quad\ \ |\qquad |\qquad\qquad |$$
$$\ \ \text{OH}\quad \text{CH}_2\text{OH}\qquad \text{OH}$$

(XIX.)

(30-membered ring)

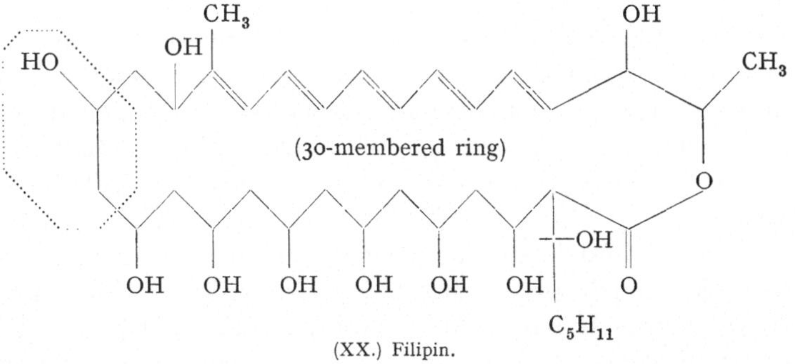

(XX.) Filipin.

[Differs from fungichromin (IX) only in the presence of the circled atoms and the absence of a homoallylic OH.]

As Djerassi et al. remark (*35*), filipin "represents the largest macrolide ring which has so far been encountered in nature".

The heptaenes, however, will certainly be found to contain larger rings than this. Models show that to "string" a strain-free chain from end to end of an all-*trans* polyene of n double bonds requires at least $2\,n + 4$ atoms—better $2\,n + 6$; but since all-*trans* heptaenes are known in which the lactone bond is intact, these must assuredly contain macrolide rings of at least $4\,n + 4$, or thirty-two atoms. [Nakano (*77*), on the same basis, more parsimoniously stipulates "at least 31".] However, $4\,n + 6$ (34 atoms) is a more realistic lower limit for the actual size of the heptaene macrolide ring.

6. Other Methylpentaenes.

Some chemical work has been reported on other methylpentaenes. Pentamycin (*123*) is clearly differentiable from filipin (its infrared spectrum

---

* Actually, the cited reference still left open the possibility that the lactone ring might be 29-membered (closed to the allylic rather than the homoallylic hydroxyl). Formula (XX) arbitrarily shows it closed as in fungichromin.

shows more hydroxyl and other differences, and it is less soluble in organic solvents), but not from fungichromin, with which its elemental analysis* and infrared spectrum** are in rather good agreement. Like fungichromin (*23*), it is cleaved by periodate (after saponification) to produce in good yield the dialdehyde (IV, p. 32) containing the methylpentaene chromophore (*121*). No reactions distinguishing pentamycin definitely from fungichromin have yet been reported. "Moldcidin B", another methylpentaene discovered in 1959 by Japanese researchers, proved on comparison to be identical with pentamycin (*83*). Another antibiotic likewise discovered by Sakamoto, moldcidin A (*97*), is unique in combining a methylpentaene chromophore with the basic nitrogen atom and carboxyl group (visible as COO^- in the infrared curve) of the unmethylated series.

7. Pimaricin.

Pimaricin is the only "normal" polyene antibiotic—i. e. one containing amine and carboxyl functions—for which a complete structure has yet been proposed. This solution, by Patrick and co-workers (*89*) at Lederle, was published as a "Communication" in 1958, thus antedating the solutions of the methylpentaenes; however, complete details have not yet appeared.

The antibiotic used was a substance isolated at Lederle, which had been identified as pimaricin by comparison with authentic material (*109*) from Holland. An empirical formula of $C_{34}H_{49}O_{14}N$ was assigned (previously given as $C_{30-32}H_{45-50}O_{13}N$) (*109*). Functional groups detected were primary amine (van Slyke), carboxyl and "a somewhat hindered keto group". The lactone present was apparently α,β-unsaturated in this case (the infrared λ_{max} of $5.84\,\mu$ shifted on perhydrogenation to $5.77\,\mu$, and the non-tetraenic absorption peak in the ultraviolet, λ_{max} 222 mμ, ε 22,000, also disappeared). Perhydrogenation required six equivalents of hydrogen, one more than the tetraene and the conjugated lactone would account for. The appearance of a new acetylatable hydroxyl group, although the ketone had not been reduced, indicated that the sixth equivalent of hydrogen had opened an *epoxide* ring. The liberation of iodine from potassium iodide in acetic acid by pimaricin confirmed the presence of an epoxide, and suggested that it was "probably adjacent to a carbonyl group". The carboxyl group likewise appeared to be adjacent to the ketone, since both in pimaricin and in perhydropimaricin

---

* Ogawa et al. (*83*) note explicitly that the empirical formula of fungichromin monohydrate, $C_{35}H_{60}O_{13}$, fits the pentamycin data well.

** The infrared curve of fungichromin has apparently never been published; but comparison of the list of infrared $\lambda\lambda_{max}$ given in reference (*118*) with the infrared curves of pentamycin (*83*) and lagosin (*13 a*) suggests that identity is possible.

it was readily decarboxylated by warm dilute sulfuric acid, but after reduction of the ketone with borohydride this facile decarboxylation could no longer be elicited. It was thus inferred that the system (XXI) was present in pimaricin.

The amine present was found to be mycosamine, rather easily detached by refluxing in methanolic hydrogen chloride. Alkali treatment yielded ammonia and 13-hydroxytetradecapentaenal (III, p. 31). The interpretation of this reaction given by the Lederle investigators (retro-aldol cleavage of the molecule and β-elimination of mycosamine) has already been mentioned. Since the carbonyl group involved in this cleavage was necessarily the same as the one activating the carboxyl and epoxide groups, the C_{28} partial structure (XXII) could now be postulated, in which only a $-C_6H_{12}O_3-$ bridge, known to contain three hydroxyls, remained undetermined.

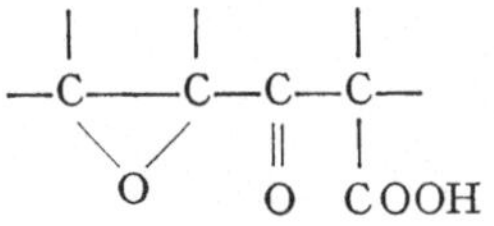

(XXI.)

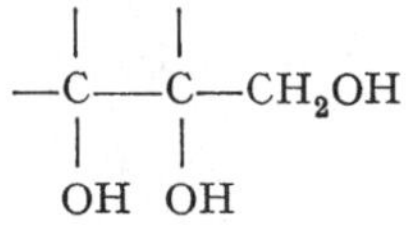

(XXIII.)

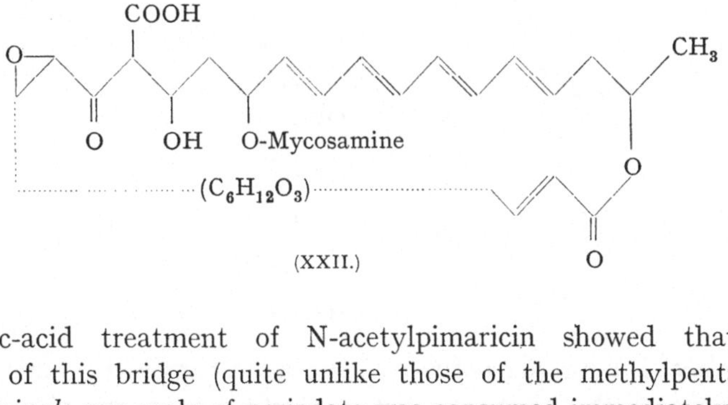

(XXII.)

Periodic-acid treatment of N-acetylpimaricin showed that the hydroxyls of this bridge (quite unlike those of the methylpentaenes) were *all vicinal*: one mole of periodate was consumed immediately, with the production of formaldehyde, and a second mole more slowly. The three hydroxyls were therefore arranged as in (XXIII). The final choice of the arrangement shown in (XXIV) below was made on the basis of a peculiar acid-catalyzed dehydration of N-acetylperhydropimaricin (with hot 1 N sulfuric acid for three minutes) to a crystalline compound of λ_{max} 280 mμ, identified as an alkyl furyl ketone. This reaction suggested the presence of a "deoxyhexose" structure in perhydropimaricin (*Chart 4*, p. 42). A three-carbon C_3H_6 bridge remained to be identified; its formulation as in (XXV) was confirmed when, on chromic-acid oxidation of perhydropimaricin, it proved possible to isolate and identify *pimelic acid*, corresponding to $C_{(1)}-C_{(7)}$.

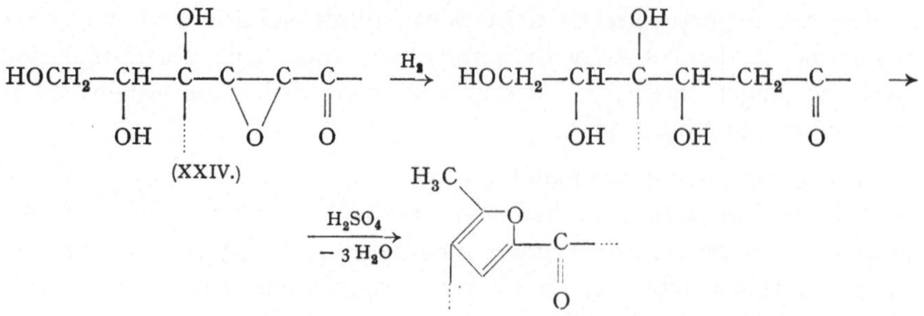

$$\text{HOCH}_2\text{—CH—C—C—C—C—} \xrightarrow{\text{H}_2} \text{HOCH}_2\text{—CH—C—CH—CH}_2\text{—C—} \rightarrow$$

(XXIV.)

$$\xrightarrow[-3\,\text{H}_2\text{O}]{\text{H}_2\text{SO}_4}$$

Chart 4. Dehydration of Perhydropimaricin.

This notable feat of deduction was, as the authors remarked, "the first complete structure determination on any of the numerous polyene antifungal antibiotics reported in the literature. It seems likely that most of these substances are macrolides of the same general type as pimaricin" (*89*).

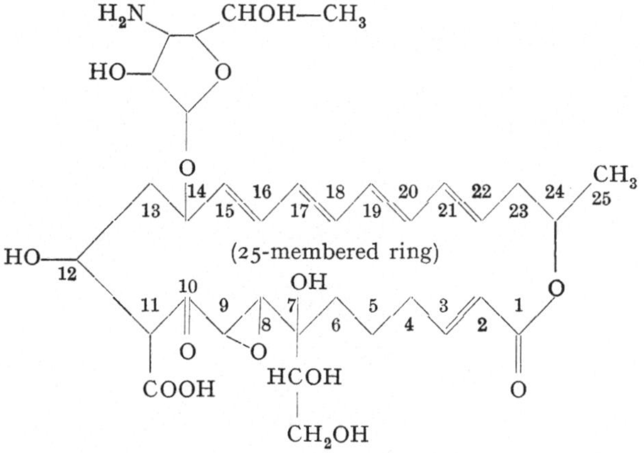

(XXV.) Pimaricin.

To account for the positive iodoform test given by pimaricin under conditions which do not hydrolyze the lactone, mycosamine was presumed to be present in the furanose form, as shown in (XXV). This and perhaps other details are not unshakably established by the evidence adduced in 1958; and it seems possible that when the complete work is published there may be some modifications in the preliminary formulation. For example, a 1,2-substituted glycerol fragment rather than the 1,1-substituted version of (XXV) does not appear inconsistent with the data and would lead to a formula (with 26-membered lactone ring) in better harmony with current biogenetic generalizations (cf. p. 51). The location of the epoxide ring also calls for more direct substantiation. A disturbing anomaly is noticeable in the circumstance that both the *saturated* lactone of the methylpentaenes and the *unsaturated* lactone of pimaricin show the same infrared absorption (at 5.83 μ), but there seems to be no objective basis for questioning either assignment.

8. Nystatin and Other Tetraenes.

The longest-known and most-used polyene antibiotic, nystatin, is not the simplest in structure: at the time of writing, its constitution is still only partially known. Even the empirical formula $C_{46}H_{75}O_{19}N$ (*38*), or perhaps $C_{46}H_{75}O_{18}N$ (*36*) is not yet entirely certain. As in the case of pimaricin, the only detachable appendages of the molecule are the acidic and basic functional groups, but neither of these can be situated like those of pimaricin, since no ketonic carbonyl group is present (*38*). As already noted, the mycosaminide linkage is much more resistant to acidolysis in nystatin (*42*); decarboxylation is also less facile, though it will occur at a temperature of 210° (*36*). The ultraviolet spectrum, like that of pimaricin, shows an additional absorption peak at short wavelengths which is not attributable to the tetraene chromophore, and which vanishes on perhydrogenation (six equivalents of hydrogen being absorbed). However, in the case of nystatin this second chromophore has been attributed (*41*) to a diene* rather than an α,β-unsaturated lactone. Since succinic acid is formed on nitric acid oxidation (*36*), it is inferred that a —CH_2—CH_2— link separates the tetraene from the diene, i. e., that the system —$(CH{=}CH)_2$—CH_2—CH_2—$(CH{=}CH)_4$— is present. This is interesting in that it can be regarded as formally derived from a *heptaene* by hydrogenation of the third double bond, which suggests that nystatin probably contains a lactone ring of "heptaene" size, viz. one of at least thirty-four atoms.

As has already been mentioned, DUTCHER (*42*) was able to deduce from the empirical formula as early as 1956 that all of the oxygen atoms, not involved in the glycoside, —COOH, and —COO— functions, must be hydroxylic, so that in this sense the molecule must be a "simpler" one than pimaricin. After several years during which no further publications on the structure of nystatin appeared, a recent preliminary report (*36*) on collaborative work at the University of Manchester, Squibb, and Stanford indicates the resumption of active progress. The "interrupted-heptaene" chromophore was verified by the isolation and identification of 2-methylheptadecanedioic acid (XXVI) as the largest fragment from nitric acid oxidation of perhydronystatin; this permitted expansion of the partial formula to (XXVII)**. As in the methyl-

---

* The low λ_{max} shows it to be *trans-trans*-1,4-disubstituted: a *cis*-configuration or a methyl substituent would produce a λ_{max} of $\geqslant$ 232 mμ.

** Alternatives in which either the tetraene or the diene terminates at the methyl group are excluded by the ultraviolet spectrum of nystatin.

pentaenes, isolation was effected by gas chromatography of the dimethyl esters, and identification primarily by mass spectroscopy.

$$CH_3$$
$$|$$
$$HOOC\!-\!CH\!-\!(CH_2)_{14}\!-\!COOH$$

(XXVI.) 2-Methylheptadecanedioic acid.

(XXVII.) Chromophore of nystatin (as of 1962).

According to periodic-acid titration, one (*38*) to three (*41*) vicinal glycol systems occur among the sixteen or seventeen hydroxyls in nystatin; as in pimaricin, one of the glycol hydroxyls is primary, since a mole of formaldehyde is formed (*41*). In the more recent work (*36*), lead tetraacetate was found to liberate, presumably from another vicinal glycol, the five-carbon fragment *tiglic aldehyde* (XXVIII). The double bond was of course not originally present, and is considered to have arisen by β-elimination of the lactone *(Chart 5)*.

(XXVIII.) Tiglic aldehyde.

Chart 5. Oxidative Fission of Nystatin.

Kuhn-Roth oxidation showed the presence of four C-methyl groups, all of which are accounted for by the above fragments (two in tiglic aldehyde, one as the branch-methyl in the C_{18}-dibasic acid, and one in mycosamine). How the moieties should be connected is as yet uncertain, but (XXIX) has been suggested (*36*), with a ring-closing chain containing only non-vicinal hydroxyls (one of which is primary) and a carboxyl, tentatively formulated as in (XXX).

It may be noted that the presence of three methyl substituents and one hydroxymethyl limits the maximum possible ring size in nystatin to thirty-six atoms.

A striking circumstance is that deep-seated chain rupture by alkali, apparently similar to that seen in pimaricin, is also observed in the case of nystatin. It has been known at least since 1955 (*41*) that ammonia

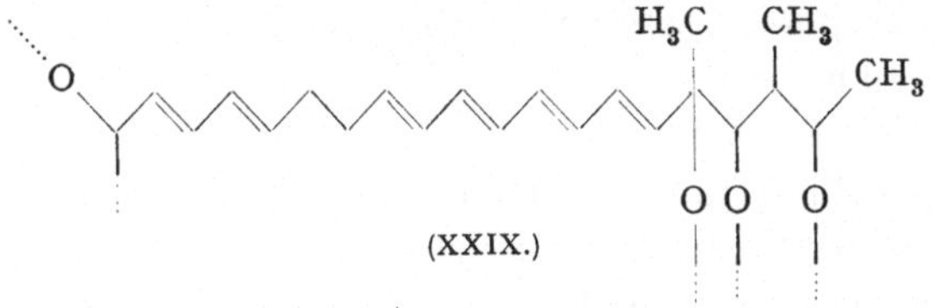

(XXIX.)

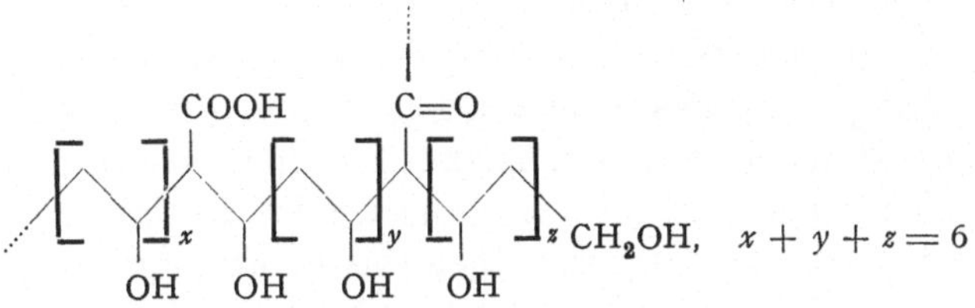

(XXX.)

$$x + y + z = 6$$

(+ mycosamine linked to some OH)

is liberated on treatment of nystatin with alkali; more recently it was reported, without details, that under these conditions the molecule is cleaved into two approximately equal fragments (36). One of these has since been demonstrated to be a pentaenal, resulting from retro-aldol fission and β-elimination, the activating group in this case being the carboxylate ion (16a).

It was also observed in 1955 that on heating with sodium hydroxide, the tetraene amphotericin A similarly releases "a volatile base" (125) (necessarily ammonia, since it originates from mycosamine); the same reaction is even more facile in its heptaene congener amphotericin B, which gives up its nitrogen even at room temperature (125). Yet the infrared spectra of these compounds (40, 125) show that neither contains any carbonyl groups other than the "normal" lactone (5.88 μ) and carboxylate ion (6.4 μ). It would appear from these examples that pimaricin-like alkaline cleavage of the chain, with β-elimination of mycosamine, must be quite a general reaction among the "normal" polyene antibiotics and that the presence of the carboxyl group is a sufficient condition for its occurrence.

9. Pentaenes and Hexaenes.

No chemical work beyond the determination of the empirical formula has yet been reported on any pentaene antibiotic of the "normal" type. Although, as we have seen, empirical formulas in this field are subject to considerable alteration on further investigation, the recurrence of the C_{37} formula [rimocidin in the tetraenes (38), PA-153 in the pentaenes] and the C_{46} formula (nystatin and perhaps amphotericin A in the tetraenes, candidin and amphotericin B in the heptaenes) may be significant of more than the preconceptions of the investigators. It certainly appears a reasonable prediction that some of these compounds will turn out to possess identical carbon skeletons, differing only in the number of double bonds.

The pentaene 2814-P was shown to yield mycosamine (*114*). The pentaene capacidin is unique among polyene antibiotics in containing a non-basic as well as a basic N atom (*20*).

None of the hexaenes has been obtained pure enough, or in sufficient quantity, to permit elemental analysis (*129*). The presence of non-hexaenic absorption peaks around 300 mμ is common to all four of them in the crude state. Vining and Taber (*129*) have shown this to be due to contamination with a tetraene congener in the case of endomycin, but there is a strong possibility that this explanation may not apply to the others, whose absorption curves suggest rather the presence of a ca. 295 mμ chromophore without sharp fine structure, possibly a trienoic ester. [The displacement of the "normal" 5.83 μ carbonyl peak to nearly 6 μ in the infrared spectrum of cryptocidin (*96*) shows unequivocally that in this compound this carbonyl group is conjugated.] The status of nitrogen (if any) in these compounds is not clear. Flavacid is reported to contain nitrogen but to be non-basic (like the tetraene unamycin); similarly, cryptocidin (*96*) is reported to contain nitrogen but to give no test for sugars, and its infrared curve confirms the absence of amine by the absence of a carboxylate ion peak. One is tempted to imagine some connection between the absence of amino nitrogen and the presence of carbonyl conjugation: possibly, these compounds contain the system —(CH=CH)$_3$—COO— where the "normal" series has —(CHOH—CH$_2$)$_3$—COO—, one hydroxyl etherified to mycosamine. The comparative rarity of the hexaene chromophore cannot, however, be accounted for by supposing that it is formed only in the presence of such a second conjugated system, since endomycin B shows only a pure hexaene spectrum (Fig. 2, p. 21).

10. Trichomycin and Other Heptaenes.

Since it is in the heptaenes that the highest antifungal potencies are found (*11*), this group commands the most interest from the practical standpoint. Unfortunately, these molecules also excel in size, complexity, insolubility, instability, and intractability. Few have even been obtained in a completely pure state. The only extensive chemical investigations yet published have been those carried out on trichomycin by Nakano (*76*) and Hattori (*51 a*). Pure crystals of the major component, i. e. trichomycin A, were first obtained in 1960, eight years after the discovery of trichomycin, by means of two 240-tube countercurrent fractionations followed by crystallization from Hosoya and Hamamura's mixture (*58*), in which the pure compound proved to be far less soluble than partially-purified material had been.

Analysis *(76)* showed two atoms of nitrogen and a molecular weight of about 1230, indicating a formula of $C_{61}H_{90-92}O_{23}N_2$ (interpreted by NAKANO as $C_{61}H_{86}O_{21}N_2$, $2\,H_2O$). Methoxyl was absent; Kuhn-Roth oxidation showed three to four C-methyls, as in nystatin; twelve to thirteen acetylatable hydroxyl or amino groups were present, and also (as estimated by consumption of sodium borohydride) four carbonyl groups. As no polyene ketone was detectable after treatment with manganese dioxide, it was judged that none of the hydroxyl groups was allylic *(77)*. From chromic acid oxidation, $O=CH—(CH=CH)_4—CHO$, apparently an over-oxidized fragment of the chromophore, was obtained in poor yield; no methylated derivatives of this dialdehyde were detectable, tending to confirm the absence of internal methyl substituents on the heptaene. Prolonged treatment with 2 N sodium hydroxide at room temperature yielded another polyenic aldehyde: λ_{max} 418 mμ (ether), reducible by borohydride to a sharply-peaked polyene, $\lambda\lambda_{max}$ 418, 394, 374 mμ (methanol). NAKANO *(77)* assumes, no doubt correctly, that this aldehyde was formed by a retro-aldol fission and β-elimination as in the case of pimaricin *(Chart 6)*:

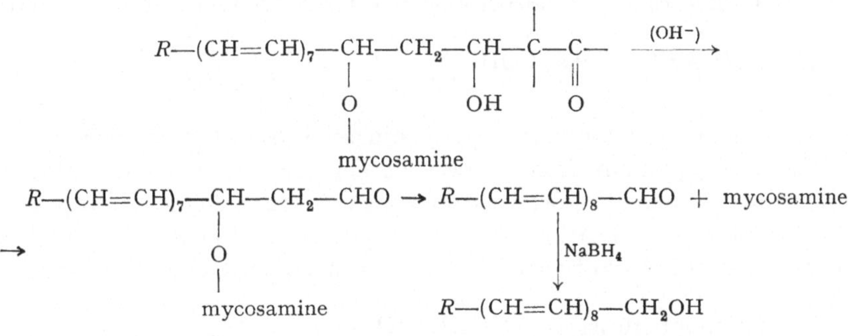

Chart 6. Alkaline Fission of Trichomycin*.

The amino sugar present in trichomycin was best obtained by acid-catalyzed alcoholysis of the N-dinitrophenyl derivative of perhydro-trichomycin, giving the crystalline N-dinitrophenyl ethylmycosaminide. Mycosamine could also be detected by paper chromatography after perhydrogenation of trichomycin in glacial acetic acid (ten to eleven

---

* Although no chemical verification of these spectrally-deduced structures was reported, they may well be correct: the spectrum of the polyene from boro-hydride reduction agrees well with that predicted for $HOCH_2—(CH=CH)_8—R$ with one internal *cis*-bond. (The relative lowness of the longest-wavelength peak shows that either an internal *cis*-bond or an internal substituent must be present.) Since the λ_{max} is ca. 4 mμ *lower* than expected for the all-*trans* isomer, the former alternative is indicated. But the presence of an internal methyl group and *two* internal *cis* double bonds would also account for the observed spectra.

equivalents taken up); this facile hydrogenolysis of mycosamine was taken as evidence that it is bound to an allylic hydroxyl. As in pimaricin (89), the positive iodoform test shown by trichomycin was ascribed to the presence of a furanosidic glycoside.

By gentle alkaline treatment, a carbonyl compound identifiable as *p-aminoacetophenone* was released, evidently by a retro-aldol cleavage. Unlike the hexanaldehyde side-chain of the methylpentaenes, this was

not originally present as the ketol $p\text{-}H_2N\text{---}C_6H_4\text{---}C(CH_3)(OH)\text{---}\overset{|}{C}\text{---}C{=}O$;

the *p*-aminobenzoyl group was found to be preexistent in trichomycin, its ultraviolet spectrum (λ_{max} ca. 325 mμ, in methanol), being revealed after saturation of the heptaene chromophore with seven moles of hydrogen. It was possible by the same criterion to rule out the presence of the $p\text{-}H_2N\text{---}C_6H_4\text{---}COCH_2CO\text{---}$ system; and the absence of $p\text{-}H_2N\text{---}C_6H_4\text{---}COCH_2COO\text{---}$ was proved by the fact that after borohydride reduction of the carbonyl group, the aminophenyl fragment was no longer detachable by alkali treatment. Nakano (77) therefore concluded that the *p*-aminoacetophenone must originate from the ketol

$$p\text{-}H_2N\text{---}C_6H_4\text{---}CO\text{---}CH_2\text{---}\overset{|}{\underset{|}{C}}OH.$$

The presence of a β-diketone (*not* conjugated to phenyl) in trichomycin was likewise apparent from ultraviolet evidence: in alkaline solution, perhydro-trichomycin showed a strong peak at 272 mμ which disappeared reversibly on neutralization; after borohydride treatment this could no longer be elicited. A positive ferric chloride test supported the identification of this chromophore as $\text{---}CO\text{---}\overset{|}{CH}\text{---}CO\text{---}$.

By periodate titration (two moles being consumed in two hours), two vicinal glycol systems were found to be present in perhydro-trichomycin. As no formic acid was detected, these cannot be formulated as $\overset{|}{\underset{|}{C}}(OH)\text{---}CH(OH)\text{---}\overset{|}{\underset{|}{C}}(OH)$. One of them was ascribed to the mycosamine moiety. For the other, the partial structure (XXXI) could be deduced (77) from the fact that after periodate fission a 237-mμ absorption peak appeared, identifiable as a substituted methacrolein* by ozonolysis which produced methylglyoxal *(Chart 7)*.

---

* This production of $R\text{---}CH{=}C(CH_3)\text{-}CHO$ by periodic acid appears analoguous to the production of tiglic aldehyde (R = methyl) from nystatin by lead tetraacetate; however, the 237-mμ compound from trichomycin could not have been tiglic aldehyde, which shows λ_{max} 227 mμ in alcohol.

$$\text{—C—CH—C—CH—}R \xrightarrow{\text{HIO}_4} \left[\text{O=CH—CH—CH—}R \right] \xrightarrow{\beta\text{-elim.}} \text{O=CH—C=CH—}R$$

(XXXI.)

or alternatively:

$$\text{—C—C—CH}_2\text{—CH—}R \xrightarrow{\text{HIO}_4} \left[\text{O=C—CH}_2\text{—CH—}R \right] \xrightarrow{\beta\text{-elim.}} \text{O=C—CH—CH—}R$$

Chart 7. Oxidative Fission of Trichomycin.

The alternate formulation shown in Chart 7 was excluded on the grounds that after acetylation of perhydrotrichomycin, the infrared spectrum shows that no free hydroxyl remains; hence the compound contains no tertiary hydroxyl.

NAKANO's findings may be summed up by the collection of partial structures shown in *Chart 8*, in which all of the functional groups have

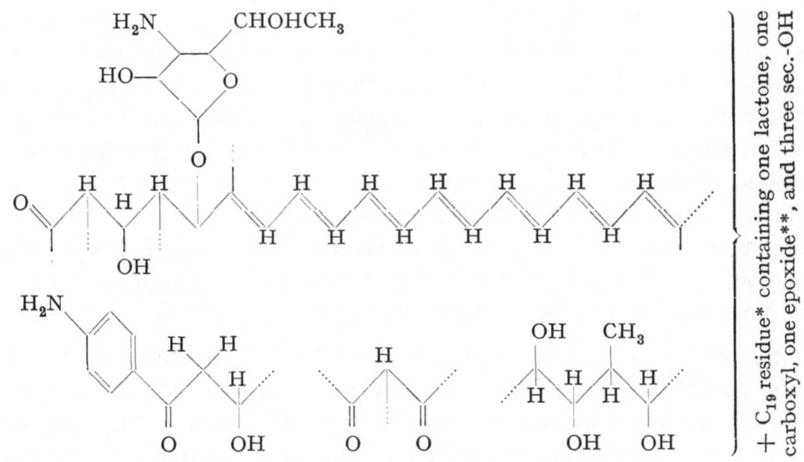

Chart 8. Partial Structures Found in Trichomycin.

been accounted for, and only their interconnection remains in doubt. Clarification of the position of the *p*-aminobenzoyl side-chain may be anticipated from further study of a crystalline oxidation product,

---

* Assuming no overlap between the fragments. If the C=O of the chromophore fragment belongs to the β-diketone, this should be C_{20}.

** The existence of an epoxide group was suggested only by a positive sodium-thiosulfate test (*94*).

probably $C_{21}H_{27}O_4N$, obtainable by chromic-acetic acid oxidation of trichomycin (*77*): this contains intact the *p*-aminophenyl moiety shown in Chart 8 (as the N-acetyl derivative), since with alkali it splits off *p*-acetaminoacetophenone, and it also contains the terminal group R—CH=CH—CHO (λ_{max} 223 mμ; ozone → glyoxal), evidently derived by β-elimination from R—CH(O—)—CH_2—CHO.

Very recently reported work by Hattori (*51a*) has led the Japanese workers to propose formula (XXXII) for the structure of trichomycin.

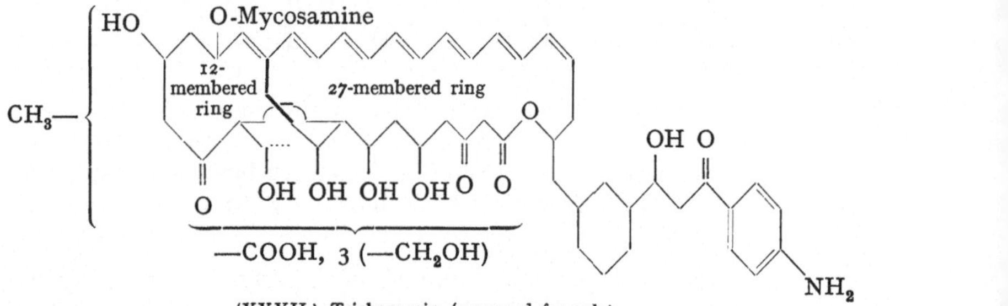

(XXXII.) Trichomycin (proposed formula).

Nothing comparable to the above work on trichomycin has yet been published for other heptaenes. On amphotericin B, nothing has appeared since the important advance reported in late 1956, when Dutcher's group (*42, 133*) first succeeded in isolating and identifying mycosamine. However, the renewed investigation of nystatin now in progress doubtless includes parallel work on amphotericin B, since the Squibb researchers who assigned a C_{46}-formula to both of these compounds must certainly have entertained the suspicion that their skeletons are identical; the infrared spectrum of amphotericin B is in fact very similar to that of nystatin.

No work has ever been published on two of the earliest and also most potent heptaenes, ascosin and candicidin, apart from paper-chromatographic comparisons which show that both crude products are mixtures apparently consisting of the same components as crude trichomycin (*3, 26, 130*). This suspicion of identity is strengthened by the finding of the Rutgers workers (*99*) that candicidin, like trichomycin, yields both mycosamine and *p*-aminoacetophenone.

Candidin was obtained crystalline by Vining (*128*) in 1956, but on more intensive countercurrent fractionation these crystals were found to be still heterogeneous, and large crystals of pure candidin were first produced at Rutgers in 1962 (*98*)—nine years after the discovery of the antibiotic. As in the case of nystatin and trichomycin, a renewed attack on the structural problem was inspired by the solutions of pimaricin and the methylpentaenes, but no results have yet appeared. It is known, however, that candidin contains mycosamine (which, in contrast to nystatin, is easily freed) (*18*), and that no periodate-reactive α-glycol groups are present (*98*). The C_{46} formula of course suggests that here too the skeleton may be the same as that of nystatin.

Perimycin (*18*) is an unusual heptaene, the only one yet encountered that lacks a carboxyl group. (The striking 6.25 μ peak in its infrared curve is not that of the carboxylate ion but some other feature, possibly aromatic amine.) Like trichomycin, it contains two primary amine groups in detachable fragments, but both of these proved to differ from those of trichomycin. The carbohydrate moiety was not mycosamine but a new amino sugar of still unknown structure, which was called "perosamine", and the fragment released by retro-aldol fission (fifteen minutes at 100° in 10% sodium hydroxide) was p-aminophenylacetone, the homologue of the p-aminoacetophenone present in trichomycin. The molecule of perimycin appears to be substantially smaller than that of trichomycin, although the suggested tentative formula, $C_{47}H_{75}O_{14}N_2$, must underestimate it somewhat, since it implies a ring size of thirty-one atoms or less.

Hamycin (*14*, *15*, *16*) produced by a *Streptomyces* strain from soil of Pimpri, India, is unique in that it is reported to contain a polypeptide moiety. At least nine amino acids were detected, yet the nitrogen content is only the usual ca. 2%, which necessitates, as the authors remark, a molecular weight "of the order of several thousands". Yet the properties of hamycin are like those of other heptaenes; its infrared spectrum (*16*) is scarcely distinguishable from those of PA-150 or amphotericin B, and on paper chromatograms (*15*) it was not separated from amphotericin B and candidin. Since the same analysts who isolated amino acids from hamycin also found indications that these and other well-known polyene antibiotics "also may contain peptide portions in their molecules" (*16*), it seems likely that some error of technique accounts for their surprising findings.

From the same laboratories it has been reported (*14*) that the antibacterial antibiotic pumilin, produced by *Bacillus pumilis* and devoid of antifungal action, most surprisingly proves to contain a heptaene chromophore like that of hamycin. Until corroboration is available, this observation should probably be viewed with reserve.

IV. Biogenetic Relationships.

The predominance of a simple two-carbon rhythm $-[-CH_2-CH(OH)-]_n$ and its dehydration product $(-CH=CH-)_n-$ in these molecules is as striking as the $[\text{—CH(OH)—}]_n$ of the carbohydrates, the $(\text{—NHCH}_2\text{CO—})_n$ of polypeptides, or the polyisoprene chains of the carotenoids. The skeleton is fundamentally a long, unbranched polymer chain such as might be formed by repetitive aldol condensation of acetaldehyde. In fact it is actually formed in this way, except that the reaction involved in the "polymerization" is an iterated acetylation (via acetyl-coenzyme A) rather than an iterated aldol condensation. In recent years a surprisingly large proportion of natural products have been shown to be formed from

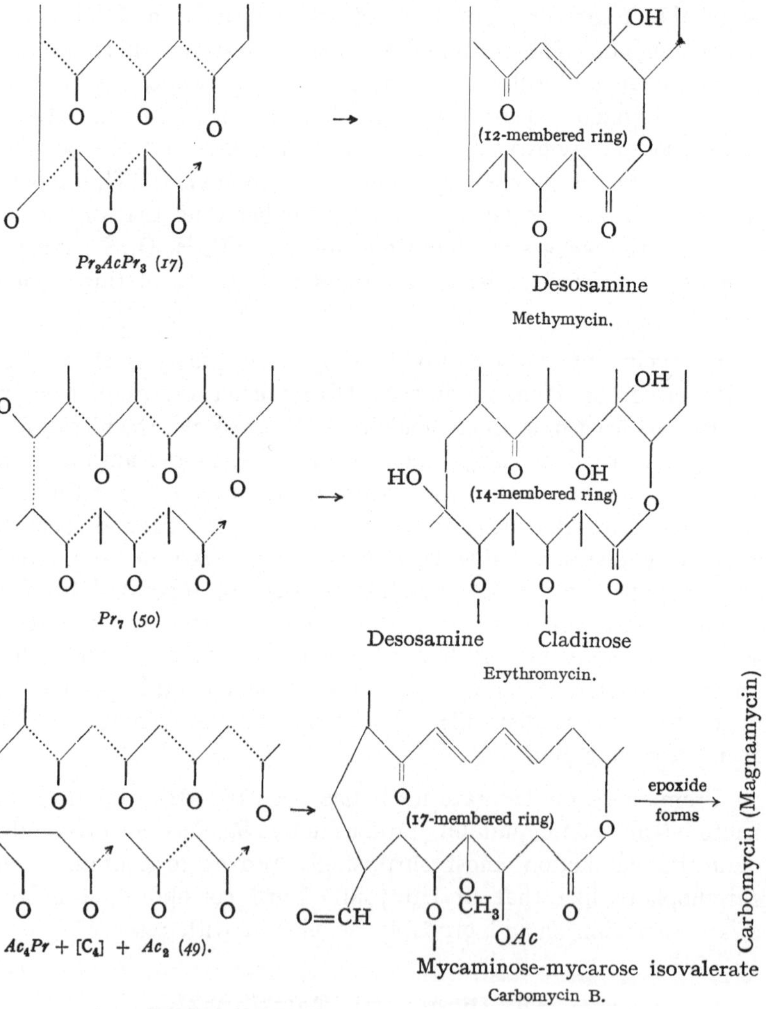

Chart 9. Biosynthesis of Polyenic and Other Macrolides.
(Ac = acetyl; Pr = propionyl.)

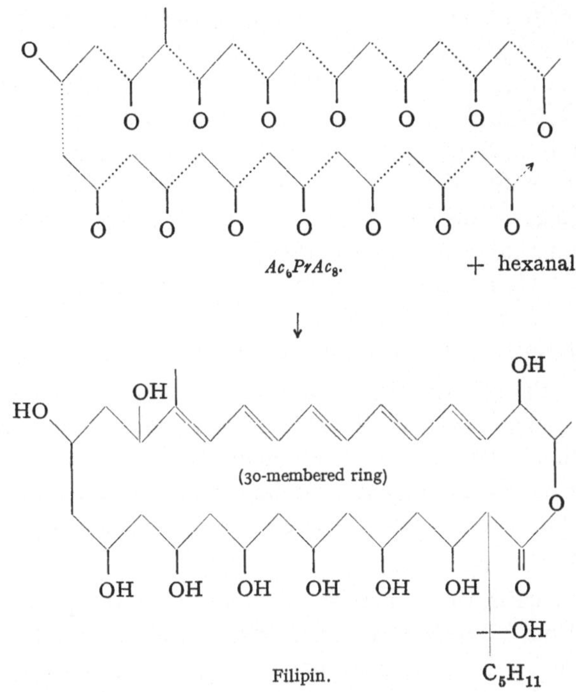

Chart. 9 (Continued).

such polyacetyl chains [see, e. g., BIRCH (*17*)], though the "seams" left by this mode of biosynthesis are seldom quite so obvious as they are in these antibiotics. The ubiquitous saturated and unsaturated straight-chain fatty acids are uncomplicated reduction products of such poly-acetate chains, as are their hyper-unsaturated relatives, the straight-chain polyenes and polyacetylenes of which so many have now been demonstrated in higher plants and in Basidiomycetous higher fungi. Since some at least of these fungal polyenynes have antibiotic properties— in fact, were originally detected in that manner—these might also come under the general rubric of "polyene antibiotics". However, these mushroom antibiotics are clearly much more closely allied to the fatty acids* and the polyacetylenes of the Compositae than to the polyene

---

* Tracer experiments by BIRCH, BU'LOCK, and others at the University of Manchester [see for example (*20a, 43a*)] have verified that the natural poly-acetylenes are biosynthesized from acetate chains in a manner essentially similar to that of fatty acids save for the conversion of —CH_2CO— to —$C{\equiv}C$— units. Polyacetylenes having an odd number of carbon atoms (e. g. the *Nocardia* anti-biotic mycomycin) almost invariably possess a terminal —$C{\equiv}CH$; it has been demonstrated that these arise by oxidation of a terminal methyl to —$C{\equiv}C$—COOH followed by decarboxylation (*43a*).

macrolides of *Streptomyces*, with which there seems to be no intergradation. As of the end of 1962, no *Streptomyces* polyene is yet known which includes any acetylenic bonds (although this feature may be expected to turn up eventually); and, more strikingly, none of simple straight-chain construction is known: all, to judge by their infrared similarities, are folded in the center of the chain and fastened together at the ends by an ester linkage. This biosynthetic "originality" of *Streptomyces* is all the more remarkable in view of the fact that a related actinomycete, *Nocardia acidophilus*, converts acetate into mycomycin, an excessively unsaturated straight-chain acid of exactly the same sort synthesized by the mushrooms and higher plants.

However, though the *Streptomyces* antifungal polyenes may seem at first sight to have little but their building blocks in common with these straight-chain products, further study may reveal a more direct kinship: in the typical carboxylated polyene macrolide there is certainly some suggestion of a cyclic "dimer" formed between two preformed acids of C_{10}–C_{20} chain length.

More obvious is the structural relationship to the previously known "broad-spectrum" *Streptomyces* antibiotics, the macrolides. These are written in *Chart 9* in a manner similar to that adopted for the polyene macrolides, with schemes illustrating their presumed precursor chains.

The interpolation of *propionate* units into the polyacetate skeleton is characteristic of all of these *Streptomyces* antibiotics*—indeed, erythromycin is synthesized from a pure heptapropionyl chain. This structural feature, which has not yet been encountered in the non-*Streptomyces* polyenynes, is present also in some (and perhaps all) of the polyenic macrolides: for example, chromophoric methyl substituents are unquestionably propionate-derived, as indicated in the hypothetical scheme for filipin (p. 53).

Biogenetic tracer experiments on nystatin by Birch and his colleagues (*36*) (almost the first reported in this field) have given more precise structural information. Neither mevalonic acid** (the precursor of polyisoprenes) nor methionine is utilized in the biosynthesis of nystatin;

---

* The tetracyclines, which are likewise products of cyclization of a polyacetyl chain, evidently differ in this respect: when branch-methyl groups occur, they are found to be supplied by methionine alkylation rather than by propionate (*44*). [On tetracyclines, see Muxfeldt and Bangert's paper on p. 80 of this Volume.]

** In 1961, Katz and Lechevalier (*64*) reported that labeled mevalonic acid was not incorporated into antimycoin A—an unexpected result at the time, since it had previously been observed (*100*) that the production of this tetraene could be stimulated by the addition of mevalonic acid to the culture medium.

For the role of mevalonic acid in the biosynthesis of rubber, cf. Bonner's paper on p. 1 of this Volume.

References, pp. 72—79.

only acetate and propionate. All three atoms of one of the labeled propionic acid molecules are present in the tiglic aldehyde produced by lead tetraacetate (see Section on nystatin, p. 43); and as this tiglal carries one-third of the total activity introduced by the labeled propionate, nystatin must incorporate three propionate units. The carboxyl group (recoverable as carbon dioxide by decarboxylation at 210°) proved to have its origin in the methyl group of one of these propionate units, not in acetate carboxyl: this significant finding in the case of nystatin is likely to be generally valid for the entire group. The other two propionate methyls are present as such—one in tiglic aldehyde, the other as the branch-methyl group of the C_{18}-dibasic acid produced by oxidation— and since this latter methyl was found to be flanked by *two* atoms derived from propionate carboxyl, two of the three propionate units must be consecutive. The tentative structure cited on p. 44 for "Nystatin", derived from a precursor chain beginning with $AcPr_2Ac_7$, was chosen by the Manchester workers so as to accommodate these biogenetic findings.

Similar experiments with carbomycin indicate that its protruding formyl group, which bears such a suggestive resemblance to the protruding carboxyl of the polyene macrolides, must be biogenetically unrelated: this aldehyde is part of a four-carbon unit whose biosynthesis utilizes neither acetate nor propionate (*49*).

The apparent "head-to-head" junction of two chain units at this point (*48*), a striking peculiarity of the carbomycin ring, would therefore appear to have no implications for the structure of the carboxylated polyenes.

V. Tables.

Table 1. Typical Tetraenes: Spectral Data.

Name and References	First three $\lambda\lambda_{max}$ (mμ) (corr. to 95% Ethanol)*			$\Delta\lambda$ From Unsubstituted Polyene		
1,3,5,7-Octatetraene (*135*)..	302.5	289.2	277.5			
2,4,6,8-Decatetraene (*80*) ..	310.8	297.0	284.5	8.3	7.8	7.0
β-Parinaric acid (*1, 65*) ...	315.5	301.0	288.0	13.0	11.8	10.5
Tetrin (*48*)..............	318.0	304.0	290.0	15.5	14.8	12.5
Antimycoin A (*93*)........	318.0	304.0	291.0	15.5	14.8	13.5
Pimaricin (*2, 34, 109*)	318.5	303.3	290.3	16.0	14.1	12.8
Protocidin (*95*)	318.5	303.5	290.5	16.0	14.3	13.0
Rimocidin (*28, 29*)........	318.5	304.0	291.5	16.0	14.8	14.0
Etruscomycin (*7*)	319.0	305.8	290.8	16.5	16.6	13.3
Akitamycin (*43*)	319.5	304.0	291.5	17.0	14.8	14.0
Amphotericin A (*40, 125*)..	319.5	304.8	291.8	17.0	15.6	14.3
PA-166 (*66*)..............	319.5	305.0	291.8	17.0	15.8	14.3
Sistomycosin (*88*)........	319.5	305.2	291.6	17.0	16.0	14.1
Unamycin (*73*)	320.0	304.8	290.8	17.5	15.6	13.3

* Where the solvent used differed from 95% ethanol, the reported $\lambda\lambda_{max}$ have been approximately "corrected to alcohol" by use of the empirical equation: $\lambda_{95\% \text{ ethanol}} \approx \lambda_{solvent} - k_{solvent} (\lambda_{solvent} - 180)/1000$.

$$k_{solvent}: \quad H_2O \quad = \quad 8 \qquad \text{Cyclohexane} = \quad 12$$
$$80\% \text{ Methanol} = -4 \qquad \text{Hexane} \quad = -7$$
$$\text{Methanol} \quad = -7 \qquad \text{Chloroform} = \quad 40$$

Although the data in Tables 1–5, are recorded to the nearest 0.1 mμ, this is actually the result of applying the above "correction" equation rather than an indication of the precision of the original spectral observations. — When more than one reference is cited, the spectral data in the Table represent an average of the values reported.

Table 2. Typical Pentaenes: Spectral Data.

Name and References	First three $\lambda\lambda_{max}$ (mμ) (corr. to 95% Ethanol)*			$\Delta\lambda$ From Unsubstituted Polyene		
1,3,5,7,9-Decapentaene (*74*)	335.3	318.7	304.5			
2,4,6,8,10-Dodecapentaene (*80*)	342.9	326.0	310.0	7.6	7.3	5.5
PA-153 (*66*).....................	350.0	332.6	317.5	14.7	13.9	13.0
Distamycin B (*104*)	351.2	334.1	319.0	15.9	15.4	14.5
St. effluvius pentaene (*70*)...........	350.7	332.6	317.5	15.4	13.9	13.0
Eurocidin (*124*)	352.2	334.1	319.0	16.9	15.4	14.5

* See footnote, Table 1.

References, pp. 72—79.

Table 3. Typical Methylpentaenes: Spectral Data.

Name and References	First three $\lambda\lambda_{max}$ (mμ) (corr. to 95% Ethanol)*			$\Delta\lambda$ From Unsubstituted Pentaene		
Cabicidin (*82*)	355.2	340.1	321.0	19.9	21.4	16.5
Filipin (*134*)	356.2	339.1	323.0	20.9	20.4	18.5
Pentamycin (*123*).................	357.2	339.1	323.0	21.9	20.4	18.5
Moldcidin A (*97*).................	358.7	339.7	324.6	23.4	21.0	20.1

* See footnote, Table 1.

Table 4. Typical Hexaenes: Spectral Data.

Name and References	First three $\lambda\lambda_{max}$ (mμ) (corr. to 95% Ethanol)*			$\Delta\lambda$ From Unsubstituted Polyene		
1,3,5,7,9,11-Dodecahexaene (*107*)	365.5	346.0	330.0			
2,4,6,8,10,12-Tetradecahexaene (*80*) ..	372.0	352.0	334.5	6.5	6.0	4.5
Mediocidin (*124*)..................	378.9	357.7	340.6	13.4	11.7	10.6
Endomycin B (*127, 129*)...........	379.0	358.0	339.0	13.5	12.0	9.0
Flavacid (*85*).....................	379.2	358.2	341.0	13.7	12.2	11.0
Cryptocidin (*96*)	380.8	358.7	341.6	15.3	12.7	11.6

* See footnote, Table 1.

Table 5. Typical Heptaenes: Spectral Data.

Name and References	First three $\lambda\lambda_{max}$ (mμ) (corr. to 95% Ethanol)*			$\Delta\lambda$ From Unsubstituted Polyene		
1,3,5,7,9,11,13-Tetradecaheptaene (*74*) .	391.4	369.4	351.2			
PA-150 (*66*)...............	398.0	376.8	358.7	6.6	7.4	7.5
Ascosin (*57*)..............	400.5	378.0	359.0	9.1	8.6	7.8
Ascosin (stereoisomerized) (*75*)	405.0	382.0	362.2	13.6	12.6	11.0
Candicidin (*75*)	401.0	378.9	359.7	9.6	9.5	8.5
Candicidin (stereoisomerized) (*75*)........	405.7	382.6	363.1	14.3	13.2	11.9
Trichomycin A (*52*)	401.6	378.4	359.3	10.2	9.0	8.1
Trichomycin A crystalline (*76*)	405.6	383.0	362.8	14.2	13.6	11.6
All-*trans*-Trichomycin A (*77*)	411.0	386.7	365.5	19.6	17.3	14.3
26/1 (*117*)................	403.0	380.0	360.0	11.6	10.6	8.8
757 (*27*)	404.0	381.0	361.0	12.6	11.6	9.8
Eurotin A (*106*)	404.7	380.0	360.0	13.3	10.6	8.8
Amphotericin B (*12*).......	407.5	383.5	364.3	16.1	14.1	13.1
Candidin (*128*)	407.5	383.5	364.3	16.1	14.1	13.1
Hamycin (*16*)	406.9	383.8	363.8	15.5	14.4	12.6
Perimycin (*18*)	407.6	384.4	362.3	16.2	15.0	11.1
Candimycin (*102*)	407.6	383.4	363.3	16.2	14.0	12.1
F-17-C (*26*)	408.0	383.0	365.0	16.6	13.6	13.8

* See footnote, Table 1.

Table 6. Tetraenes: Physical

Name and References	Ultraviolet Spectrum $\lambda\lambda_{max}$ (mμ) a*	$[\alpha]_D$	Infrared Peaks (μ) >C=O Reg. 9–15 μ Reg.	Crystal Form and m. p.
Nystatin (36, 38, 39, 41, 54)	(Solvent?) 318.5 78.0 304.0 85.0 291.0 57.0 230.0 30.0	— 10 (AcOH) + 12 (DMF) + 21 (C$_6$H$_5$N) — 7 (0.1 N-HCl/MeOH)	5.87 10.03 6.37 11.8	Fine needles d. > 160°
Rimocidin (28, 29, 38, 39)	(In 80% MeOH) 318.0 92.0 303.5 100.0 291.0 65.0	+ 75 (Sulfate in MeOH) + 116 (base in C$_6$H$_5$N)	5.82 9.93 6.13 11.24 6.35 11.8 12.4	Sulfate m. p. 151° base d. 110°
Chromin (131, 132)	(Solvent?) 320.0 96.0 305.0 109.0 292.5 76.0		5.9 9.9 6.4 10.9	Fine needles d. 145–150°
Sistomycosin (88)	(In H$_2$O) 320.5 29.5 306.0 34.0 292.5 25.5		6.34	Light yellow microcrystalline > 230°
Amphotericin A (38, 40, 46, 125)	(In MeOH) 318.5 80.0 304.0 87.0 291.0 58.0 230.0 30.0	+ 93 (AcOH) + 163 (C$_6$H$_5$N) + 136 (DMF) — 10 (0.1 N-HCl/MeOH)	5.88 9.91 6.45 10.08 11.77 11.95	Crystals m. p. ca. 210° d. gradual decomp. > 180°
7071-RP (30)	(Solvent?) 318.0 78.3 304.0 86.3 291.0 56.2	+ 90 (MeOH) + 80 (C$_6$H$_5$N)	Differs from rimocidin, nystatin and chromin	Colorless platelets d. 275–280°
Protocidin (8, 95)	(In 80% MeOH) 318.0 63.0 303.0 70.0 290.0 46.0	+ 37.5 (C$_6$H$_5$N, 16°)	5.85 6.10 9.95 6.28 11.2	Pale yellow needles m. p. > 200°
Etruscomycin (Lucensomycin) (6, 7)	(In MeOH) 318.0 117.0 305.0 139.0 290.0 85.0 222.0 45.0	+ 296 (C$_6$H$_5$N) + 50 (0.1 N-HCl/MeOH)	5.83 9.95 6.1 11.2 6.3 11.85	White crystals d. > 150°
Endomycin A (Helixin A) (129)	Tetraene spectrum present			Non-crystalline

* $a = 0.1 \ E^{1\%}_{1\,cm}$ (4). DMF = dimethylformamide.

References, pp. 72—79.

and Chemical Properties.

% C	Elemental Analysis % H and Empirical Formula	% N	Neutralization Equivalent as Acid	as Base	Other Chemical Information
58.4	8.18 $C_{46}H_{75}O_{18-19}N$	1.6	950	955	Violet → blue in conc. H_2SO_4; no—OMe or—NMe; neg. 2,4-DNHP, $FeCl_3$; pos. Molisch. Also see text
57.65	(Sulfate) 7.82 $C_{37}H_{59}O_{13}N$	1.81	826–840	889	Red-brown in conc. H_2SO_4 Separable by paper chrom. from nystatin and amphotericin A but not from rimocidin or chromin
58.2	7.81	2.29	positive		Pos. Fehling; neg. Molisch, ninhydrin, $FeCl_3$, Sakaguchi
					Cherry-red → chocolate in conc. H_2SO_4; pos. Benedict, Molisch; neg. $FeCl_3$
60.32	8.39 $C_{44}H_{75}O_{17}N$	1.72	positive	915	See text
58.3	8.0	1.65		859	
	$C_{29}H_{45}O_{13}N$		positive	positive	Brown-black in conc. H_2SO_4; pos. Fehling; neg. Molisch, $FeCl_3$ ninhydrin, anthrone
59.57	8.05 $C_{36}H_{57}O_{14}N$	2.04	positive	731	Brown-red in conc. H_2SO_4; weak Molisch, neg. $FeCl_3$
					R_f data given (*129*)

(Table 6, continued.)

Name and References	Ultraviolet Spectrum $\lambda\lambda_{max}$ (mμ) a*		$[\alpha]_D$	Infrared Peaks (μ) >C=O Reg. 9–15 μ Reg.		Crystal Form and m. p.
Pimaricin (Tennecetin) (*2, 21, 34, 89, 109*)	(In MeOH) 317.5 102.0 302.5 110.0 289.5 71.0 220.0 30.0		+ 180 ± 5 (Me$_2$SO$_4$)	5.84 6.12 6.36	9.95 11.3 11.9	Colorless **needles**, d. ca. 200°
PA-166 (*66*)	(In 80% MeOH) 319.0 99.0 304.0 110.0 291.0 71.5 223.0 28.5		+ 275 (C$_6$H$_5$N) + 257 (DMF) + 191 (0.1 N-HCl/DMF)	5.8 6.4	9.95 11.3	Colorless needles **d. > 260°**
Akitamycin (*43, 105*)	(In 80% MeOH) 319.0 64.0 303.5 70.0 291.0 60.0		+ 88 (MeOH, pH 9) + 158 (DMF)	5.85 6.09 6.25 6.43	9.9 10.85 11.05 11.9 12.45	White needles d. 180°
Fungicidin A-94 (*87*)	(Solvent?) 318.5 304.5 291.0			5.80 6.03 6.25	9.8	Small needles
St. gilvosporeus Tetraene (*10*)	(In MeOH) 317.0 112.0 302.0 120.0 289.0 76.0 220.0 33.0		+ 248 (DMF)	5.8 6.35	9.95 10.6 11.2 11.8	White, crystalline
Unamycin (*73, 119*)	(In MeOH) 319.0 87.5 304.0 101.0 290.0 64.3		−92 (80% MeOH) + 79 (DMF)	5.9 6.1	9.9 10.1 10.7 11.2 11.5 11.7 12.3	Needles m. p. 165°
Tetrin (*48*)	(In EtOH) 318.0 304.0 290.0					Crystalline; sinters > 150° d. 160°

References, pp. 72—79.

Elemental Analysis			Neutralization Equivalent		Other Chemical Information
% C	% H and Empirical Formula	% N	as Acid	as Base	
$58.53 \pm .3$	$7.32 \pm .17$ $C_{34}H_{49}O_{14}N$	$2.12 \pm .14$	positive	positive	Wine-red in conc. H_2SO_4; pos. van Slyke, N, iodoform. Also see text
59.6	7.66 $C_{35}H_{53}O_{14}N$	2.00	positive	708	Violet in conc. H_2SO_4; pos. Fehling, ninhydrin, 2,4-DNHP; C—Me (three)
57.26	7.68	1.64			Yellow → green, purple, black in conc. H_2SO_4; weak Molisch
	$C_{46}H_{77}O_{19}N$				
58	7.5 $C_{31-34}H_{47-51}O_{13-14}N$	2	positive	positive	K salt prepared; picrate prepared
52.24	7.77	1.74	positive	0	Red-brown → black in conc. H_2SO_4; pos. Molisch; neg. Millon, Fehling, Tollens
	$C_{31}H_{59}O_{12}N$				Blue in conc. H_2SO_4

Table 7. Pentaenes: Physical

Name and References	Ultraviolet Spectrum $\lambda\lambda_{max}$ (mμ) a*	$[\alpha]_D$	Infrared Peaks (μ) >C=O Reg.	9—15 μ Reg.	Crystal Form and m. p.
Eurocidin (5, 79, 82, 124)	(In MeOH) 351.0 333.0 318.0	—200 (0.1 N-HCl) +22 (0.1 N-NaOH)			Pale yellow columnar crystals > 300°
No. 83, No. 90 (5)	(Solvent?) 350.0 330.0 315.0				
Fungichromatin (118)	(Solvent?) 350.0 333.0 318.0				Crystalline
Aliomycin (62)	(Solvent?) 351.0 330.0 321.0				Yellow-brown powder
St. effluvius Pentaene (40)	(In 80% MeOH) 350.0 113.8 332.0 113.5 317.0	+253 (AcOH)	5.83 6.36	9.94 10.35 11.25 11.8 12.65	Pale yellow needles d. > 250°
PA-153 (66)	(In 80% MeOH) 349.0 144.5 332.0 140.0 317.0 84.5	+398 (C_6H_5N) +296 (DMF) +353 (0.1 N-HCl/DMF)	5.8 6.4	9.9 11.3 11.8	Colorless needles > 260°
Capacidin (20)	(Solvent?) 350.0 93.6 332.0 103.9 318.0 77.3	—34.5 (MeOH) —45 (C_6H_5N) —54 (DMF) —49 (60% Acetone)	5.9 6.16 6.32 6.56	9.95 10.25 11.9	Mustard-yellow powder indef. m. p.
2814-P (113, 114)	(In MeOH) 351.0 333.0 317.0	0 (C_6H_5N) +32 (DMF) 0 (0.1 N-HCl)	5.82 6.36		Yellow non-crystalline
Distamycin B (104)	(Solvent?) 350.0 333.0 318.0				Non-crystalline

* See footnote to Table 6.

References, pp. 72—79.

and Chemical Properties.

Elemental Analysis			Neutralization Equivalent		Other Chemical Information
% C	% H and Empirical Formula	% N	as Acid	as Base	
57.97	8.17	1.65	positive	positive	Red-purple in conc. H_2SO_4; pos. Fehling; neg. biuret, Molisch, Seliwanoff; no S, hal.
					Red-purple in conc. H_2SO_4
		N present			Red-purple in conc. H_2SO_4; pos. Fehling; weak Molisch, Seliwanoff; pos. S; no hal.
58.36	8.24	2.04	positive	positive	Red-violet in conc. H_2SO_4; violet Carr-Price; pos. Fehling; neg. ninhydrin, alkali soluble
59.94	8.29 $C_{37}H_{61}O_{14}N$	1.88	positive	736	Violet in conc. H_2SO_4; pos. ninhydrin, Fehling, 2,4-DNHP; C—Me (three); neg. N—Me, OMe, OAc, S, hal.
61.52	8.05 $C_{54}H_{85}O_{18}N_2$	2.32		1035	Red-purple in conc. H_2SO_4; emerald green in HCl; emerald green $\rightarrow$ sapphire blue in H_3PO_4; cherry-red in NaOH; $FeCl_3$ neg.; ninhydrin red-violet; amino: 1.03%
57.52	8.23	1.76	positive	positive	Red-violet in conc. H_2SO_4; contains mycosamine. Separable on paper from eurocidin and capacidin
		N present			Violet in conc. H_2SO_4

Table 8. Methylpentaenes: Physical

Name and References	Ultraviolet Spectrum $\lambda\lambda_{max}$ (mμ) a*	$[\alpha]_D$	Infrared Peaks (μ) $>$C=O Reg.	9–15 μ Reg.	Crystal Form and m. p.
Fungichromin (*22, 23, 118*)	(Solvent?) 356.5 146.0 338.5 155.0 322.5 96.0		5.85 6.10 6.30	9.90 10.25 11.17 11.82 12.42 13.10 13.83	Pale yellow cryst. m. p. 205–210°
Filipin (*13, 35, 47, 134*)	(In MeOH) 355.0 133.0 338.0 136.0 322.0 91.0	—148 (MeOH)	5.83	9.95 11.9	Needles m. p. 195–205°
Eurocidin-group antibiotic (*72*)	(Solvent?) 356.5 16.4 338.5 17.9 322.5 12.5				White powder
St. sanguineus Methylpentaene (*69*)	(Solvent?) 355.0 156.3 338.0 156.8 322.0 95.4	—175 (Solvent?)			Colorless plates or yellow needles m. p. 248–250°
Pentamycin (Moldcidin B) (*83, 121, 123*)	(In MeOH) 356.0 150.0 338.0 145.0 322.0 90.0	—224–230 (C_6H_5N)	5.80 6.1	9.95 11.82	Pale yellow needles d. 226–9°
Lagosin (Glaxo A-246) (*11, 31, 32, 33, 132*)	(In MeOH) 356.0 148.0 338.0 149.0 322.0 93.0	—160 (MeOH)	5.83 6.08	9.93 10.26 11.16 11.82 12.42 13.09 13.85	Crystalline m. p. 230–240°
Cabicidin (*82*)	(In MeOH) 354.0 339.0 320.0	—135 (MeOH 34°)			Colorless or pale yellow columns 225°
Moldcidin A (*9, 97*)	(In 80% MeOH) 358.0 83.0 339.0 85.0 324.0 52.0		5.81 6.35	9.95 11.8	Pale yellow needles d. 180–230°

* See footnote to Table 6.

References, pp. 72—79.

and Chemical Properties.

Elemental Analysis			Neutralization Equivalent		Other Chemical Information
% C	% H and Empirical Formula	% N	as Acid	as Base	
60.93	8.65 $C_{35}H_{58}O_{12} \cdot H_2O$	o	o	o	Violet $\rightarrow$ blue in conc. H_2SO_4; Tollens? See text
63.99	8.82 $C_{37}H_{62}O_{12}$	o	o	o	Blue-violet in conc. H_2SO_4; pos. Molisch; neg. ninhydrin, $FeCl_3$, 2,4-DNPH. See text
62.05	9.15	o	o	o	
61.16	8.97	o	o	o	Deep purple in conc. H_2SO_4; Tollens?; neg. Fehling; $FeCl_3$
61.25	9.12 $C_{35}H_{58}O_{12}$	o	o	o	See text
	$C_{35}H_{60}O_{13}$	o			
56.05	8.75 $C_{42}H_{81}O_{19}N$	1.52			Pos. ninhydrin; neg. Fehling, Molisch, $FeCl_3$

Table 9. Hexaenes: Physical

Name and References	Ultraviolet Spectrum $\lambda\lambda_{max}$ (mμ) a^*		Infrared Peaks (μ) $>$C=O Reg. 9–15 μ Reg.		Crystal Form and m. p.
Flavacid (67, 112, 127, 85)	(In EtOH) 379.2 358.2 341.0	63.5 63.0 41.0			Light yellow powder d. 200° Na salt d. 102–105°
Mediocidin (84, 124)	(In MeOH) 377–378 356–357 339–340				Yellow powder
Endomycin B (Helixin B) (127, 129)	(In EtOH) 379.0 358.0 339.0	25.0 24.5 17.0			
Cryptocidin (96)	(In 80% MeOH) 380.0 358.0 341.0	61.0 57.0 37.0	6.0 6.2	10.0 10.4	Crystalline d. 100–115°

Table 10. Heptaenes: Physical

Name and References	Ultraviolet Spectrum $\lambda\lambda_{max}$ (mμ) a^*		$[\alpha]_D$	Infrared Peaks (μ) $>$C=O Reg. 9–15 μ Reg.		Crystal Form and m. p.
Ascosin (40, 57, 75, 127)	(In MeOH) 399.0 376–377 358.0	54.0 64.0 48.0	+ 12 (DMF) — 13 (0.1 N-HCl/MeOH)	5.87 6.15 6.25 6.40	7.98	Yellow-brown powder
Trichomycin A (52, 59, 60, 76, 77)	(In MeOH) 404.0 382.0 361.5	75.0 86.3 62.5		5.77 5.86 6.12 6.27 6.38	9.99 11.8 12.97–13.17 14.16	Light yellow plates > 320°
Candicidin (68, 75, 99, 127, 130)	(In EtOH) 401 379 360			5.87 6.15 6.25 6.40		
Candidin (99, 111, 128, 130)	(In EtOH) 407.5 383.5 364.0	191.0 173.0 98.5	+ 205 (AcOH) + 363 (DMF)	5.83 5.87 6.18 6.41	9.88 10.35 11.9	Golden yellow needles 180°

* See footnote to Table 6.

and Chemical Properties.

% C	Elemental Analysis % H and Empirical Formula	% N	Neutralization Equivalent as Acid	as Base	Other Chemical Information
61.57	7.77 (ash, 2.20)	1.06	positive		Pos. ninhydrin, $FeCl_3$; neg. Fehling; no amino sugar (98)
			positive	positive	R_f values given
	Present		positive		Deep blue in conc. H_2SO_4; neg. ninhydrin, $FeCl_3$, Fehling, Molisch, S, hal.

and Chemical Properties.

% C	Elemental Analysis % H and Empirical Formula	% N	Neutralization Equivalent as Acid	as Base	Other Chemical Information
			positive	positive	Blue in conc. H_2SO_4
60.06	7.62 $C_{61}H_{86}O_{21}N_2 \cdot 2H_2O$	2.28	1230	positive	Blue $\to$ violet in conc. H_2SO_4; no S, hal.; pos. $FeCl_3$; neg. Fehling; Molisch?; also see text
		1.6–2.16	positive	positive	Blue in conc. H_2SO_4; contains —COOH pK_a ca. 6.2; no allylic —OH; contains mycosamine and p-$NH_2C_6H_4COCH_3$
60.20	8.24 $C_{46}H_{75}O_{17}N$	1.52	922	928	Blue in conc. H_2SO_4; contains C—Me (six), mycosamine; gives N-acetyl deriv.; neg. 2,4-DNPH; no cleavage by HIO_4; U. V. spectrum unaltered by I_2 and light

(Table 10, continued.)

Name and References	Ultraviolet Spectrum $\lambda\lambda_{max}$ (mμ) $a*$	$[\alpha]_D$	Infrared Peaks (μ) >C=O Reg. 9—15 μ Reg.		Crystal Form and m. p.
Candimycin (*102, 26*)	(In MeOH) 406.0 382.0 362.0				
St. abikoensis Heptaene (*120*)	(In MeOH) 399—403 378—380 358—360				Yellow powder
Amphotericin B (*12, 37, 40, 42, 46, 125*)	(In MeOH) 406.0 189.0 382.0 167.0 363.0 98.0	+ 238 (DMF) + 333 (acidic DMF) — 33.5 (0,1 N-HCl/MeOH)	5.83 6.34	9.88 10.2 11.08 11.29 11.75	Yellow crystals > 170°
Aureofacin (*61*)	(Solvent?) 402.0 380.0 359.0				Yellow-brown powder
757 (*24, 26, 27*)	(In EtOH) 404.0 381.0 361.0				Light yellow powder
Heptamycin (*55*)	(In EtOH) 403.0 380.0 362.0				
PA-150 (*66*)	(In 80% MeOH) 397.0 89.5 376—377 103.3 358.0 73.0	+ 294 (C$_6$H$_5$N) + 148 (DMF) — 34 (0.1 N-HCl/DMF) Sodium salt: — 2590 (H$_2$O)	5.87 6.14 6.26 6.36	10.00 11.82 13.1	Yellow crystals
AYF (Ayfactin) (*63, 19*) A	(In aqueous dimethylacet-amide) 409.0 383.0 52.6 363.0				Brown crystals
B	(In aqueous dimethylacet-amide) 409.0 383.0 55.6 363.0		5.9 6.1—6.2 6.3 6.4	10.05 11.82 13.1	Dark yellow crystals

References, pp. 72—79.

Elemental Analysis			Neutralization Equivalent		Other Chemical Information
% C	% H and Empirical Formula	% N	as Acid	as Base	
57.17	8.18	1.70			Neg. Molisch, Fehling, $FeCl_3$ Differs from candidin on paper
			positive		Neg. Molisch, Tollens, $FeCl_3$, S
57.59	8.0 $C_{46}H_{73-75}O_{18-20}N$	1.7	Perhydro 929–959 deriv. 970 Saponification equ. 490		Blue-purple in conc. H_2SO_4; neg. $FeCl_3$, —OMe; pos. Molisch. Contains mycosamine; also see text
	Present				Blue-purple in conc. H_2SO_4; neg. $FeCl_3$, S; weak Molisch
			positive		Differs on paper from trichomycin, candidin, amphotericin
			positive		
62.03	7.83 $C_{54}H_{82}O_{18}N_2$	2.73	positive	positive	Blue in conc. H_2SO_4; pos. Fehling, 2,4-DNHP; —C—Me (four) neg. —OMe, —N—Me; ninhydrin?
62.45	7.64 $C_{25}H_{35-36}O_7N$	2.8	positive		
62.55	7.86 $C_{25}H_{35-36}O_7N$	2.65	positive		

(Table 10, continued.)

Name and References	Ultraviolet Spectrum $\lambda\lambda_{max}$ (mμ) a^*	$[\alpha]_D$	Infrared Peaks (μ) $\geq$C=O Reg. 9–15 μ Reg.	Crystal Form and m. p.
26/1 (*81, 115, 117*)	(In EtOH) 403.0 380.0 360.0			Yellow
Antifungin 4915 (*51*)	(In 70% MeOH- 20% *Ac*OH) 407.5 63.6 379.5 76.4 360.0 54.0			Yellow crystals d. 100°
Eurotin A (*106*)	(In 80% EtOH) 405.0 380.0 360.0			Yellow brown powder
F-17-C (*26*)	(In EtOH) 408.0 383.0 365.0			Yellow powder
AE-56 (*108*)	(In H$_2$O) 412.0 388.0 365.0 322.0			Yellow amorph.
Perimycin (*18, 86, 99, 137, 138*)	(In MeOH) 406.0 383.0 100.0 361.0		5.83 9.90 6.05 11.80 6.25 12.05 6.50	Golden-yellow amorph. decomp. on heat
2814-H (*114*)	(In MeOH) 406.0 384.0 362.0	+ 353 (C$_6$H$_5$N) + 347 (DMF)	5.82 6.36	Yellow solid
Hamycin (*14, 15, 16*)	(In 80% MeOH) 406.0 91.8 383.0 96.2 363.0	+ 216 (C$_6$H$_5$N)	5.8 9.96 6.1 11.8 6.3 6.4	Golden yellow powder d. > 160°

References, pp. 72—79.

Elemental Analysis			Neutralization Equivalent		Other Chemical Information
% C	% H	% N	as Acid	as Base	
	and Empirical Formula				
			positive	positive	Violet in conc. H_2SO_4. Two active fractions separable on paper chromatography. Amphoteric
63.6 (Ash: 1.3)	7.8	2.8			Blue in conc. H_2SO_4; neg. Tollens, DNPH, $FeCl_3$
	N present			positive	Deep blue in conc. H_2SO_4; neg. Fehling, Tollens
			positive	positive	Deep blue in conc. H_2SO_4
					Blue in conc. H_2SO_4
63.19	8.28 $C_{47}H_{74-76}O_{14}N_2$	3.38		948	No —COOH; one amino N as perosamine, other as p-$H_2NC_6H_4$; C—Me (four); see text
56.9	8.15	1.92		positive	Blue in conc. H_2SO_4; mycosamine present
59.5	8.3	2.2	positive	positive	Deep blue in conc. H_2SO_4; HCl hydrolysis yields amino acids; see text

References.

1. AHLERS, N. H. E., R. A. BRETT and N. G. McTAGGART: An Infra-Red Study of the *cis-* and *trans*-Isomers of Some C_{18} Fatty Acids. J. Appl. Chem. **3**, 433 (1953).

2. *American Cyanamid Co.:* Pimaricin, An Antifungal Substance. British Patent 846933 (1960).

3. AMMANN, A. and D. GOTTLIEB: Paper Chromatography of Antifungal Antibiotics. Appl. Microbiol. **3**, 181 (1955).

4. *Analytical Chemistry*, Editors: Spectrometry Nomenclature. Analyt. Chemistry **34**, 1852 (1962).

5. ARAI, T., Y. MORITA and Y. TAKAMIZAWA: Studies on Antifungal Substances Extracted from the Mycelia of Streptomyces. V. On Certain Characteristics of Antifungal Substances of Mycelial Origin. J. Antibiotics (Tokyo) **7 A,** 169 (1954).

6. ARCAMONE, F., C. BERTAZZOLI, G. CANEVAZZI, A. DI MARCO, M. GHIONE e A. GREIN: La etruscomicina, nuovo antibiotico antifungino prodotto dallo *Streptomyces lucensis* N. SP. Giorn. Microbiol. **4**, 119 (1957).

7. ARCAMONE, F. e M. PEREGO: Isolamento e caratterizzazione di un nuovo antibiotico, la etruscomicina. Ann. chim. (Roma) **49**, 345 (1959).

8. ARIMA, N., J. SAKAMOTO and E. OKAMOTO: Protocidin, an Antibiotic Substance. Japanese Patent 8648 (1960) [Chem. Abstr. **55**, 7758 (1961)].

9. ARISHIMA, N. and J. SAKAMOTO: Moldcidin A, Antibiotic Substance. Japanese Patent 1148 (1961) [Chem. Abstr. **55**, 21476 (1961)].

10. BACKUS, E. J. and M. DANN: Biological Fungicides. German Patent 1056785 (1959).

11. BALL, S., C. J. BESSELL and A. MORTIMER: The Production of Polyenic Antibiotics by Soil Streptomycetes. J. Gen. Microbiol. **17**, 96 (1957).

12. BARTNER, E., H. ZINNES, R. A. MOE and J. S. KULESZA: Studies on a New Solubilized Preparation of Amphotericin B. Antibiotics Annu. **1957–1958**, 53.

13. BERKOZ, B. and C. DJERASSI: Macrolide Antibiotics. IX. Filipin. Proc. Chem. Soc. (London) **1959**, 316.

13a. BESSELL, C. J., W. K. ANSLOW, A. M. MORTIMER, D. L. FLETCHER and A. RHODES: Antifungal Antibiotic Production. U. S. Patent 3013947 (1961).

14. BHATE, D. S.: Chemistry of Polyene Antibiotics. In: T. S. GORE, B. S. JOSHI, S. V. SUNTHANKAR and B. D. TILAK, Recent Progress in the Chemistry of Natural and Synthetic Colouring Matters, p. 341. New York: Academic Press. 1962.

15. BHATE, D. S. and S. P. ACHARYA: Purification of Hamycin. Hindustan Antibiotics Bull. **5**, 16 (1962).

16. BHATE, D. S., G. R. AMBEKAR, K. K. BHATNAGAR and R. K. HULYALKAR: Hamycin. II. Isolation and Chemical Properties. Hindustan Antibiotics Bull. **3**, 139 (1961).

16a. BIRCH, A. J.: Aspects of the Biosynthesis and Structures of Some Antibiotics. Lecture, North Jersey Section, Amer. Chem. Soc., March 26, 1963.

17. BIRCH, A. J., E. PRIDE, R. W. RICKARDS, P. J. THOMSON, J. D. DUTCHER, D. PERLMAN and C. DJERASSI: Biosynthesis of Methymycin. Chem. and Ind. **1960**, 1245.

18. BOROWSKI, E., C. P. SCHAFFNER and H. LECHEVALIER: Perimycin, a Novel Type of Heptaene Antifungal Antibiotic. Antimicrobial Agents Annu. **1960**, 532.

19. *Bristol Laboratories, Inc.:* Ayfactin. British Patent 796982 (1958).

20. BROWN, R. and E. L. HAZEN: Capacidin, A New Member of the Polyene Antibiotic Group. Antibiot. & Chemother. **10**, 702 (1960).

20a. Bu'Lock, J. D., D. C. Allport and W. B. Turner: The Biosynthesis of Polyacetylenes. Part III. Polyacetylenes and Triterpenes in *Polyporus anthracophilus*. J. Chem. Soc. (London) 1961, 1654.

21. Burns, J. and D. F. Holtman: Tennecetin: A New Antifungal Antibiotic. General Considerations. Antibiot. & Chemother. 9, 398 (1959).

22. Cope, A. C., R. K. Bly, E. P. Burrows, O. J. Ceder, E. Ciganek, B. T. Gillis, R. F. Porter and H. E. Johnson: Fungichromin: Complete Structure and Absolute Configuration at C_{26} and C_{27}. J. Amer. Chem. Soc. 84, 2170 (1962).

23. Cope, A. C. and H. E. Johnson: Fungichromin. Determination of the Structure of the Pentaene Chromophore. J. Amer. Chem. Soc. 80, 1504 (1958).

24. Craveri, R. and G. Giolitti: Isolation and Study of an Antifungal Antibiotic. Ann. Microbiol. 7, 81 (1956).

25. Craveri, R., A. M. Lugli, B. Sgarzi and G. Giolitti: Distribution of Antibiotic-Producing Streptomycetes in Italian Soils. Antibiot. & Chemother. 10, 306 (1960).

26. Craveri, R., O. L. Shotwell, R. G. Dworschack, T. G. Pridham and R. W. Jackson: Antibiotics Against Plant Disease. VII. The Antifungal Heptaene Component (F-17-C) Produced by *Streptomyces cinnamomeus* f. *azacoluta*. Antibiot. & Chemother. 10, 430 (1960).

27. Craveri, R. and U. Veronesi: Effect on Mitosis in *Allium cepa* of an Antifungal Antibiotic produced by *Streptomyces* sp. Exptl. Cell Res. 11, 560 (1956).

28. Davisson, J. W., F. W. Tanner, Jr., A. C. Finlay and J. H. Kane: Rimocidin and Methods for its Recovery. U. S. Patent 2963401 (1960).

29. Davisson, J. W., F. W. Tanner, Jr., A. C. Finlay and I. A. Solomons: Rimocidin, A New Antibiotic. Antibiot. & Chemother. 1, 289 (1951).

30. Despois, R., S. Pinnert-Sindico, L. Ninet et J. Preud'homme: Trois antibiotiques de groupes différents produits par une même souche de *Streptomyces*. Giorn. Microbiol. 2, 76 (1956).

31. Dhar, M. L., V. Thaller and M. C. Whiting: A New Type of Macrolide Antibiotic. Proc. Chem. Soc. (London) 1958, 148.

32. — — — The Structures of Lagosin and Filipin. Proc. Chem. Soc. (London) 1960, 310.

33. Dhar, M. L., V. Thaller, M. C. Whiting, R. Ryhage, S. Ställberg-Stenhagen and E. Stenhagen: The Carbon Skeleton of Lagosin (Antibiotic A-246). Proc. Chem. Soc. (London) 1959, 154.

34. Divekar, P. V., J. L. Bloomer, J. F. Eastham, D. F. Holtman and D. A. Shirley: The Isolation of Crystalline Tennecetin and the Comparison of this Antibiotic with Pimaricin. Antibiot. & Chemother. 11, 377 (1961).

35. Djerassi, C., M. Ishikawa, H. Budzikiewicz, J. N. Shoolery and L. F. Johnson: The Structure of the Macrolide Antibiotic Filipin. Tetrahedron Letters 1961, No. 12, 383.

36. Djerassi, C., P. C. Seidel, J. Westley, A. J. Birch, C. W. Holzapfel, R. W. Rickards, J. D. Dutcher and R. Thomas: Biosynthetic Approaches to Structure Determination: Nystatin. Abstr. Papers, 142nd Meeting, Amer. Chem. Soc., Sept. 1962, p. 5 P.

37. Donovick, R., B. A. Steinberg, J. D. Dutcher and J. Vandeputte: Biological and Chemical Characteristics of Amphotericin B. Giorn. Microbiol. 2, 147 (1956).

38. Dutcher, J. D.: The Chemistry of Nystatin and Related Antibiotics. Monographs on Therapy, Squibb Inst. Med. Res., New Brunswick, N. J. 2, No. 1, 87 (1957).

39. DUTCHER, J. D., G. BOYACK and S. FOX: The Preparation and Properties of Crystalline Fungicidin (Nystatin). Antibiotics Annu. 1953–1954, 191.

40. DUTCHER, J. D., W. GOLD, J. F. PAGANO and J. VANDEPUTTE: Amphotericin B and its Salts. U. S. Patent 2908611 (1959).

41. DUTCHER, J. D., D. R. WALTERS and O. P. WINTERSTEINER: Studies of the Chemical Properties and Structure of Nystatin (Mycostatin). In: T. H. Sternberg and V. D. Newcomer, Therapy of Fungus Diseases, p. 168. Boston: Little, Brown. 1955.

42. DUTCHER, J. D., M. B. YOUNG, J. H. SHERMAN, W. HIBBITS and D. R. WALTERS: Chemical Studies on Amphotericin B. I. Preparation of the Hydrogenation Product and Isolation of Mycosamine, An Acetolysis Product. Antibiotics Annu. 1956–1957, 866.

43. FUJITA, H.: Chemical Study of Akitamycin, An Antifungal Antibiotic. J. Antibiotics (Tokyo) 12 B, 297 (1959).

43a. GARDNER, J. N., G. LOWE and G. READ: Chemistry of the Higher Fungi. Part XII. The Enzymic Decarboxylation of an $\alpha\beta$-Acetylenic Acid. J. Chem. Soc. (London) 1961, 1532.

44. GATENBECK, S.: The Biosynthesis of Oxytetracycline. Biochem. Biophys. Res. Commun. 6, 422 (1962).

45. GILNER, D. and P. R. SRINIVASAN: The Biosynthesis of Magnamycin, a Macrolide Antibiotic. Biochem. Biophys. Res. Commun. 8, 299 (1962).

46. GOLD, W., H. A. STOUT, J. F. PAGANO and R. DONOVICK: Amphotericins A and B, Antifungal Antibiotics Produced by a Streptomycete. I. In-Vitro Studies. Antibiotics Annu. 1955–1956, 579.

47. GOTTLIEB, D., A. AMMANN and H. E. CARTER: A New Antifungal Agent, Filipin. Plant Disease Reporter 39, 219 (1955).

48. GOTTLIEB, D. and H. L. POTE: Tetrin, an Antifungal Antibiotic. Phytopathology 50, 817 (1960).

49. GRISEBACH, H. und H. ACHENBACH: Zur Biogenese der Makrolide. Über die Herkunft der Aldehydgruppe des Magnamycins. Tetrahedron Letters 13, 569 (1962).

50. GRISEBACH, H., H. ACHENBACH und U. C. GRISEBACH: Zur Biogenese des Erythromycins. Naturwiss. 47, 206 (1960).

51. HAGEMANN, G., G. NOMINE and L. PÉNASSE: Antifungin 4915. German Patent 1053738 (1959).

51a. HATTORI, K.: Studies on Trichomycin. VII. Structure of C_{21} Aldehyde. VIII. Chemical Structure of Trichomycin A. J. Antibiotics (Tokyo) 15 B, 37 (1962) [Chem. Abstr. 58, 5532 (1963)].

52. HATTORI, K., H. NAKANO, M. SEKI and Y. HIRATA: Studies on Trichomycin. IV. J. Antibiotics (Tokyo) 9 A, 176 (1956).

53. HAUSSER, K. W., R. KUHN und G. SEITZ: Lichtabsorption und Doppelbindung. V. Über die Absorption von Verbindungen mit konjugierten Kohlenstoffdoppelbindungen bei tiefer Temperatur. Z. physik. Chem. 29 B, 391 (1935).

54. HAZEN, E. L. and R. BROWN: Two Antifungal Agents Produced by a Soil Actinomycete. Science (Washington) 112, 423 (1950).

55. HENIS, Y., N. GROSSOWICZ and M. ASCHNER: Heptamycin, an Antifungal and Antiprotozoal Antibiotic. Bull. Res. Council Israel 6 E, VII (1957).

56. HICKEY, R. J.: The Antagonism Between the Antifungal Antibiotic, Ascosin, and Some Long-Chain, Unsaturated Fatty Acids. Arch. Biochem. Biophys. 46, 331 (1953).

57. HICKEY, R. J., C. J. CORUM, P. H. HIDY, I. R. COHEN, U. F. B. NAGER and E. KROPP: Ascosin, An Antifungal Antibiotic Produced by a Streptomycete. Antibiot. & Chemother. 2, 472 (1952).

58. HOSOYA, S. and N. HAMAMURA: A New Solvent for the Purification of the Antifungal Antibiotics Belonging to the Trichomycin Group. J. Antibiotics (Tokyo) 9 A, 129 (1956).

59. HOSOYA, S., N. KOMATSU, M. SOEDA and Y. SONODA: Trichomycin, a New Antibiotic Produced by *Streptomyces hachijoensis* with Trichomonadicidal and Antifungal Activity. Japan J. Exp. Med. 22, 595 (1952) [Chem. Abstr. 48, 3444 (1952)].

60. HOSOYA, S., N. KOMATSU, M. SOEDA, T. YUNAGUCHI and Y. SONODA: Trichomycin, A New Antibiotic With Trichomonicidal and Antifungal Activities. J. Antibiotics (Tokyo) 5 A, 564 (1952).

61. IGARASHI, M., K. OGATA and A. MIYAKE: An Antifungal Substance Produced by *Streptomyces aureofaciens*. J. Antibiotics (Tokyo) 9 B, 79 (1956) [Chem. Abstr. 53, 20 251 (1959)].

62. — — — Streptomyces Pentaene Group Substances. I. An Antifungal Antibiotic Produced by *Streptomyces acidomyceticus*. J. Antibiotics (Tokyo) 9 B, 101 (1956) [Chem. Abstr. 53, 22 224 (1959)].

62 a. JONES, E. R. H.: Pedler Lecture: Polyacetylenes. Proc. Chem. Soc. (London) **1960**, 199.

63. KAPLAN, M. A., B. HEINEMANN, I. MYDLINSKI, F. H. BUCKWALTER, J. LEIN and I. R. HOOPER: An Antifungal Antibiotic (AYF) Produced by a Strain of *Streptomyces aureofaciens*. Antibiot. & Chemother. 8, 491 (1958).

64. KATZ, E. and H. LECHEVALIER: Nutritional Control of Antibiotic Synthesis as Applied to Actinomycin and the Polyene Antifungal Antibiotics. In: Recent Advances in Botany, p. 657. Toronto: Univ. of Toronto Press. 1961.

65. KAUFMANN, H. P. und R. K. SUD: Zur Stereochemie der vierfach konjugiert-ungesättigten Parinärsäuren. Chem. Ber. 92, 2797 (1959).

66. KOE, B. K., F. W. TANNER, Jr., K. V. RAO, B. A. SOBIN and W. D. CELMER: PA 150, PA 153, and PA 166: New Polyene Antifungal Antibiotics. Antibiotics Annu. **1957–1958**, 897.

67. KUROYA, M.: Antibiotic Substance, Flavacid. Japanese Patent 8547 (1954) [Chem. Abstr. 50, 9696 (1956)].

68. LECHEVALIER, H., R. F. ACKER, C. T. CORKE, C. M. HAENSELER and S. A. WAKSMAN: Candicidin, A New Antifungal Antibiotic. Mycologia 45, 155 (1953).

69. LINDNER, F., J. SCHMIDT-THOMÉ, K. KÜHN, G. NESEMANN, A. STEIGLER and K. H. WALLHÄUSSER: Antifungal Antibiotic. German Pat. 1 017 329 (1957) [Chem. Abstr. 55, 898 (1961)].

70. LINDNER, F., J. SCHMIDT-THOMÉ, K. H. WALLHÄUSSER and H. WEIDEN-MÜLLER: Fungicidal Antibiotic. German. Pat. 1 012 430 (1957) [Chem. Abstr. 54, 6045 (1960)].

71. LOEB, J. N., P. K. BROWN and G. WALD: *Cis-trans* Isomerism and Steric Hindrance. Nature (London) 184, 617 (1959).

72. MAEDA, K., K. OI, H. KOSAKA, E. LIN WANG and H. UMEZAWA: Simultaneous Production of an Actinomycin and a Eurocidin-Group Antibiotic. J. Antibiotics (Tokyo) 9 A, 125 (1956).

73. MATSUOKA, M. and H. UMEZAWA: Unamycin, An Antifungal Substance Produced by *Streptomyces fungicidicus*. J. Antibiotics (Tokyo) 13 A, 114 (1960).

74. MEBANE, A. D.: 1,3,5,7,9-Decapentaene and 1,3,5,7,9,11,13-Tetradeca-heptaene. J. Amer. Chem. Soc, 74, 5227 (1952).

75. Mebane, A. D. and W. Oroshnik: Unpublished observations.
76. Nakano, H.: Studies on Trichomycin. V. J. Antibiotics (Tokyo) 14 A, 68 (1961).
77. — Studies on Trichomycin. VI. J. Antibiotics (Tokyo) 14 A, 72 (1961).
78. Nakano, H., K. Hattori, M. Seki and Y. Hirata: Trichomycin. III. J. Antibiotics (Tokyo) 9 A, 172 (1956).
79. Nakazawa, K.: Streptomycetes. III. Eurocidin, An Antibiotic Produced by S. albireticuli. J. Agr. Chem. Soc. Japan 29, 650 (1955) [Chem. Abstr. 50, 5830 (1956)].
80. Nayler, P. and M. C. Whiting: Researches on Polyenes. Part III. The Synthesis and Light Absorption of Dimethylpolyenes. J. Chem. Soc. (London) 1955, 3037.
81. Nefelova, M. V. and I. N. Pozmogova: Culturing Actinomycetes Producing Antibiotics Toxic to Fungi. Mikrobiologiya 29, 856 (1960) [Chem. Abstr. 55, 14586 (1961)].
82. Ogata, K., S. Igarashi and Y. Nakao: Culture Medium for Separation of Bacteria from Mixtures of Bacteria, Yeasts, and Molds. Japanese Patent 9245 (1958) [Chem. Abstr. 53, 6545 (1959)].
83. Ogawa, H., T. Ito, S. Inoue and M. Nishio: Chemical Study on Moldcidin B and its Identification with Pentamycin. J. Antibiotics (Tokyo) 13 A, 353 (1960).
84. Okami, Y., R. Utahara, S. Nakamura and H. Umezawa: Studies on Antibiotics from Actinomycetes. IX. J. Antibiotics (Tokyo) 7 A, 98 (1954).
85. Oroshnik, W., L. C. Vining, A. D. Mebane and W. A. Taber: Polyene Antibiotics. Science (Washington) 121, 147 (1955).
86. Oswald, E. J., R. J. Reedy and W. A. Randall: An Antifungal Agent, 1968, Produced by a New Streptomyces Species. Antibiotics Annu. 1955–1956, 236.
87. Pao, Ch'in-Chu, P.-Y. Kung and J.-S. Ts'ai: Fungicidin A-94. II. Preparation and Chemical Characteristics. K'o Hsüeh T'ung Pao 24, 825 (1959) [Chem. Abstr. 54, 20054 (1960)].
88. Parke, Davis & Co.: Sistomycosin. British Patent 712547 (1954).
89. Patrick, J. B., R. P. Williams, C. F. Wolf and J. S. Webb: Pimaricin. I. Oxidation and Hydrolysis Products. II. The Structure of Pimaricin. J. Amer. Chem. Soc. 80, 6688, 6689 (1958).
90. Pledger, R. A. and H. Lechevalier: Survey of the Production of Polyenic Substances by Soil Streptomycetes. Antibiotics Annu. 1955–1956, 249.
91. Rabinowitch, E. and L. F. Epstein: Polymerization of Dyestuffs in Solution. Thionine and Methylene Blue. J. Amer. Chem. Soc. 63, 69 (1941).
92. Rao, K. V., W. S. Marsh and A. L. Garretson: Biologically Active Substances. German Patent 1052065 (1959) [Chem. Abstr. 55, 18009 (1961)].
93. Raubitscheck, F., R. F. Acker and S. A. Waksman: Production of an Antifungal Agent of the Fungicidin Type by S. aureus. Antibiot. & Chemother. 2, 179 (1952).
93a. Richardson, A. C.: The Synthesis and Stereochemistry of Mycaminose. Proc. Chem. Soc. (London) 1961, 430.
94. Ross, W. C. J.: The Reactions of Certain Epoxides in Aqueous Solution. J. Chem. Soc. (London) 1950, 2257.
95. Sakamoto, J. M. J.: Étude sur antibiotique antifongique. I. La protocidine, un nouvel antibiotique produit par les Streptomycete. J. Antibiotics (Tokyo) 10 A, 128 (1957).
96. — Étude sur antibiotique antifongique. II. La cryptocidine, un nouvel antibiotique produit par les Streptomycete. J. Antibiotics (Tokyo) 12 A, 21 (1959).

97. SAKAMOTO, J. M. J.: Étude sur antibiotique antifongique. III. La moldcidine A, un nouvel antibiotique produit par les Streptomycete. J. Antibiotics (Tokyo) **12 A**, 169 (1959).

98. SCHAFFNER, C. P.: Personal communication.

99. SCHAFFNER, C. P. and E. BOROWSKI: Biologically Active N-Acyl Derivatives of Polyene Macrolide Antifungal Antibiotics. Antibiot. & Chemother. **11**, 724 (1961).

100. SCHAFFNER, C. P., I. D. STEINMAN, R. S. SAFFERMAN and H. LECHEVALIER: Role of Inorganic Salts and Mevalonic Acid in the Production of a Tetraenic Antifungal Antibiotic. Antibiotics Annu. **1957–1958**, 869.

101. SCHNABEL, E., H. NÖTHER and H. KUHN: Monomers, Dimers, and Tetramers in Solutions of Phthalocyanine Sulphonates. In: T. S. GORE et al., Recent Progress in the Chemistry of Natural and Synthetic Colouring Matters, p. 561. New York: Academic Press. 1962.

102. SHIBATA, M., M. HONJO, Y. TOKUI and K. NAKAZAWA: On a New Antifungal and Antiyeast Substance, Candimycin, Produced by a *Streptomyces*. J. Antibiotics (Tokyo) **7 B**, 168 (1954).

103. SHIBATA, Y., K. TANABE, A. MIYAKE, Y. TOKUI, S. YOSHISHIRO, J. KANEKO, S. FUJII, K. OKI, I. TADOKORO and K. NAKAZAWA: Enteromycin, a New Antibiotic. Japanese Patent 4995 (1956) [Chem. Abstr. **52**, 9529 (1958)].

103a. SMAKULA, A.: Über physikalische Methoden im chemischen Laboratorium. XXII. Lichtabsorption und chemische Konstitution. Angew. Chem. **47**, 657 (1934).

104. Società Farmaceutici Italia: The Fermentative Production of the Antibiotics Distamycine and Distacyne. British Patent 872734 (1961).

105. SOEDA, M. and H. FUJITA: Akitamycin, An Antifungal Antibiotic. J. Antibiotics (Tokyo) **12 B**, 293 (1959) [Chem. Abstr. **54**, 14359 (1960)].

106. — — An Antifungal Antibiotic, Eurotin A. J. Antibiotics (Tokyo) **12 B**, 368 (1959) [Chem. Abstr. **54**, 16644 (1960)].

107. SONDHEIMER, F., D. A. BEN-EFRAIM and R. WOLOVSKY: Unsaturated Macrocyclic Compounds. XVII. The Prototropic Rearrangement of Linear 1,5-Enynes to Conjugated Polyenes. The Synthesis of a Series of Vinylogs of Butadiene. J. Amer. Chem. Soc. **83**, 1675 (1961).

108. STARON, T. et A. FAIVRE-AMIOT: Isolement et propriétés physicochimiques d'un nouvel antibiotique antifongale. C. R. hebd. Séances Acad. Sci. **250**, 1730 (1960).

109. STRUYK, A. P., I. HOETTE, G. DROST, J. M. WAISVISZ, T. VAN EEK and J. C. HOOGERHEIDE: Pimaricin, A New Antifungal Antibiotic. Antibiotics Annu. **1957–1958**, 878.

110. TABER, W. A. and L. C. VINING: A Comparison of the Antifungal Antibiotics Candidin, Ascosin, Candicidin, and Trichomycin. Bacteriol. Proc. (Soc. Amer. Bacteriologists) **54**, 86 (1954).

111. TABER, W. A., L. C. VINING and S. A. WAKSMAN: Candidin, A New Antifungal Antibiotic Produced by *Streptomyces viridoflavus*. Antibiot. & Chemother. **4**, 455 (1954).

112. TAKAHASHI, I.: Studies on Antibiotic Substances from Actinomycetes. XXVII. A New Antifungal Substance, Flavacid. J. Antibiotics (Tokyo) **6 A**, 117 (1953).

113. THRUM, H.: Eine neue, von einer Spezies der *Streptomyces reticuli*-Gruppe gebildete Antibiotikakombination. Naturwiss. **46**, 87 (1959).

114. — Stoffwechselprodukte und Antibiotika von *Streptomyces* I A 2814, einem *Streptomyces-Netropsis*-Stamm. Planta Med. **8**, 376 (1960).

115. Tsyganov, V. A., P. N. Golyakov, A. M. Bezborodov, U. P. Namestnikova, G. V. Khopko, S. N. Solov'ev, M. A. Malyshkina and L. O. Bol'shakova: Antifungal Antibiotic 26/1. Antibiotiki 4, 21 (1959) [Chem. Abstr. 53, 19167 (1959)].

116. Tsyganov, V. A., P. N. Golyakov, S. N. Solov'ev, B. G. Belen'kii and A. I. Filippova: Antibiotic Substances of the Polyene Type. I. The Biological Properties of Actinomycetes Producing Polyene Antibiotics. II. Physico-chemical Properties of Polyene Antibiotics. Sbornik Nauch. Trudov Leningrad. Nauch-Issledovatel. Inst. Antibiotikov 2, 6, 13 (1960) [Chem. Abstr. 55, 17751, 22703 (1961)].

117. — — — — — Antibiotic Properties and Systematic Position of Some Actinomycetes of the *Globisporus* Group. II. Trudy Inst. Mikrobiol., Akad. Nauk S. S. S. R. 1960, 182 [Chem. Abstr. 55, 10585 (1961)].

118. Tytell, A. A., F. J. McCarthy, W. P. Fisher, W. A. Bolhofer and J. Charney: Fungichromin and Fungichromatin: New Polyene Antifungal Agents. Antibiotics Annu. 1954–1955, 716.

119. Umezawa, H. and M. Nagase: Unamycin, A New Antifungal Substance. Japanese Patent 3392 (1959) [Chem. Abstr. 53, 20708 (1959)].

120. Umezawa, H., Y. Okami and R. Kihara: New Antibiotic Substance Obtained from *Streptomyces abikoensis*. Japanese Patent 9346 (1955) [Chem. Abstr. 51, 18495 (1957)].

121. Umezawa, S., S. Nakada and H. Onuma: Isolation of 2-Methyl-2,4,6,8,10-dodecapentaenedial from the Oxidative Degradation of Pentamycin. J. Antibiotics (Tokyo) 11 A, 273 (1958).

122. Umezawa, S., M. Ooka and S. Shiozu: Pentamycin, A New Antibiotic. Japanese Patent 6000 (1959) [Chem. Abstr. 53, 22764 (1959)].

123. Umezawa, S., Y. Tanaka, M. Ooka and S. Shiotzu: A New Antifungal Antibiotic, Pentamycin. J. Antibiotics (Tokyo) 11 A, 26 (1958).

124. Utahara, R., Y. Okami, S. Nakamura and H. Umezawa: A New Antifungal Substance, Mediocidin, and Other Antifungal Substances of *Streptomyces* with Three Characteristic Absorption Maxima. J. Antibiotics (Tokyo) 7 A, 120 (1954).

125. Vandeputte, J., J. L. Wachtel and E. T. Stiller: Amphotericins A and B, Antifungal Antibiotics Produced by a Streptomycete. II. The Isolation and Properties of the Crystalline Amphotericins. Antibiotics Annu. 1955–1956, 587.

126. Vaněk, Z., L. Doležilová and Z. Řeháček: Formation of a Mixture of Antibiotic Substances Including Antibiotics of a Polyene Character, by Strains of Actinomycetes Freshly Isolated from Soil Samples. J. Gen. Microbiol. 18, 649 (1958).

127. Vining, L. C.: The Polyene Antifungal Antibiotics. Hindustan Antibiot. Bull. 3, 37 (1960).

128. Vining, L. C. and W. A. Taber: Preparation and Properties of Crystalline Candidin. Canad. J. Chem. 34, 1163 (1956).

129. — — Separation of Endomycins A and B, and their Identification as Members of the Polyene Group of Antifungal Antibiotics. Canad. J. Chem. 35, 1461 (1957).

130. Vining, L. C., W. A. Taber and F. J. Gregory: The Candidin-Candicidin Group of Antifungal Antibiotics. Antibiotics Annu. 1954–1955, 980.

130a. Von Saltza, M. H., J. Reid, J. D. Dutcher and O. Wintersteiner: Nystatin. II. Stereochemistry of Mycosamine. J. Amer. Chem. Soc. 83, 2785 (1961).

131. WAKAKI, S., S. AKANABE, K. HAMADA and T. ASAHINA: Antifungal Substances Produced by Actinomycetes. Antibiotic from Strain C-6. J. Antibiotics (Tokyo) **5**, 677 (1952).

132. WAKAKI, S., K. HAMADA, S. AKANABE and T. ASAHINA: Studies of an Antifungal Antibiotic from Streptomyces, IV. On the Physico-Chemical Properties of Chromin. J. Antibiotics (Tokyo) **6 A**, 145 (1953); **6 B**, 247 (1953).

133. WALTERS, D. R., J. D. DUTCHER and O. WINTERSTEINER: Structure of Mycosamine. J. Amer. Chem. Soc. **79**, 5076 (1957).

134. WHITFIELD, G. B., T. D. BROCK, A. AMMANN, D. GOTTLIEB and H. E. CARTER: Filipin, an Antifungal Antibiotic: Isolation and Properties. J. Amer. Chem. Soc. **77**, 4799 (1955).

135. WOODS, G. F. and L. H. SCHWARTZMAN: 1,3,5,7-Octatetraene. J. Amer. Chem. Soc. **71**, 1396 (1949).

136. WOODWARD, R. B.: The Structure and Biogenesis of the Macrolides, a New Class of Natural Products. In: Festschrift Arthur Stoll, p. 524. Basel: Birkhäuser. 1957.

137. WOOLDRIDGE, W. E.: Fungicidal and Antibiotic Substances from *Streptomyces aminophilus*. German Patent 1000966 (1957) [Chem. Abstr. **54**, 23177 (1960)].

138. — Antifungal Antibiotic. British Patent 828792 (1960).

139. YAJIMA, T.: On the Classification of Antifungal Antibiotics. J. Antibiotics (Tokyo) **8 A**, 189 (1955).

140. ZECHMEISTER, L.: *Cis-trans* Isomeric Carotenoids, Vitamins A, and Arylpolyenes, pp. 30—44. Vienna: Springer-Verlag, and New York: Academic Press. 1962.

(Received, March 18, 1963.)

Die Chemie der Tetracycline.

Von **H. Muxfeldt** und **R. Bangert**, Madison, Wisconsin.

Inhaltsübersicht.

                                                     Seite

I. Einleitung 80

II. Konstitutionsaufklärung 82

    1. Terramycin 82

        Alkalischer Abbau 83

        Saurer Abbau 87

        Reduktiver Abbau 90

    2. Aureomycin 91

    3. 6-Desmethyl-tetracycline 96

    4. 5a,11a-Dehydro-7-chlor-tetracyclin 96

    5. 2-Acetyl-2-descarboxamido-tetracycline 97

III. Weitere chemische Eigenschaften 98

    1. Reaktionen am $C_{(2)}$ 99

    2. Reaktionen am $C_{(4)}$ 100

    3. Reaktionen am $C_{(6)}$ 102

    4. Reaktionen am $C_{(11a)}$ und $C_{(12a)}$ 108

IV. Biogenese der Tetracycline 113

V. Versuche zur Synthese von Tetracyclinen 115

Literaturverzeichnis 116

I. Einleitung.

Tetracycline sind blaßgelb gefärbte, kristallisierte Antibiotika, die von verschiedenen *Streptomyces*-Stämmen produziert werden und sich nicht nur durch besonders vorteilhafte chemotherapeutische Eigenschaften auszeichnen (*14, 48, 35, 73*), sondern zugleich eine neue Klasse strukturell interessanter Naturstoffe darstellen, deren chemisches und physikalisches Verhalten Gegenstand eingehender Untersuchungen verschiedener Laboratorien während des letzten Jahrzehnts gewesen ist.

Seit **Duggar** (*15*) im Jahre 1948 das Aureomycin aus Kulturlösungen von *Streptomyces aureofaciens* isolierte, sind bislang neun weitere Vertreter dieser Stoffklasse aus natürlichen Streptomyceten oder Mutanten tetracyclin-bildender Mikroorganismen gewonnen worden. Von diesen· war das 1950 in den Laboratorien der Chas. Pfizer and Co. isolierte

Terramycin (*18*) die erste Verbindung, deren Konstitution in den genannten Laboratorien gemeinsam mit Woodward vollständig aufgeklärt werden konnte (*27*). Wie kurz darauf von dem gleichen Arbeitskreis sowie auch in den Lederle-Laboratorien gezeigt wurde (*69, 74*), sind Aureomycin und Terramycin sehr nahe verwandte Verbindungen, die beide als Substitutionsprodukte des ebenfalls natürlich vorkommenden Tetracyclins (I) anzusehen sind; Terramycin ist ein 5-Hydroxy-tetracyclin (II) und Aureomycin (III) das 7-Chlor-tetracyclin.

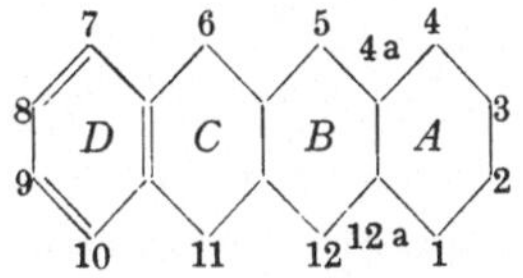

Bezifferung (*27*).

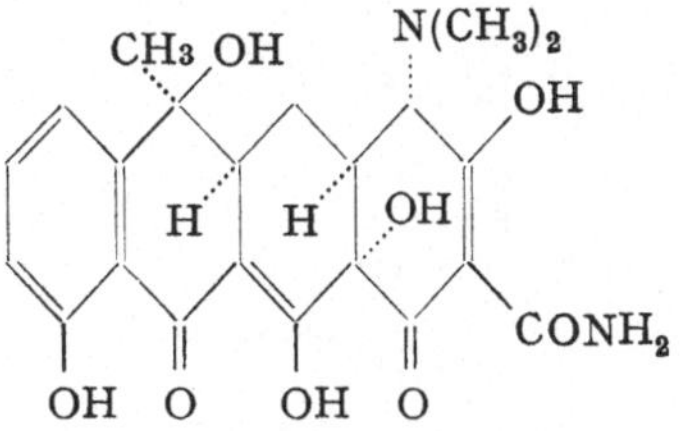

(I.) Tetracyclin.

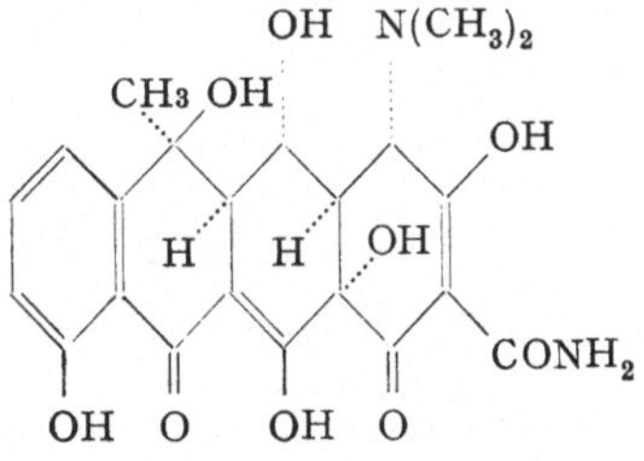

(II.) Terramycin.

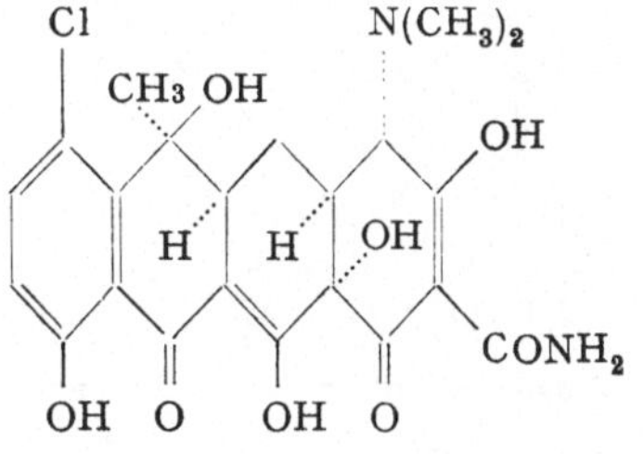

(III.) Aureomycin.

Bereits die während der Konstitutionsaufklärung des Terramycins (II) erarbeiteten experimentellen Befunde haben es seinerzeit erlaubt, Aussagen über die Konfiguration dieser sechs Asymmetriezentren enthaltenden Verbindung zu machen, welche zu dem durch Formel (IV) wiedergegebenen Vorschlag führten (*27*).

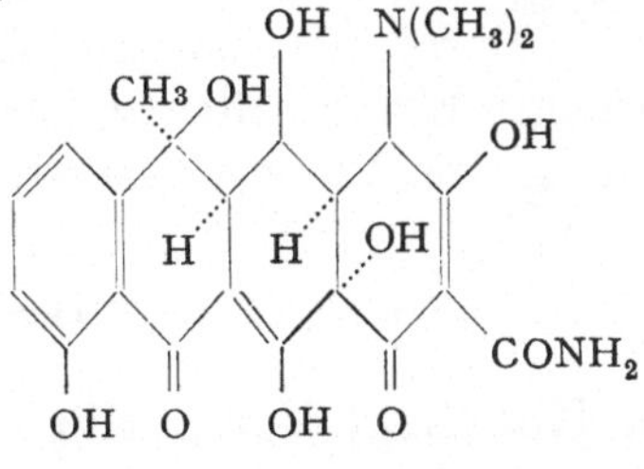

(IV.)

Eine kürzlich ausgeführte Röntgenstruktur-Analyse hat diese Aussagen zum größten Teil bestätigt und die durch Formel (II) dargestellte Konfiguration ergeben (*72*). Weiterhin konnte durch eine Röntgen-

struktur-Analyse des Aureomycins (III) gezeigt werden (*22*), daß Aureomycin (III) und Terramycin (II) nicht nur strukturell sehr ähnlich sind,
sondern darüber hinaus an allen gemeinsamen Asymmetriezentren die
gleiche Konfiguration besitzen, was auch für alle übrigen aus Streptomyceten isolierten Tetracycline (V—IX) als sehr wahrscheinlich angesehen
werden darf.

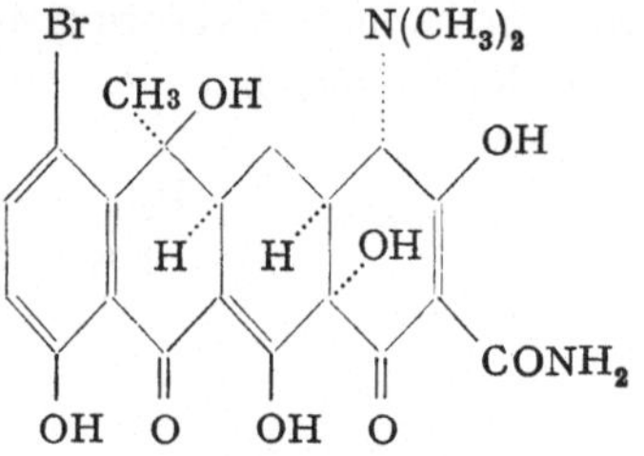

(V.) 7-Brom-tetracyclin (*13, 64*).

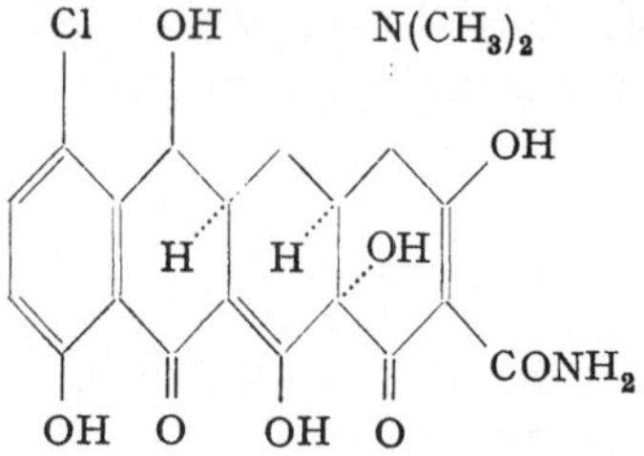

(VI.) 6-Desmethyl-tetracyclin (H statt Cl) (*44*).
(VII.) 6-Desmethyl-7-chlor-tetracyclin (*44*).

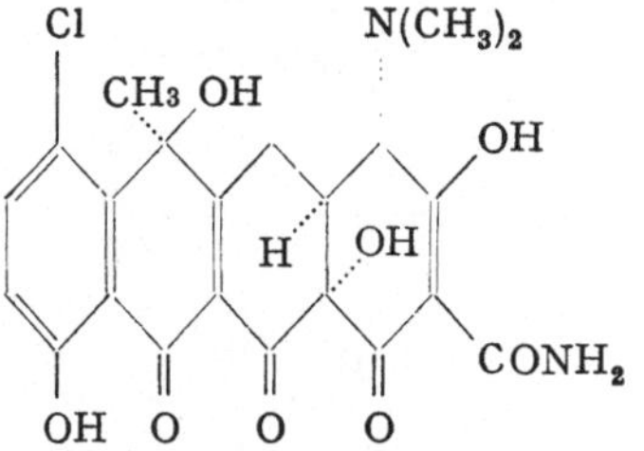

(VIII.) 5a,11a-Dehydro-7-chlor-tetracyclin (*41*).

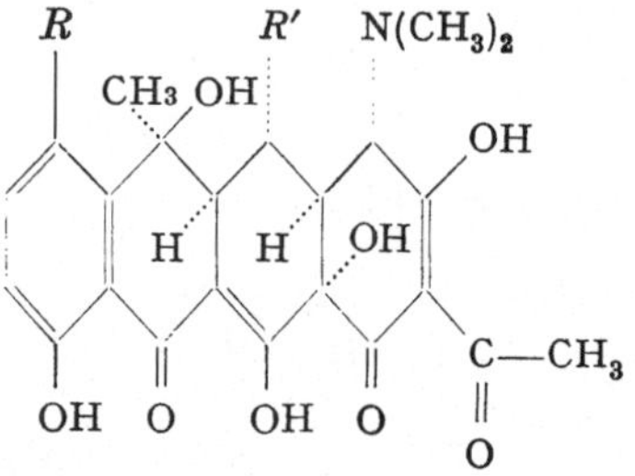

(IXa.) R = H; R' = OH,
5-Hydroxy-descarboxamido-2-acetyl-tetracyclin (*26*).
(IXb.) R = Cl; R' = H,
7-Chlor-descarboxamido-2-acetyl-tetracyclin (*46*).
(IXc.) R = R' = H,
Descarboxamido-2-acetyl-tetracyclin (*46*).

Die im letzten Jahrzehnt auf dem Gebiet der Chemie der Tetracycline
durchgeführten Untersuchungen lassen sich in drei verschiedene Arbeits-
richtungen einteilen: Isolierung und Konstitutionsaufklärung neuer
Tetracycline, Darstellung von Abbau- und Umwandlungsprodukten
natürlich vorkommender Tetracycline mit dem Ziel, Verbindungen mit
neuen chemotherapeutischen Eigenschaften zu gewinnen und schließlich
Versuche zur Totalsynthese von Tetracyclinen oder biologisch wirksamen
Abkömmlingen derselben.

Über Actinomycine s. Brockmann (*8a*), Van Tamelen (*72a*).

II. Konstitutionsaufklärung.

1. Terramycin (*27*).

Terramycin kann durch Extraktion von Kulturlösungen des Actinomyceten *Streptomyces rimosus* isoliert werden (*18, 60*). Es kristallisiert
normalerweise als Hydrat, jedoch kann es in kristallisierter, wasserfreier

Form gewonnen werden, wenn man das Hydrat in siedendem Toluol löst und das Kristallwasser durch azeotrope Destillation entfernt. Das wasserfreie Kristallisat besitzt die Summenformel $C_{22}H_{24}O_9N_2$. Von den zwei Stickstoff-Funktionen ist eine basisch, denn Terramycin bildet beispielsweise ein Hydrochlorid. Potentiometrische Titration des Hydrochlorids ergibt pKa-Werte von 3,27, 7,32 und 9,11 (70). Aus dem UV-Absorptionsspektrum ließen sich anfänglich keine Rückschlüsse auf die Konstitution der Verbindung ziehen, und aus dem IR-Spektrum war lediglich die Abwesenheit isolierter Carbonyl-gruppen erkennbar, da keine Absorption zwischen $5\,\mu$ und $6\,\mu$ vorhanden ist. Da Terramycin mit Diazomethan einen Dimethyl-äther und mit Essigsäureanhydrid ein Diacetat bildet, sind von den neun Sauerstoff-Funktionen zwei sauer und mindestens zwei alko-holisch. Von den zwei Stickstoff-Funktionen liegt die eine in Form einer Carboxamid-gruppe vor, was sich dadurch nach-

$$C_{18}H_9O_4 \quad \begin{cases} -CH_3 \\ -OH \\ -OH \end{cases} \text{sauer} \\ \begin{cases} -OH \\ -OH \end{cases} \text{alkoholisch} \\ -N(CH_3)_2 \\ -CONH_2$$

(X.) Terramycin (Teilformel).

weisen ließ, daß bei Behandlung mit Benzolsulfonsäurechlorid ein Benzolsulfonylnitril, $C_{21}H_{21}O_8N_2(SO_2C_6H_5)(CN)$, mit charakteristischer Ultrarot-Absorption bei $4,5\,\mu$ entsteht. Läßt man auf Terramycin Natronlauge einwirken, so entsteht 1 Mol Dimethylamin. Daher liegt die basische Stickstoff-Funktion als Dimethylaminogruppe vor. Schließlich zeigt die C-Methyl-Bestimmung nach KUHN und ROTH das Vorhandensein von mindestens einer C-Methylgruppe an. Die aufgeführten Fakten lassen sich durch die Teilformel (X) zusammenfassen.

Weiterer Einblick in die Konstitution und schließlich die gesamte Struktur wurden durch die Deutung der Ergebnisse von alkalischem, saurem und reduktivem Abbau erhalten.

Alkalischer Abbau.

Den Bedingungen einer Alkalischmelze ausgesetzt, lieferte Terramycin Salicylsäure (XI) und 3-Hydroxy-benzoesäure (XII).

(XI.) (XII.)

Milde Einwirkung von Natronlauge auf Terramycin ergibt eine Verbindung $C_{13}H_{12}O_6$, welche Terracinsäure genannt wurde und der die Konstitution (XIII) zugeordnet werden konnte (56). Läßt man dagegen

6*

Natronlauge in Gegenwart von Zink auf Terramycin einwirken, so entstehen zwei von (XIII) verschiedene Verbindungen. Das eine Abbauprodukt ist das Phthalid (XIV), und die zweite Verbindung wurde Terranaphthol genannt. Terranaphthol ist ein Derivat des 1,8-Dihydroxynaphthalins, denn es besitzt ein dieser Verbindung sehr ähnliches UV-Spektrum und erhöht wie 1,8-Dihydroxy-naphthalin die Acidität von Borsäure um 2,8 pH-Einheiten, wenn man 1 Mol dieser Verbindung in 12,5 Molen einer 0,36 M-wäßrig-alkoholischen Borsäurelösung löst. Terranaphthol wurde schließlich als das 1,8-Dihydroxy-3-hydroxymethyl-4-methyl-naphthalin (XV) erkannt, das unter den Bedingungen einer Alkalischmelze in Terranaphthoesäure (XVI) übergeht.

(XIII.) Terracinsäure.

(XIV.)

(XV.) Terranaphthol.

(XVI.) Terranaphthoesäure.

Von diesen Ergebnissen war die Isolierung der Terracinsäure (XIII) das überraschendste und zugleich aufschlußreichste. Es war sehr unwahrscheinlich, daß Terracinsäure eine Teilstruktur des Terramycins darstellt, da sich sein Indanon-Carbonyl im IR-Spektrum des Terramycins hätte zu erkennen geben müssen. Für die Deutung der Entstehung von Terracinsäure (XIII) aus Terramycin in Natronlauge und Abwesenheit eines reduzierenden Agens waren zwei weitere Fakten von wesentlicher Bedeutung. Einmal fällt Terracinsäure (XIII) als Racemat an, obgleich diese Verbindung neben einem äquilibrierbaren Asymmetriezentrum ein weiteres Asymmetriezentrum enthält, das unter der Einwirkung von Natronlauge unter den angewandten Reaktionsbedingungen nicht äquilibrierbar sein sollte. Auffallend war weiterhin, daß bei reduzierendem alkalischem Abbau das Phthalid (XIV) mit einer Sauerstoff-Funktion in Nachbarstellung zur Methylgruppe entsteht, während unter nicht reduzierenden Bedingungen Terracinsäure (XIII) entsteht, der diese

Sauerstoff-Funktion fehlt. Die Genese aller Alkali-Abbauprodukte wurde schließlich in eleganter Weise durch die Postulierung der Teilstruktur (XVII) für Terramycin gedeutet. So ist die Entstehung des Terranaphthols (XV) aus einer Verbindung der Teilformel (XVII) einleuchtend, wenn man annimmt, daß unter alkali-katalysierter Spaltung der Bindungen b und d ein Zwischenprodukt (XVIII) entsteht, das dann unter β-Eliminierung von Wasser und Reduktion in Terranaphthol (XV) übergeht.

(XVII.) Terramycin (Teilformel).

(XVIII.)

Nimmt man dagegen an, daß die Bindungen d und a zuerst gespalten werden, so wird die Entstehung des Phthalids (XIV) einleuchtend, da das Spaltprodukt (XIX) durch Retro-Aldolreaktion in das Hydroxyphthalid (XX) übergehen dürfte, welches dann durch Zink zum Phthalid (XIV) reduziert werden kann.

(XIX.)

(XX.)

Schließlich läßt sich auch die Entstehung der Terracinsäure logisch deuten, wenn man eine Spaltung der Bindungen d, c und a annimmt. Das dann entstehende Spaltstück (XXI) könnte durch β-Eliminierung von Wasser die Verbindung (XXII) ergeben, dessen Aldehydfunktion den Aromaten elektrophil substituiert, wobei (XXIII) resultieren sollte, welches schließlich unter prototroper Isomerisierung in Terracinsäure (XIII) übergehen dürfte. Eine derartige Entstehung der Terracinsäure erklärt sowohl die Tatsache, daß Terracinsäure als Racemat anfällt, als auch die Abwesenheit einer Sauerstoff-Funktion in Nachbarstellung zur Methylgruppe.

$$\text{(XXI.)} \qquad \text{(XXII.)}$$

$$\text{(XXIII.)}$$

Eine wesentliche Stütze für die Teilformel (XVII) fand sich, als man aus den Mutterlaugen des nicht-reduzierenden Alkaliabbaus Iso-des-carboxyterracinsäure (XXIV) isolierte. Diese Verbindung sollte man als Nebenprodukt erwarten, wenn der Aldehyd (XXII) intermediär auftritt und den Aromaten in ortho-Stellung zur Hydroxygruppe elektrophil substituiert, wobei das Trienon (XXV) entstehen sollte, welches dann durch Decarboxylierung und prototrope Isomerisierung in (XXIV) übergehen dürfte.

$$\text{(XXIV.) Iso-descarboxyterracinsäure.} \qquad \text{(XXV.)}$$

Schließlich konnte an dieser Stelle auch das UV-Spektrum des Terramycins zur Erhärtung der Teilformel (XVII) herangezogen werden, denn

$$\text{(XXVI.) 2-Acetyl-8-hydroxy-tetralon.} \qquad \text{(XXVII.) Terramycin (Teilformel).}$$

Literaturverzeichnis: SS. 116—120.

Terramycin besitzt einen Chromophor, der Ähnlichkeiten mit dem Chromophor des 2-Acetyl-8-hydroxy-tetralons (XXVI) aufweist.

Den Ergebnissen des Alkaliabbaus zufolge besitzt Terramycin demnach die Teilformel (XXVII).

Saurer Abbau.

Unter milder Einwirkung von Mineralsäuren in hydroxylfreiem Medium liefert Terramycin eine um 1 Mol Wasser ärmere, Anhydro-terramycin ($C_{22}H_{22}O_8N_2$) genannte Verbindung. Um diesen Befund zu deuten, lag es nahe anzunehmen, daß die tertiäre Hydroxygruppe in der Teilformel (XXVII) eliminiert wird. Diese Annahme ließ sich in überraschend einfacher Weise bestätigen, da Anhydro-terramycin ein Elektronenspektrum besitzt, das dem der Modellsubstanz (XXVIII) sehr ähnlich ist. Dem Anhydro-terramycin kommt demnach die Teilformel (XXIX) zu.

(XXVIII.)   (XXIX.) Anhydro-terramycin (Teilformel).

Läßt man auf Terramycin Mineralsäure in hydroxylhaltigem Medium einwirken, so entsteht kein Anhydro-terramycin (XXIX), sondern ein Gemisch zweier sehr ähnlicher, isomerer Verbindungen $C_{22}H_{22}O_8N_2$, welche α- und β-Apo-terramycin genannt wurden. Das gleiche Gemisch erhält man auch, wenn man auf Anhydro-terramycin (XXIX) Mineralsäuren in hydroxylhaltigem Medium oder Basen einwirken läßt. Der Übergang von Anhydro-terramycin (XXIX) in die Apo-terramycine ist von einer erheblichen Änderung des Elektronenspektrums begleitet, so daß man eine Änderung im Chromophor von (XXIX) postulieren mußte. Näheren Einblick in die Konstitution der Apo-terramycine ergab wiederum ein spektroskopischer Vergleich mit einer Modellsubstanz: Die Absorptionsspektren der Apo-terramycine sind dem des Naphthalin-Derivates (XXXII)

(XXX.) (XXXI.) α- und β-Apo-terramycine (Teilformel).   (XXXII.)

sehr ähnlich. Man konnte daher die Hypothese aufstellen, daß den α- und β-Apo-terramycinen die Teilformeln (XXX) bzw. (XXXI) zukommen, worauf auch das Vorhandensein einer Bande bei 5,75 μ im IR-Spektrum hinwies. Die beiden Verbindungen müssen sich demzufolge in dem noch unbekannten Teil der Molekel unterscheiden.

Diese Annahme konnte einerseits weiter erhärtet werden durch den Befund, daß die Apo-terramycine unter den Bedingungen einer Alkalischmelze Terranaphthoesäure (XVI, S. 84) liefern und zum anderen durch die Ergebnisse eines energischeren Säureabbaus von Terramycin, Anhydroterramycin (XXIX) und den Apo-terramycinen (XXX und XXXI). Dabei entstehen je nach den angewandten Reaktionsbedingungen zunächst Terrinolid ($C_{20}H_{15}O_8N$) und Dimethylamin oder Descarboxamido-terrinolid ($C_{19}H_{14}O_7$). Terrinolid und Descarboxamido-terrinolid liefern Pentamethyläther, Pentaacetate und Pentatosylate. Der Pentamethyläther des Descarboxamido-terrinolids ergibt bei Oxydation mit Salpetersäure die Verbindung (XXXIII). Näheren Einblick in die

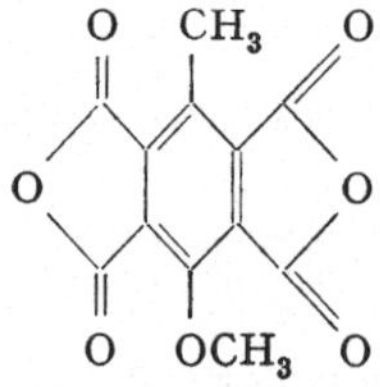

(XXXIII.)

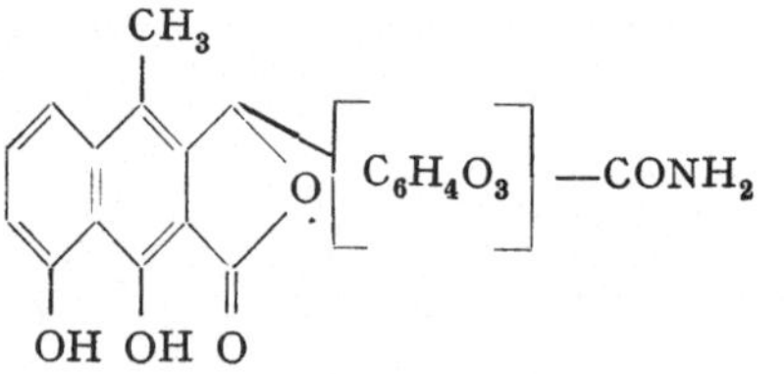

(XXXIV.) Terrinolid (Teilformel).

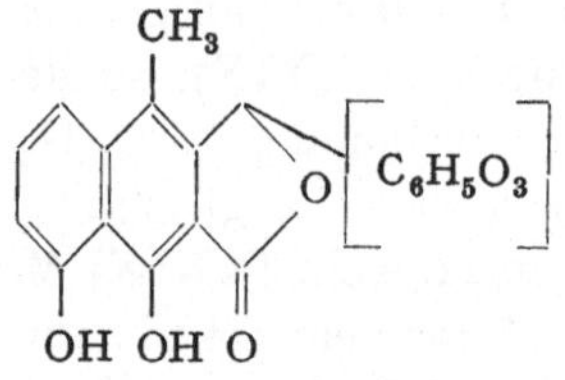

(XXXV.) Descarboxamido-terrinolid (Teilformel).

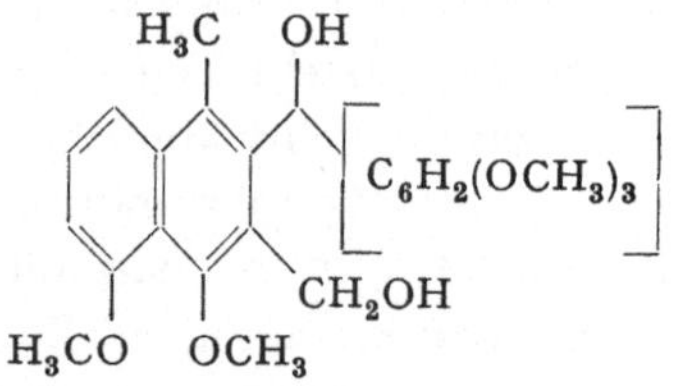

(XXXVI.)

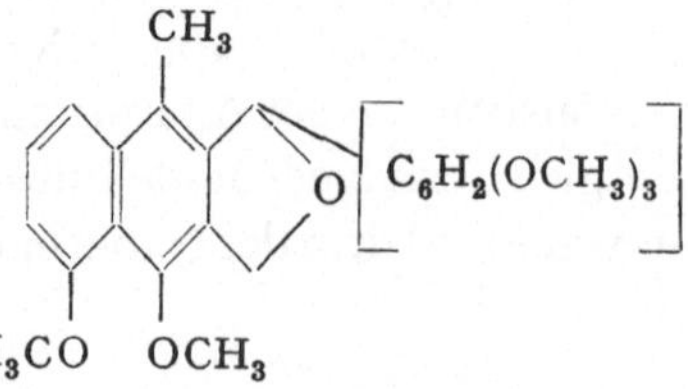

(XXXVII.)

Strukturen von Terrinolid und Descarboxamido-terrinolid, für die sich die Teilformeln (XXXIV) und (XXXV) ergeben, hat die Reduktion des Descarboxamido-terrinolid-pentamethyläthers geliefert. Dabei entsteht ein Diol (XXXVI), das mit Säure in den cyclischen Äther (XXXVII)

verwandelt werden kann. Da das Absorptionsspektrum des Diols (XXXVI) demjenigen eines Gemisches äquivalenter Mengen von Terranaphthol (XV, S. 84) und 1,2,4-Trihydroxy-benzol sehr ähnlich ist, kann man die Teilformel für Descarboxamido-terrinolid zu (XXXVIII) erweitern. Daß Terrinolid die Teilformel (XXXIX) besitzt, ergibt sich aus einem Vergleich der pK_a-Werte von Terrinolid und Descarboxamido-terrinolid.

(XXXVIII.) Descarboxamido-terrinolid (Teilformel).   (XXXIX.) Terrinolid (Teilformel).

Descarboxamido-terrinolid besitzt zusätzlich zu einem pK_a-Wert von 4,7, welcher dem Dihydroxy-naphthalin-System zuzuordnen ist, einen zweiten bei 10,2. Der zweite pK_a-Wert des Terrinolids liegt dagegen bei 7,5, woraus sich ergibt, daß die Carboxamidgruppe des Terrinolids das 1,2,4-Trihydroxy-benzol-System um 2,7 pK_a-Einheiten saurer macht. Dies ist nur möglich, wenn Terrinolid die Teilformel (XXXIX) besitzt, in der die Carboxamidgruppe zwischen zwei phenolischen Hydroxygruppen steht.

Da α- und β-Apo-terramycin (XXX und XXXI) bei der Alkalischmelze 2,5-Dihydroxy-benzochinon liefern und unter energischer Säureeinwirkung unter Eliminierung von Dimethylamin in Terrinolid übergehen, müssen sie zwangsläufig Derivate eines Dihydro-benzols sein. Da ihre Hydrochloride weiterhin drei titrierbare saure Gruppierungen enthalten (α-Apo-terramycin, pK_a 4,0, 5,1 und 8,4; β-Apo-terramycin, pK_a 3,6, 5,2 und 7,8), kann die dritte zusätzliche saure Funktion nur einem α- oder β-Diketon-System zukommen. Von diesen beiden grundsätzlichen Möglichkeiten kann ein α-Diketon von vornherein ausgeschlossen werden, da man die dritte Sauerstoff-Funktion, welche zwangsläufig als Hydroxygruppe vorliegen muß, in β-Stellung zu einer der Carbonylgruppen formulieren müßte, um den Teilformeln für Terrinolid und Descarboxamido-terrinolid (XXXVIII und XXXIX) Rechnung zu tragen.

(XL.) Apo-terramycine (Teilformel).   (XLI.) Anhydro-terramycin (Teilformel).
$X = OH$, $Y = N(CH_3)_2$ oder $X = N(CH_3)_2$, $Y = OH$.

Einer solchen Formulierung steht jedoch die relativ große Säurestabilität der Apo-terramycine entgegen. Die Apo-terramycine müssen daher β-Diketone mit einer Hydroxygruppe und einer Dimethylaminogruppe jeweils in α-Stellung zu einer der Carbonylgruppen sein, wodurch sich für sie die Teilformel (XL) ergibt. Da Anhydro-terramycin unter den Bedingungen einer Spaltung von β-Diketonen in die Apo-terramycine übergeht, ergibt sich für diese Verbindung die Teilformel (XLI).

Eine Entscheidung, wo die Hydroxygruppe und die Dimethylaminogruppe in (XLI) zu lokalisieren sind und wie das Ringgerüst des Terramycins aussieht, hat sich aus den Ergebnissen des reduktiven Abbaus ergeben.

Reduktiver Abbau.

Läßt man auf Terramycin Zink in Eisessig einwirken, so entsteht ein Desdimethylamino-desoxy-terramycin. Unter schonenderer Einwirkung von Zink in Eisessig wird dagegen lediglich die Dimethylaminogruppe reduktiv entfernt. Das so entstehende Desdimethylamino-terramycin hat ein UV-Spektrum, das mit dem des Terramycins nahezu identisch ist. Desdimethylamino-desoxy-terramycin dagegen besitzt ein Spektrum, das

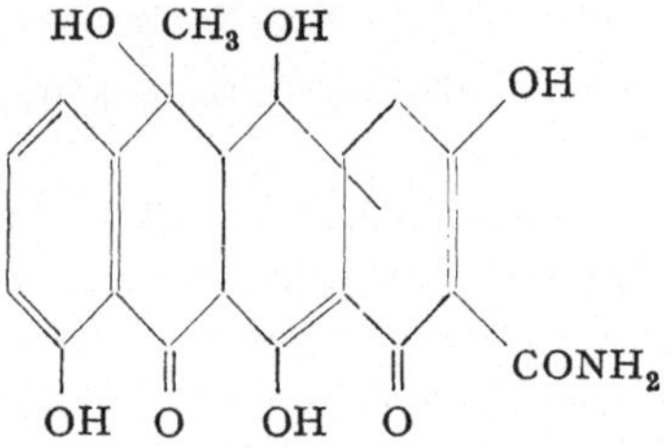

(XLII.) 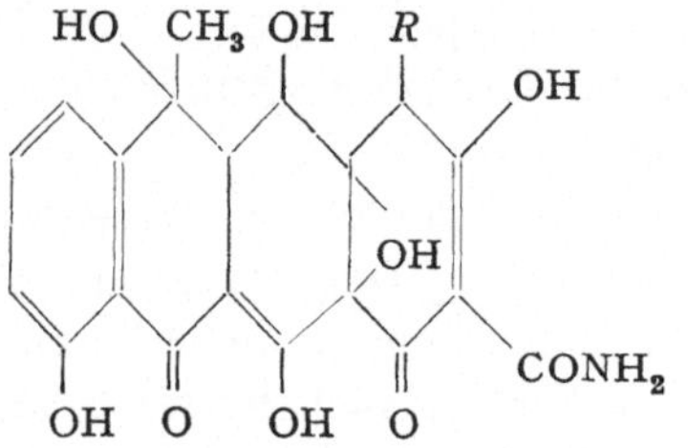   (XLIII.) $R = H.$
(XLIV.) $R = N(CH_3)_2.$

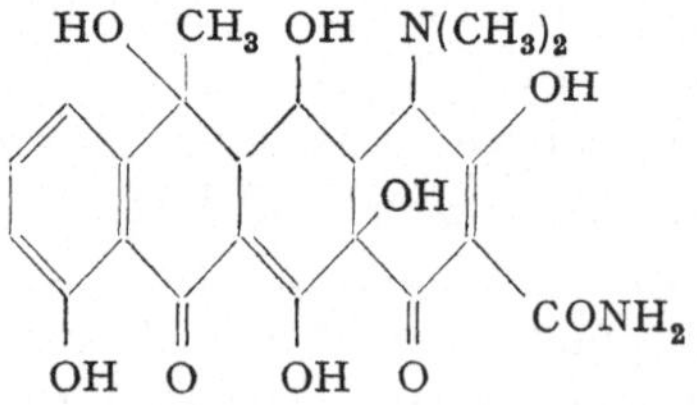

(XLV.) Terramycin.

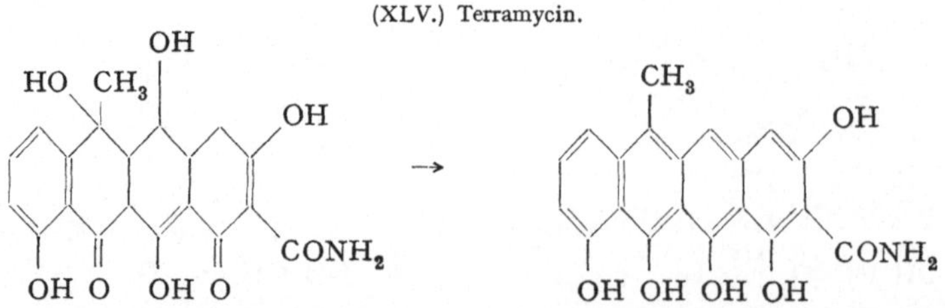

(XLVI.) Desdimethylamino-desoxyterramycin. (XLVII.) Desdimethylamino-terrarubein.

Literaturverzeichnis: SS. 116—120.

nicht einem über alle drei Ringe konjugierten Chromophor entspricht. Demzufolge muß es ein Tautomeres, wie durch Teilformel (XLII) dargestellt, sein. Daraus ergibt sich zwangsläufig für Desdimethylamino-terramycin die Teilformel (XLIII) und für Terramycin die Teilformel (XLIV). Daß dem Terramycin die Konstitution (XLV) zukommt, folgt schließlich aus folgenden Fakten: Desdimethylamino-desoxy-terramycin (XLVI) verliert bei Säurebehandlung 2 Mol Wasser und geht dabei in das rote, voll-aromatische Desdimethylamino-terrarubein (XLVII) über, welches bei der Zinkstaubdestillation Tetracen liefert.

2. Aureomycin.

Nachdem die Konstitution des Terramycins in den Laboratorien der Chas. Pfizer and Co. aufgeklärt worden war, war es relativ einfach, auch dem erstmals in den Lederle-Laboratorien isolierten (*15*) Aureomycin die Konstitution (XLVIII, S. 92) zuzuordnen (*69*).

Diese Struktur-Zuordnung gründet sich auf folgende in den Laboratorien der Chas. Pfizer and Co. erarbeiteten Befunde: Aureomycin hat die Summenformel $C_{22}H_{23}O_8N_2Cl$ und unterscheidet sich daher vom Terramycin ($C_{22}H_{24}O_9N_2$) durch das Fehlen einer Sauerstoff-Funktion und das Vorhandensein von einem Chloratom. Bei der Alkalischmelze von Aureomycin entsteht u. a. 5-Chlor-salicylsäure (*33*). Das UV-Spektrum des Aureomycins ist dem des Terramycins sehr ähnlich, und die UV-Spektren werden identisch, wenn man aus dem Aureomycin das Chlor durch katalytische Hydrierung entfernt, wobei Tetracyclin (IL) entsteht. Dadurch ist bewiesen, daß Aureomycin und Terramycin praktisch den gleichen Chromophor besitzen, und es blieb lediglich die Frage zu beantworten, welche der drei alkoholischen Hydroxygruppen des Terramycins dem Aureomycin fehlt. Da Aureomycin unter milder Einwirkung von Zink in Eisessig Desdimethylamino-aureomycin (L) liefert, das den gleichen Chromophor wie Aureomycin besitzt, ist die Stellung der Dimethylaminogruppe bewiesen. Das Vorhandensein einer 12a-Hydroxy-gruppe ließ sich ebenfalls nachweisen, denn Aureomycin liefert bei der Behandlung mit Zink in Eisessig unter energischeren Bedingungen Desdimethylamino-12a-desoxy-aureomycin (LI), und diese Reaktion ist von einer ähnlichen Änderung des Absorptionsspektrums begleitet wie beim Terramycin, so daß Desdimethylamino-12a-desoxy-aureomycin als das durch Formel (LI) dargestellte Tautomere identifiziert werden kann. Die Anwesenheit einer tertiären Hydroxygruppe am $C_{(6)}$ gibt sich durch mehrere Reaktionen zu erkennen. So geht Desdimethylamino-12a-desoxy-aureomycin in Desdimethylamino-12a-desoxy-isoaureomycin (LII) über (IR-Absorption des Phthalidcarbonyls bei 5,72 μ), das bei Pyrolyse 3-Methyl-4-chlor-7-hydroxy-phthalid (LIII) liefert.

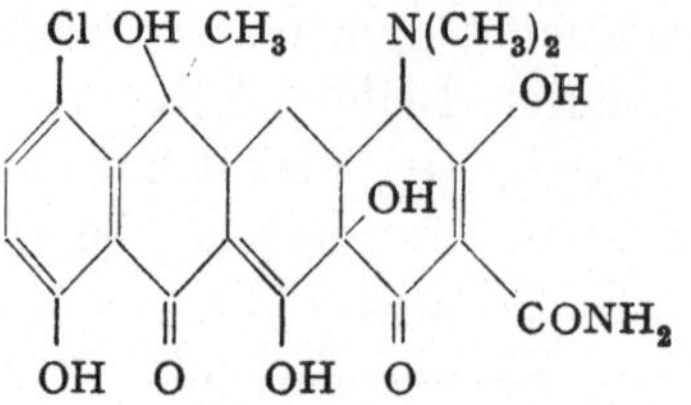

(XLVIII.) Aureomycin.

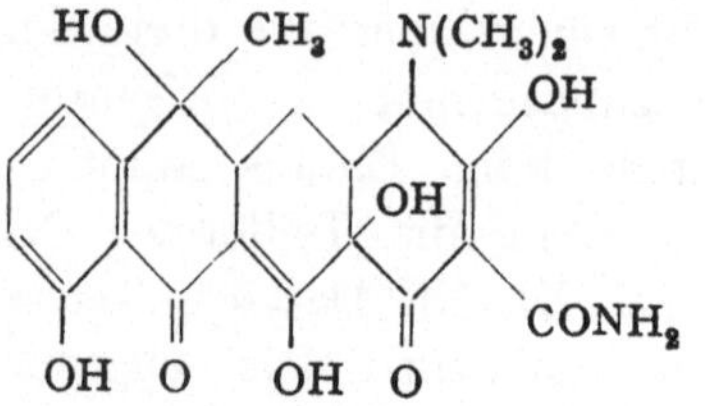

(IL.) Tetracyclin.

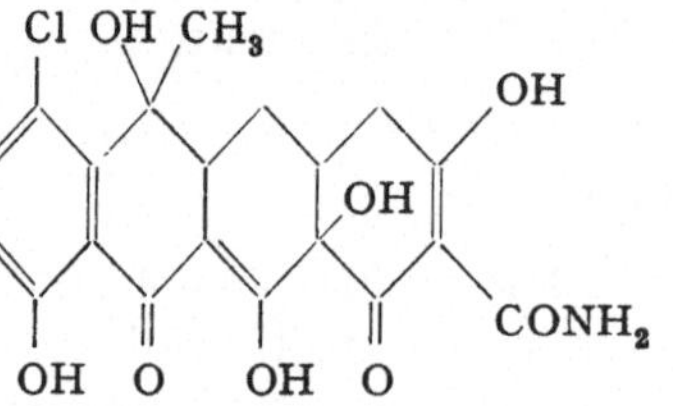

(L.) Desdimethylamino-aureomycin.

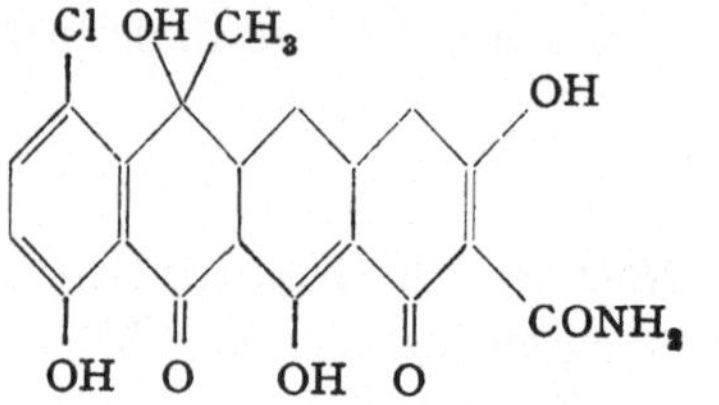

(LI.) Desdimethylamino-12a-desoxy-aureomycin.

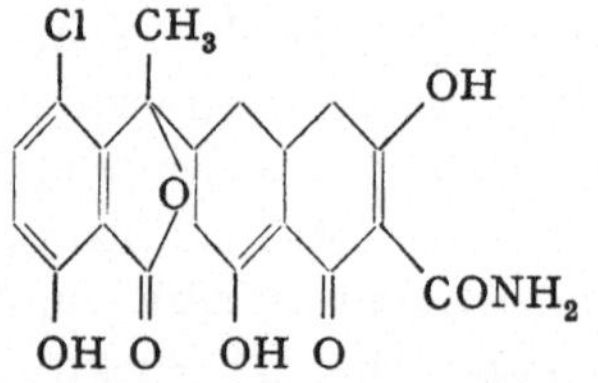

(LII.) Desdimethylamino-12a-desoxy-isoaureomycin.

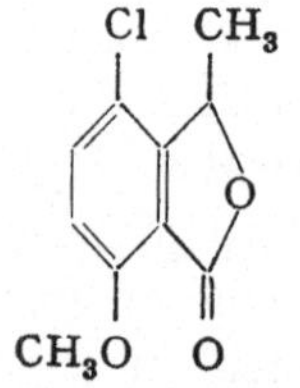

(LIII.) 3-Methyl-4-chlor-7-hydroxy-phthalid.

Wenn man von der Absorptionskurve von (LII) diejenige von (LIII) subtrahiert, so resultiert eine Kurve, die fast identisch ist mit einer Kurve, welche man erhält, wenn man die Absorptionskurve des 3-Methyl-7-hydroxy-phthalids (XIV, S. 84) von derjenigen des Desdimethylamino-12a-desoxy-isoterramycins (LVII) subtrahiert. Diese Verbindung entsteht in analoger Weise wie (LII), wenn man auf Desdimethylamino-12a-desoxy-terramycin (XLVI) unter schonenden Bedingungen Alkali einwirken läßt (27).

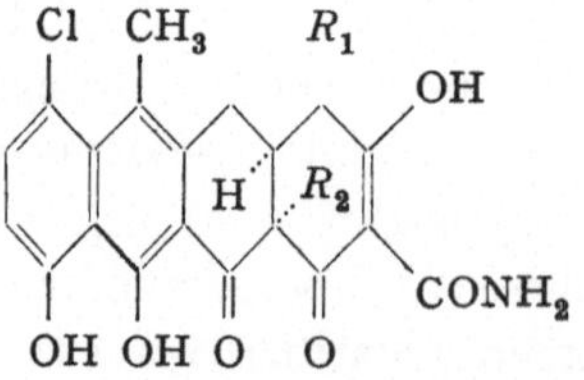

(LIV.) $R_1 = N(CH_3)_2$, $R_2 = OH$.
(LV.) $R_1 = H$, $R_2 = OH$.
(LVI.) $R_1 = R_2 = H$.

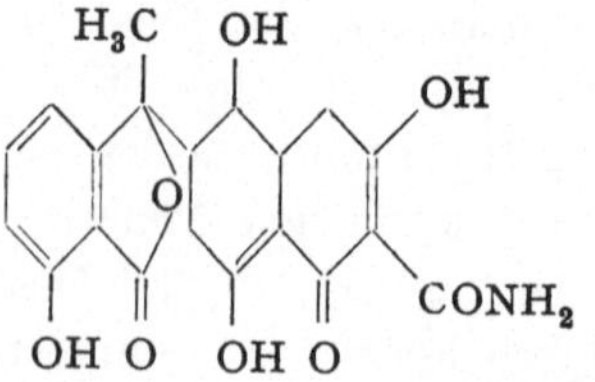

(LVII.) Desdimethylamino-12a-desoxy-isoterramycin.

Da sich weiterhin Aureomycin (XLVIII), Desdimethylamino-aureomycin (L) und Desdimethylamino-12a-desoxy-aureomycin (LI) in die

Literaturverzeichnis: SS. 116—120.

entsprechenden Anhydro-Verbindungen (LIV), (LV) und (LVI) durch Säurebehandlung überführen lassen, ist das Vorhandensein der 6-Hydroxygruppe im Aureomycin eindeutig bewiesen. Schließlich läßt sich auch noch die Carboxamidgruppe dadurch nachweisen, daß Aureomycin bei Behandlung mit p-Tosylchlorid in Pyridin in Aureomycin-nitril mit charakteristischer IR-Absorption bei 4,6 μ übergeht. Da außerdem Desdimethylamino-12a-desoxy-anhydro-aureomycin (LVI) bei der Zinkstaubdestillation Tetracen liefert, ist die Konstitution (XLVIII) für das Aureomycin bewiesen.

Unabhängig von den hier aufgeführten Ergebnissen wurden in den Lederle-Laboratorien Resultate erzielt, die es ermöglichten, Aureomycin die Konstitution (XLVIII) oder (LVIII) zuzuordnen (77).

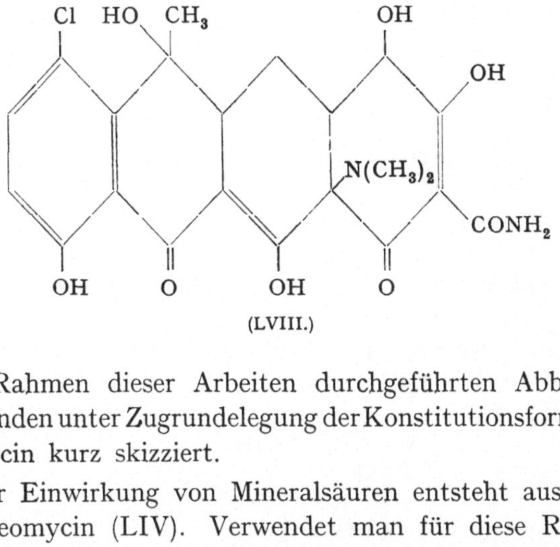

(LVIII.)

Die im Rahmen dieser Arbeiten durchgeführten Abbaureaktionen seien im folgenden unter Zugrundelegung der Konstitutionsformel (XLVIII) für Aureomycin kurz skizziert.

Unter der Einwirkung von Mineralsäuren entsteht aus Aureomycin Anhydro-aureomycin (LIV). Verwendet man für diese Reaktion Jodwasserstoffsäure, so entsteht Anhydro-tetracyclin (LIX) (74). Unter milder Einwirkung von Alkali auf Aureomycin (XLVIII) wird dieses in Iso-aureomycin (LX) überführt, während unter energischeren Bedingungen Desdimethylamino-aureomycinsäure (LXI) entsteht. Wird Desdimethylamino-aureomycinsäure weiter mit Natronlauge und Sauerstoff behandelt, so wird es zur Dicarbonsäure (LXVIIIa, S. 95) und zu dem Derivat (LXII) des Cyclopentan-1,3-dions weiter abgebaut. Aus (LXII) konnte mit Bromwasserstoffsäure das Cyclopentan-1,2,4-trion (LXIII) gewonnen werden, das bei Behandlung mit Jodwasserstoffsäure das bis dahin unbekannte Cyclopentan-1,3-dion (LXIV) lieferte (75, 76).

Mit konz. Schwefelsäure kann man Desdimethylamino-aureomycinsäure (LXI) zu dem Tetralon-Derivat (LXV) cyclisieren, welches Aureonamid genannt wurde. Unter Verseifungsbedingungen verliert Aureonamid seine Carboxamidgruppe und geht in Aureon (LXVI) über. Wird

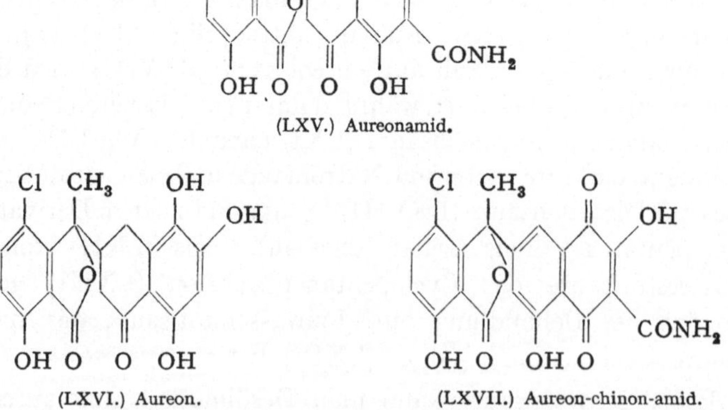

(LIX.) Anhydro-tetracyclin.

(LX.) Iso-aureomycin.

(LXI.) Desdimethylamino-aureomycinsäure.

(LXII.)

(LXIII.) Cyclopentan-1,2,4-trion.

(LXIV.) Cyclopentan-1,3-dion.

es dagegen mit 5 n-Natronlauge und Sauerstoff behandelt, so liefert es Aureon-chinon-amid (LXVII) (76).

(LXV.) Aureonamid.

(LXVI.) Aureon.

(LXVII.) Aureon-chinon-amid.

Näheren Einblick in die Entstehung der Desdimethylamino-aureomycinsäure (LXI) lieferte die Behandlung von Aureomycin mit 5 n-

Natronlauge in Gegenwart von Reduktionsmitteln. Dabei entstehen zwei isomere Verbindungen (LXIX), die α- und β-Aureomycinsäure genannt wurden.

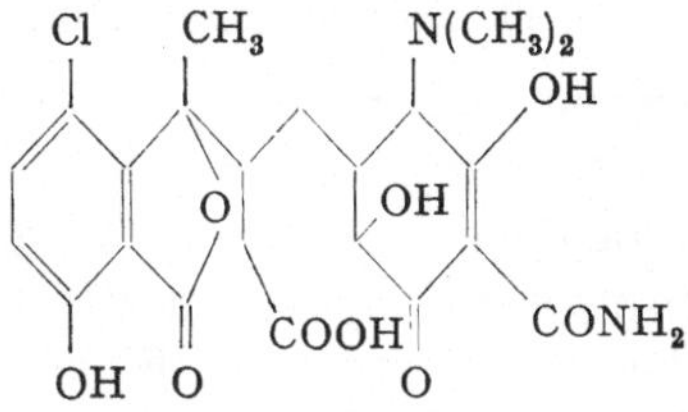

(LXIX.) α- und β-Aureomycinsäure.

(LXIX) geht mit 5 n-Natronlauge in Abwesenheit eines Reduktionsmittels und Gegenwart katalytischer Mengen Luft in Desdimethylamino-aureomycinsäure (LXI) über — eine Reaktion, die so gedeutet werden muß, daß das α-Ketol-System in (LXIX) durch katalytische Mengen Sauerstoff ins Triketon (LXX) übergeht, welches dann leicht Dimethylamin eliminieren kann, wobei ein Chinon des Typs (LXXI) gebildet wird, das sodann wieder das α-Ketol (LXIX) dehydrieren kann und dabei zu Desdimethylamino-aureomycinsäure reduziert wird (30).

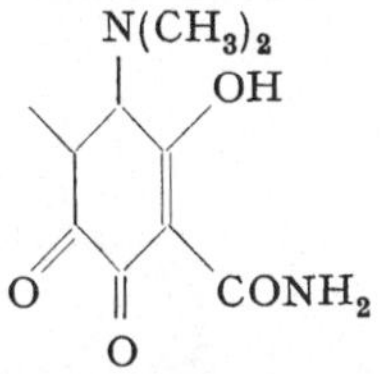

(LXX.)

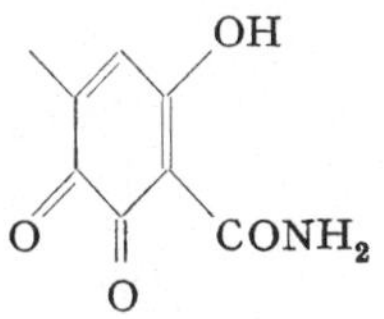

(LXXI.)

Bei der Oxydation eines Methyläthers des Aureomycins mit Kaliumpermanganat entstehen 3-Methoxy-6-chlor-phthalsäure und die Phthalid-Derivate (LXXII), (LXXIII) und (LXVIIIb) (31).

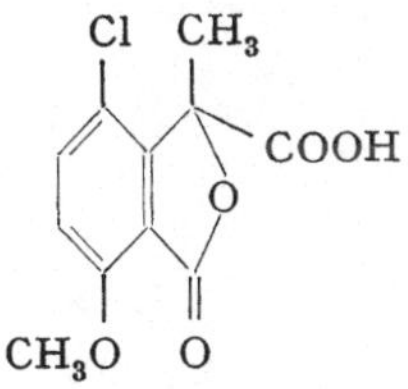

(LXXII.)

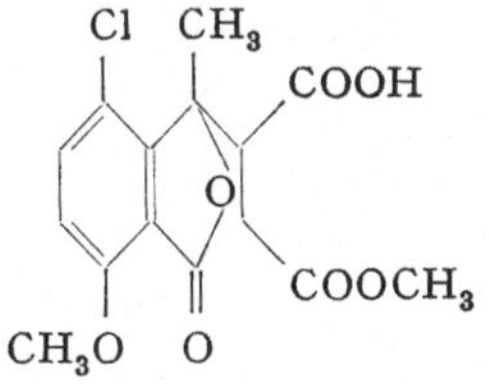

(LXXIII.)

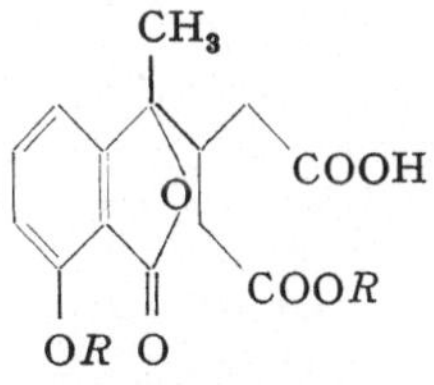

(LXVIIIa.) $R = H.$
(LXVIIIb.) $R = CH_3.$

Die Verbindungen (LXXII) und (LXXIII) wurden auch durch Synthese gewonnen (34).

7-Brom-tetracyclin (V, S. 82) entsteht, wenn man *Streptomyces aureofaciens* in chloridionen-freiem und bromidionen-haltigem Medium wachsen läßt (*13*). Auf seine Konstitution wurde in Analogie zur Konstitution des Aureomycins geschlossen.

3. 6-Desmethyl-tetracycline.

6-Desmethyl-tetracyclin (VI, S. 82) und 6-Desmethyl-7-chlor-tetracyclin (VII) wurden aus einer Mutante von *Streptomyces aureofaciens* isoliert (*44*). Die Verbindungen liefern unter den Bedingungen einer Kuhn-Roth-Oxydation keine Essigsäure, und (VII) kann durch katalytische Hydrierung in (VI) überführt werden. Bei Einwirkung von Mineralsäure entstehen die entsprechenden Anhydro-Verbindungen (LXXIVa) und (LXXIVb), deren Absorptionsspektren denen von Anhydro-aureomycin (LIV) bzw. Anhydrotetracyclin (LIX) sehr ähnlich sind (*78*). Bei der Oxydation von (VII) mit Natronlauge und Luft wird die β-(4-Chlor-7-hydroxy-phthalid-3)-glutarsäure (LXXV) gebildet und daneben entsteht das Cyclopentan-dion-Derivat (LXII) in völliger Analogie zu entsprechenden Reaktionen des Aureomycins. Zink und Eisessig reduzieren (VII) zum Desdimethylamino-6-desmethyl-12a-desoxy-7-chlor-tetracyclin (LXXVI), welches unter der Einwirkung von Alkali zu (LXXVII) isomerisiert wird. Diese Verbindung liefert unter Pyrolysebedingungen das Phthalid (LXXVIII), das mit einem synthetisch gewonnenen Präparat identisch ist (*6*).

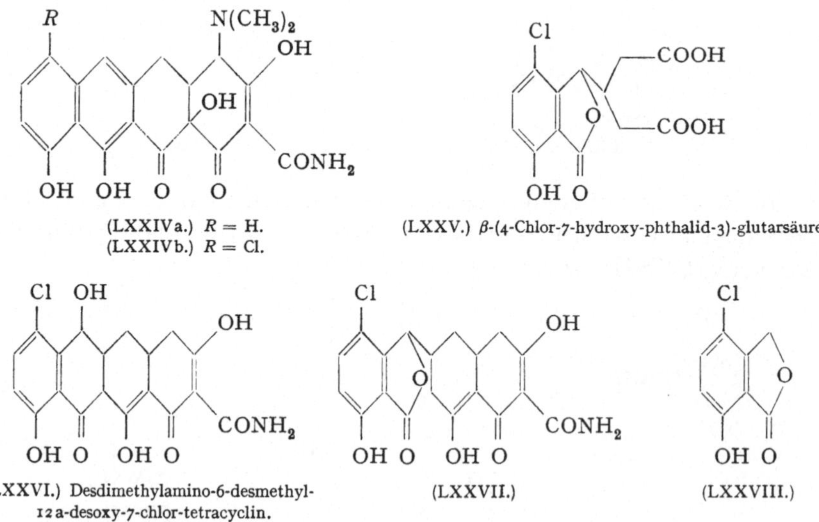

(LXXIVa.) R = H.
(LXXIVb.) R = Cl.

(LXXV.) β-(4-Chlor-7-hydroxy-phthalid-3)-glutarsäure.

(LXXVI.) Desdimethylamino-6-desmethyl-12a-desoxy-7-chlor-tetracyclin.

(LXXVII.)

(LXXVIII.)

4. 5a,11a-Dehydro-7-chlor-tetracyclin.

Die Verbindung (VIII, S. 82) wurde ebenfalls aus einer Mutanten von *Streptomyces aureofaciens* isoliert (*41*). Da es bei katalytischer Hydrierung in Tetracyclin (I, S. 81) und eine mit Tetracyclin isomere Verbindung (LXXIX) übergeht, ist die Identität des Ringsystems und der Stellung aller mit Tetracyclin gemeinsamen Substituenten bewiesen. Die Konstitution von (LXXIX) ergibt sich aus der Tatsache, daß (LXXIX) bei Säurebehandlung Anhydro-tetracyclin (LIX, S. 94) liefert, woraus sich zugleich die Konstitution (VIII, S. 82) für das Dehydro-7-chlor-tetracyclin ergibt.

(LXXIX.)

5. 2-Acetyl-2-descarboxamido-tetracycline.

Über die Isolierung des ersten Vertreters dieser Reihe, des 2-Acetyl-2-descarboxamido-5-hydroxy-tetracyclins (IXa, S. 82), wurde im Jahre 1960 berichtet (26). Diese Verbindung besitzt ein UV-Spektrum, das im langwelligen Bereich dem des Terramycins (II, S. 81) ähnlich ist, sich im kurzwelligen Bereich jedoch erheblich von dem Absorptions-spektrum des Terramycins unterscheidet. Da die langwellige Absorption ausschließlich der Absorption der Ringe B, C und D des Terramycins zukommt (27), müssen die Unterschiede zwischen diesen beiden Ver-bindungen im Ring A liegen. Den analytischen Daten zufolge unter-scheidet sich die neue Verbindung vom Terramycin durch das Fehlen einer Carboxamidgruppe und das Vorhandensein einer Acetylgruppe, welche sich auch dadurch zu erkennen gibt, daß bei der Hydrolyse Essig-säure gebildet wird und eine C-Methyl-Bestimmung zwei C-Methyl-gruppen anzeigt. (IXa) besitzt weiterhin sehr ähnliche pK_a-Werte wie Terramycin von pK_a 3,3, 7,1 und 9,2. Im IR-Spektrum ist im Gegensatz zum Terramycin unterhalb von 6 μ eine Absorptionsbande bei 5,92 μ erkennbar. Daß (IXa) Hydroxygruppen in 5- und 6-Stellung besitzt, ließ sich dadurch nachweisen, daß aus (IXa) unter der Einwirkung von Alkali Terracinsäure (XIII, S. 84) und unter energischer Säureeinwirkung Descarboxamido-terrinolid (LXXXI) gebildet werden. Die Stellung der zusätzlichen Acetylgruppe läßt sich schließlich mit Hilfe des UV-Spektrums von (IXa) festlegen: Subtrahiert man von den in saurem und alkalischem

(LXXX.) 2-Acetyl-dimedon.   (LXXXI.) Descarboxamido-terrinolid.

Medium aufgenommenen Absorptionskurven von (IXa) diejenigen des 2-Acetyl-8-hydroxy-tetralons, so resultieren Kurven, die den gleichen Verlauf wie die Kurven des 2-Acetyl-dimedons (LXXX) aufweisen.

Die Verbindung (IXa) wurde aus einer Mutante von *Streptomyces rimosus* isoliert, und in analoger Weise ließen sich aus einer Mutante von *St. aureofaciens* 2-Acetyl-2-descarboxamido-tetracyclin (IXc, S. 82) und 2-Acetyl-2-descarboxamido-7-chlor-tetracyclin (IXb) gewinnen (46). Die Verbindung (IXb) kann durch katalytische Hydrierung in (IXc) übergeführt werden, welches ein (IXa) ähnliches Absorptionsspektrum besitzt. Wie (IXa), zeigt auch (IXc) eine IR-Bande unterhalb $6\,\mu$ bei $5{,}95\,\mu$. Es besitzt zwei C-Methylgruppen und geht bei Säurebehandlung in die Anhydro-Verbindung (LXXXII) über, wodurch die Stellung der Hydroxygruppe in 6-Stellung nachgewiesen ist. Auf die Stellung des Chlors in (IXb) wurde in Analogie zur Stellung des Chlors im Aureomycin (III, S. 81) geschlossen.

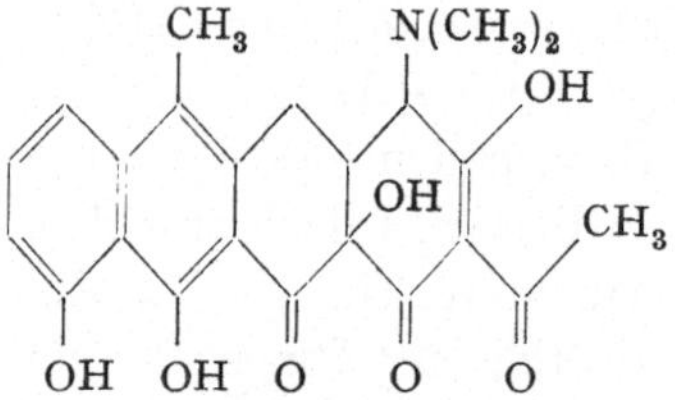

(LXXXII.)

III. Weitere chemische Eigenschaften.

Auch nach der Konstitutionsaufklärung der Tetracycline hat man sich eingehend mit dem Studium ihrer chemischen Eigenschaften be-

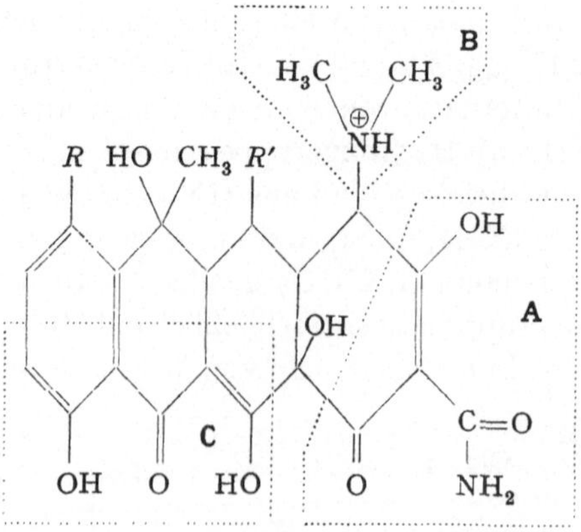

(LXXXIII.)

		A	B	C
(LXXXIIIa.) Tetracyclin-hydrochlorid. $R = R' = H$	pK:	3,30	7,68	9,69
(LXXXIIIb.) Aureomycin-hydrochlorid. $R = Cl,\ R' = H$	pK:	3,30	7,44	9,27
(LXXXIIIc.) Terramycin-hydrochlorid. $R = H,\ R' = OH$	pK:	3.27	7,32	9,11

Literaturverzeichnis: SS. 116—120.

schäftigt, wobei man vorwiegend das Ziel verfolgte, näheren Einblick in die Zusammenhänge zwischen Konstitution und biologischer Aktivität zu erhalten.

In einer ausführlichen Untersuchung konnte gezeigt werden, daß die drei pK_a-Werte von Tetracyclin-hydrochlorid (LXXXIIIa), Aureomycin-hydrochlorid (LXXXIIIb) und Terramycin-hydrochlorid (LXXXIIIc) den in Formel (LXXXIII) dargestellten Systemen A, B und C zuzuordnen sind (70).

1. Reaktionen am $C_{(2)}$.

Es konnte gezeigt werden, daß Anhydro-aureomycin-nitril, welches sich aus Aureomycin und Mesylchlorid in Pyridin sowie anschließende Behandlung mit Mineralsäuren gewinnen läßt, bei der Einwirkung von Isobutylen in konz. Schwefelsäure das Substitutionsprodukt (LXXXIV) liefert (67). Relativ große praktische Bedeutung haben Verbindungen erlangt, welche man erhält, wenn man sekundäre Amine und Formaldehyd auf Tetracycline einwirken läßt (65). So entsteht z. Beisp. aus Tetracyclin, Pyrrolidin und Formaldehyd das N-Pyrrolidinomethyl-tetracyclin (LXXXV) (65, 19).

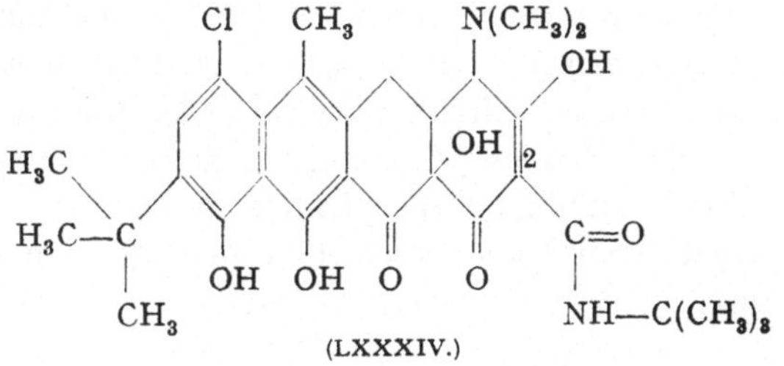

(LXXXIV.)

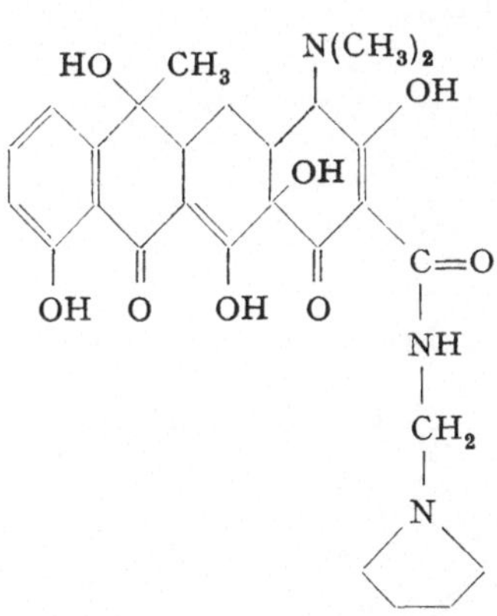

(LXXXV.) N-Pyrrolidinomethyl-tetracyclin.

Weiterhin läßt sich Tetracyclin leicht mit 9-Xanthenol in (LXXXVI) überführen (19).

7*

(LXXXVI.)

2. Reaktionen am $C_{(4)}$.

Bereits relativ früh erkannte man, daß Tetracycline unter dem Einfluß verschiedener Puffer zwischen pH 2 und pH 6 eine reversible Isomerisierung erleiden (*12*), und man nannte die Isomerisierungsprodukte anfänglich Quatrimycine (*12*). Es konnte jedoch bald gezeigt werden, daß es sich bei diesen Verbindungen um 4-epi-Tetracycline handelt (*68, 36, 37*). So konnte z. Beisp. 4-epi-Tetracyclin in Anhydro-4-epi-tetracyclin (LXXXIX) überführt werden, welches verschieden von Anhydro-tetracyclin (LIX, S. 94) und unter Äquilibrierungsbedingungen in dies umwandelbar ist. Damit ist bewiesen, daß die Isomerisierung im Ring *A* eintreten muß. Da unter den schonenden Reaktionsbedingungen eine Isomerisierung am $C_{(4a)}$ und $C_{(12a)}$ ausgeschlossen ist, bleibt nur das $C_{(4)}$ als Isomerisierungszentrum übrig. Das konnte weiterhin dadurch erhärtet werden, daß Aureomycin (III, S. 81) und 4-epi-Aureomycin (LXXXVII) bei Behandlung mit Mesylchlorid in Pyridin die Nitrile (XCa) bzw. (XCb) ergeben, welche verschieden und nicht mehr ineinander überführbar sind.

(LXXXVII.) 4-epi-Aureomycin.

(LXXXIX.) Anhydro-4-epi-tetracyclin.

(XCa.) $R = N(CH_3)_2$, $R' = H$.
(XCb.) $R = H$. $R' = N(CH_3)_2$.

Literaturverzeichnis: SS. 116—120.

Ein eindeutiger Beweis dafür, daß das $C_{(4)}$ das Isomerisierungszentrum ist, konnte über quartäre Derivate geführt werden. So liefern Aureomycin (III, S. 81), 4-epi-Aureomycin (LXXXVII), Tetracyclin (I, S. 81) und 4-epi-Tetracyclin quartäre Trimethylammoniumjodide (XCIa, XCIb, XCIc und XCId), welche bei 15-minütiger Einwirkung von Zink in 50%iger Essigsäure Desdimethylamino-aureomycin (XCIe) bzw. Desdimethylamino-tetracyclin (XCIf) liefern. Da gezeigt werden konnte, daß unter diesen Reaktionsbedingungen keine Epimerisierung eintritt, muß das $C_{(4)}$ das Zentrum der Isomerisierung sein (36, 37).

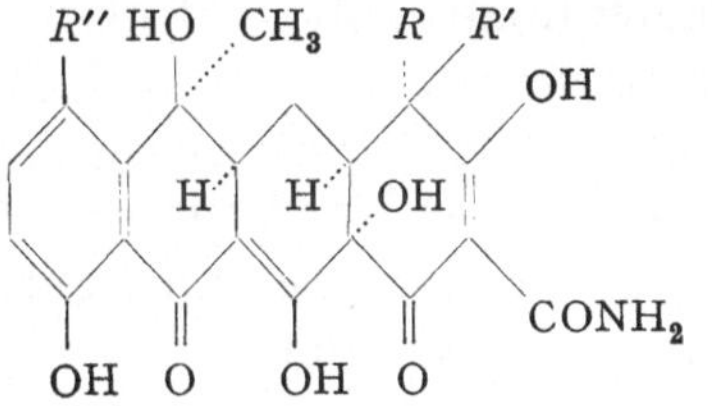

(XCIa.) $R = \overset{\oplus}{N}(CH_3)_3\overset{\ominus}{J}$; $R' = R'' = H$.

(XCIb.) $R = R'' = H$; $R' = \overset{\oplus}{N}(CH_3)_3\overset{\ominus}{J}$.

(XCIc.) $R = \overset{\oplus}{N}(CH_3)_3\overset{\ominus}{J}$; $R' = H$; $R'' = Cl$.

(XCId.) $R = H$; $R' = \overset{\oplus}{N}(CH_3)_3\overset{\ominus}{J}$; $R'' = Cl$.

(XCIe.) $R = R' = H$; $R'' = Cl$.

(XCIf.) $R = R' = R'' = H$.

Terramycin (II, S. 81) wird bei der Einwirkung von Methyljodid vollständig abgebaut (5). Als Reaktionsprodukte konnten Tetramethyl-

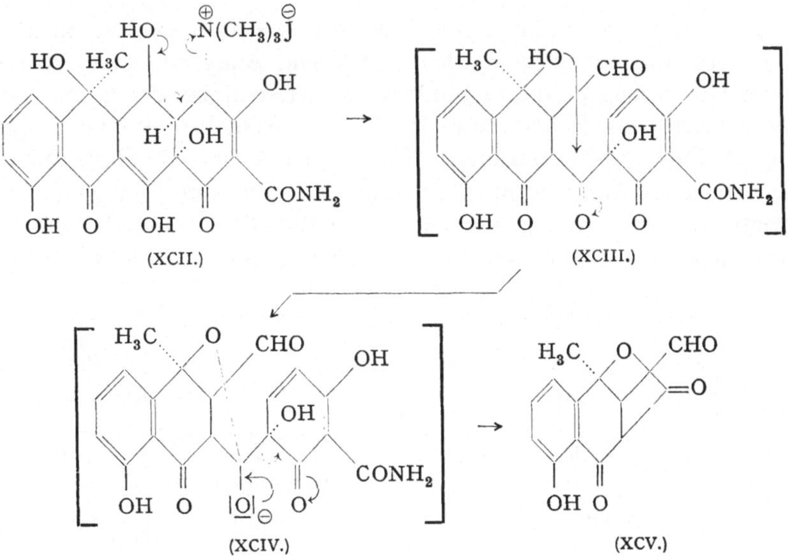

(XCII.)

(XCIII.)

(XCIV.)

(XCV.)

ammoniumjodid und der Aldehyd (XCV) isoliert werden. Wahrscheinlich entsteht diese Verbindung über Terramycin-jodmethylat (XCII) auf dem durch die Formeln (XCII—XCV) skizzierten Weg (*10*).

3. Reaktionen am $C_{(6)}$.

Die Hydroxygruppe am $C_{(6)}$ der Tetracycline läßt sich hydrogenolytisch entfernen (*71, 40*). Dabei entsteht aus Tetracyclin (I, S. 81) 6-Desoxy-6-epi-tetracyclin (XCVI a), während Terramycin (II) 6-Desoxy-6-epi-terramycin (XCVI b) ergibt (*49, 55, 62*). Die katalytische Hydrierung führt demnach nicht unter Konfigurationserhalt zu beispielsweise (XCVI c), sondern unter Konfigurationsumkehr zu (XCVI b). Schließlich konnte durch katalytische Hydrierung von 6-Desmethyl-tetracyclin (VI, S. 82) 6-Desmethyl-6-desoxy-tetracyclin (XCVI d) dargestellt werden. Diese Verbindung ist die einfachste bisher bekannte Substanz der Tetracyclin-Reihe mit voller biologischer Aktivität.

(XCVI a.) $R' = CH_3$; $R = R'' = H$.
(XCVI b.) $R = H$; $R' = CH_3$; $R'' = OH$.
(XCVI c.) $R = CH_3$; $R' = H$; $R'' = OH$.
(XCVI d.) $R = R' = R'' = H$.

Bei diesen Reaktionen sind die 6-Desoxy-tetracycline nicht die Hauptprodukte, denn es werden vorwiegend Anhydro-tetracycline und deren Hydrierungsprodukte gebildet. So entsteht bei der katalytischen Hydrierung von Tetracyclin (I, S. 81) Anhydro-tetracyclin (LIX, S. 94), 6-Desoxy-6-epi-tetracyclin (XCVI a) sowie die Verbindung (XCVII), die man auch erhält, wenn man Anhydro-tetracyclin hydriert (*71*). Es konnte weiterhin gezeigt werden (*40*), daß (XCVIII a) und (XCVIII b) sowie (IC a) und (IC b) Zwischenprodukte bei der Bildung von (XCVII) sind.

(XCVII.)

(XCVIII a.) $R = H$; $R' = OH$.
(XCVIII b.) $R = OH$; $R' = H$.

Literaturverzeichnis: SS. 116—120.

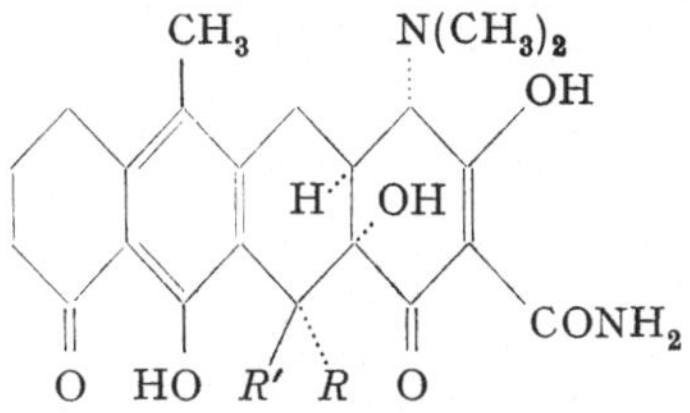

$$\text{(ICa.)} \quad R = H; \quad R' = OH.$$
$$\text{(ICb.)} \quad R = OH; \quad R' = H.$$

6-Desoxy-tetracycline sind erheblich stabiler gegen Säure und Alkali als die natürlich vorkommenden Tetracycline. Da sie, wie bereits erwähnt, volle biologische Aktivität besitzen, hat man versucht, eine große Zahl von Substitutions- und Umwandlungsprodukten dieser Verbindungen darzustellen, mit dem Ziel, Verbindungen mit höheren oder modifizierten biologischen Aktivitäten zu gewinnen.

So liefert die Nitrierung von 6-Desoxy-6-desmethyl-tetracyclin (XCVI d) mit Natriumnitrat in konz. Schwefelsäure (7, 58) oder 70%iger Salpetersäure in konz. Schwefelsäure (2) ein Gemisch von 7-Nitro-6-desoxy-6-desmethyl-tetracyclin (C) *(Tabelle 1)* und 9-Nitro-6-desoxy-6-desmethyl-tetracyclin (CXIV). Die Konstitution dieser Verbindungen folgte einerseits aus ihren Absorptionsspektren und konnte weiterhin wie folgt bewiesen werden: Bei der Behandlung von 6-Desmethyl-7-chlor-tetracyclin (VII, S. 82) mit Tritium und Palladium entsteht 6-Desmethyl-tetracyclin-7-^{3}H (CXXX), das bei Hydrierung mit Palladium 6-Desmethyl-6-desoxy-tetracyclin-7-^{3}H (CXXXI) gibt (38, 1). Wenn diese Verbindung nitriert und anschließend die zwei Nitrierungsprodukte auf ihren Tritiumgehalt untersucht werden, so zeigt die 7-Nitro-Verbindung (C) keinen Gehalt an Tritium mehr, während die 9-Nitro-Verbindung (CXIV) noch Tritium enthält.

Diese Technik der Konstitutionszuordnung wurde auch in allen anderen, im folgenden beschriebenen elektrophilen Substitutionsreaktionen angewandt. Die so gewonnenen Nitro-Verbindungen haben als Ausgangsmaterial für weitere Umwandlungen gedient. Ihre katalytische Hydrierung ergab die entsprechenden Amino-Verbindungen (CI) und (CXV) (7, 58), die in die entsprechenden N-Formyl-Verbindungen (CIV) und (CXVI) umgewandelt werden konnten. Weiterhin liefern die Amino-Verbindungen (CI) und (CXV) mit Butylnitrit die Diazoniumsalze (CII) und (CXVII), die als Ausgangsmaterial für verschiedene nucleophile Substitutionsreaktionen benutzt worden sind. So erhält man aus ihnen mit Natriumazid die Azide (CIII) und (CXVIII) sowie mit Natrium-äthyl-xanthogenat die Äthoxy-thiocarbonyl-thio-Verbindungen (CV) und (CXIX) (25). Photolyse des Sulfat-hydrochlorides von (CII) in Eisessig liefert drei Produkte: 7-Chlor-6-desmethyl-6-desoxy-tetra-

Tabelle 1. Partialsynthetische Derivate von 6-Desoxy-tetracyclinen.

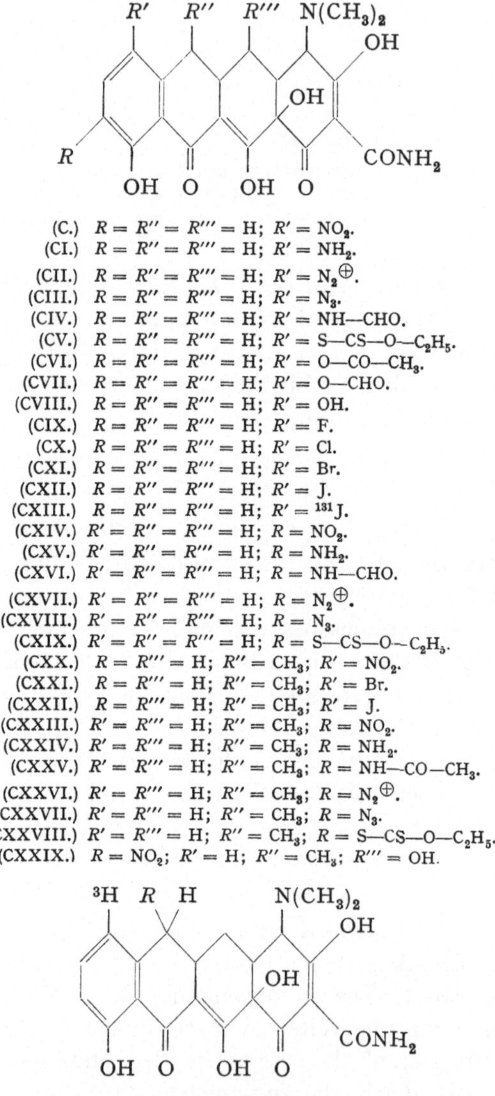

(C.) $R = R'' = R''' = H; R' = NO_2$.
(CI.) $R = R'' = R''' = H; R' = NH_2$.
(CII.) $R = R'' = R''' = H; R' = N_2{}^\oplus$.
(CIII.) $R = R'' = R''' = H; R' = N_3$.
(CIV.) $R = R'' = R''' = H; R' = NH—CHO$.
(CV.) $R = R'' = R''' = H; R' = S—CS—O—C_2H_5$.
(CVI.) $R = R'' = R''' = H; R' = O—CO—CH_3$.
(CVII.) $R = R'' = R''' = H; R' = O—CHO$.
(CVIII.) $R = R'' = R''' = H; R' = OH$.
(CIX.) $R = R'' = R''' = H; R' = F$.
(CX.) $R = R'' = R''' = H; R' = Cl$.
(CXI.) $R = R'' = R''' = H; R' = Br$.
(CXII.) $R = R'' = R''' = H; R' = J$.
(CXIII.) $R = R'' = R''' = H; R' = {}^{131}J$.
(CXIV.) $R' = R'' = R''' = H; R = NO_2$.
(CXV.) $R' = R'' = R''' = H; R = NH_2$.
(CXVI.) $R' = R'' = R''' = H; R = NH—CHO$.
(CXVII.) $R' = R'' = R''' = H; R = N_2{}^\oplus$.
(CXVIII.) $R' = R'' = R''' = H; R = N_3$.
(CXIX.) $R' = R'' = R''' = H; R = S—CS—O~C_2H_5$.
(CXX.) $R = R''' = H; R'' = CH_3; R' = NO_2$.
(CXXI.) $R = R''' = H; R'' = CH_3; R' = Br$.
(CXXII.) $R = R''' = H; R'' = CH_3; R' = J$.
(CXXIII.) $R' = R''' = H; R'' = CH_3; R = NO_2$.
(CXXIV.) $R' = R''' = H; R'' = CH_3; R = NH_2$.
(CXXV.) $R' = R''' = H; R'' = CH_3; R = NH—CO—CH_3$.
(CXXVI.) $R' = R''' = H; R'' = CH_3; R = N_2{}^\oplus$.
(CXXVII.) $R' = R''' = H; R'' = CH_3; R = N_3$.
(CXXVIII.) $R' = R''' = H; R'' = CH_3; R = S—CS—O—C_2H_5$.
(CXXIX.) $R = NO_2; R' = H; R'' = CH_3; R''' = OH$.

(CXXX.) $R = OH$.   6-Desmethyl-tetracyclin-^{3}H.
(CXXXI.) $R = H$.   6-Desmethyl-6-desoxy-tetracyclin-^{3}H.

cyclin (CX), 6-Desmethyl-6-desoxy-tetracyclin (XCVI d) und 7-Acetoxy-
6-desmethyl-6-desoxy-tetracyclin (CVI) (24). Führt man die Photolyse
in Ameisensäure aus, so entsteht als Hauptprodukt 7-Formyloxy-6-
desmethyl-6-desoxy-tetracyclin (CVII) (24). Sowohl (CVI) als (CVII)
lassen sich zu 7-Hydroxy-6-desmethyl-6-desoxy-tetracyclin (CVIII) ver-
seifen. Bei der Photolyse des Fluorborates von (CII) konnte 7-Fluor-

Literaturverzeichnis: SS. 116—120.

6-desmethyl-6-desoxy-tetracyclin (CIX) erstmals dargestellt werden. 7-Chlor-6-desmethyl-6-desoxy-tetracyclin (CX) ist durch katalytische Hydrierung von 7-Chlor-6-desmethyl-tetracyclin (VII, S. 82) mit einem Rhodiumkatalysator dargestellt worden (*39*), und die entsprechende 7-Brom-Verbindung (CXI) entsteht bei der Bromierung von 6-Desmethyl-6-desoxy-tetracyclin (XCVI d) mit N-Brom-succinimid in konz. Schwefelsäure. In gleicher Weise kann auch 7-Jod-6-desmethyl-6-desoxy-tetracyclin (CXII) und 7-^{131}Jod-6-desmethyl-6-desoxy-tetracyclin (CXIII) bei Verwendung der entsprechenden Jod-succinimide gewonnen werden.

Die letztgenannte Verbindung soll eine gewisse Bedeutung für die Diagnostik und Lokalisierung von Tumoren besitzen, da sie in Tumorgewebe angereichert wird (*23*).

Werden Bromierung oder Jodierung nicht in konz. Schwefelsäure durchgeführt, so entstehen keine 7-Brom-Derivate, sondern 11a-Brom-Verbindungen, welche sich in konz. Schwefelsäure in 7-Brom-Verbindungen umlagern (*25*). Es konnte gezeigt werden, daß dies eine intermolekulare Umlagerung ist, denn wenn dem Reaktionsmedium α-Naphthol zugesetzt wurde, erhielt man weniger als 10% an 7-Brom-tetracyclinen, da die Bromkationen, welche durch Protonisierung der 11a-Brom-Verbindungen freigesetzt wurden, von α-Naphthol abgefangen wurden.

Die substituierten 6-Desoxy-tetracycline (CXIV—CXXIX) wurden in gleicher Weise wie die Verbindungen (C—CXIII) dargestellt (*7*, *58*, *25*, *2*).

Interessante Ergebnisse haben Chlorierungsversuche in wäßrigem Medium ergeben. 6-Desoxy-6-desmethyl-tetracyclin (XCVI d, S. 102) liefert beim Umsatz mit N-Chlor-succinimid in Wasser 11a-Chlor-6-desoxy-6-desmethyl-tetracyclin (CXXXII). Tetracyclin (I, S. 81) und Terramycin (II) reagieren ähnlich, doch liegen die entsprechenden Chlorierungsprodukte (CXXXIII) und (CXXXIV) als Hemiketal vor (*3*). Alle 11a-Chlor-Verbindungen können durch Reduktion mit Dithionit oder durch katalytische Hydrierung wieder in die entsprechenden Tetracycline zurückverwandelt werden. Mit heißer methanolischer Salzsäure wird 11a-Chlor-tetracyclin-6,12-hemiketal (CXXXIII) nicht dehydratisiert, sondern zum Chlor-isotetracyclin (CXXXV) isomerisiert. Der Befund, daß (CXXXIII) und (CXXXIV) als Hemiketale vorliegen, ist ein eindeutiger Beweis für die Stereochemie am $C_{(5a)}$ und $C_{(6)}$ in Tetracyclinen. Weiterhin haben (CXXXIII) und (CXXXIV) als Ausgangsmaterial für die Gewinnung von 6-Methylen-tetracyclinen gedient, welche entstehen, wenn man (CXXXIII) und (CXXXIV) mit Fluorwasserstoffsäure zunächst in die 6-Methylen-11a-chlor-Verbindungen (CXXXVI) und (CXXXVII) überführt und diese mit Natriumdithionit reduziert. 5-Hydroxy-6-methylen-11a-chlor-tetracyclin (CXXXVII) läßt sich außerdem mit N-Chlor-succinimid in flüssiger Fluorwasserstoffsäure in 5-Hydroxy-6-methylen-7,11a-dichlor-tetracyclin (CXXXVIII) überführen, welches bei Reduktion mit Natriumdithionit 5-Hydroxy-6-methylen-7-chlor-tetracyclin (CXLI) ergibt.

(CXXXII.) 11a-Chlor-6-desoxy-6-desmethyl-tetracyclin.

(CXXXIII.) $R = $ H.
(CXXXIV.) $R = $ OH.

(CXXXV.) Chlor-isotetracyclin.

(CXXXVI.) $R = R' = $ H.
(CXXXVII.) $R = $ H; $R' = $ OH.
(CXXXVIII.) $R = $ Cl· $R' = $ OH.

(CXXXIX.) $R = R' = $ H.
(CXL.) $R = $ H;. $R' = $ OH.
(CXLI.) $R = $ Cl: $R' = $ OH.

(CXLII.) 11a-Fluor-tetracyclin-6,12-hemiketal.

11a-Fluor-tetracycline lassen sich durch Reaktion von Tetracyclinen mit Fluorperchlorat gewinnen (61). Sie liegen ebenfalls als Hemiketale vor, sofern sie in 6-Stellung eine Hydroxygruppe tragen. So erhält man z. Beisp. aus Tetracyclin (I, S. 81) 11a-Fluor-tetracyclin-6,12-hemi-ketal (CXLII), und dieses kann mit Fluorwasserstoffsäure in 6-Methylen-11a-fluor-tetracyclin (CXLIII) umgewandelt werden. Diese Verbindung ergibt bei katalytischer Hydrierung 6-Desoxy-tetracyclin (CXLIV), das verschieden von dem bei katalytischer Hydrierung von Tetracyclin entstehenden 6-Desoxy-6-epi-tetracyclin (XCVIa, S. 102) ist. Aus Terramycin (II) kann man so in analoger Weise 6-Desoxy-terramycin (CXLV) gewinnen.

Literaturverzeichnis: SS. 116—120.

(CXLIII.) 6-Methylen-11a-fluor-tetracylin.

(CXLIV.) $R =$ H. 6-Desoxy-tetracyclin.
(CXLV.) $R =$ OH. 6-Desoxy-terramycin.

Weitere 11a-Fluor-tetracyclin-6,12-hemiketale, welche mit Fluor-perchlorat gewonnen wurden, sind die Verbindungen (CXLVI—CIL). Alle diese Hemiketale unterscheiden sich von anderen Tetracyclinen charakteristisch durch ihre erheblich höhere Säurestabilität. Sie sind z. Beisp. in siedender methanolischer Salzsäure vollständig stabil. Weiter-hin läßt sich das Fluor nur unter relativ energischen Bedingungen durch katalytische Hydrierung entfernen, im Gegensatz zu den 11a-Fluor-tetracyclinen (CL—CLII), welche man aus 6-Desoxy-tetracyclinen mit Fluorperchlorat erhalten kann.

(CXLVI.) $R =$ CH$_3$; $R' =$ OH; $R'' =$ N(CH$_3$)$_2$.
(CXLVII.) $R =$ CH$_3$; $R' = R'' =$ H.
(CXLVIII.) $R =$ CH$_3$; $R' =$ OH; $R'' =$ H.
(CIL.) $R = R' =$ H; $R'' =$ N(CH$_3$)$_2$.

(CL.) $R = R' =$ H; $R'' =$ N(CH$_3$)$_2$.
(CLI.) $R =$ CH$_3$; $R' =$ OH; $R'' =$ N(CH$_3$)$_2$.
(CLII.) $R = R' = R'' =$ H.

6-Methylen-tetracycline sind für zahlreiche Substitutionen am $C_{(6)}$ benutzt worden (59). So entsteht bei radikalischer Addition von Benzyl-mercaptan an 6-Methylen-tetracyclin (CXXXIX) oder an 6-Methylen-terramycin (CXL) die Verbindung (CLIII) bzw. (CLIV). Die Konfigu-ration am $C_{(6)}$ der Additionsprodukte ergibt sich aus der Tatsache, daß bei Entschwefelung mit Raney-Nickel 6-Desoxy-tetracyclin (CXLIV) bzw.

6-Desoxy-terramycin (CXLV) entsteht (*62*). Die Gewinnung von 6-Desoxy-tetracyclin aus 11a-Fluor-6-methylen-tetracyclin (CXLIII) sowie aus der Verbindung (CLIII) ist als Beweis für die Konfiguration am $C_{(6)}$ angesehen worden (*62*).

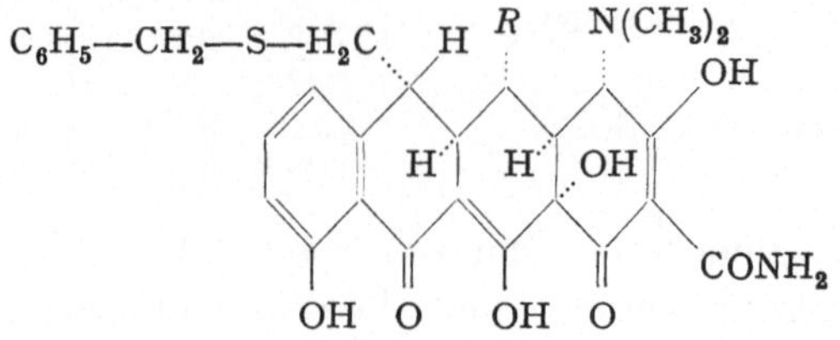

(CLIII.) $R =$ H.
(CLIV.) $R =$ OH.

Eine sehr interessante Reaktion am $C_{(6)}$ des Anhydro-aureomycins (LIV, S. 92) ist die licht-katalysierte Oxydation dieser Verbindung zu dem Hydroperoxyd (CLV), das sich durch katalytische Hydrierung mit Palladium in Alkohol in 5a,11a-Dehydro-7-chlor-tetracyclin (VIII, S. 82) überführen läßt (*63*). Da diese Verbindung bereits vorher in Tetracyclin (I, S. 81) umgewandelt worden ist (*41*), ist durch diese Reaktionsschritte die Möglichkeit gegeben, Anhydro-aureomycin in Tetracyclin zu überführen.

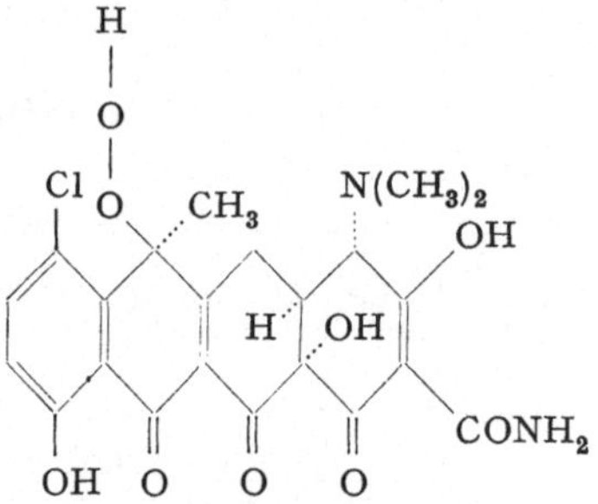

(CLV.)

4. Reaktionen am $C_{(11\,a)}$ und $C_{(12\,a)}$.

Das $C_{(11a)}$ von Tetracyclinen kann, wie bereits am Beispiel der 6-Desoxy-tetracycline gezeigt wurde, elektrophil substituiert werden. So erhält man beim Umsatz einer Suspension von Desdimethylamino-tetracyclin (XCIf, S. 101) in Chloroform mit N-Brom-succinimid bei Zimmertemperatur Desdimethylamino-11a-brom-tetracyclin (CLVI). Im Gegensatz dazu liefert die Bromierung der Verbindung (XCIf, S. 101) mit N-Brom-succinimid in Eisessig Desdimethylamino-9-brom-anhydro-tetracyclin (CLVII), das auch entsteht, wenn man (CLVI) mit Eisessig und Bromwasserstoffsäure behandelt. Obgleich die Konstitution dieser Verbindung nicht eindeutig bewiesen ist, kann man sie als sehr wahr-

Literaturverzeichnis: SS. 116—120.

scheinlich ansehen, da bei der Chlorierung von Desdimethylamino-anhydro-tetracyclin (CLX) ein im aromatischen Kern substituiertes Chlor-Derivat gebildet wird, welches nicht identisch mit Desdimethyl-

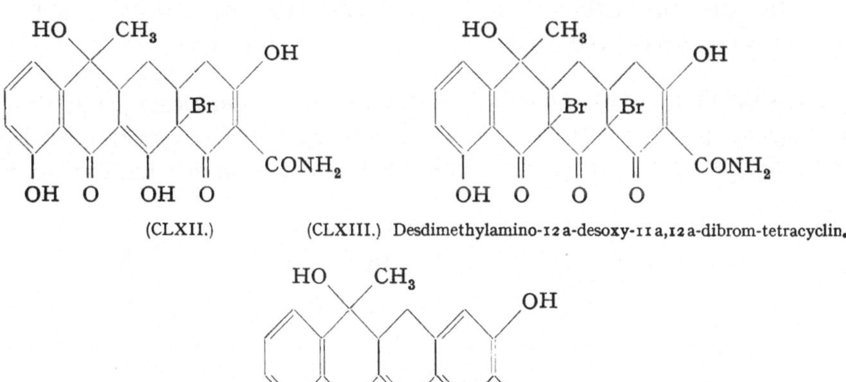

(CLVI.) Desdimethylamino-11a-brom-tetracylin.

amino-anhydro-aureomycin (LV, S. 92) und demzufolge wahrscheinlich 9-Chlor-desdimethylamino-anhydro-tetracyclin (CLVIII) ist (*21*).

(CLVII.) $R = Br$: $R' = OH$.
(CLVIII.) $R = Cl$; $R' = OH$.
(CLIX.) $R = R' = Br$.
(CLX.) $R = H$; $R' = OH$.

(CLXI.) Desdimethylamino-12a-desoxy-tetracyclin.

Die Bromierung von Desdimethylamino-aureomycin (L, S. 92) hat Produkte mit nicht vollständig geklärter Konstitution ergeben.

Umsatz von Desdimethylamino-12a-desoxy-tetracyclin (CLXI) mit zwei Äquivalenten N-Brom-succinimid liefert Desdimethylamino-12a-

(CLXII.)

(CLXIII.) Desdimethylamino-12a-desoxy-11a,12a-dibrom-tetracyclin.

(CLXIV.) 4a,12a-Dehydro-desdimethylamino-12a-desoxy-tetracyclin.

desoxy-11a,12a-dibrom-tetracyclin (CLXIII), während mit einem Mol N-Brom-succinimid die 12a-Stellung selektiv angegriffen wird, wobei die Verbindung (CLXII) resultiert. Beim Behandeln von (CLXIII) mit Bromwasserstoffsäure entsteht die Anhydro-Verbindung (CLIX). Aus (CLXII) kann man mit schwachen Basen sehr leicht Bromwasserstoffsäure eliminieren und so 4a,12a-Dehydro-desdimethylamino-12a-desoxy-tetracyclin (CLXIV) gewinnen (21).

Eine selektive reduktive Entfernung der Hydroxygruppe in 12a-Stellung läßt sich mit zwei Methoden erreichen. So erhält man aus Tetracyclin (I, S. 81) unter gleichzeitiger Epimerisierung am $C_{(4)}$ (52) 12a-Desoxy-4-epi-tetracyclin (CLXV), wenn man Tetracyclin mit Zink in Ammoniak behandelt (20). Das gleiche Reaktionsprodukt entsteht bei der katalytischen Hydrierung von 12a-O-Formyl-tetracyclin (CLXVIII), welches sich durch Einwirkung eines Gemisches von Ameisensäure und Essigsäureanhydrid auf eine Lösung von Tetracyclin in Pyridin gewinnen läßt (4).

Die Konstitution des 12a-Desoxy-4-epi-tetracyclins (CLXV) ergibt sich aus verschiedenen Fakten. In Natriumtetraboratlösung besitzt die Verbindung ein gegenüber Tetracyclin um mehr als 100 mμ bathochrom verschobenes Absorptionsspektrum, das der Absorption eines Borat-komplexes eines über alle vier Ringe durchkonjugierten Enolsystems entspricht, was das Fehlen der 12a-Hydroxygruppe anzeigt. Weiterhin

(CLXV.) 12a-Desoxy-4-epi-tetracyclin.

(CLXVI.)

geht (CLXV) in methanolischer Salzsäure in ein 12a-Desoxy-anhydro-tetracyclin (CLXVI) über, was zeigt, daß die 6-Hydroxygruppe nicht reduktiv entfernt wurde (20). Schließlich erhält man bei katalytischer

(CLXVII.)

(CLXVIII.) 12a-O-Formyl-tetracyclin.

Literaturverzeichnis: SS. 116—120.

Hydroxylierung des 12a-Desoxy-4-epi-tetracyclins mit Platin und Sauerstoff in Dimethylformamid kein Tetracyclin, sondern 4-epi-Tetracyclin (CLXIX) (52). Eine katalytische Hydroxylierung zu Tetracyclin gelingt mit Natriumnitrit und Luft (28). Weiterhin kann man diese Reaktion auch durch mikrobiologische Oxydation erzielen (29). Als Nebenprodukte der katalytischen Oxydationen entstehen auch in 11a-Stellung hydroxylierte Verbindungen (52, 28).

(CLXIX.) 4-epi-Tetracyclin.

Bei Versuchen, 12a-Desoxy-4-epi-tetracyclin (CLXV) mit Perbenzoesäure zu hydroxylieren, wurde lediglich Desdimethylamino-4a,12a-dehydro-12a-desoxy-tetracyclin (CLXIV) in 20%iger Ausbeute isoliert (20). Diese Verbindung soll durch Eliminierung von Hydroxylamin aus dem N-Oxyd (CLXVII) entstehen (20). Sie ist weiterhin durch Erhitzen von 12a-Desoxy-4-epi-tetracyclin (CLXV) mit Methyljodid, Propylenoxyd und Tetrahydrofuran zugänglich (20).

4a,12a-Dehydro-tetracyclin (CLXX) selbst entsteht dagegen bei Pyrolyse von 12a-O-Formyl-tetracyclin (CLXVIII) in siedendem Toluol (4). Diese Reaktion kann als Beweis für die *cis*-Verknüpfung der Ringe *A* und *B* angesehen werden. Einwirkung von Säure auf diese Verbindung

(CLXX.) 4a,12a-Dehydro-tetracyclin.

(CLXXI.) Terrarubein.

ergibt das bereits vorher aus Terramycin (II, S. 81) gewonnene Terrarubein (CLXXI) (27, 4).

In analoger Weise wie Tetracyclin (I) konnte 6-Desmethyl-6-desoxy-tetracyclin (XCVI d, S. 102) in die 12a-O-Formyl-Verbindung (CLXXII) überführt werden, welche bei katalytischer Hydrierung 6-Desmethyl-6,12a-didesoxy-tetracyclin (CLXXIV) ergibt. Diese Verbindung ist als Racemat auch totalsynthetisch aufgebaut und mit Cerchlorid und Sauerstoff in racemisches 6-Desoxy-6-desmethyl-tetracyclin verwandelt worden (*11*). Bei einer Pyrolyse geht (CLXXII) in 4a,12a-Dehydro-6-desmethyl-6-desoxy-tetracyclin (CLXXIII) (*4*) über.

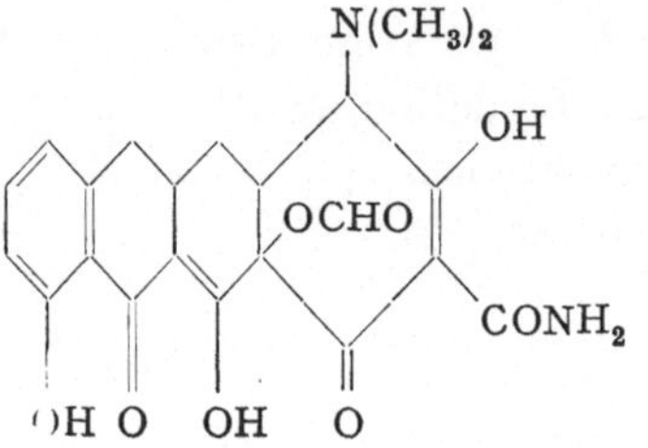

(CLXXII.)

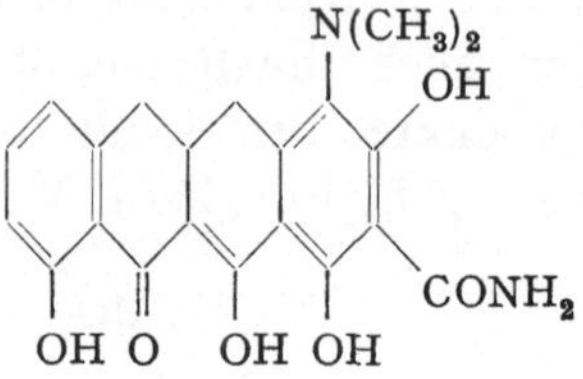

(CLXXIII.) 4a,12a-Dehydro-6-desmethyl-6-desoxy-tetracyclin.

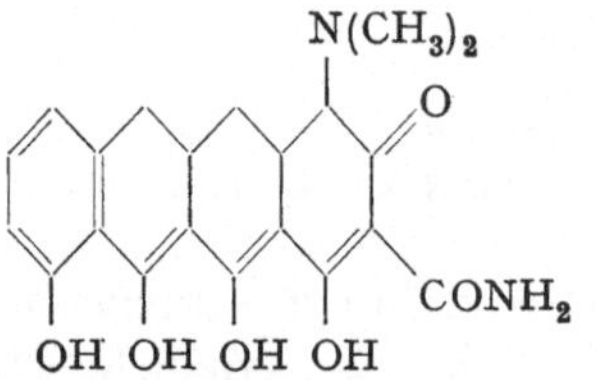

(CLXXIV.) 6-Desmethyl-6,12a-didesoxy-tetracyclin.

Im Rahmen von Versuchen zur Totalsynthese von Anhydro-tetra-cyclinen wurden Versuche unternommen, den Desdimethylamino-12a-desoxy-anhydro-aureomycin-monomethyläther (CLXXV) und das entsprechende Abbauprodukt (CLXXVI) des Tetracyclins in 12a-Stellung zu hydroxylieren. Zwar erhält man bei Bromierung von (CLXXV) und (CLXXVI) mit N-Brom-succinimid in guter Ausbeute 12a-Brom-verbindungen, wie z. B. (CLXXXIII), doch läßt sich das Brom nicht durch nucleophile Substitution gegen eine Acetoxygruppe austauschen. Oxydation von (CLXXV) und (CLXXVI) mit Perbenzoesäure ergibt ebenfalls in guter Ausbeute 12a-Hydroxyverbindungen, doch erhält man die Verbindungen (CLXXXI) und (CLXXXII) mit der Konfiguration von 12a-epi-Tetracyclinen als Hauptprodukt. Die Monomethyläther des Desdimethyl-amino-anhydro-aureomycins (CLXXVII) und des Des-dimethylamino-anhydro-tetracyclins (CLXXVIII) werden nur in sehr

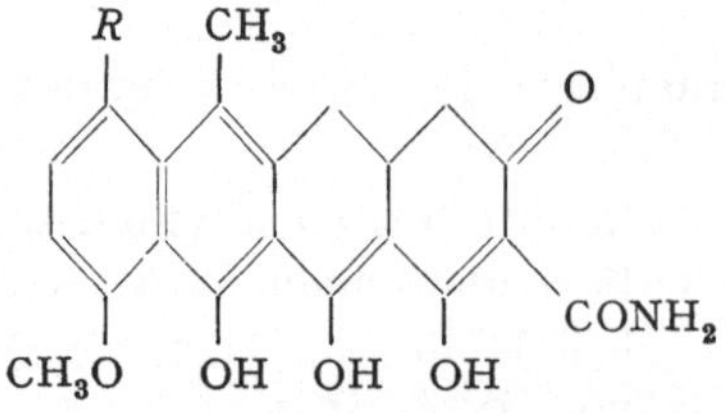

(CLXXV.) R = Cl.
(CLXXVI.) R = H.

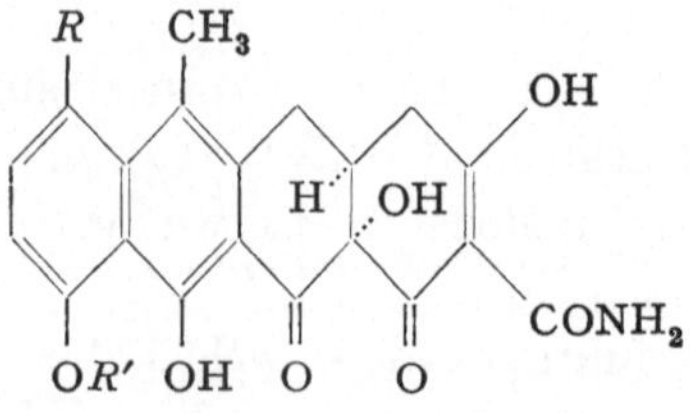

(CLXXVII.) R = Cl; R′ = CH₃.
(CLXXVIII.) R = H; R′ = CH₃.
(CLXXIX.) R = Cl; R′ = H.
(CLXXX.) R = R′ = H.

Literaturverzeichnis: SS. 116—120.

geringer Ausbeute gebildet. Sie lassen sich jedoch durch Chromatographie reinigen und in kristallisierter Form isolieren sowie zu Desdimethyl-amino-anhydro-aureomycin (CLXXIX) bzw. Desdimethylamino-anhydro-tetracyclin (CLXXX) entmethylieren (*49, 53, 54*). In sehr viel besserer

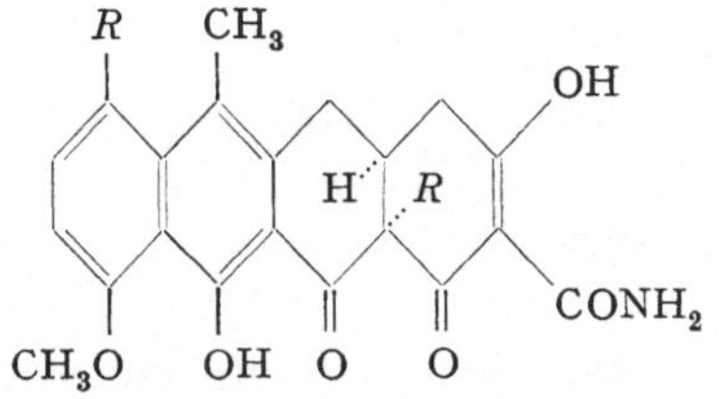

(CLXXXI.) $R = $ Cl; $R' = $ OH.
(CLXXXII.) $R = $ H; $R' = $ OH.
(CLXXXIII.) $R = $ H; $R' = $ Br.

Ausbeute erhält man dagegen Desdimethylamino-anhydro-tetracyclin, wenn man die entsprechende 12a-Desoxy-Verbindung mit Platin und Sauerstoff behandelt (*49, 52*).

IV. Biogenese der Tetracycline.

Wird *Str. rimosus* in Gegenwart von Natriumacetat-^{14}C kultiviert, so entsteht ein radioaktives Terramycin (II, S. 81) (*66*),

Abbauversuche von SNELL et al. (*66*) haben gezeigt, daß zumindest die C-Atome 5 bis 12 (vgl. CLXXXIV), wahrscheinlich aber auch 12a und 1 nach dem Essigsäureschema miteinander verknüpft werden. Für die übrigen Kohlenstoffatome des Ringgerüstes (2 bis 4) sowie die Carboxamidgruppe soll Glutaminsäure ein biogenetischer Vorläufer sein (*66*). Markierungsversuche von GATENBECK (*18a*) haben dagegen gezeigt, daß Glutaminsäure kein notwendiger biogenetischer Vorläufer der Tetra-cycline ist, sondern daß das Ringgerüst der Tetracycline vollkommen aus Essigsäure- bzw. Malonsäureeinheiten aufgebaut und daß die Carboxamidgruppe sekundär inkorporiert wird. Schließlich hat sich auch noch zeigen lassen, daß die Methylgruppen der Dimethylaminogruppe

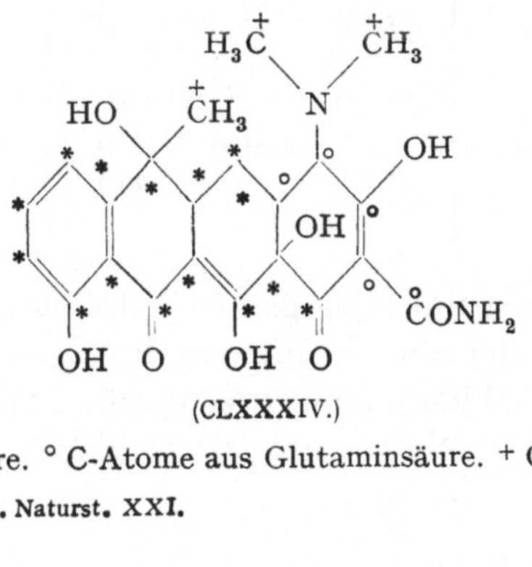

(CLXXXIV.)

* C-Atome aus Essigsäure. ° C-Atome aus Glutaminsäure. + C-Atome aus Methionin.

sowie die Methylgruppe am $C_{(6)}$ des Terramycins radioaktiv werden und alle die gleiche Aktivität besitzen, wenn man dem Fermentierungsmedium Methionin-^{14}C zusetzt (CLXXXIV).

Bemerkenswerterweise greifen Sulfonamide selektiv in den Methioninstoffwechsel ein, denn *Streptomyces aureofaciens*, der normalerweise Aureomycin (III, S. 81) produziert, bildet bei Zusatz von Sulfonamiden, wie z. Beisp. Sulfaguanidin, als Hauptprodukt 6-Desmethyl-7-chlortetracyclin (VII, S. 82). Sulfonamide vermögen also den Einbau der Methylgruppe am $C_{(6)}$ zu verhindern, nicht aber den der Methylgruppen am Stickstoff (57).

Als Halogenquelle für den Aufbau halogenhaltiger Tetracycline können Chlor- und Bromionen dienen. So produzieren z. Beisp. bestimmte Stämme von *St. aureofaciens* in Gegenwart von Chlorionen Aureomycin (III), in Gegenwart von Bromionen 7-Bromtetracyclin (V, S. 82) und in Abwesenheit von Halogenionen Tetracyclin (*13, 64*).

Die Endstufen der Biogenese von Tetracyclinen sind seit kurzem genau bekannt. Es konnte gezeigt werden, daß zahlreiche Varianten von *St. aureofaciens* Anhydro-tetracycline in Tetracycline verwandeln können, und die in *Tabelle 2* verzeichneten biologischen Umwandlungen ließen sich realisieren (*42*).

Tabelle 2. Biologische Umwandlungen von Tetracyclin-Abbauprodukten in Tetracycline.

Ausgangsmaterial	*Streptomyces*	Reaktionsprodukt
Anhydro-tetracyclin	*St. aureofaciens* BC 41	Tetracyclin
Anhydro-aureomycin	*St. aureofaciens* BC 41	Aureomycin
Anhydro-terramycin	*St. aureofaciens* V 828	Terramycin
Anhydro-desmethyl-tetracyclin	*St. aureofaciens* S 2242	Desmethyl-tetracyclin
Anhydro-7-chlor-desmethyl-tetracyclin	*St. aureofaciens* S 2311	7-Chlor-desmethyl-tetracyclin
Anhydro-2-acetyl-descarboxamido-aureomycin	*St. aureofaciens* S 2242	2-Acetyl-descarboxamido-tetracyclin
Anhydro-aureomycin	*St. aureofaciens* S 1308	5a,11a-Dehydro-tetracyclin
Anhydro-tetracyclin	*St. rimosus* T 1686 B	Terramycin
Anhydro-tetracyclinnitril	*St. aureofaciens* T 219	Tetracyclin-nitril
12a-Desoxy-anhydro-tetracyclin	*St. aureofaciens* T 219	12a-Desoxy-tetracyclin

Nicht angegriffen wurden dagegen Anhydro-aureomycin und Anhydro-6-desmethyl-tetracyclin durch *St. rimosus* T 1686 B sowie Anhydro-4-epi-tetracyclin, Anhydro-4-epi-aureomycin und Desdimethylamino-anhydro-tetracyclin durch *St. aureofaciens* T 219.

Literaturverzeichnis: SS. 116—120.

Diese Ergebnisse zeigen, daß alle Anhydro-tetracycline, welche sich direkt von natürlichen Tetracyclinen herleiten, wahrscheinlich biogenetische Vorstufen darstellen. Dagegen sind Anhydro-tetracyclin-nitril und 12a-Desoxy-anhydro-tetracyclin zwar hydratisierbar, jedoch werden sie nicht in natürlich vorkommende Tetracycline umgewandelt.

St. aureofaciens S 1308 bildet aus Anhydro-aureomycin nicht Aureomycin (III, S. 81), sondern 5a,11a-Dehydro-aureomycin (VIII, S. 82), das von Streptomyceten, welche Aureomycin produzieren, in dieses umgewandelt werden kann (*45*). Die Mutante S 1308 ist also nicht in der Lage, diesen letzten Schritt durchzuführen. Die Ursache dafür ist das Fehlen der Eigenschaft, einen Cosynthesefaktor zu produzieren, der diese Reaktion katalysiert und von Aureomycin produzierenden Streptomyceten aufgebaut und aus solchen in kristallisierter Form isoliert werden kann (*47*). Wenn dieser Cosynthesefaktor einer Kulturlösung von *St. aureofaciens* S 1308 zugesetzt wird, so wird nicht mehr 5a,11a-Dehydro-aureomycin, sondern Aureomycin gebildet. Man darf daher annehmen, daß bei der Biogenese der Tetracycline zunächst Anhydro-tetracycline gebildet werden, welche sodann zunächst zu 5a,11a-Dehydro-tetracyclinen oxydiert und dann zu den entsprechenden Tetracyclinen reduziert werden. Bei der Biogenese des Terramycins (II) muß noch eine weitere Oxydation am $C_{(5)}$ vorgeschaltet sein, da *St. rimosus* Anhydro-tetracyclin in Terramycin umwandeln kann.

Verschiedene Mutanten tetracyclin-bildender Streptomyceten, welche nicht mehr in der Lage sind, Tetracycline zu produzieren, bilden Tetracycline, wenn man sie im Gemisch miteinander kultiviert. Dieser Vorgang ist als biologische Cosynthese bezeichnet worden (*38*).

Aus der Mutante S 652 von *St. aureofaciens* konnten relativ große Mengen (+)-*trans*-2,3-Dihydro-3-hydroxy-anthranilsäure isoliert werden (*43*). Da diese Mutante keine Tetracycline mehr produziert, lag es nahe anzunehmen, daß diese neue Aminosäure ein biogenetischer Vorläufer von Tetracyclinen sei. Dies hat sich jedoch nicht bestätigen lassen (*43*).

V. Versuche zur Synthese von Tetracyclinen.

Versuche zur Synthese von Tetracyclinen sind seit etwa fünf Jahren in verschiedenen Laboratorien im Gange. So wurden 1959 das Desdimethylamino - desmethyl - 12 a - desoxy - 7 - chlor - anhydro - tetracyclin (CLXXXV) (*8, 32*) und im gleichen Jahr das Desdimethylamino-anhydro-aureomycin (CLXXXVI) (*49, 50, 53, 54*) synthetisiert.

Ein Jahr später gelangen die Synthesen des Desdimethylamino-6,12a-didesoxy-6-desmethyl-7-chlor-tetracyclins (CLXXXVII) (*16, 17*) und der

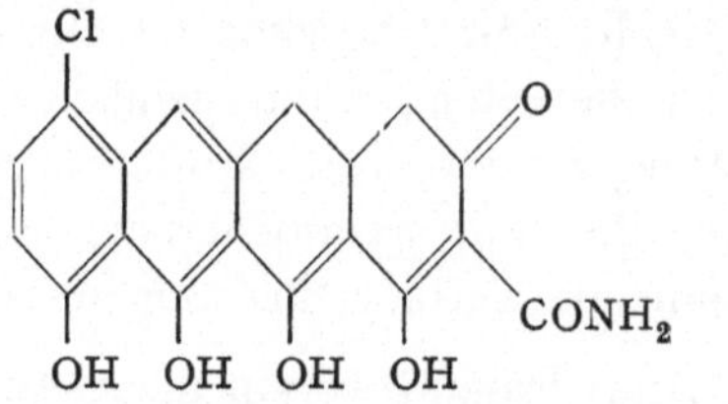

(CLXXXV.) Desdimethylamino-desmethyl-12a-desoxy-
7-chlor-anhydro-tetracyclin.

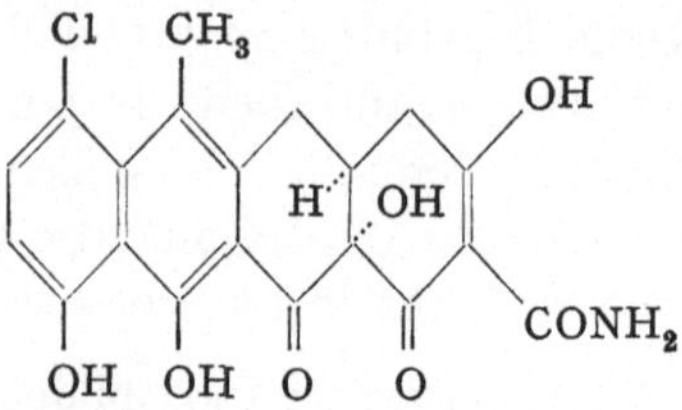

(CLXXXVI). Desdimethylamino-anhydro-
aureomycin.

tetracyclischen Verbindung (CLXXXVIII) (*49, 55*), der in diesem Jahre
der Aufbau von (CIXC) folgte (*51*).

Das erste Abbauprodukt eines natürlich vorkommenden Tetracyclins
mit voller biologischer Aktivität und zugleich das einfachste Abbauprodukt mit voller biologischer Aktivität, das 6-Desoxy-6-desmethyl-
tetracyclin (CXC), wurde kürzlich in den Laboratorien der Chas. Pfizer
and Co. in Zusammenarbeit mit Woodward synthetisiert (*11*).

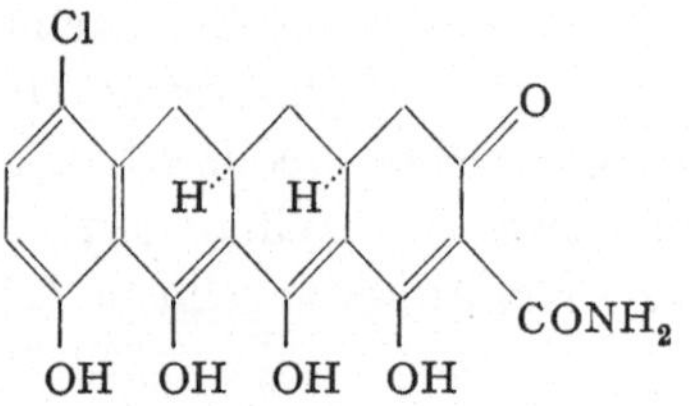

(CLXXXVII.) Desdimethylamino-
6,12a-didesoxy-6-desmethyl-7-chlor-tetracyclin.

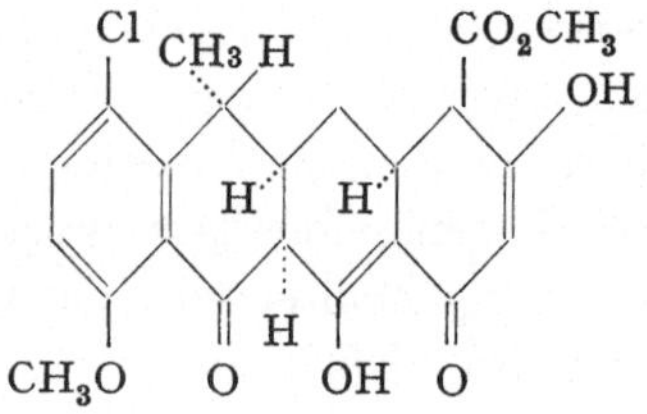

(CLXXXVIII.)

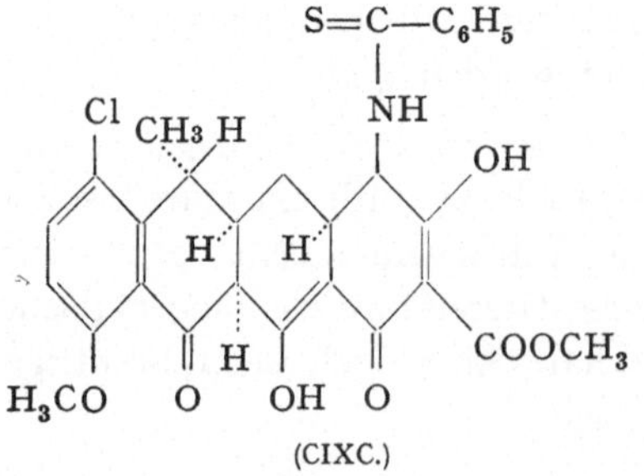

(CIXC.)

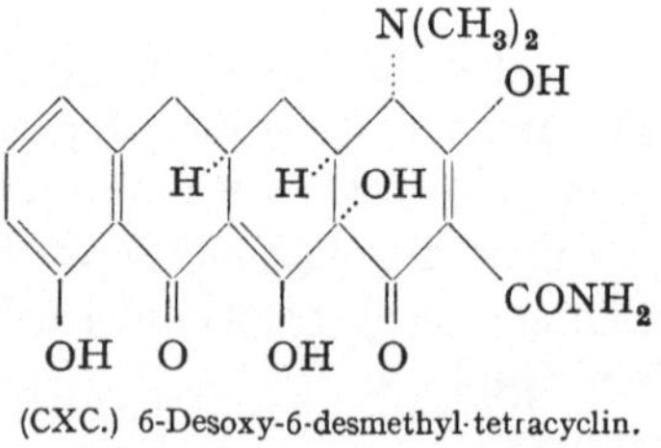

(CXC.) 6-Desoxy-6-desmethyl-tetracyclin.

Da sich die synthetischen Arbeiten noch in voller Entwicklung befinden, sollen
sie zu einem späteren Zeitpunkt gesondert zusammenfassend beschrieben werden.

Literaturverzeichnis.

1. André, T. and S. Ullberg: Radioactive Tetracycline. J. Amer. Chem. Soc. **79**,
494 (1957).

2. Beereboom, J. J., J. J. Ursprung, H. H. Rennhard and C. R. Stephens:
Further 6-Deoxytetracycline Studies: Effect of Aromatic Substituents on
Biological Activity. J. Amer. Chem. Soc. **82**, 1003 (1960).

3. BLACKWOOD, R. K., J. J. BEEREBOOM, H. H. RENNHARD, M. SCHACH VON WITTENAU and C. R. STEPHENS: 6-Methylenetetracyclines. I. A New Class of Tetracycline Antibiotics. J. Amer. Chem. Soc. **83**, 2773 (1961).

4. BLACKWOOD, R. K., H. H. RENNHARD and C. R. STEPHENS: Some Transformations at the 12a-Position in the Tetracycline Series. J. Amer. Chem. Soc. **82**, 5194 (1960).

5. BOOTHE, J. H., G. E. BONVICINO, C. W. WALLER, J. P. PETISI, R. W. WILKINSON and R. W. BROSCHARD: Chemistry of the Tetracycline Antibiotics. I. Quaternary Derivatives. J. Amer. Chem. Soc. **80**, 1654 (1958).

6. BOOTHE, J. H., A. GREEN, J. P. PETISI, R. G. WILKINSON and C. W. WALLER: Demethyltetracyclines. Synthesis of a Degradation Product. J. Amer. Chem. Soc. **79**, 4564 (1957).

7. BOOTHE, J. H., J. J. HLAVKA, J. P. PETISI and J. L. SPENCER: 6-Deoxytetracyclines. I. Chemical Modification by Electrophilic Substitution. J. Amer. Chem. Soc. **82**, 1253 (1960).

8. BOOTHE, J. H., A. S. KENDE, T. L. FIELDS and R. G. WILKINSON: Total Synthesis of Tetracyclines. I. ($\pm$)-Dedimethylamino-12a-deoxy-6-demethylanhydrochlorotetracycline. J. Amer. Chem. Soc. **81**, 1006 (1959).

8a. BROCKMANN, H.: Die Actinomycine. Fortschr. Chem. organ. Naturst. **18**, 1 (1960).

9. BROSCHARD, R. W., A. C. DORNBUSH, S. GORDON, B. L. HUTCHINGS, A. R. KOHLER, G. KRUPKA, S. KUSHNER, D. V. LEFIMINE and C. PIDACKS: Aureomycin, a New Antibiotic. Science (Washington) **109**, 199 (1949).

10. CONOVER, L. H.: Progress in the Chemistry of Oxytetracycline and Related Compounds. Chem. Soc. (London), Spec. Publ. No. 5, p. 73 (1956).

11. CONOVER, L. H., K. BUTLER, J. D. JOHNSTON, J. J. KORST and R. B. WOODWARD: The Total Synthesis of 6-Demethyl-6-deoxy-tetracycline. J. Amer. Chem. Soc. **84**, 3222 (1962).

12. DOERSCHUK, A. P., B. A. BITLER and J. R. D. McCORMICK: Reversible Isomerizations in the Tetracycline Family. J. Amer. Chem. Soc. **77**, 4687 (1955).

13. DOERSCHUK, A. P., J. R. D. McCORMICK, J. J. GOODMAN, S. A. SZUMSKI, J. A. GROWICH, P. A. MILLER, B. A. BITLER, E. R. JENSEN, M. MATRISHIN, M. A. PETTY and A. S. PHELPS: Biosynthesis of Tetracyclines. I. The Halide Metabolism of *Streptomyces aureofaciens* Mutants. The Preparation and Characterization of Tetracycline, 7-Chloro[36]-tetracycline and 7-Bromotetracycline. J. Amer. Chem. Soc. **81**, 3069 (1959).

14. DOWLING, H. F.: Tetracycline. New York: Medical Encyclopedia, Inc. 1955.

15. DUGGAR, B. M.: Aureomycin—a New Antibiotic. Ann. New York Acad. Sci. **51**, 175 (1948).

16. FIELDS, T. L., A. S. KENDE and J. H. BOOTHE: Total Synthesis of Tetracyclines. II. Stereospecific Synthesis of ($\pm$)-Dedimethylamino-6-demethyl-6,12a-dideoxy-7-chlorotetracycline. J. Amer. Chem. Soc. **82**, 1250 (1960).

17. — — — Total Synthesis of Tetracyclines. V. The Stereospecific Elaboration of the Tetracycline Ring System. J. Amer. Chem. Soc. **83**, 4612 (1961).

18. FINLAY, A. C., G. L. HOBBY, S. Y. PLAN, P. P. REGNA, J. B. ROUTIEN, D. B. SEELEY, G. M. SHULL, B. A. SOBIN, I. A. SOLOMONS, J. W. VINSON and J. H. KANE: Terramycin, a New Antibiotic. Science (Washington) **111**, 85 (1950).

18a. GATENBECK, S.: The Biosynthesis of Oxytetracycline. Biochem. and Biophys. Res. Comm. **6**, 422 (1961–62).

19. GOTTSTEIN, W. J., W. F. MINOR and L. C. CHENEY: Carboxamido Derivatives of the Tetracyclines. J. Amer. Chem. Soc. **81**, 1198 (1959).

20. GREEN, A. and J. H. BOOTHE: Chemistry of the Tetracycline Antibiotics. III. 12a-Deoxy-tetracycline. J. Amer. Chem. Soc. **82**, 3950 (1960).

21. Green, A., R. G. Wilkinson and J. H. Boothe: Chemistry of the Tetracycline Antibiotics. II. Bromination of Dedimethylamino-tetracyclines. J. Amer. Chem. Soc. **82**, 3946 (1960).

22. Hirokawa, S., Y. O. Okaya, F. M. Lovell and R. Pepinsky: Abstracts Amer. Cryst. Assoc. Meeting, Cornell Univ., Juli 1959, p. 44.

23. Hlavka, J. J. and D. A. Buyske: Radioactive 7-Iodo-6-deoxy-tetracycline in Tumor Tissue. Nature (London) **186**, 1064 (1960).

24. Hlavka, J. J., H. Krazinski and J. H. Boothe: The 6-Deoxytetracyclines. IV. A Photochemical Displacement of a Diazonium Group. J. Organ. Chem. (USA) **27**, 3674 (1962).

25. Hlavka, J. J., A. Schneller, H. Krazinski and J. H. Boothe: The 6-Deoxytetracyclines. III. Electrophilic and Nucleophilic Substitution. J. Amer. Chem. Soc. **84**, 1426 (1962).

26. Hochstein, F. A., M. Schach von Wittenau, F. W. Tanner, Jr. and K. Murai: 2-Acetyl-2-decarboxamido-oxytetracycline. J. Amer. Chem. Soc. **82**, 5934 (1960).

27. Hochstein, F. A., C. R. Stephens, L. H. Conover, P. P. Regna, R. Pasternack, P. N. Gordon, F. J. Pilgrim, K. J. Brunings and R. B. Woodward: The Structure of Terramycin. J. Amer. Chem. Soc. **75**, 5455 (1953).

28. Holmlund, C. E., W. W. Andres and A. J. Shay: Chemical Hydroxylation of 12a-Deoxy-tetracycline. J. Amer. Chem. Soc. **81**, 4748 (1959).

29. — — — Microbiological Hydroxylation of 12a-Deoxy-tetracycline. J. Amer. Chem. Soc. **81**, 4750 (1959).

30. Hutchings, B. L., C. W. Waller, R. W. Broschard, C. F. Wolf, P. W. Fryth and J. H. Williams: Degradation of Aureomycin. V. Aureomycinic Acid. J. Amer. Chem. Soc. **74**, 4980 (1952).

31. Hutchings, B. L., C. W. Waller, S. Gordon, R. W. Broschard, C. F. Wolf, A. A. Goldman and J. H. Williams: Degradation of Aureomycin. J. Amer. Chem. Soc. **74**, 3710 (1952).

32. Kende, A. S., T. L. Fields, J. H. Boothe and S. Kushner: Total Synthesis of Tetracyclines. IV. Synthesis of an Anhydrotetracycline Derivative. J. Amer. Chem. Soc. **83**, 439 (1961).

33. Kuhn, R. und K. Dury: 6-Acetyl-salicylsäure. Chem. Ber. **84**, 848 (1951).

34. Kushner, S., J. H. Boothe, J. Morton II, J. Petisi and J. H. Williams: Synthesis of Degradation Products of Aureomycin. J. Amer. Chem. Soc. **74**, 3710 (1952).

35. Lepper, M. H.: Terramycin. New York: Medical Encyclopedia, Inc. 1956.

36. McCormick, J. R. D., S. M. Fox, L. L. Smith, B. A. Bitler, J. Reichenthal, V. E. Origoni, W. H. Muller, R. Winterbottom and A. P. Doerschuk: On the Nature of the Reversible Isomerizations Occurring in the Tetracycline Family. J. Amer. Chem. Soc. **78**, 3547 (1956).

37. — — — — — — — — — Studies of the Reversible Epimerization Occurring in the Tetracycline Family. The Preparation, Properties and Proof of Structure of Some 4-epi-Tetracyclines. J. Amer. Chem. Soc. **79**, 2849 (1957).

38. McCormick, J. R. D., U. Hirsch, N. O. Sjolander and A. P. Doerschuk: Cosynthesis of Tetracyclines by Pairs of *Streptomyces aureofaciens* Mutants. J. Amer. Chem. Soc. **82**, 5006 (1960).

39. McCormick, J. R. D. und E. R. Jensen: Deutsches Bundespat. 1082905 [Chem. Abstr. **56**, 4703 (1962)].

40. McCormick, J. R. D., E. R. Jensen, P. A. Miller and A. P. Doerschuk: The 6-Deoxytetracyclines. Further Studies on the Relationship between Structure and Antibacterial Activity in the Tetracycline Series. J. Amer. Chem. Soc. **82**, 3381 (1960).

41. McCormick, J. R. D., P. A. Miller, J. A. Growich, N. O. Sjolander and A. P. Doerschuk: Two New Tetracycline-related Compounds: 7-Chloro-5a(11a)-Dehydrotetracycline and 5a-epi-Tetracycline. A New Route to Tetracycline. J. Amer. Chem. Soc. **80**, 5572 (1958).

42. McCormick, J. R. D., P. A. Miller, S. Johnson, N. Arnold and N. O. Sjolander: Biosynthesis of the Tetracyclines. IV. Biological Rehydration of the 5a,6-Anhydrotetracyclines. J. Amer. Chem. Soc. **84**, 3023 (1962).

43. McCormick, J. R. D., J. Reichenthal, U. Hirsch and N. O. Sjolander: Biosynthesis of the Tetracyclines. III. A New Amino Acid from *Streptomyces aureofaciens*: (+)-*trans*-2,3-Dihydro-3-hydroxyanthranilic Acid. J. Amer. Chem. Soc. **84**, 3711 (1962).

44. McCormick, J. R. D., N. O. Sjolander, U. Hirsch, E. R. Jensen and A. P. Doerschuk: A New Family of Antibiotics: The Demethyl-tetracyclines. J. Amer. Chem. Soc. **79**, 4561 (1957).

45. McCormick, J. R. D., N. O. Sjolander, U. Hirsch, P. A. Miller, N. H. Arnold and A. P. Doerschuk: The Biological Reduction of 7-Chloro-5a(11a)-dehydrotetracycline to 7-Chloro-tetracycline by *Streptomyces aureofaciens*. J. Amer. Chem. Soc. **80**, 6460 (1958).

46. Miller, M. W. and F. A. Hochstein: Isolation and Characterization of Two New Tetracycline Antibiotics. J. Organ. Chem. (USA) **27**, 2525 (1962).

47. Miller, P. A., N. O. Sjolander, S. Nalesnyk, N. Arnold, S. Johnson, A. P. Doerschuk and J. R. D. McCormick: Cosynthetic Factor I, a Factor Involved in Hydrogen Transfer in *Streptomyces aureofaciens*. J. Amer. Chem. Soc. **82**, 5002 (1960).

48. Musselman, M. M.: Aureomycin. New York: Medical Encyclopedia, Inc. 1956.

49. Muxfeldt, H.: Synthesen in der Tetracyclinreihe. Angew. Chem. **74**, 443 (1962). Syntheses in the Tetracycline Series. Angew. Chem. Internat. Ed. **1**, 372 (1962).

50. — Synthese tetracyclischer Abbauprodukte von Anhydrotetracyclinen. Chem. Ber. **92**, 3122 (1959).

51. — Synthese eines Terramycin-Bausteins. Angew. Chem. **74**, 825 (1962).

52. Muxfeldt, H., G. Buhr und R. Bangert: Katalytische Hydroxylierung von 12a-Desoxy-tetracyclinen. Angew. Chem. **74**, 213 (1962). Catalytic Hydroxylation of 12a-Deoxy-tetracyclines. Angew. Chem. Internat. Ed. **1**, 157 (1962).

53. Muxfeldt, H. und A. Kreutzer: Synthese des Desdimethylamino-anhydroaureomycins. Chem. Ber. **94**, 881 (1961).

54. — — Ab- und Aufbaureaktionen in der Anhydro-tetracyclin-Reihe. Naturwiss. **46**, 204 (1959).

55. Muxfeldt, H., W. Rogalski und K. Striegler: Aufbau des Ringsystems von 6-Desoxy-tetracyclinen. Chem. Ber. **95**, 2581 (1962).

56. Pasternack, R., L. H. Conover, A. Bavley, F. A. Hochstein, G. B. Hess and K. J. Brunings: Terramycin. III. Structure of Terracinoic Acid, an Alkaline Degradation Product. J. Amer. Chem. Soc. **74**, 1928 (1952).

57. Perlman, D., L. J. Heuser, J. B. Semar, W. R. Frazier and J. A. Boska: Process for Biosynthesis of 7-Chloro-6-demethyltetracycline. J. Amer. Chem. Soc. **83**, 4481 (1961).

58. Petisi, J., J. L. Spencer, J. J. Hlavka and J. H. Boothe: 6-Deoxytetracyclines. II. Nitrations and Subsequent Reactions. J. Med. Pharm. Chem. **5**, 538 (1962).

59. *Chas. Pfizer and Co.*, Medical Research Laboratories, Groton, Conn., Privatmitteilung.

60. Regna, P. P., I. A. Solomons, K. Murai, A. E. Timreck, K. J. Brunings and W. A. Lazier: The Isolation and General Properties of Terramycin and Terramycin Salts. J. Amer. Chem. Soc. **73**, 4211 (1951).

61. Rennhard, H. H., R. K. Blackwood and C. R. Stephens: Fluorotetracyclines. I. Perchloryl Fluoride Studies in the Tetracycline Series. J. Amer. Chem. Soc. **83**, 2775 (1961).

62. Schach von Wittenau, M., J. J. Beereboom, R. K. Blackwood and C. R. Stephens: 6-Deoxytetracyclines. III. Stereochemistry at C.6. J. Amer. Chem. Soc. **84**, 2645 (1962).

63. Scott, A. I. and C. T. Bedford: Simulation of the Biosynthesis of Tetracyclines. A Partial Synthesis of Tetracycline from Anhydroaureomycin. J. Amer. Chem. Soc. **84**, 2271 (1962).

64. Sensi, P., G. A. De Ferrari, G. G. Gallo und G. Rolland: 7-Bromtetracyclin. Farmaco (Pavia), Ed. sci. **10**, 337 (1955).

65. Siedel, W., F. Soeder und F. Lindner: Die Aminomethylierung der Tetracycline. Zur Chemie des Reverin®. Münchner Med. Wochenschr. **17**, 661 (1958).

66. Snell, J. F., A. J. Birch and P. L. Thomson: The Biosynthesis of Tetracycline Antibiotics. J. Amer. Chem. Soc. **82**, 2402 (1960).

67. Stephens, C. R.: Abstracts of papers, 109th Meeting Amer. Chem. Soc., Dallas, Texas, 1956, p. 18 M.

68. Stephens, C. R., L. H. Conover, P. N. Gordon, F. C. Pennington, R. L. Wagner, K. J. Brunings and F. J. Pilgrim: Epitetracycline. The Chemical Relationship between Tetracycline and "Quatrimycin". J. Amer. Chem. Soc. **78**, 1515 (1956).

69. Stephens, C. R., L. H. Conover, R. Pasternack, F. A. Hochstein, W. T. Moreland, P. P. Regna, F. J. Pilgrim, K. J. Brunings and R. B. Woodward: The Structure of Aureomycin. J. Amer. Chem. Soc. **76**, 3568 (1954).

70. Stephens, C. R., K. Murai, K. J. Brunings and R. B. Woodward: Acidity Constants of the Tetracycline Antibiotics. J. Amer. Chem. Soc. **78**, 4155 (1956).

71. Stephens, C. R., K. Murai, H. H. Rennhard, L. H. Conover and K. J. Brunings: Hydrogenolysis Studies in the Tetracycline Series. 6-Deoxytetracyclines. J. Amer. Chem. Soc. **80**, 5324 (1958).

72. Takeuchi, Y. and M. J. Buerger: The Crystal Structure of Terramycin Hydrochloride. Proc. Nat. Acad. Sci. (USA) **46**, 1366 (1960).

72a. Van Tamelen, E. E.: Structural Chemistry of Actinomycetes Antibiotics. Fortschr. Chem. organ. Naturst. **16**, 90 (1958).

73. Vonderbank, H.: Aureomycin und Achromycin. Arzneimittel-Forsch., 6. Beiheft, 1956.

74. Waller, C. W., B. L. Hutchings, R. W. Broschard, A. A. Goldman, W. J. Stein, C. F. Wolf and J. H. Williams: Degradation of Aureomycin. VII. Aureomycin and Anhydroaureomycin. J. Amer. Chem. Soc. **74**, 4981 (1952).

75. Waller, C. W., B. L. Hutchings, C. F. Wolf, R. W. Broschard, A. A. Goldman and J. H. Williams: Degradation of Aureomycin. III. 3,4-Dihydroxy-2,5-dioxocyclopentane-1-carboxamide. J. Amer. Chem. Soc. **74**, 4978 (1952).

76. Waller, C. W., B. L. Hutchings, A. A. Goldman, C. F. Wolf, R. W. Broschard and J. H. Williams: Degradation of Aureomycin. IV. Des-dimethylaminoaureomycinic Acid. J. Amer. Chem. Soc. **74**, 4979 (1952).

77. Waller, C. W., B. L. Hutchings, C. F. Wolf, A. A. Goldman, R. W. Broschard and J. H. Williams: Degradation of Aureomycin. VI. Isoaureomycin and Aureomycin. J. Amer. Chem. Soc. **74**, 4981 (1952).

78. Webb, J. S., R. W. Broschard, D. B. Cosulich, W. J. Stein and C. F. Wolf: Demethyltetracyclines. Structure Studies. J. Amer. Chem. Soc. **79**, 4563 (1957).

79. Wilkinson, R. G., T. L. Fields and J. H. Boothe: Total Synthesis of Tetracyclines. III. Synthesis of a T icyclic Model System. J. Organ. Chem. (USA) **26**, 637 (1961).

(Eingelaufen am 12. Dezember 1962.)

Anthracyclinone und Anthracycline.
(Rhodomycinone, Pyrromycinone und ihre Glykoside.)

Von **Hans Brockmann**, Göttingen.

Mit 7 Abbildungen.

Inhaltsübersicht.

	Seite
I. Einleitung	122
II. Isolierung der Anthracyclinone und Anthracycline	123
1. Gewinnung der ε-Pyrromycinon-glykoside und Pyrromycinone	124
Cinerubin A und B	124
Pyrromycin und Pyrromycinone	124
2. Gewinnung der Rhodomycinone, Iso-rhodomycinone und ihrer Glykoside	125
Trennung von Rhodomycinon/Iso-rhodomycinon-Gemischen	126
Trennung von Rhodomycinen und Iso-rhodomycinen	127
III. Die Anthracyclinone	127
1. Vorbemerkungen zur Struktur der Anthracyclinone	127
2. Zur Konstitutionsermittlung der Anthracyclinone	130
Die Aufklärung des Chromophors	131
Die Anellierung des alicyclischen Ringes	133
Die Substituenten an Ring A	134
Schreibweise und Bezifferung der Anthracyclinon-Formeln	137
3. Konstitution der Anthracyclinone	137
A. Iso-rhodomycinone	137
ε-Iso-rhodomycinon	137
ζ-Iso-rhodomycinon	139
β-Iso-rhodomycinon	140
B. Rhodomycinone	140
ε-Rhodomycinon	140
ζ-Rhodomycinon	141
β-Rhodomycinon	141
γ-Rhodomycinon	145
δ-Rhodomycinon	145
C. Pyrromycinone	146
η-Pyrromycinon	146
ε-Pyrromycinon	151
ζ-Pyrromycinon	151
D. Aklavinone	154
Aklavinon	154
7-Desoxy-aklavinon	154

Seite

4. Die KMR-Spektren der Anthracyclinone.......................... 155

5. Zur Stereochemie der Anthracyclinone........................... 160

6. Zur Biogenese der Anthracyclinone.............................. 163

IV. Die Anthracycline.. 170

    1. Die Zucker der Anthracycline 171
        Rhodosamin 171
        2-Desoxy-*L*-fucose 173
        Rhodinose 173

    2. Anthracycline des ε-Pyrromycinons............................. 174
        Pyrromycin 174
        Cinerubine... 175
        Rutilantine 176

    3. Anthracycline der Rhodomycinone 176
        Rhodomycin A 176
        Rhodomycin B 177
        γ-Rhodomycine... 177
        Iso-rhodomycin A... 178
        Antibiotica der Mycetin-Violarin-Gruppe 179

    4. Anthracycline des Aklavinons................................. 179
        Aklavin 179

Literaturverzeichnis 179

I. Einleitung.

Die in den letzten Jahren aus verschiedenen *Streptomyces*-Arten isolierten Rhodomycinone, Iso-rhodomycinone und Pyrromycinone sind gelbrote oder rote, optisch aktive Farbstoffe, die man ihrem Kohlenstoffgerüst nach als Derivate des 7,8,9,10-Tetrahydro-tetracenchinons-(5,12) und dem chromophoren Teil ihres Moleküls nach als Anthrachinonfarbstoffe ansehen kann. Da sie eine Gruppe neuer, nahe verwandter Naturfarbstoffe bilden, von denen bereits vierzehn Vertreter bekannt sind, hat man kürzlich für sie den in diesem Bericht verwendeten Sammelnamen *Anthracyclinone* vorgeschlagen (*14*). Er soll zum Ausdruck bringen, daß das Kohlenstoffgerüst wie bei den Tetracyclinen aus linear anellierten Sechsringen besteht, von denen drei einem Anthrachinon-Ringsystem angehören.

Die in Wasser schwer löslichen Anthracyclinone liegen im Kulturmedium und Mycel der *Streptomyces*-Arten ganz oder größtenteils als wasserlösliche, basische, ein oder mehrere Zuckerreste enthaltende Glykoside vor, für die dem neuen Namen ihrer Aglykone entsprechend die Bezeichnung *Anthracycline* vorgeschlagen wurde (*14*). Einige von ihnen sind antibiotisch hochwirksam und zwei hemmen das Wachstum von Krebszellen. Ihrer Giftigkeit wegen sind sie bisher medizinisch nicht verwendet worden.

Literaturverzeichnis: SS. 179—182.

Die Rhodomycinone und Iso-rhodomycinone sowie ihre Glykoside, die Rhodomycine und Iso-rhodomycin A, ferner η-Pyrromycinon und ζ-Pyrromycinon haben BROCKMANN und Mitarb. aufgefunden und eingehend bearbeitet (*3*). ε-Pyrromycinon und seine Glykoside sind unabhängig voneinander von PRELOG, KELLER-SCHIERLEIN und Mitarb. sowie von BROCKMANN und Mitarb. untersucht worden. An der Konstitutionsaufklärung der Pyrromycinone beteiligten sich später auch OLLIS und SUTHERLAND, die in jüngster Zeit Aklavinon und Desoxy-aklavinon, zwei neue Vertreter der Anthracyclinone, entdeckt haben (*48*).

Der folgende Bericht faßt zusammen, was bisher über Anthracyclinone und Anthracycline bekanntgeworden ist.

II. Isolierung der Anthracyclinone und Anthracycline.

Rhodomycinone und Iso-rhodomycinone sowie ihre Glykoside, die Rhodomycine und Iso-rhodomycine, werden von *Streptomyces purpurascens* [LINDENBEIN (*47*)] produziert, die Pyrromycinone und ihre Glykoside, das Pyrromycin, Cinerubin A und Cinerubin B, von *Str. antibioticus* WAKSMAN et HENRICI, *Str. niveoruber*, *Str. galileus* [PRELOG, KELLER-SCHIERLEIN und Mitarb. (*40*)] sowie *Str. DOA 1205* [BROCKMANN und LENK (*22, 23*)].

Bisher hat man in Kulturen von *Str. purpurascens*-Stämmen keine Pyrromycinone oder Pyrromycinonglykoside gefunden und umgekehrt bei Pyrromycinon bildenden Stämmen keine Rhodomycinone, Iso-rhodomycinone oder deren Glykoside. Ob die Aklavine und Aklavinon erzeugenden *Streptomyces*-Arten noch andere Anthracyclinone oder Anthracycline synthetisieren, geht aus den bisherigen kurzen Literatur-

Tabelle 1. Anthracyclinone und Anthracycline.

Anthracyclinone.

β-*Rhodomycinon*	β-*Iso-rhodomycinon*	ε-*Pyrromycinon*	*Aklavinon*
γ-*Rhodomycinon*	γ-*Iso-rhodomycinon*	ζ-*Pyrromycinon*	*7-Desoxy-aklavinon*
δ-*Rhodomycinon*		η-*Pyrromycinon*	
ε-*Rhodomycinon*	ε-*Iso-rhodomycinon*		
ζ-*Rhodomycinon*	ζ-*Iso-rhodomycinon*		

Anthracycline.

Glykoside des β-Rhodomycinons	Glykosid des β-Iso-rhodomycinons	Glykoside des ε-Pyrromycinons	Glykoside des Aklavinons	Glykoside des γ-Rhodomycinons
Rhodomycin A	*Iso-rhodomycin A*	*Pyrromycin*	*Aklavine*	γ-*Rhodomycin I*
Rhodomycin B		*Cinerubin A*		γ-*Rhodomycin II*
		Cinerubin B		γ-*Rhodomycin III*
		Rutilantine		γ-*Rhodomycin IV*

angaben nicht hervor. *Tabelle 1* zeigt die bisher bekannten Anthracyclinone und Anthracycline.

Die Anthracycline sind basische Glykoside, in denen ein Mol eines Anthracyclinons mit einem oder zwei Mol der Tridesoxy-dimethylamino-hexose Rhodosamin (LXXVIII, S. 171) verknüpft ist. Manche enthalten außerdem noch stickstoff-freie Zucker, die sehr leicht durch Säure abgespalten werden. Bei der Isolierung solcher Anthracycline muß daher auf jede Anwendung von Säure verzichtet werden.

Die Anthracycline, ausgenommen solche, die nur ein Mol Rhodosamin enthalten, lösen sich leicht in Wasser. Alle sind amphoter: basisch durch die Dimethylaminogruppe des Rhodosamins, schwach sauer durch die phenolischen Hydroxygruppen des Anthracyclinons. Ihre basische Eigenschaft hat man benutzt, um sie als Salze von den Anthracyclinonen abzutrennen, wobei zur Salzbildung auch Kationenaustauscher, z. B. Lewatit KSN (*20*) oder Carboxymethylcellulose (*54*), dienten.

Die Isolierung der Anthracyclinone und Anthracycline ist einfach, wenn ein Mikroorganismen-Stamm neben wenigen anderen überwiegend nur ein Anthracyclinon und dessen Glykoside bildet. Das ist bei den Pyrromycinonen und ihren Glykosiden der Fall; ferner beim Aklavinon und seinen Glykosiden.

1. Gewinnung der ε-Pyrromycinon-glykoside und Pyrromycinone.

Cinerubin A und B.

Cinerubin A und B haben PRELOG, KELLER-SCHIERLEIN und Mitarb. (*40*) aus Kulturen von *Streptomyces antibioticus* WAKSMAN et HENRICI isoliert.

Den Verdampfungsrückstand vom Acetonauszug des Mycels vereinigte man mit dem Äthylacetatauszug der Kulturlösung und entzog dieser Lösung die Cinerubine mit verd. Essigsäure. Aus der auf pH 9,0 gebrachten, wäßrigen Phase wurden die Cinerubine mit Chloroform ausgeschüttelt und aus der eingeengten Chloroformlösung mit Petroläther gefällt. Ihre Trennung gelang durch fraktionierte Gegenstromverteilung im System Tetrachlorkohlenstoff/Methanol/Wasser 6,7 : 5,7 : 1.

Bequemer ist die Chromatographie aus Chloroform/Aceton an feinem, mit Borsäure aktiviertem Kieselgel [BROCKMANN jr. (*36*)], bei der sich beigemengte Pyrromycinone sowie Pyrromycin gut abtrennen. Reihenfolge der Zonen von oben nach unten: 1. Pyrromycin; 2. Cinerubin A; 3. Cinerubin B; 4. Pyrromycinone.

Papierchromatographisch trennen sich die Cinerubine im System: 20%ige Lösung von Formamid in Aceton (stationäre Phase)/Benzol-Cyclohexan (2 : 1) (mobile Phase) (*40*).

Pyrromycin und Pyrromycinone.

Das folgende *Schema 1* zeigt die Isolierung des Pyrromycins und der Pyrromycinone aus *Streptomyces DOA 1205* [BROCKMANN und LENK (*22*)]. Nach Extraktion mit Aceton enthielt das Mycel noch

Literaturverzeichnis: SS. 179—182.

beträchtliche Farbstoffmengen, die sich mit salzsäurehaltigem Aceton ausziehen ließen. Dabei und bei der gemeinsamen Verarbeitung von saurem und neutralem Acetonauszug wurden auch die von diesem Stamm gebildeten Cinerubine zu Pyrromycin abgebaut.

Schema 1. Isolierung von Pyrromycin und Pyrromycinonen (*22*).

Kulturlösung	*Mycel*
Butanolauszug verdampft, Rückstand verteilt zwischen Chloroform und Wasser. Rückstand der Chloroformlösung: *Fraktion B*	Extrahiert mit Aceton und Aceton + 0,5% HCl. Neutralisierter Extrakt verdampft: *Fraktion M*

Fraktion B + Fraktion M

mit Petroläther ausgezogen:

Rückstand (Fraktion P) erschöpfend mit Äther extrahiert:	*Lösung* Fette, wenig Farbstoffe
Ätherauszug (Pyrromycinone) Trennung der Pyrromycinone durch Chromatographie an Kieselgel:	*Ätherunlösliche Fraktion X* Enthält u. a. das wasserlösliche *Pyrromycin*

Zone 1: ε-Pyrromycinon
Zone 2: ζ-Pyrromycinon
Zone 3: η-Pyrromycinon

Isolierung von Pyrromycin-hydrochlorid (*23*).

Fraktion X erschöpfend mit Chloroform extrahiert:

Rückstand Enthält Verunreinigungen	*Lösung* Im Vakuum verdampft, Rückstand mit Wasser extrahiert
Rückstand Pyrromycinone	*Lösung* Im Vakuum verdampft, Rückstand mit Aceton extrahiert. Aus Acetonextrakt krist. *Pyrromycin-hydrochlorid*

2. Gewinnung der Rhodomycinone, Iso-rhodomycinone und ihrer Glykoside.

Die meisten *Str. purpurascens*-Stämme synthetisieren mehrere Rhodomycinone, Iso-rhodomycinone und ihre Glykoside nebeneinander, deren Mengenverhältnis recht unterschiedlich sein kann. Ein bestimmtes Rhodomycinon bzw. Iso-rhodomycinon oder dessen Glykosid zu gewinnen, kann daher weniger eine präparative als eine mikrobiologische Aufgabe sein, deren Lösung davon abhängt, ob man einen Stamm findet, der die gewünschte Verbindung in ausreichendem Umfange erzeugt [vgl. dazu FROMMER (*41, 42*)].

Ebenso wie die Pyrromycinon bildenden *Streptomyces*-Arten gibt auch *Str. purpurascens* seine Rhodomycine und Iso-rhodomycine in begrenztem Umfang an die Kulturlösung ab, die Hauptmenge bleibt meistens im Mycel. Ein Teil davon läßt sich mit Aceton herauslösen, ein anderer erst, wenn das Aceton 1—3% Säure enthält, und der Rest mitunter nur mit heißem, säurehaltigem Aceton. Durch die Säure werden die stickstoff-freien Zucker der Glykoside und zum Teil auch ihr Rhodosamin abgespalten, ein Verlust, der belanglos ist, so lange es allein auf die Gewinnung der Aglykone oder ihrer nur Rhodosamin enthaltenden Glykoside ankommt. Eine unter diesem Gesichtspunkt vorgenommene Aufarbeitung von Mycel zeigt das folgende *Schema 2 (20)*.

Schema 2. Abtrennung der Rhodomycinone und Iso-rhodomycinone aus dem Mycel von *Str. purpurascens.*

Extraktion des Mycels mit kaltem, 2% Schwefelsäure enthaltendem 80%igem Aceton.

Acetonauszug		*Mycel*
Im Vakuum eingeengt, wäßriger Rückstand auf pH 3,0 gebracht und mit Äther geschüttelt		Mit Aceton-konz. Salzsäure (9 : 1) heiß extrahiert
		Auszug
		Mit Ammoniak neutralisiert, im Vakuum eingeengt, wäßriger Rückstand auf pH 3,0 eingestellt und mit Äther extrahiert
Ätherphase	*Wasserphase*	
Mit n-NaOH ausgeschüttelt, Lauge angesäuert: Rhodomycinone und Iso-rhodomycinone	Enthält die Rhodomycine und Iso-rhodomycine	*Ätherauszug*
		enthält Hauptteil der Rhodomycinone und Iso-rhodomycinone

Trennung von Rhodomycinon/Iso-rhodomycinon-Gemischen.

Die Rhodomycinone lassen sich im Ring-Papierchromatogramm im System Benzol-Formamid trennen [BROCKMANN und PATT (27)]. Ihre Zonen bezeichnet man in der Reihenfolge der R_F-Werte alphabetisch mit kleinen griechischen Buchstaben, die gleichzeitig zur Kennzeichnung der in den einzelnen Zonen enthaltenen Rhodomycinone dienen (Tabelle 1, S. 123). In gleicher Weise werden die Iso-rhodomycinone getrennt und bezeichnet. Ihre R_F-Werte laufen denen der Rhodomycinone parallel. Jedem Iso-rhodomycinon entspricht ein Rhodomycinon mit gleichem R_F-Wert; oder anders ausgedrückt: Rhodomycinone und Iso-rhodomycinone mit gleichem griechischen Buchstaben bilden Paare, die papierchromatographisch in Benzol/Formamid nicht zu trennen sind.

Statt Benzol/Formamid kann man Tetralin/Eisessig/Wasser 10 : 10 : 1 verwenden. Eine Trennung der Rhodomycinon/Iso-rhodomycinon-Paare gelingt auch in diesem System nicht; man erkennt die Anwesenheit von Iso-rhodomycinonen lediglich an einem schmalen, karmoisinroten, inneren Rand der Zonen.

Das gleiche Bild wie im Papierchromatogramm zeigen Rhodomycinon/Iso-rhodomycinon-Gemische bei der Adsorption aus acetonhaltigem Benzol an Kieselgel-Säulen [BOLDT (2)] oder bei der Verteilungschromatographie an Cellulose-Säulen im System Benzol:Ligroin/Eisessig:Wasser 2:8/10:1 [BROCKMANN und BROCK-MANN jr. (11)]. Das zweite Verfahren arbeitet erheblich schneller und ist daher bequemer.

Literaturverzeichnis: SS. 179—182.

Rhodomycinon/Iso-rhodomycinon-Paare präparativ zu trennen, gelingt durch Chromatographie an Polyamidpulver (BASF) aus Methanol/7%igem wäßrigen Pyridin (*11*) [BROCKMANN jr. (*36*)], ein Verfahren, das auch für Dünnschicht-chromatogramme brauchbar ist.

Trennung von Rhodomycinen und Iso-rhodomycinen.

Die ersten Versuche zur Isolierung von Rhodomycinen wurden mit Präparaten durchgeführt, die unter Verwendung von Säure hergestellt waren und daher keine Glykoside mit stickstoff-freien (durch Säure leicht abspaltbaren) Zuckern enthielten. Durch fraktionierte Gegenstromverteilung in Butanol/m/15 Phosphatpuffer pH 6,0 ließen sie sich in zwei Fraktionen (A und B) zerlegen. Aus der A-Fraktion trennte man durch präparative Papierchromatographie (Papierpack) Rhodomycin A und Iso-rhodomycin A als kristallisierte Hydrochloride und Perchlorate ab; die B-Fraktion lieferte kristallisiertes Rhodomycin B-hydrochlorid [BROCKMANN und PATT (*27*)]. Zu dieser Zeit waren nur β-Rhodomycinon und β-Iso-rhodomycinon (damals noch ohne griechische Buchstaben bezeichnet), die Aglykone von Rhodomycin A und B bzw. Iso-rhodomycin, bekannt.

Als sich später herausstellte, daß es mehrere Rhodomycinone und Iso-rhodomycinone gibt, von denen jedes mehrere Glykoside bilden kann, war vorauszusehen, daß Versuche zur Isolierung weiterer und komplizierterer Rhodomycine oder Iso-rhodomycine nur Erfolg haben würden, wenn dafür Stämme zur Verfügung stehen, die bevorzugt oder ausschließlich die Glykoside *eines* Rhodomycinons oder Iso-rhodomycinons erzeugen.

Der erste Stamm, der diese Bedingung erfüllt, war einer, der fast ausschließlich γ-Rhodomycinon-glykoside produzierte. Das unter sehr schonenden, jede Hydrolyse vermeidenden Bedingungen aus seinem Mycel und seiner Kulturlösung isolierte Rhodomycingemisch trennte sich bei Verteilungschromatographie an Cellulose-Säulen (Butanol/Phosphatpuffer pH 5,8) in vier Rhodomycine auf, die der Anzahl ihrer Zuckerreste entsprechend durch römische Ziffern gekennzeichnet werden [BROCKMANN und WAEHNELDT (*31*)].

III. Die Anthracyclinone.

1. Vorbemerkungen zur Struktur der Anthracyclinone.

Alle Anthracyclinone sind Derivate eines 7,8,9,10-Tetrahydro-tetracen-chinons-(5,12) (Formel I b), in dem ein zum Anthrachinon-Ringsystem β-ständiges C-Atom des alicyclischen Ringes *A* mit einer Äthyl- und Hydroxygruppe verknüpft ist. Diese Stammverbindung (I a) der Anthracyclinone muß der Nomenklatur-Vorschrift nach als 8-Hydroxy-8-äthyl-

7,8,9,10-tetrahydro-tetracenchinon-(5,12) bezeichnet werden. Sie ist dem entgegen in (Ia) und allen Formeln dieser Zusammenfassung als 9-Hydroxy-9-äthyl-derivat bezeichnet, weil in den meisten Originalarbeiten den Anthracyclinonformeln eine nach (Ia) geschriebene Stammverbindung zugrunde liegt (vgl. dazu S. 137).

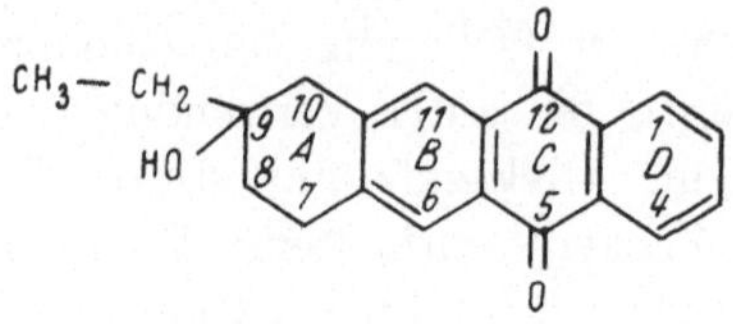

(Ia.) Stammverbindung der Anthracyclinone.

(Ib.) CH_3—CH_2, HO = H.

Die einzelnen Anthracyclinone unterscheiden sich durch verschiedenartige Substitution der Stammverbindung (Ia). Substituenten sind entweder nur Hydroxyle oder Hydroxyle und *eine* Carbomethoxygruppe. Träger von Hydroxygruppen können sein: a) in Ring A $C_{(7)}$ und $C_{(10)}$, b) im Anthrachinon-Ringsystem der Stammverbindung die den Chinoncarbonylen benachbarten C-Atome 1, 4, 6 und 11. Ist eine Carbomethoxygruppe vorhanden, steht sie in allen Fällen an $C_{(10)}$.

Die in Tabelle 1 (S. 123) angeführten Gruppen der Anthracyclinone, die *Iso-rhodomycinone, Rhodomycinone, Pyrromycinone* und *Aklavinone*, unterscheiden sich durch die Zahl der α-Hydroxygruppen, die mit dem Anthrachinon-Ringsystem der Stammverbindung (Ia) verbunden sind. Die Iso-rhodomycinone enthalten vier α-Hydroxygruppen, die Rhodomycinone und Pyrromycinone drei und die Aklavinone zwei.

Für vier α-Hydroxyle ist in (Ia) nur *eine* Anordnung möglich. Drei α-Hydroxyle dagegen können vier verschiedene Stellungen einnehmen, von denen die in (IVa) bzw. (VIa) wiedergegebenen in den Rhodomycinonen (ausgenommen δ-Rhodomycinon) bzw. Pyrromycinonen vorliegen. Für zwei α-Hydroxyle gibt es sechs verschiedene Stellungen. Nur eine dieser Möglichkeiten, die der Formel (VIIIa), ist in den beiden Aklavinonen realisiert.

So wie (Ia) als Stammverbindung aller Anthracyclinone gelten darf, können (IIb), (IVb), (VIb) und (VIIIb) als Stammverbindungen der verschiedenen Anthracyclinongruppen aufgefaßt werden. Von ihnen leiten sich die einzelnen Vertreter jeder Gruppe durch unterschiedliche Substitution an $C_{(7)}$ und $C_{(10)}$ ab.

Der Hydroxyanthrachinon-Teil des Moleküls ist der Chromophor der Anthracyclinone. Zahl und Stellung seiner Hydroxygruppen haben entscheidenden Einfluß auf das Absorptionsspektrum und damit auf die Farbe der Anthracyclinone in organischen Solvenzien, konz. Schwefelsäure und wäßrigem Alkali.

Literaturverzeichnis: SS. 179—182.

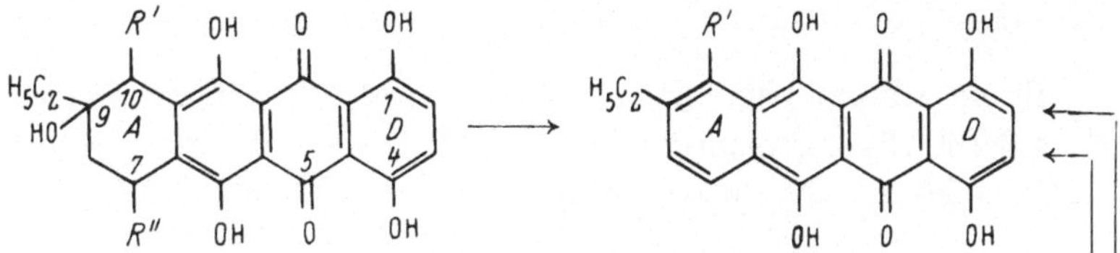

(IIa.) Iso-rhodomycinone.
(IIb.) R', $R'' = $ H.
(IIc.) R', R'', H_5C_2, OH an C(9) = H.

(IIIa.)
(IIIb.) H_5C_2, $R' = $ H.

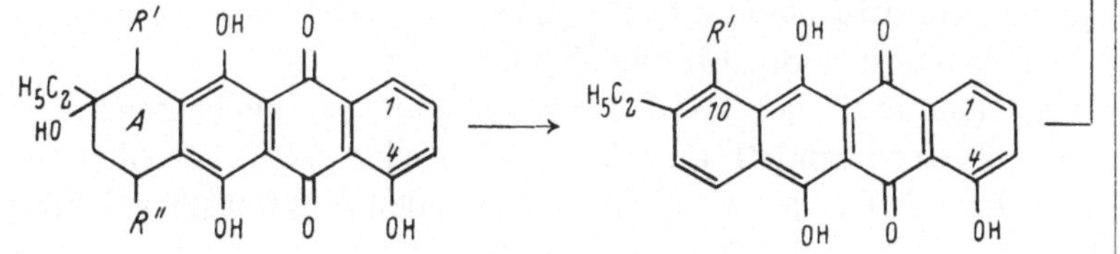

(IVa.) Rhodomycinone.
(IVb.) R', $R'' = $ H.
(IVc.) R', R'', H_5C_2, OH an C(9) = H.

(Va.)
(Vb.) H_5C_2, $R' = $ H.

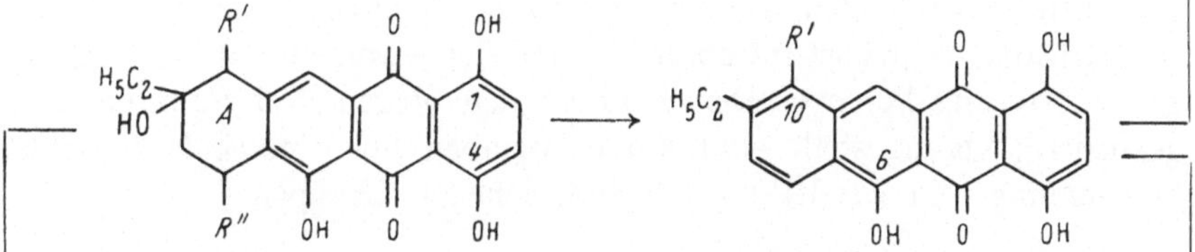

(VIa.) Pyrromycinone.
(VIb.) R', $R'' = $ H.
(VIc.) R', R'', H_5C_2, OH an C(9) = H.

(VIIa.)
(VIIb.) H_5C_2, $R' = $ H.

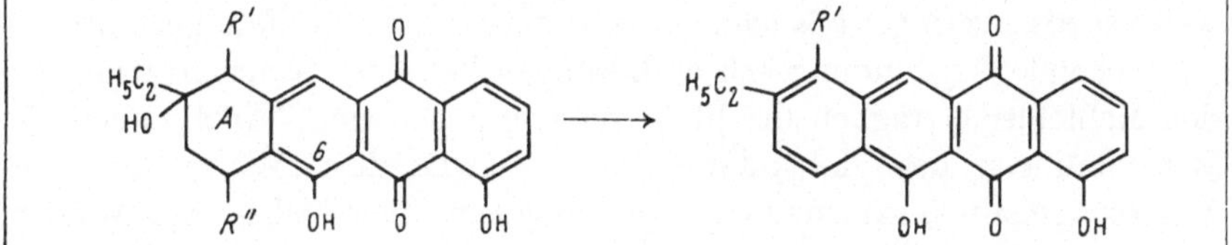

(VIIIa.) Aklavinon.
(VIIIb.) R', $R'' = $ H.
(VIIIc.) R', R'', H_5C_2, OH an C(9) = H.

(IXa.)
(IXb.) H_5C_2, $R' = $ H.

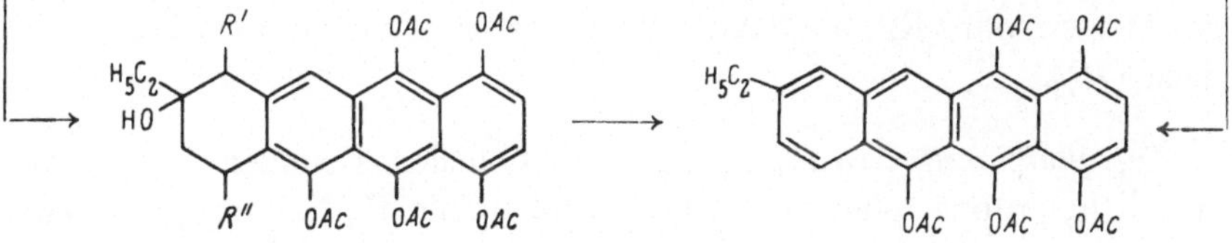

(X.) $Ac = CH_3 \cdot CO$.

(XI.)

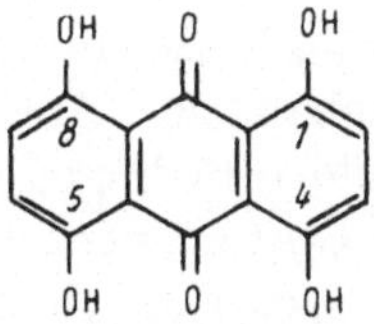

(XII.)

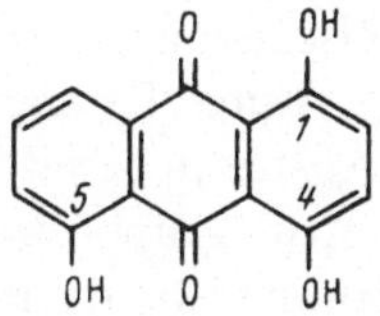

(XIII.)

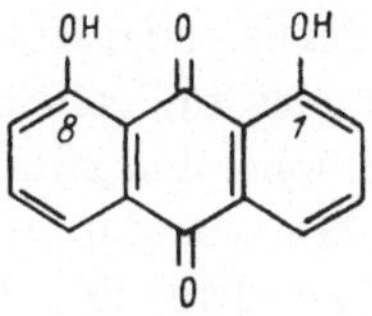

(XIV.)

Formelübersicht 1.

Fortschritte d. Chem. org. Naturst. XXI.

Ihrem Chromophor nach sind die Anthracyclinone Derivate von 1,4,5,8-Tetrahydroxy-anthrachinon (XII) ,1,4,5-Trihydroxy-anthrachinon (XIII) oder 1,8-Dihydroxy-anthrachinon (Chrysazin) (XIV).

2. Zur Konstitutionsermittlung der Anthracyclinone.

Wie aus der kurzen Übersicht des vorigen Abschnittes hervorgeht, ist zur Konstitutionsermittlung eines Anthracyclinons erforderlich:

1. Aufklärung des Chromophors.

2. Nachweis, daß an den Chromophor ein alicyclischer Sechsring linear, entsprechend (IIa), (IVa), (VIa) oder (VIIIa) anelliert ist.

3. Ermittlung der zu Ring A gehörenden funktionellen Gruppen und ihrer Stellung.

Soweit die Verfahren zur Lösung dieser Aufgaben bei den verschiedenen Anthracyclinonen die gleichen waren, werden sie, um Wiederholungen zu vermeiden, vorweg in diesem Abschnitt besprochen, bevor im nächsten die Konstitutionsermittlung der einzelnen Anthracyclinone zur Erörterung kommt. Der Stoff ist den obigen drei Punkten nach gegliedert; vorangestellt sind kurze Angaben über Molekulargewichtsbestimmung und Ermittlung der funktionellen Gruppen.

Bestimmung des Molekulargewichtes. Wie bei vielen Naturstoffen versagen auch bei den Anthracyclinonen die klassischen Methoden der Molekulargewichtsbestimmung oder liefern Werte mit zu großen Fehlern, weil wegen zu geringer Löslichkeit die Voraussetzung für diese Verfahren — ausreichende Verdünnung echter Lösungen bei einer Konzentration, die den Meßfehler erträglich macht — nicht gegeben sind. Auch chemische Bestimmungen des Molekulargewichtes, z. B. durch Ermittlung des Methoxylgehaltes, sind nicht genau genug, um über Mehr- oder Mindergehalt an einer CH_2-Gruppe sicher entscheiden zu können. Das Verfahren der Wahl ist die Massenspektroskopie, die erst zum Einsatz kommen konnte, als die Aufklärung der Anthracyclinone nahezu abgeschlossen war [REED, OLLIS, BROCKMANN jr. unveröffentlicht; vgl. dazu (*36*)].

Die funktionellen Gruppen. An funktionellen Gruppen liegen vor: eine Äthylgruppe, phenolische und alkoholische Hydroxygruppen, zwei Chinon-carbonyle und in den meisten Anthracyclinonen eine Carbomethoxygruppe. Aus der Äthylgruppe wird bei der Kuhn-Roth-Oxydation etwa 0,9 Mol eines Gemisches aus Propionsäure und Essigsäure (*39*). Propionsäure allein entsteht beim Permanganat-Abbau der Anthracyclinone, und zwar in guter Ausbeute bereits unter milden Bedingungen; ein Hinweis darauf, daß die Äthylgruppe mit einem C-Atom verknüpft ist, an dem eine Hydroxygruppe steht. Im KMR-Spektrum (S. 155) erkennt man die Äthylgruppe an zwei Signalen mit den relativen Intensitäten 3 bzw. 2 und den τ-Werten 8,8—8,9 bzw. 8,1—8,4.

Kurze Einwirkung von Acetanhydrid-Pyridin verwandelt die Anthracyclinone in gelbe Acetate mit einer OH-Bande bei 3640—3500 cm^{-1} (2,75—2,85 μ) (Tetrachlorkohlenstoff, Nujol), die eine schwer acetylierbare und demnach tertiäre Hydroxygruppe anzeigt.

Aus der starken Farbaufhellung beim Acetylieren ergibt sich die Anwesenheit von Hydroxygruppen, die Chinon-sauerstoffatomen benachbart und daher bathochrom sind; desgleichen aus der starken bathochromen Verschiebung der Absorptionsbanden und der damit verbundenen Farbvertiefung beim Verestern der Anthracyclinone mit Acetborsäure (*40, 20, 22*).

Eine Methoxygruppe, bei deren Verseifung eine Carboxygruppe frei wird, sowie eine Carbonylbande bei 1740 cm^{-1} (5,75 μ) zeigt das Vorliegen einer Carbomethoxygruppe.

Sind beide Chinon-sauerstoffatome mit α-Hydroxylen durch Wasserstoffbrücken verbunden, so liegt ihre Bande im IR-Spektrum bei 1600 cm^{-1} (6,25 μ). Die Bande verschiebt sich nach 1680 cm^{-1} (5,95 μ), wenn die Wasserstoffbrücken durch Acetylierung der Hydroxyle aufgehoben werden. Im Aklavinon hat die nicht chelierte Chinon-carbonylgruppe eine Bande bei 1675 cm^{-1} (5,98 μ) (Nujol) und die chelierte eine bei 1623 cm^{-1} (6,16 μ) (*43*).

Die Aufklärung des Chromophors.

Ob eine Verbindung ein Hydroxy-anthrachinon bzw. dessen Derivat ist, läßt sich durch reduzierende Acetylierung und spektroskopische Untersuchung des Reduktionsproduktes feststellen. Denn Hydroxyanthrachinone und ihre Derivate gehen bei reduzierender Acetylierung in Acetoxy-anthracene über, die das gleiche charakteristische UV-Spektrum haben wie Anthracen [BROCKMANN und BUDDE (*18*)]. Die Anwendung dieses Verfahrens auf die Anthracyclinone, z. B. (VIa) → (X) [PRELOG, KELLER-SCHIERLEIN und Mitarb. (*40*)], hat gezeigt, daß sie Anthrachinonderivate sind.

Ob ein Anthracyclinon ein Derivat von (XII), (XIII) oder (XIV) ist, ergibt der spektroskopische Vergleich mit diesen Verbindungen, die sich in ihren Absorptionskurven und daher auch in der Farbe ihrer Lösungen charakteristisch unterscheiden.

(XII) löst sich in organischen Solvenzien karmoisinrot, in konz. Schwefelsäure und wäßrigem Alkalihydroxyd rein blau. Die Ähnlichkeit seines in Cyclohexan besonders charakteristischen Absorptionsspektrums mit dem der Iso-rhodomycinone, die *Abb. 1* am Beispiel des ε-Iso-rhodomycinons (*8*) demonstriert, beweist, daß die Iso-rhodomycinone Derivate von (XII) sind, oder anders ausgedrückt, daß (XII) ihr Chromophor ist.

Die Farbe von (XIII) ist in organischen Solvenzien rotgelb, in konz. Schwefelsäure und wäßrigem Alkalihydroxyd violett. Die Ähnlichkeit seines Spektrums mit dem der Rhodomycinone und Pyrromycinone (ausgenommen η-Pyrromycinon) —

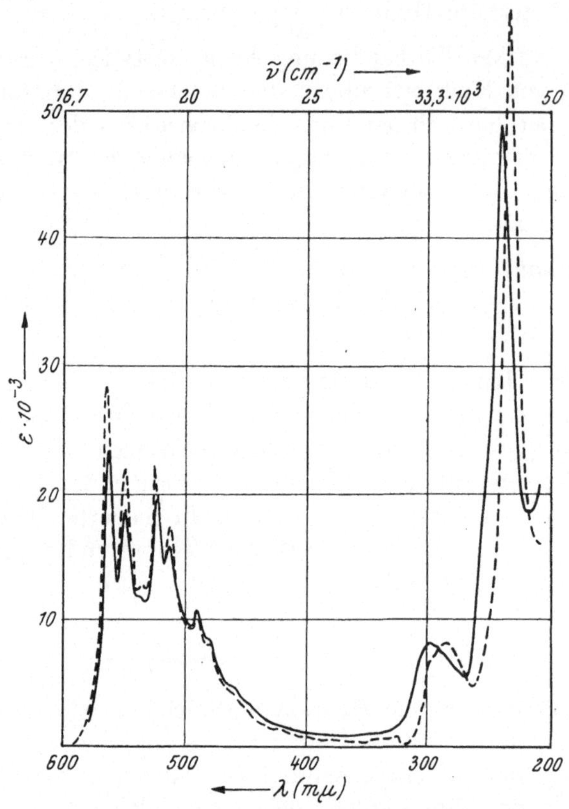

Abb. 1. Absorptionskurven von ε-Iso-rhodomycinon ——————— (λ_{max} 563, 549, 524, 513, 490, 299, 242 mμ) und 1,4,5,8-Tetrahydroxy-anthrachinon (XII) ——————— (λ_{max} 563, 548, 524, 513, 490, 285, 235 mμ) in Cyclohexan. [Aus: Chem. Ber. **94**, 2174 (1961).] In Cyclohexan sind beide Verbindungen so schwer löslich, daß gesättigte Lösungen gemessen werden mußten. Aus ihnen können sich kleine Mengen Farbstoff kolloidal ausscheiden und dadurch die Extinktionswerte fälschen. Dieser Nachteil wurde in Kauf genommen, weil die Kurven in Cyclohexan besonders charakteristisch sind und das für den Vergleich wichtiger war als genaue Extinktionswerte.

in *Abb. 2* am Beispiel des ε-Rhodomycinons (*11*) gezeigt — beweist, daß die Rhodomycinone und Pyrromycinone (ausgenommen η-Pyrromycinon) Derivate von (XIII) sind bzw. (XIII) ihr Chromophor ist.

(XIV) löst sich in organischen Solvenzien gelb, in konz. Schwefelsäure und wäßrigem Alkalihydroxyd rot. Im sichtbaren Gebiet ist sein Absorptionsspektrum weniger charakteristisch als das von (XII) oder (XIII). Dennoch läßt sich, wenn man das UV-Gebiet mit heranzieht, durch Vergleich von Aklavinon mit (XIV) beweisen, daß dieses der Aklavinon-Chromophor ist (*43*).

Literaturverzeichnis: SS. 179—182.

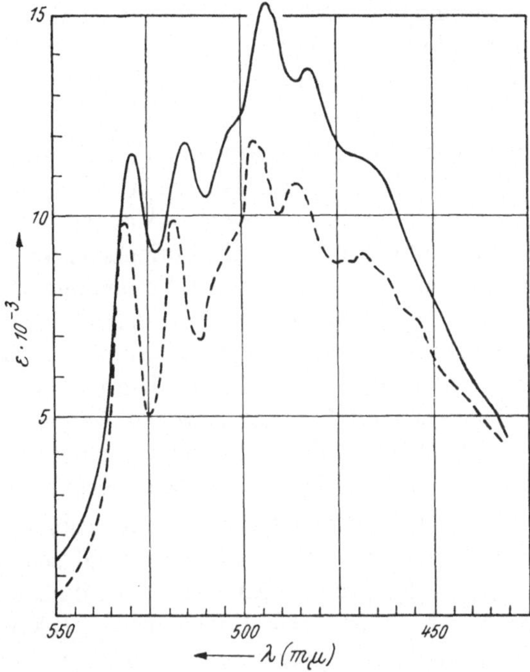

Abb. 2. Absorptionskurven in Cyclohexan von ε-Rhodomycinon ——————— (λ_{max} 529, 515, 493, 483 mμ) und 1,4,5-Trihydroxy-anthrachinon (XIII) ——————— (λ_{max} 530, 517, 495, 483 mμ). [Aus: Chem. Ber. **94**, 2681 (1961).]

Die Anellierung des alicyclischen Ringes.

Daß dem Hydroxyanthrachinon-Chromophor der Anthracyclinone ein alicyclischer Sechsring linear anelliert ist, ergibt sich aus dessen Aromatisierung, die, wie unten gezeigt, durch Abspaltung von zwei Molekülen Wasser bzw. einem Molekül Wasser und anschließende Dehydrierung leicht zu erreichen ist. Bei dieser Aromatisierung entstehen nämlich in guter Ausbeute Verbindungen, die sich a) durch das oben erwähnte Verfahren von BROCKMANN und BUDDE (*18*), b) durch das Absorptionsspektrum ihrer Acetate [BROCKMANN und MÜLLER (*25*)] und c) durch die rote Küpe ihrer gelben Acetate [BROCKMANN, BOLDT und NIEMEYER (*9*)] als Derivate von Hydroxy-tetracenchinonen identifizieren lassen. Zum gleichen Ergebnis führt die Zinkstaubdestillation der Anthracyclinone, die Sublimate mit dem charakteristischen Absorptionsspektrum des Tetracens liefert (*20*).

Aus den symmetrisch gebauten Chromophoren (XII) bzw. (XIV) der Iso-rhodomycinone und Aklavinone kann durch lineare Anellierung eines alicyclischen Sechsringes nur *ein* Hydroxy-7,8,9,10-tetrahydrotetracenchinon [(IIc) bzw. (VIIIc)] entstehen. Anellierung an den un-symmetrischen Chromophor (XIII) der Pyrromycinone und Rhodomycinone dagegen führt, je nachdem, auf welcher Seite anelliert wird, zu *zwei* Hydroxy-7,8,9,10-tetrahydro-tetracenchinonen, nämlich (IVc) und (VIc).

Aromatisierung von Ring A verwandelt (IVc) in das Trihydroxy-tetracenchinon (Vb), während (VIc) bei dieser Reaktion in das Trihydroxy-tetracenchinon (VIIb) übergeht. (Vb) und (VIIb) unterscheiden sich charakteristisch in ihren Absorptionsspektren [Brockmann und Müller (*25*)]. Um zu sehen, ob eine Verbindung (IVc) oder ein Derivat davon bzw. (VIc) oder ein Derivat davon ist, braucht man nur ihren Ring A zu aromatisieren und zu prüfen, ob das Spektrum des entstandenen Trihydroxy-tetracenchinons dem von (Vb) oder (VIIb) ähnlich ist. Mit diesem Verfahren ist bewiesen worden, daß sich die Pyrromycinone von (VIc) und die Rhodomycinone (ausgenommen δ-Rhodomycinon) von (IVc) ableiten [Brockmann und Mitarb. (*3*)].

Das aus Aklavinon durch Aromatisierung seines Ringes A gewonnene Hydroxy-tetracenchinon-derivat hat man auf Grund seines Spektrums als Derivat von (IXb) formuliert [Ollis und Sutherland (*43*)]; der spektroskopische Vergleich mit (IXb) fehlt noch. (IXb) ist erst in jüngster Zeit zugänglich geworden [Brockmann und Brockmann jr. (*16*)].

Die Konstitution der aus Iso-rhodomycinonen entstehenden Aromatisierungsprodukte (IIIa) ($R' = COOCH_3$ oder OH) ist durch Vergleich mit (IIIb) [Brockmann und Müller (*25*)] und Derivaten von (IIIb) [Brockmann und Wimmer (*35*)] gesichert.

Die Substituenten an Ring A.

Die Substituenten, die, wie im vorhergehenden Abschnitt gezeigt, die Hydroxy-7,8,9,10-tetrahydro-tetracenchinone (IIc), (IVc), (VIc) und (VIIIc) zu Anthracyclinonen machen, sind in den Kombinationen der *Tabelle 2* vertreten.

Aussagen über die Stellung der Substituenten ergeben sich aus folgenden Befunden und Überlegungen: Fehlt die Carbomethoxygruppe, so können drei oder zwei alkoholische Hydroxygruppen vorhanden sein. In beiden Fällen lassen sich durch Säure zwei in Form von Wasser abspalten, wobei aus dem Anthracyclinon (Anthrachinon-derivat) ein Tetracenchinon-derivat wird. Sind *drei* alkoholische Hydroxyle anwesend, so enthält Ring A nach der Aromatisierung *eine* phenolische Hydroxygruppe. Demnach gehören alle nichtphenolischen Hydroxyle des Ausgangsmaterials zu Ring A.

Von den drei bzw. zwei alkoholischen Hydroxygruppen ist eine schwer acetylierbar und somit tertiär. Sie steht daher am gleichen

Tabelle 2. Substituenten-Kombinationen bei Anthracyclinonen.

Anthracyclinon	CH_3—CH_2—	CH_3OOC—	HO-alkohol.	HO-alkohol., als H_2O abspaltbar
IIa: R', $R'' =$ HO	1	—	3	2
IIa: $R' =$ HO $R'' =$ H	1	—	2	2
IIa: $R' =$ CH_3OOC $R'' =$ HO	1	1	2	2
IIa: $R' =$ CH_3OOC $R'' =$ H	1	1	1	1
IVa, VIa: $R' =$ CH_3OOC $R'' =$ HO	1	1	2	2
IVa, VIa: $R' =$ CH_3OOC $R'' =$ H	1	1	1	1
VIIIa: $R' =$ CH_3OOC $R'' =$ HO	1	1	2	2
VIIIa: $R' =$ CH_3OOC $R'' =$ H	1	1	1	

C-Atom wie die Äthylgruppe, wofür auch spricht, daß bei der Permanganat-Oxydation des betreffenden Anthracyclinons leicht und in guter Ausbeute Propionsäure entsteht (*20*, *22*). Die Äthylgruppe gehört also zu Ring *A*.

Liegt eine Carbomethoxygruppe und *eine* alkoholische Hydroxygruppe vor, so ist diese schwer acetylierbar und demnach tertiär. Durch Erwärmen mit Säure wird sie in Form von Wasser eliminiert. Aus dem Reaktionsprodukt entsteht beim Dehydrieren ein Hydroxy-tetracenchinon. Hydroxy- und Äthylgruppe stehen demnach an Ring *A* und — da aus der Äthylgruppe durch Oxydation leicht Propionsäure wird — am gleichen C-Atom.

Sind außer einer Carbomethoxygruppe *zwei* Hydroxygruppen anwesend, so werden *beide* beim Erwärmen mit Säure unter Aromatisierung aus Ring *A* in Form von Wasser eliminiert. Beide gehören demnach zu Ring *A*. Eine von ihnen, die acetylierbare, wird durch katalytische Hydrierung in alkalischem Medium, z. B. Äthanol/Triäthanolamin, entfernt, was sich spektroskopisch dadurch zu erkennen gibt, daß die langwelligen Absorptionsmaxima des Hydrierungsproduktes in Piperidin (nicht aber in neutralen Solvenzien) um 7—8 mμ kürzerwellig liegen als die des Ausgangsmaterials (*8*, *12*). Eine derartige Verschiebung ist nur verständlich, wenn die reduktiv entfernbare Hydroxygruppe an einem dem aromatischen Ringsystem benachbarten C-Atom steht.

Zusammenfassend ergibt sich aus den vorstehenden Befunden:
a) Alle in Tabelle 2 angeführten Substituenten gehören zu Ring *A*.
b) Äthyl- und tertiäre Hydroxygruppe stehen am gleichen C-Atom und
c) eines der dem Anthrachinon-System benachbarten C-Atome kann eine durch katalytische Hydrierung entfernbare Hydroxygruppe tragen.

Dagegen sagen die Befunde nichts darüber aus, mit welchen C-Atomen des Ringes A die Äthyl- und Carbomethoxygruppe verbunden sind, und erst recht nichts darüber, ob sie bei allen Anthracyclinonen die gleiche Stellung einnehmen.

Hätten Äthyl- und Carbomethoxygruppe in allen Anthracyclinonen, bezogen auf das Anthrachinon-Ringsystem, die gleiche Anordnung, so würden sich die Hydroxy-tetracenchinon-derivate, die durch Aromatisierung von Ring A aus den Anthracyclinonen mit $R' = CH_3OOC$ entstehen, nur dadurch unterscheiden, daß die α-C-Atome des Anthrachinon-Ringsystems mit vier, drei oder zwei Hydroxygruppen besetzt sind (IIIa, Va, VIIa, IXa); Unterschiede, die verschwinden, wenn man bei denen mit weniger als vier α-Hydroxygruppen die noch nicht substituierten α-C-Atome hydroxylieren würde. Eine derartige α-Hydroxylierung ist bei den aus Pyrromycinonen und Rhodomycinonen ($R' = CH_3OOC$) durch Aromatisierung ihres Ringes A entstehenden Derivaten von (Vb) bzw. (VIIb) gelungen [Brockmann und Lenk (22); Brockmann und Brockmann jr. (12)]. Die dabei erhaltenen Derivate von (IIIb) waren identisch mit dem aus Iso-rhodomycinonen ($R' = CH_3OOC$) gewonnenen Hydroxy-tetracenchinon-derivat (IIIa).

Damit war bewiesen, daß in den Iso-rhodomycinonen (mit $R' =$ $= CH_3OOC$), Rhodomycinonen (mit $R' = CH_3OOC$) und Pyrromycinonen die Äthyl- und Carbomethoxygruppe in Ring A, bezogen auf das Anthrachinon-Ringsystem, die gleiche Stellung haben — ein Ergebnis, das die nahe Verwandtschaft der drei Verbindungsklassen aufgedeckt und zum anderen ihre Strukturaufklärung erleichtert hat. Denn nunmehr genügte es, die Stellung der Äthyl- und Carbomethoxygruppe in *einem* der drei Hydroxy-tetracenchinone (IIIa), (Va) oder (VIIa) zu ermitteln, um zugleich ihre Stellung in den beiden anderen und damit auch in den zugehörigen, eine Carbomethoxygruppe enthaltenden Anthracyclinonen (IIa), (IVa) und (VIa) ($R' = CH_3OOC$) zu beweisen.

Diese Aufgabe ist zuerst für die Äthylgruppe gelöst worden, und zwar nicht bei einem der Aromatisierungsprodukte (IIIa), (Va), (VIIa), (IXa), sondern bei einem Anthracyclinon selber. Wie im nächsten Abschnitt gezeigt, ließ sich durch oxydativen Abbau eines aus ε-Iso-rhodomycinon erhaltenen Reduktionsproduktes beweisen, daß die Äthylgruppe des ε-Iso-rhodomycinons mit einem zum Chromophor β-ständigen C-Atom des Ringes A verbunden ist; womit diese Stellung nach dem vorstehend Gesagten zugleich auch für die Rhodomycinone mit Carbomethoxygruppe und die Pyrromycinone bewiesen war.

Den Nachweis, daß die Carbomethoxygruppe der Äthylgruppe *und* dem Ringsystem benachbart an einem α-C-Atom des Ringes A steht, brachte, wie ebenfalls im nächsten Abschnitt geschildert, der oxydative Abbau des aus den Pyrromycinonen durch Aromatisierung von Ring A

entstehenden Hydroxy-tetracenchinons (VIIa). Denn dabei wurde Benzoltetracarbonsäure-1,2,3,4 (XLVI, S. 149) gefaßt.

Damit war bei den Pyrromycinonen und — dank ihrer oben geschilderten strukturellen Verknüpfung mit den Rhodomycinonen und Iso-rhodomycinonen (VIIa → IIIa sowie Va → IIIa) — auch bei den eine Carbomethoxygruppe enthaltenden Rhodomycinonen und Iso-rhodomycinonen die Stellung der Äthyl- und Carbomethoxygruppe in bezug auf das Anthrachinon-Ringsystem bekannt. Bezogen auf die Hydroxygruppen des Anthrachinon-Ringsystems dagegen blieben bei den Rhodomycinonen und Pyrromycinonen zwei Möglichkeiten offen: a) Stellung wie in (IVa) bzw. (VIa), oder b) Äthylgruppe an $C_{(8)}$ und Carbomethoxygruppe an $C_{(7)}$; eine Isomerie, die bei den Iso-rhodomycinonen mit $R' = CH_3OOC$ entfällt, weil bei ihnen die α-Hydroxygruppen des Chromophors symmetrisch angeordnet sind. Bei den Pyrromycinonen hat sich eine eindeutige Entscheidung zwischen den beiden Formeln treffen lassen, bei den Rhodomycinonen noch nicht (vgl. SS. 141, 145).

Schreibweise und Bezifferung der Anthracyclinon-Formeln.

Im Zusammenhang mit der eben erwähnten Isomerie soll hier — ergänzend zu dem, was darüber in Abschn. III, 1 (S. 127) gesagt ist — die im vorstehenden und folgenden gewählte Schreibweise und Bezifferung der Anthracyclinon-Formeln begründet werden.

Als man fand, daß sich die Pyrromycinone vom 1,4,6-Trihydroxy-7,8,9,10-tetrahydro-tetracenchinon-(5,12) (VIc) ableiten, hat man den im Laufe der Strukturaufklärung benutzten Partialformeln der Pyrromycinone zunächst (VIc) zugrunde gelegt. Das war auch dann noch berechtigt, als man schon wußte, daß die Äthylgruppe in Ring A eine β-Stellung zum Chromophor einnimmt, denn für diese blieb noch zwischen $C_{(8)}$ und $C_{(9)}$ zu entscheiden. Als schließlich die Entscheidung zugunsten von $C_{(9)}$ der Formel (VIc) fiel, war die der Formel (Ia) bzw. (VIc) entsprechende Schreibweise in Veröffentlichungen von PRELOG und Mitarb. (*40*) sowie BROCKMANN und Mitarb. bereits verschiedentlich verwendet worden; desgleichen in einer gemeinsamen Arbeit dieser Autoren mit OLLIS, GORDON und SUTHERLAND (*17*).

Ferner haben BROCKMANN und Mitarb. auch bei den Rhodomycinonen und Iso-rhodomycinonen, um die Ähnlichkeit mit den Pyrromycinonen deutlich zu machen, die Schreibweise mit der Äthylgruppe an $C_{(9)}$ verwendet. Es besteht daher kein Grund, in dieser Zusammenfassung von der bisherigen Gepflogenheit abzugehen.

3. Konstitution der Anthracyclinone.

A. *Iso-rhodomycinone.*

ε-Iso-rhodomycinon.

ε-Iso-rhodomycinon, $C_{22}H_{20}O_{10}$ (XVa), dunkelrote Kristalle vom Schmp. 227—229° (Zers.), wurde von BROCKMANN und FRANCK (*20*) isoliert und von BROCKMANN und BOLDT (*6, 7*) aufgeklärt. Es verliert beim Kochen mit Jodwasserstoffsäure seine Carbomethoxygruppe und zwei alkoholische Hydroxyle, die alle drei durch Wasserstoff ersetzt werden. Das rote, kristallisierte Reduktionsprodukt $C_{20}H_{18}O_6$ gibt sich durch sein Absorptionsspektrum als Derivat des 1,4,5,8-Tetrahydroxy-anthrachinons (XII, S. 129) zu erkennen, liefert beim oxydativen Abbau β-Äthyl-adipinsäure (XVIII), hat demnach die Konstitution (XVI) und beweist damit, daß die Äthylgruppe des ε-Iso-rhodomycinons mit einem zum Chromophor (XII) β-ständigen C-Atom des Ringes A verknüpft ist.

Beim Erhitzen auf 250° oder in besserer Ausbeute bei kurzem Kochen mit Eisessig/Bromwasserstoffsäure verwandelt sich ε-Iso-rhodomycinon unter Abspaltung von 2 Mol. Wasser in kristallisiertes, rotes Bisanhydro-ε-iso-rhodomycinon, $C_{22}H_{16}O_8$, das durch sein Absorptionsspektrum eindeutig als Derivat des 1,4,6,11-Tetrahydroxy-tetracenchinons-(5,12) (XVIIb) (25) charakterisiert ist. Es erwies sich als identisch mit η-Iso-pyrromycinon, das man durch vorsichtige Oxydation mit Mangandioxid/Schwefelsäure aus η-Pyrromycinon erhält [Brockmann und Lenk (22)] und das, wie unten gezeigt, die Konstitution (XVIIa) hat. Damit war für Bisanhydro-ε-iso-rhodomycinon die Formel (XVIIa) bewiesen. Für ε-Iso-rhodomycinon wurde auf Grund dieses Ergebnisses sowie der Tatsache, daß eine seiner alkoholischen Hydroxygruppen tertiär und die andere durch Hydrierung unter Bildung einer in Piperidin kürzerwellig absorbierenden Desoxyverbindung $C_{22}H_{20}O_9$ abgespalten wird, die Konstitutionsformel (XVa) abgeleitet.

Daß die Absorptionsmaxima der Desoxy-Verbindung in Piperidin um 6—7 mμ kürzerwellig sind als die des ε-Iso-rhodomycinons, kann streng genommen nur als Beweis dafür gelten, daß die reduktiv abspaltbare Hydroxygruppe des ε-Iso-rhodomycinons α-ständig zum Anthrachinon-Ringsystem d. h. mit $C_{(7)}$ oder $C_{(10)}$ verknüpft ist. Die in (XVa) formulierte Stellung an $C_{(7)}$ ließ sich zunächst nur damit begründen, daß sie mit der Acetathypothese (vgl. S. 163) besser in Einklang steht. Bewiesen wird sie ebenso wie die Stellung der anderen Substituenten des Ringes A durch das KMR-Spektrum des ε-Iso-rhodomycinons (S. 156).

Die Absorptionsmaxima des Bisanhydro-ε-iso-rhodomycinons (XVIIa) sind in Cyclohexan um 4—5 mμ längerwellig als die von (XVIId). Diese Differenz ist der bathochromen Wirkung der Carbomethoxygruppe zuzuschreiben, denn die Absorptionskurve des Methylesters (XXIb), der aus (XIX) und (XX) durch Friedel-Crafts-Kondensation synthetisierten Säure (XXIa), ist in Cyclohexan bis auf 1—2 mμ deckungsgleich mit der von (XVIIa) [Brockmann und Wimmer (35)].

Die Carbomethoxygruppe kann sich, behindert durch das *peri*-ständige Hydroxyl an $C_{(11)}$, nicht coplanar zum Ringsystem einstellen. Auf Mesomerie mit dem Ringsystem ist ihre bathochrome Wirkung daher nicht zurückzuführen. Vielmehr dürfte eine durch die sterische Behinderung bewirkte Verdrillung des Ringes A dafür verantwortlich sein. Dafür spricht, daß die bathochrome Wirkung entfällt, wenn, wie im η-Pyrromycinon (XLa), die *peri*-ständige OH-Gruppe fehlt. Dieser Befund kann, einerlei, ob seine Deutung richtig ist, bei der Aufklärung

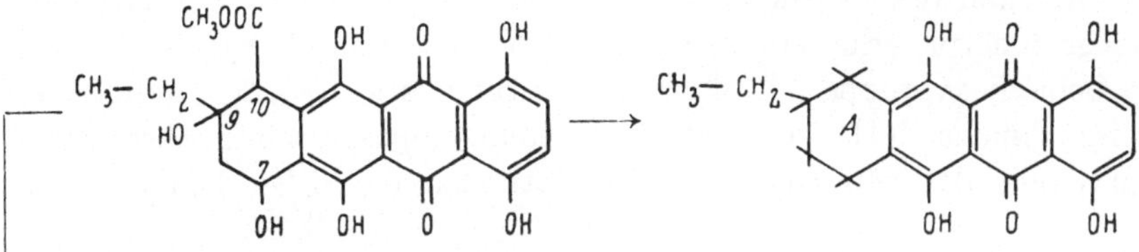

(XVa.) ε-Iso-rhodomycinon.
(XVb.) OH an C(7) = H.

(XVI.)

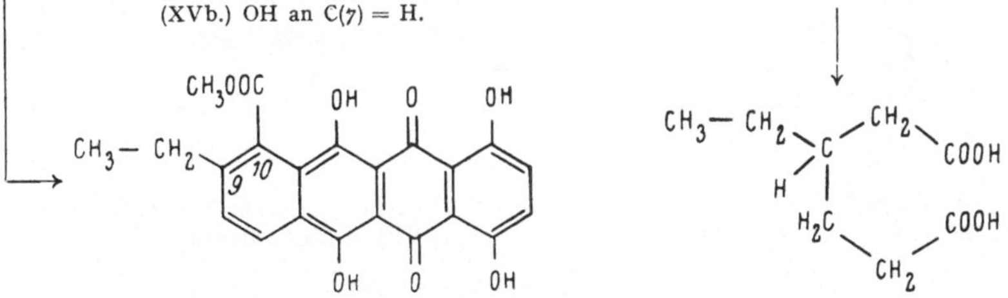

(XVIIa.) Bisanhydro-ε-iso-rhodomycinon, identisch
  mit η-Iso-pyrromycinon.
(XVIIb.) CH₃—CH₂, CH₃OOC = H.
(XVIIc.) CH₃—CH₂ = H.
(XVIId.) CH₃OOC = H.

(XVIII.) β-Äthyl-adipinsäure.

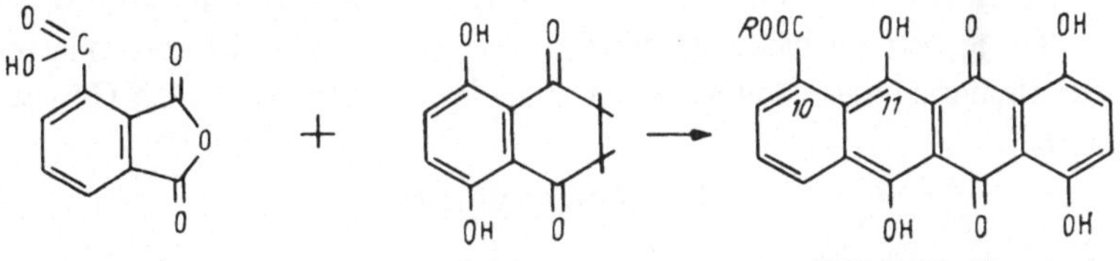

(XIX.)

(XX.)

(XXIa.) R = H.
(XXIb.) R = CH₃.
(XXIc.) ROOC = H.

Formelübersicht 2.

unbekannter, eine Carbomethoxygruppe enthaltender Hydroxy-tetracenchinone genutzt werden. Absorbiert die betreffende Verbindung längerwellig als ihre Stammverbindung (ohne Carbomethoxygruppe), so deutet dies darauf hin, daß die Carbomethoxygruppe in *peri*-Stellung zu einem Hydroxyl steht.

ζ-Iso-rhodomycinon.

ζ-Iso-rhodomycinon $C_{22}H_{20}O_9$ [BROCKMANN und BOLDT (7, 8)], rote, bei 258 bis 260° schmelzende Kristalle, stimmt im Schmp., Misch-Schmp., Spektrum des sichtbaren sowie des IR- und UV-Gebietes und im R_F-Wert mit dem roten, durch katalytische Hydrierung erhaltenen Desoxyderivat des ε-Iso-rhodomycinons überein und hat demnach wie dieses die Konstitution (XVb) (S. 139).

β-Iso-rhodomycinon.

β-Iso-rhodomycinon (BROCKMANN und RODE, unveröffentlicht) ist das Aglykon des Iso-rhodomycins A (S. 178) und bisher nur in kleiner Menge amorph erhalten worden. Es enthält laut IR-Spektrum keine Carbomethoxygruppe. Daß es ein Derivat des 1,4,5,8-Tetrahydroxy-anthrachinons (XII) ist, zeigt die Übereinstimmung seiner Absorptions-kurve mit der von (XII). Kurzes Erhitzen des β-Iso-rhodomycinons

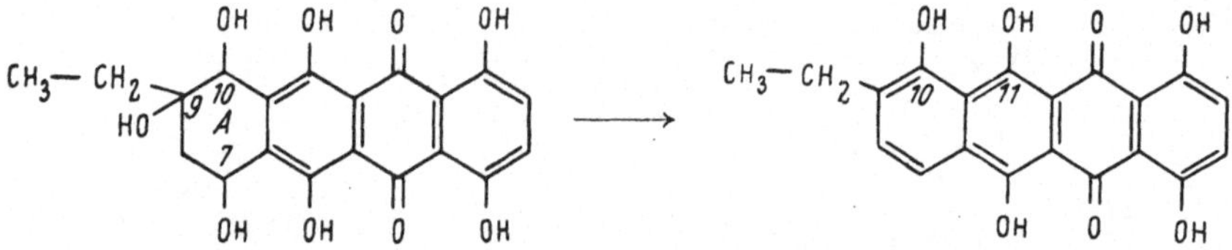

(XXIIa.) β-Iso-rhodomycinon.
(XXIIb.) OH an C(7) = H, γ-Iso-rhodomycinon.

(XXIIIa.) Bisanhydro-β-iso-rhodomycinon.
(XXIIIb.) CH_3—CH_2 = H.

mit Säure liefert ein rotes Hydroxy-tetracenchinon, dessen charakteristi-sches Absorptionsspektrum mit dem des 1,4,6,10,11-Pentahydroxy-tetracenchinons (XXIIIb) [BROCKMANN und WIMMER (35)] deckungs-gleich ist. Aus diesen Befunden und in Analogie zum β-Rhodomycinon läßt sich für β-Iso-rhodomycinon die Formel (XXIIa) ableiten, die durch das KMR-Spektrum bestätigt wird. Für γ-Iso-rhodomycinon, das bisher nur in kleiner Menge isoliert wurde, ist die Konstitution (XXIIb) wahr-scheinlich.

B. Rhodomycinone.

ε-Rhodomycinon.

ε-Rhodomycinon, $C_{22}H_{20}O_9$ (XXIVa, b, S. 142), rote Kristalle vom Schmp. 210°, wurde von BROCKMANN und FRANCK (20) aus *Streptomyces purpurascens*-Kulturen isoliert und in seiner Struktur von BROCKMANN und BROCKMANN jr. (11, 12) aufgeklärt. Es verwandelt sich bei kurzem Erhitzen mit Eisessig/Bromwasserstoffsäure unter Abspaltung von 2 Mol. Wasser in kristallisiertes, rotes Bisanhydro-ε-rhodomycinon, $C_{22}H_{16}O_7$, das sich durch sein Absorptionsspektrum als Derivat des Trihydroxy-tetracenchinons (Vb, S. 129) zu erkennen gibt. Seine langwelligen Absorptionsmaxima sind in Cyclohexan gegen die von (Vb) um 5 mμ nach Rot verschoben. Da eine gleiche spektroskopische Differenz bei (XXIc) und (XXIb) vorliegt und hier eindeutig auf die zu einem Hydroxyl des Ringes B peri-ständige Carbomethoxygruppe des Ringes A zurückzuführen ist, konnte man die im Vergleich zu (Vb) längerwellige

Absorption des Bisanhydro-ε-rhodomycinons als Hinweis darauf ansehen, daß dessen Carbomethoxygruppe an Ring A in *peri*-Stellung zu einer der beiden Hydroxyle des Ringes B steht. Daß dies zutrifft und daß die Äthylgruppe der Carbomethoxygruppe benachbart ist, zeigte die α-Hydroxylierung des Bisanhydro-ε-rhodomycinons mit Mangandioxid/Schwefelsäure. Denn dabei entstand eine dunkelrote, kristallisierte Verbindung, die identisch mit Bisanhydro-ε-iso-rhodomycinon (XVIIa) und somit nach (XXVc) zu formulieren ist.

Für Bisanhydro-ε-rhodomycinon war damit α-Stellung seiner Carbomethoxygruppe und β-Stellung seines Äthylrestes an Ring A bewiesen, aber noch nicht entschieden, ob ihre Stellung zur Hydroxygruppe des Ringes D nach (XXVa) oder (XXVb) zu formulieren ist. Dieses Ergebnis und die Tatsache, daß ε-Rhodomycinon zwei alkoholische Hydroxyle, ein sekundäres und ein tertiäres, besitzt, die sich ebenso verhalten wie die des ε-Iso-rhodomycinons (XVa), führte für ε-Rhodomycinon zu den beiden Formeln (XXIVa) und (XXIVb), von denen (XXIVa) besser in das Biogenese-Schema der Anthracyclinone (S. 163) paßt und daher den Vorzug verdient. Ihr entsprechend kommt der durch Hydrierung aus ε-Rhodomycinon entstehenden Desoxyverbindung die Formel (XXVIa) zu. Die in (XXIVa) bzw. (XXVIa) angegebene Struktur des Ringes A steht in Einklang mit den KMR-Spektren der beiden Verbindungen (S. 155).

ζ-Rhodomycinon.

ζ-Rhodomycinon, $C_{22}H_{20}O_8$, feuerrote Kristalle vom Schmp. 274 bis 275°, wurde von BOLDT (*2*) neben anderen Rhodomycinonen aus *Str. purpurascens*-Kulturen abgetrennt. Es stimmt im Schmp., Misch-Schmp., Absorptions- und IR-Spektrum in Analysenzahlen und R_F-Werten mit dem roten Hydrierungsprodukt des ε-Rhodomycinons überein und hat demnach wie dieses die Konstitution (XXVIa) oder weniger wahrscheinlich (XXVIb) [BROCKMANN und BROCKMANN jr. (*12*)].

β-Rhodomycinon.

β-Rhodomycinon, $C_{20}H_{18}O_8$, rote Kristalle vom Schmp. 225° [BROCKMANN und FRANCK (*20*)], wurde von BROCKMANN und BOLDT (*6*) sowie BROCKMANN, BOLDT und NIEMEYER (*9*) aufgeklärt. Es ist das am längsten bekannte Anthracyclinon und das Aglykon der Rhodomycine A und B (SS. 176, 177). Von seinen sechs Hydroxygruppen verliert β-Rhodomycinon drei beim Kochen mit Jodwasserstoffsäure; zwei spalten sich in Form von Wasser ab, die dritte wird durch Wasserstoff ersetzt. Das rote, kristallisierte, im Hochvakuum sublimierbare „β-Rhodomycinon-HJ-Produkt" $C_{20}H_{14}O_5$ (*6*) enthält noch die Äthylgruppe des β-Rhodomycinons, stimmt im Spektrum mit (Vb) überein und ist somit ein

(XXIVa.) ε-Rhodomycinon.
(XXIVb.) OH an C(1), H an C(4).

(XXVa.) Bisanhydro-ε-rhodomycinon.
(XXVb.) OH an C(1), H an C(4).
(XXVc.) OH an C(1).

(XXVIa.) ζ-Rhodomycinon.
(XXVIb.) OH an C(1), H an C(4).

(XXVIIa.) ζ-Rhodomycinon-HBr-Produkt.
(XXVIIb.) OH an C(1), H an C(4).

(XXVIIIa.) β-Rhodomycinon.
(XXVIIIb.) OH an C(7), C(9), C(10) = H.
(XXVIIIc.) OH an C(1), H an C(4).
(XXVIIId.) OH an C(1), H an C(4);
OH an C(7), C(9), C(10) = H.

(XXIXa.) Bisanhydro-β-rhodomycinon.
(XXIXb.) OH an C(1), H an C(4).
(XXIXc.) OH an C(7), H an C(10).
(XXIXd.) OH an C(7), H an C(10);
OH an C(1), H an C(4).

(XXXa.) γ-Rhodomycinon.
(XXXb.) OH an C(1), H an C(4).

(XXXIa.) Descarbomethoxybisanhydro--ε-
rhodomycinon (β-Rhodomycinon-
HJ-Produkt; Bisanhydro-γ-rhodo-
mycinon).
(XXXIb.) OH an C(1), H an C(4).

Formelübersicht 3.

Äthylderivat von (Vb). Über die Stellung der Äthylgruppe haben folgende
Reaktionen Auskunft gegeben.

Aus ζ-Rhodomycinon (XXVIa oder XXVIb, S. 142) entsteht bei
längerem Kochen mit Eisessig/Bromwasserstoffsäure unter Verseifung,
Decarboxylierung und Eliminierung von 1 Mol. Wasser eine als ζ-Rhodomycinon-HBr-Produkt bezeichnete Verbindung (XXVIIa) oder (XXVIIb),
die beim Erhitzen mit Palladium-Kohle zu Descarbomethoxy-bisanhydro-

ε-rhodomycinon (XXXIa bzw. XXXIb) dehydriert wird [BROCKMANN und BROCKMANN jr. (*11*)]. Diese Verbindung erwies sich als identisch mit dem „β-Rhodomycinon-HJ-Produkt" (*6*), das somit ebenfalls nach (XXXIa) oder (XXXIb) zu formulieren ist. Damit war gezeigt, daß die Äthylgruppe des β-Rhodomycinons, bezogen auf die Hydroxygruppe in Ring *D*, die gleiche Stellung hat wie im ε-Rhodomycinon und demnach β-Rhodomycinon ein Trihydroxyderivat von (XXVIIIb) oder (XXVIIId) ist.

Da von den drei durch Jodwasserstoffsäure entfernbaren Hydroxylen des β-Rhodomycinons zwei unter Aromatisierung von Ring *A* in Form von Wasser eliminiert werden, müssen sie zu Ring *A* gehören.

Über die Stellung der dritten durch Jodwasserstoffsäure reduktiv entfernbaren Hydroxygruppe erhielt man Auskunft, als man die Wasserabspaltung aus β-Rhodomycinon mit konz. Salzsäure durchführte. Da Salzsäure nicht reduzierend wirkt, enthält das unter ihrer Einwirkung entstehende Bisanhydro-β-rhodomycinon $C_{20}H_{14}O_6$ noch das durch Jodwasserstoffsäure reduktiv entfernbare Hydroxyl des β-Rhodomycinons und unterscheidet sich dadurch vom β-Rhodomycinon-HJ-Produkt [BROCKMANN und NIEMEYER (*26*)].

Dieser Unterschied gibt sich im Spektrum dadurch zu erkennen, daß die langwelligen Absorptionsmaxima des Bisanhydro-β-rhodomycinons in Cyclohexan gegen die des β-Rhodomycinon-HJ-Produktes um 23 bzw. 18 mμ nach Rot verschoben sind. Die Hydroxygruppe, um die sich Bisanhydro-β-rhodomycinon vom β-Rhodomycinon-HJ-Produkt unterscheidet, wirkt somit ungewöhnlich stark bathochrom. Das aber ist nur möglich — wie ein spektroskopischer Vergleich mit verschiedenen Hydroxy-tetracenchinonen gezeigt hat [BROCKMANN und WIMMER (*35*)] —, wenn sie an Ring *A* in *peri*-Stellung zu einer Hydroxygruppe des Ringes *B* steht, d. h. wenn dem Bisanhydro-β-rhodomycinon eine der Formeln (XXIXa, b, c, d) zukommt. Damit war bewiesen, daß die drei aliphatischen Hydroxyle des β-Rhodomycinons zu Ring *A* gehören. Da eines von ihnen schwer acetylierbar, daher tertiär ist und demnach am gleichen C-Atom steht wie die Äthylgruppe, kamen für die drei Hydroxyle zunächst die Stellungen (XXXII), (XXXIII) und (XXXIV) in Betracht.

Von den drei Hydroxygruppen wird eine bei Hydrierung des β-Rhodomycinons in alkalischem Medium durch Wasserstoff ersetzt. Die Absorptionsmaxima der dabei entstehenden roten Desoxyverbindung $C_{20}H_{18}O_7$, die ihrem Absorptionsspektrum nach noch den Chromophor (XIII, S. 129) des β-Rhodomycinons enthält, liegen in Piperidin um 10 mμ kürzerwellig als die des β-Rhodomycinons. Die durch Hydrierung abspaltbare Hydroxygruppe des β-Rhodomycinons steht demnach dem Chromophor (XIII) benachbart an $C_{(7)}$ oder $C_{(10)}$.

Kurzes Erwärmen mit Eisessig/Bromwasserstoffsäure verwandelt das Desoxy-derivat in eine rote, kristallisierte Bisanhydro-Verbindung $C_{20}H_{14}O_5$, die, wie zu erwarten, mit β-Rhodomycinon-HJ-Produkt (XXXIa) identisch ist. Durch katalytische Hydrierung und anschließende Wasserabspaltung erreicht man somit in zwei Arbeitsgängen das gleiche Derivat wie bei der Jodwasserstoffbehandlung des β-Rhodomycinons in einem.

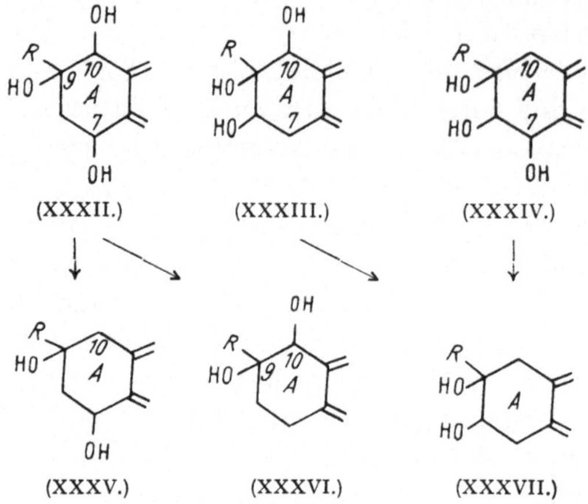

(XXXII.) (XXXIII.) (XXXIV.)

(XXXV.) (XXXVI.) (XXXVII.)

$R = CH_3{-}CH_2.$

Formelübersicht 4.

Auf Grund dieser Befunde kamen für Ring A der Desoxyverbindung die Formeln (XXXV), (XXXVI) und (XXXVII) in Frage. Eine Entscheidung zwischen (XXXV) und andererseits (XXXVI) bzw. (XXXVII) war vom Verhalten der Desoxyverbindung gegenüber Perjodsäure zu erwarten, das wegen ihrer Unlöslichkeit in Wasser nur in wasserhaltigen organischen Solvenzien untersucht werden konnte. Am besten eignete sich 90%ige Essigsäure.

Dabei trat jedoch die Schwierigkeit auf, daß a) der Abbau erst mit annehmbarer Geschwindigkeit verlief, wenn die Reaktionslösung an Perjodsäure 0,1 molar war und b) 90%ige Essigsäure nur etwa 0,005 Mol Desoxyverbindung im Liter löst. Um eine 0,005 molare Lösung der Desoxyverbindung an Perjodsäure 0,1 molar zu machen, muß man ihr diese Form einer 0,3 molaren Lösung zusetzen. Bedingungen, unter denen der Meßfehler einer Perjodsäure-Titration unerträglich groß wird.

Deshalb blieb nichts anderes übrig, als das Verhalten der Desoxyverbindung gegen Perjodsäure papierchromatographisch zu kontrollieren und mit dem von ε-Rhodomycinon und ε-Pyrromycinon zu vergleichen, in deren Ring A die Hydroxyle der Formel (XXXV) entsprechend angeordnet sind. Unter Bedingungen, unter denen die beiden Vergleichspräparate völlig intakt blieben, wurde die Desoxyverbindung vollständig abgebaut, womit Formel (XXXV) widerlegt und gezeigt war, daß die beiden Hydroxyle in Ring A wie in (XXXVI) oder (XXXVII) an benachbarten C-Atomen stehen.

Literaturverzeichnis: SS. 179—182.

Die Entscheidung zwischen (XXXVI) und (XXXVII) brachte das in Trifluoressigsäure aufgenommene KMR-Spektrum der Desoxyverbindung (S. 156). Hätte ihr Ring A die Konstitution (XXXVII), so müßten sich die beiden zum Anthrachinon-Ringsystem α-ständigen Methylengruppen durch ein Signal mit einem τ-Wert von 7,0 und der relativen Intensität 4 (4 Protonen) zu erkennen geben. Die Desoxyverbindung zeigt jedoch ein Signal mit $\tau = 7{,}17$ und der relativen Intensität 2 (2 Protonen); und außerdem eins mit $\tau = 8{,}0$ und der relativen Intensität 2, das einer zum Anthrachinon-Ringsystem β-ständigen Methylengruppe zugeordnet werden muß. Damit ist für Ring A der Desoxyverbindung die Konstitution (XXXVI) bewiesen und für Ring A des β-Rhodomycinons die Konstitution (XXXII), die das KMR-Spektrum des β-Rhodomycinons (S. 156) bestätigt hat. Von den beiden Formeln (XXVIIIa) und (XXVIIIc), die nunmehr noch für β-Rhodomycinon in Betracht kommen, verdient (XXVIIIa) den Vorzug, weil sie mit dem Biogenese-Schema der Anthracyclinone (S. 163) besser in Einklang steht als (XXVIIIc).

γ-Rhodomycinon.

γ-Rhodomycinon, $C_{20}H_{18}O_7$, rote Kristalle, die sich bei 230—240° zersetzen, wurde von BOLDT als Begleiter des β-Rhodomycinons aus *Str. purpurascens*-Kulturen isoliert [BROCKMANN, BOLDT und NIEMEYER (9)]. Es ist das Aglykon der γ-Rhodomycine (S. 177) und erwies sich als identisch mit der durch Hydrierung aus β-Rhodomycinon entstehenden Desoxyverbindung. γ-Rhodomycinon hat daher entweder Formel (XXXa) oder (XXXb), von denen (XXXa) mit der Acetathypothese besser in Einklang steht als (XXXb).

δ-Rhodomycinon.

δ-Rhodomycinon, $C_{22}H_{20}O_9$, wurde von BOLDT (2) in Rhodomycinongemischen verschiedener *Str. purpurascens*-Kulturen aufgefunden und von BROCKMANN und BROCKMANN jr. (*13, 14*) in roten Kristallen vom Schmp. 195—197° isoliert sowie in seiner Struktur aufgeklärt. Es ist isomer mit ε-Rhodomycinon, seinem Spektrum nach wie dieses ein Derivat des 1,4,5-Trihydroxy-anthrachinons (XIII) und enthält ebenfalls eine Carbomethoxygruppe.

Kurzes Erhitzen mit Eisessig/Bromwasserstoffsäure verwandelt δ-Rhodomycinon in kristallisiertes, rotes Bisanhydro-δ-rhodomycinon, das durch sein Spektrum ebenso wie η-Pyrromycinon (XLa) als Derivat des Trihydroxy-tetracenchinons (VIIb, S. 129) charakterisiert ist. Von η-Pyrromycinon (XLa) unterscheidet es sich im IR-Spektrum und in den R_F-Werten.

Ebenso wie beim Bisanhydro-ε-rhodomycinon (XXVa bzw. XXVb) und Bisanhydro-ε-iso-rhodomycinon (XVIIa) sind die Absorptionsmaxima des Bisanhydro-δ-rhodomycinons in Cyclohexan, verglichen mit denen der Stammverbindung (VIIb), um 4—5 mμ nach Rot verschoben, was wie bei (XXVa) bzw. (XXVb) und (XVIIa) anzeigt, daß die Carbomethoxygruppe in Ring *A* in *peri*-Stellung zu einer Hydroxygruppe des Ringes *B* steht. Danach hat Bisanhydro-δ-rhodomycinon — vorausgesetzt, die Äthylgruppe ist wie bei allen anderen Anthracyclinonen

(XXXVIII.) δ-Rhodomycinon.

(XXXIX.) Bisanhydro-δ-rhodomycinon.

der Carbomethoxygruppe benachbart — die Konstitution (XXXIX). Für δ-Rhodomycinon ergibt sich daraus die Formel (XXXVIII), wenn man annimmt, daß die beiden in Form von Wasser eliminierbaren Hydroxyle in Ring *A* so angeordnet sind wie bei den anderen Anthracyclinonen. Daß dies zutrifft, hat das KMR-Spektrum des δ-Rhodomycinons (S. 156) gezeigt.

Wenn man wie in Abschn. III, 1 (S. 129) die Rhodomycinone als Derivate von (IVb) und die Pyrromycinone als Derivate von (VIb) definiert, gehört δ-Rhodomycinon einem neuen Anthracyclinon-Typ an. Da es aber nur gemeinsam mit Rhodomycinonen und Iso-rhodomycinonen vorkommt und auch die Biogenese wahrscheinlich über gemeinsame Vorstufen verläuft, rechnet man es vorläufig zu den Rhodomycinen (*14*).

C. *Pyrromycinone.*

η-*Pyrromycinon.*

η-Pyrromycinon, $C_{22}H_{16}O_7$ (XLa), rote Kristalle vom Schmp. 236 bis 237°, im Hochvakuum sublimierbar, ist zuerst von Brockmann, Costa Plá und Lenk (*19*) aus Mycel und Kulturlösung verschiedener *Streptomyces*-Stämme isoliert und a) durch Zinkstaubdestillation, b) durch spektroskopischen Vergleich der acetylierten Leukoverbindung *(Abb. 3)*, c) durch das Absorptionsspektrum seines Acetates *(Abb. 4)* als Tetracenchinon-derivat und durch spektroskopischen Vergleich mit 1,4,6-Trihydroxy-tetracenchinon-(5,12) (XLb) *(Abb. 5)* als Carbomethoxy-äthylderivat von (XLb) identifiziert worden.

η-Pyrromycinon entsteht aus ε-Pyrromycinon unter Abspaltung von 2 Mol. Wasser beim Erhitzen auf 180—220° [Prelog, Keller-Schierlein und Mitarb. (*40*); BROCKMANN und LENK (*22*)], oder in besserer Ausbeute durch kurzes Erwärmen mit Eisessig/Bromwasserstoffsäure (*22*).

Die gelbe Lösung des η-Pyrromycinons in Acetanhydrid wird auf Zugabe von Pyroboracetat blau, mit roter Fluoreszenz und scharfen Absorptionsmaxima. Beim Kochen wird die Lösung karmoisinrot und die Maxima verschieben sich nach Rot (*22*). Die starke Farbvertiefung bei Zugabe von Pyroboracetat ist charakteristisch für Hydroxychinone, in denen jeder Chinon-carbonylgruppe mindestens eine Hydroxygruppe benachbart ist. Die beim Erwärmen eintretende hypsochrome Verschiebung der Absorptionsmaxima zeigt, daß mehr als zwei zu Chinoncarbonylen α-ständige Hydroxyle vorliegen (*22*, *40*).

α-Hydroxylierung mit Mangandioxid/Schwefelsäure verwandelt η-Pyrromycinon in η-Isopyrromycinon (*22*), das, wie sich später zeigte, mit Bisanhydro-ε-iso-rhodomycinon (XVIIa) identisch ist [BROCKMANN und BOLDT (*8*)]. Zu dieser Zeit war die Stellung der Äthylgruppe in (XVIIa) bekannt, nicht aber die der Carbomethoxygruppe. Die Identität der beiden Verbindungen bewies daher zunächst nur, daß in den Pyrromycinonen die Äthylgruppe an Ring *A* an einem β-C-Atom steht und die Carbomethoxygruppe die gleiche Stellung einnimmt wie bei den Rhodomycinonen und Iso-rhodomycinonen.

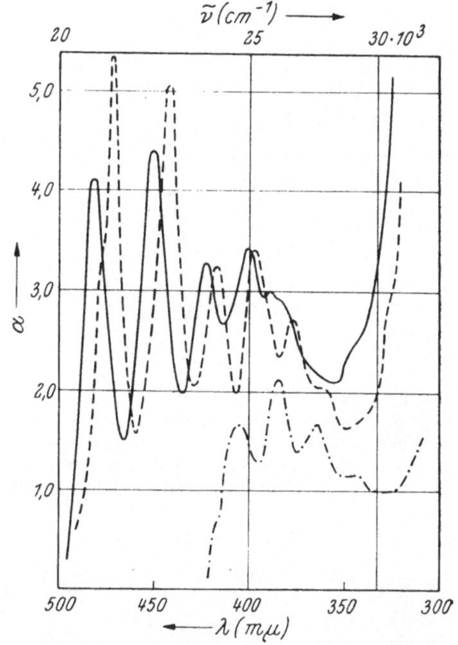

Abb. 3. Absorptionskurven in Cyclohexan. Reduzierend acetyliertes η-Pyrromycinon ——————— (λ_{max} 482, 452, 424, 400 mμ). Reduzierend acetyliertes 6,11-Diacetoxy-tetracenchinon ————— (λ_{max} 472, 442, 418, 398, 378 mμ). Reduzierend acetyliertes ζ-Pyrromycinon —·—·—· (λ_{max} 406, 384, 364 mμ). [Aus: Chem. Ber. **92**, 1880 (1959).]

Verseifung von η-Pyrromycinon liefert η-Pyrromycinonsäure (XLe), die auch bei alkalischer Verseifung des ε-Pyrromycinons entsteht (*40*) und beim Erhitzen allein (*22*) oder mit Kupferpulver in Chinolin (*40*), oder auch durch Erhitzen ihres Calciumsalzes (*22*) in Descarbomethoxy-η-pyrromycinon übergeht. Aus diesem bildet sich durch α-Hydroxylierung mit Mangandioxid/Schwefelsäure Descarbomethoxy-η-iso-pyrromycinon [BROCKMANN und LENK (*22*)], dessen Konstitution (XLIII) später durch Synthese bewiesen wurde [BROCKMANN und WIMMER (*34*, *35*)]. Damit ergaben sich für Descarbomethoxy-η-pyrromycinon die Formeln (XLIIa) und (XLIIb) (*22*), von denen (XLIIa) mit den Vorstellungen über die Biogenese der Anthracyclinone (S. 163) besser in Einklang steht als

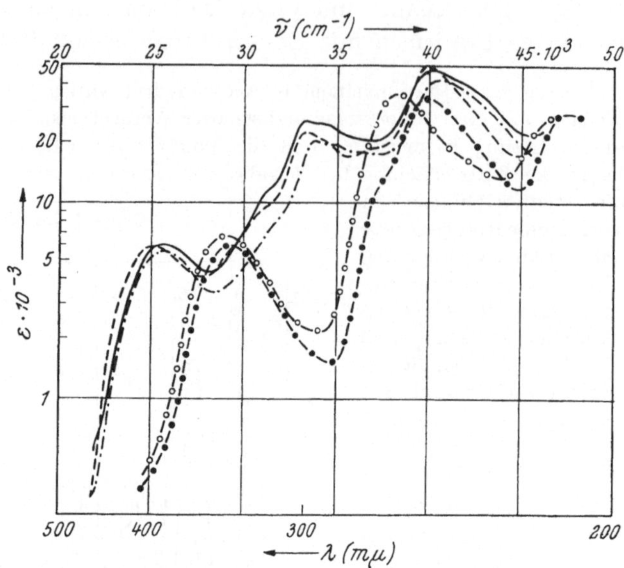

Abb. 4. Absorptionskurven in Methanol. η-Pyrromycinon-acetat ——————— (λ_max 390, 299, 248 mμ). Descarbomethoxy-η-pyrromycinon-acetat —————— (λ_max 400, 300, 250 mμ). 1,4,6-Tri-acetoxy-tetracenchinon —·—·—·— (λ_max 392, 295, 247 mμ). ζ-Pyrromycinon-acetat o — o — o — (λ_max 344, 262 mμ). 1,4,5-Triacetoxy-anthrachinon — ● — ● — ● (λ_max 342, 251, 210 mμ). [Aus: Chem. Ber. 92, 1880 (1959).]

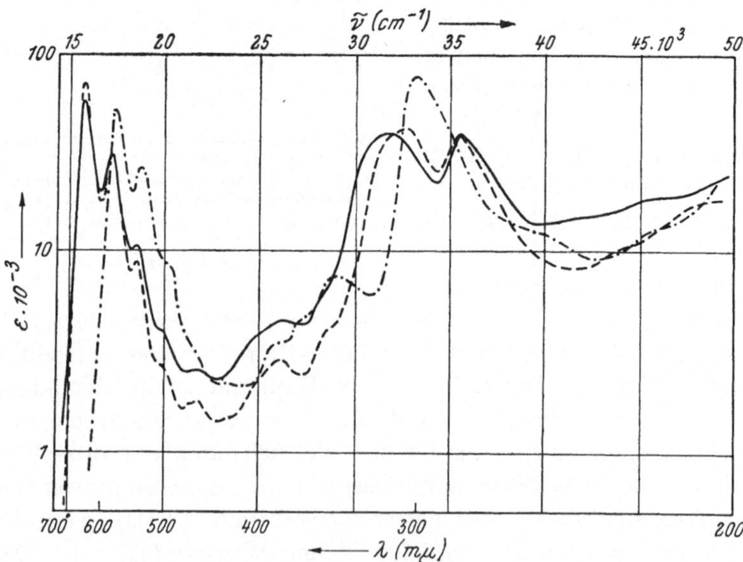

Abb. 5. Absorptionskurven in konz. Schwefelsäure. η-Pyrromycinon ——————— (λ_max 635, 582, 380, 312, 281 mμ). 1,4,6-Trihydroxy-tetracenchinon —————— (λ_max 636, 583, 385, 305, 282 mμ), 1,6,11-Trihydroxy-tetracenchinon —·—·—·— (λ_max 575, 533, 345, 301 mμ). [Aus: Chem. Ber. 92 1880 (1959).]

Literaturverzeichnis: SS. 179—182.

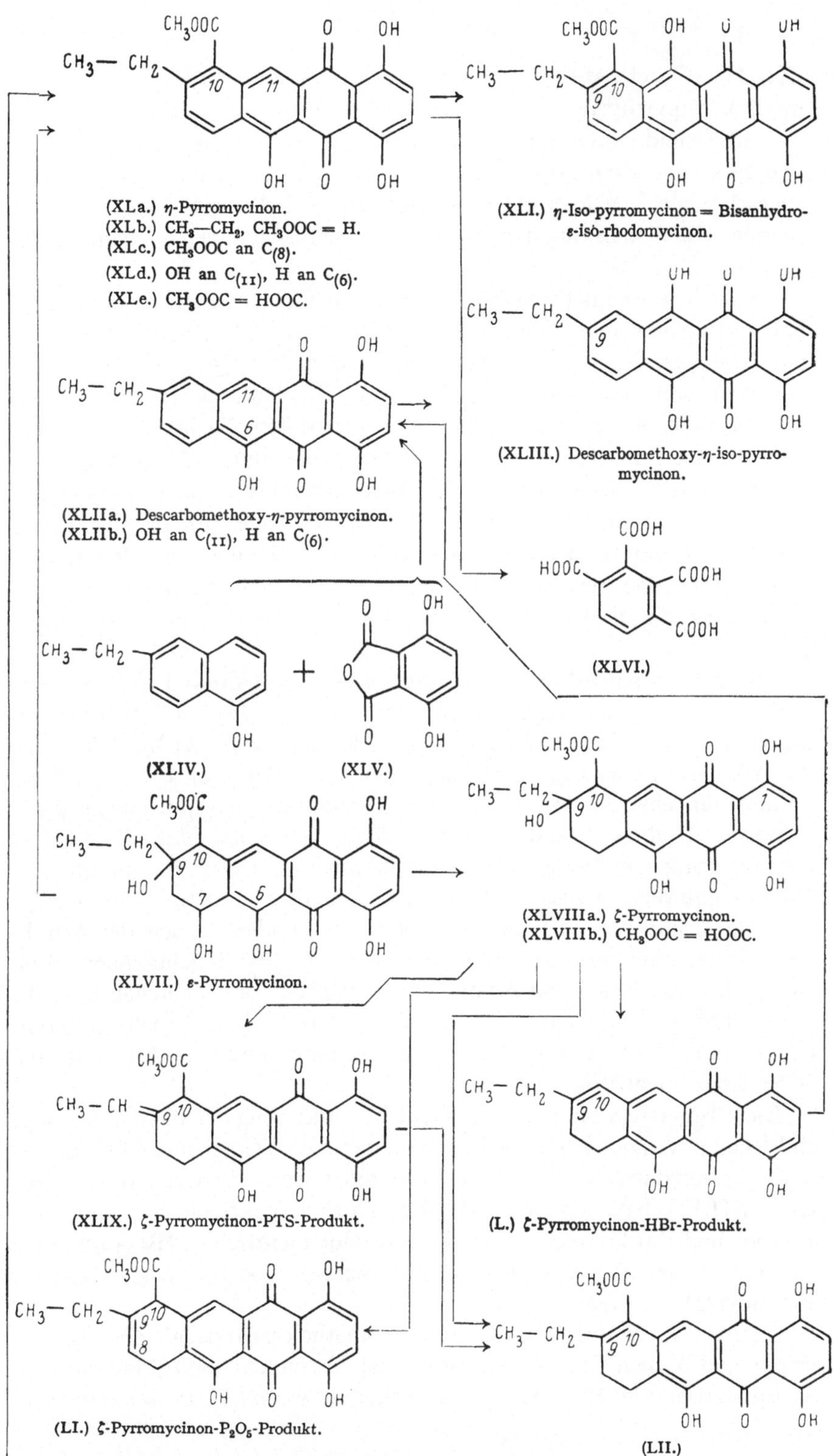

(XLa.) η-Pyrromycinon.
(XLb.) CH₃—CH₂, CH₃OOC = H.
(XLc.) CH₃OOC an C₍₈₎.
(XLd.) OH an C₍₁₁₎, H an C₍₆₎.
(XLe.) CH₃OOC = HOOC.

(XLI.) η-Iso-pyrromycinon = Bisanhydro-ε-isó-rhodomycinon.

(XLIIa.) Descarbomethoxy-η-pyrromycinon.
(XLIIb.) OH an C₍₁₁₎, H an C₍₆₎.

(XLIII.) Descarbomethoxy-η-iso-pyrromycinon.

(XLIV.)   (XLV.)   (XLVI.)

(XLVII.) ε-Pyrromycinon.

(XLVIIIa.) ζ-Pyrromycinon.
(XLVIIIb.) CH₃OOC = HOOC.

(XLIX.) ζ-Pyrromycinon-PTS-Produkt.

(L.) ζ-Pyrromycinon-HBr-Produkt.

(LI.) ζ-Pyrromycinon-P₂O₅-Produkt.

(LII.)

Formelübersicht 5.

(XLIIb). Durch Abbau mit konz. Salpetersäure wurde aus Descarbomethoxy-η-pyrromycinon Benzol-tricarbonsäure-(1,2,4) erhalten (*15*).

Auf Grund dieser und einiger anderer Befunde haben Brockmann und Lenk (*22*) dem η-Pyrromycinon zunächst die Formel (XLc, S. 149) zugeschrieben. Auf Nachbarstellung der Äthyl- und Carbomethoxygruppe schloß man aus der schweren Verseifbarkeit der Carbomethoxygruppe.

Daß die Carbomethoxygruppe nicht an $C_{(8)}$, sondern an $C_{(10)}$ steht, und η-Pyrromycinon somit die Konstitution (XLa) hat, ergab sich aus Untersuchungen von Ollis und Sutherland (*50*). Sie gewannen aus dem von Asheshov aufgefundenen Rutilantinen (S. 176) durch Hydrolyse einen roten, *Rutilantinon* genannten Farbstoff, aus dem beim Kochen in Toluol mit Toluolsulfosäure unter Abspaltung von 2 Mol. Wasser Bisanhydro-rutilantinon entstand. Diese Verbindung stimmte in ihrem Absorptionsspektrum (Cyclohexan) praktisch mit 1,4,6-Trihydroxy-tetracenchinon-(5,12) (XLb) überein, und das gleiche fand man bei den Acetaten. Von (XLb) unterschied sich die Bisanhydroverbindung durch den Besitz einer Carbomethoxy- und einer Äthylgruppe.

Da die Carbomethoxygruppe mit 2 n-Alkalihydroxyd schwer verseifbar war, wurde angenommen, daß sie von zwei o-Substituenten flankiert ist, dementsprechend an $C_{(7)}$ oder $C_{(10)}$ von (XLb) steht, und die Äthylgruppe zu $C_{(8)}$ bzw. $C_{(9)}$ gehört. Und daß man keine Anhaltspunkte für eine Lactonisierung der Carboxygruppe mit einem *peri*-ständigen Hydroxyl fand, schien dafür zu sprechen, daß die Carbomethoxygruppe mit $C_{(10)}$ der Stammverbindung (XLb) verknüpft ist. Damit ergab sich für Bisanhydro-rutilantinon Formel (XLa). Ihr zufolge ist Ring *A* mit vier vicinalen C-Atomen verbunden; denen der Äthyl- und Carbomethoxygruppe sowie den beiden zu Ring *B* gehörenden. Bei energischer Oxydation von Bisanhydro-rutilantinon war demnach Benzol-tetracarbonsäure-(1,2,3,4) (XLVI) zu erwarten. Tatsächlich konnten Ollis und Sutherland (XLVI) als Abbauprodukt fassen und als Methylester identifizieren.

Diese Ergebnisse sind der Konstitutionsaufklärung der Pyrromycinone und Rhodomycinone zugute gekommen, denn ein Vergleich der Präparate sowie der oxydative Abbau des η-Pyrromycinons zu Benzol-tetracarbonsäure (XLVI) hat gezeigt, daß Bisanhydro-rutilantinon mit η-Pyrromycinon und Rutilantinon mit ε-Pyrromycinon identisch ist [Brockmann, Brockmann jr., Gordon, Keller-Schierlein, Lenk, Ollis, Prelog und Sutherland (*17*)].

Damit war bewiesen, daß die Carbomethoxygruppe a) der Äthylgruppe und b) dem Ring *B* benachbart ist. Argument dafür, daß sie zur Hydroxygruppe des Ringes *B* so angeordnet ist wie in (XLa), war zunächst nur das Ausbleiben der Lactonbildung und die Beobachtung, daß η-Pyrromycinon leichter verseift wird als ε-Iso-rhodomycinon, dessen Ring B Träger eines zur Carbomethoxygruppe *peri*-ständigen Hydroxyls ist (*15*). Eindeutig bewies den aus diesen Befunden gezogenen Schluß erst die Synthese des Descarbomethoxy-η-pyrromycinons (XLIIa) durch Friedel-Crafts-Kondensation von 5-Hydroxy-2-äthyl-naphthalin (XLIV) mit 3,6-Dihydroxy-phthalsäureanhydrid (XLV) [OLLIS, SUTHERLAND und VEAL (*51*)].

ε-Pyrromycinon.

ε-Pyrromycinon, $C_{22}H_{20}O_9$ (XLVII), gelbrote Kristalle, ist das Aglykon der Cinerubine A und B sowie des Pyrromycins (S. 174). Es wurde von PRELOG, KELLER-SCHIERLEIN und Mitarb. (*40*) aus Hydrolysaten von Cinerubin A und B und von BROCKMANN und LENK (*22*) aus *Streptomyces*-Kulturen sowie aus Hydrolysaten von Pyrromycin isoliert und eingehend untersucht.

Durch sein Absorptionsspektrum und das seiner acetylierten Leukoverbindung sowie durch seine Farbreaktionen mit Pyroboracetat (gelbe Lösung in Acetanhydrid wird zunächst blauviolett und beim Erwärmen rotviolett) gibt sich ε-Pyrromycinon als Derivat des 1,4,5-Trihydroxy-anthrachinons (XIII, S. 129) zu erkennen. Oxydation des ε-Pyrromycinons mit Chromsäure nach KARRER (*39*) oder mit Kaliumpermanganat in Pyridin liefert ein Gemisch aus Propionsäure und Essigsäure.

Beim Erhitzen auf 180—220° (*40, 22*) oder bei kurzem Kochen mit Eisessig/Bromwasserstoffsäure (*22*) verwandelt sich ε-Pyrromycinon unter Eliminierung von 2 Mol. Wasser in η-Pyrromycinon (XLa). Dieser Befund zusammen mit der Tatsache, daß von den beiden alkoholischen Hydroxylen des Farbstoffes die eine schwer acetylierbar und daher tertiär ist und die andere bei katalytischer Hydrierung durch Wasserstoff ersetzt wird, ergab sich für ε-Pyrromycinon die Formel (XLVII) (*22, 15, 17*). Die Struktur des Ringes A wurde durch das KMR-Spektrum des ε-Pyrromycinons bestätigt (S. **157**).

ζ-Pyrromycinon.

ζ-Pyrromycinon, $C_{22}H_{20}O_8$ (XLVIIIa), feuerrote Kristalle vom Schmp. 216°, im Hochvakuum sublimierbar, wurde von BROCKMANN und LENK (*21, 22*) aus *Streptomyces*-Kulturen isoliert und als Carbomethoxyäthylderivat des 1,4,6-Trihydroxy-7,8,9,10-tetrahydro-tetracenchinons-(5,12) (VIc) erkannt. Seine Aufklärung [BROCKMANN und LENK (*24*)] gelang, nachdem die Konstitution des η-Pyrromycinons und ε-Pyrromycinons bekannt war. Dabei waren die beiden folgenden Reaktionen (*22*) entscheidend: a) Erhitzen von ζ-Pyrromycinon mit Palladium-Kohle liefert η-Pyrromycinon (XLa), womit die Stellung der kohlenstoffhaltigen Substituenten in Ring A bewiesen war. b) Katalytische Hydrierung von ε-Pyrromycinon (XLVII) in Triäthanolamin/Äthanol gibt ζ-Pyrromycinonsäure (XLVIIIb), die mit Diazomethan zu ζ-Pyrromycinon methyliert wurde [Brockmann und Brockmann jr. (*15, 12*)]. Da die langwelligen Absorptionsmaxima des ζ-Pyrromycinons in Piperidin (nicht dagegen in Cyclohexan) um 4 mμ kürzerwellig sind als die des ε-Pyrromycinons, ist bei der Hydrierung die an $C_{(7)}$ stehende Hydroxygruppe des ε-Pyrromycinons (XLVII) durch Wasserstoff ersetzt worden. Somit ergibt sich für ζ-Pyrromycinon die Konstitution (XLVIIIa). Die Struktur von Ring *A* wird durch das KMR-Spektrum bestätigt.

Außer ζ-Pyrromycinonsäure (XLVIIIb) erhielt man bei der katalytischen Hydrierung des ε-Pyrromycinons in Triäthanolamin/Äthanol η-Pyrromycinon. Das basische Lösungsmittel wirkt demnach in zweierlei Weise. In der Hauptsache verseift es — offenbar schon vor der Abspaltung des $C_{(7)}$-Hydroxyls — die Carbomethoxygruppe. Und in geringerem Umfange andererseits katalysiert es die zur Aromatisierung von Ring *A* führende Eliminierung von 2 Mol. Wasser. In dem dabei entstehenden η-Pyrromycinon (XLa) ist die Carbomethoxygruppe coplanar mit der benachbarten Äthylgruppe, durch sie daher sterisch behindert und infolgedessen schwerer verseifbar als im ε-Pyrromycinon (XLVII), in dem die beiden Gruppen nicht in einer Ebene liegen.

Hydriert man unter gleichen Bedingungen ε-Iso-rhodomycinon (XVa) und ε-Rhodomycinon (XXIVa), so wird deren Carbomethoxygruppe durch das basische Lösungsmittel nicht verändert. Vergleichende Verseifungsversuche haben gezeigt, daß ε-Iso-rhodomycinon (XVa) und ε-Rhodomycinon (XXIVa) erheblich langsamer reagieren als ε-Pyrromycinon (XLVII) (*12*). Zweifellos ist das auf eine Behinderung durch die *peri*-Hydroxygruppe zurückzuführen, die dem ε-Pyrromycinon fehlt.

Ebenso wie aus ζ-Rhodomycinon (XXVIa) wird auch aus ζ-Pyrromycinon bei längerem Kochen mit Eisessig/Bromwasserstoffsäure unter Verseifung und Decarboxylierung 1 Mol. Wasser eliminiert. Das dabei entstehende „ζ-Pyrromycinon-HBr-Produkt" kann mit Palladium/Kohle zu Descarbomethoxy-η-pyrromycinon (XLIIa) dehydriert werden. Auf Grund dieser Reaktionen und seiner — verglichen mit ζ-Pyrromycinon (XLVIIIa) — längerwelligen Absorptionsmaxima und größeren ε_{max}-Werte kommt dem ζ-Pyrromycinon-HBr-Produkt die Formel (L) zu.

Die tertiäre Hydroxygruppe des ζ-Pyrromycinons (XLVIIIa) läßt sich auch ohne gleichzeitige Decarboxylierung in Form von Wasser abspalten. Der Formel (XLVIIIa) nach war vorauszusehen, daß diese Wasserabspaltung in drei Richtungen vor sich gehen kann, je nachdem, ob sich der Wasserstoff an $C_{(10)}$, $C_{(8)}$ oder am α-C-Atom der Äthylgruppe an der Eliminierung beteiligt. Alle drei dabei zu erwartenden Anhydro-ζ-pyrromycinone konnten kristallisiert gewonnen werden [Brockmann und Lenk (*22, 24*)].

Erhitzen des ζ-Pyrromycins (XLVIIIa) auf 230° liefert unter Abspaltung von 1 Mol. Wasser eine braunrote Verbindung $C_{22}H_{18}O_7$ vom Schmp. 206—207°, deren Absorptionsmaxima in Cyclohexan, Piperidin und konz. Schwefelsäure gegen die des ζ-Pyrromycinons nach Rot verschoben sind; ein Beweis, daß die durch Wasserabspaltung entstandene Doppelbindung mit dem Anthrachinon-Ringsystem konjugiert ist. Die Anhydroverbindung hat daher die Konstitution (LII). Mit Palladium/Kohle läßt sie sich erwartungsgemäß zu η-Pyrromycinon (XLa) dehydrieren.

Ein anderes, feuerrotes Anhydro-ζ-pyrromycinon vom Schmp. 158 bis 159°, dessen Absorptionsspektrum praktisch mit dem des ζ-Pyrromycinons übereinstimmt, erhält man durch kurzes Erhitzen von ζ-Pyrromycinon mit Phosphorpentoxyd. Wie das KMR-Spektrum zeigt, liegt in diesem „ζ-Pyrromycinon-P_2O_5-Produkt" die Doppelbindung zwischen $C_{(8)}$ und $C_{(9)}$, so daß ihm Formel (LI) zukommt.

Ein drittes Anhydro-ζ-pyrromycinon entsteht beim Kochen von ζ-Pyrromycinon in Toluol mit p-Toluolsulfosäure. Sein Absorptionsspektrum stimmt zwischen 550 und 325 mμ mit dem des „ζ-Pyrromycinon-P_2O_5-Produktes" überein, und die IR-Spektren beider unterscheiden sich nur im „finger print"-Gebiet. Durch Erwärmen mit Pyridin/Triäthylamin läßt sich das mit p-Toluolsulfosäure erhaltene „ζ-Pyrromycinon-PTS-Produkt" in das Anhydro-ζ-pyrromycinon (LII) überführen. Da die Konstitution der beiden anderen Anhydroverbindungen bekannt ist, bleibt für das PTS-Produkt nur die Formel (XLIX) übrig.

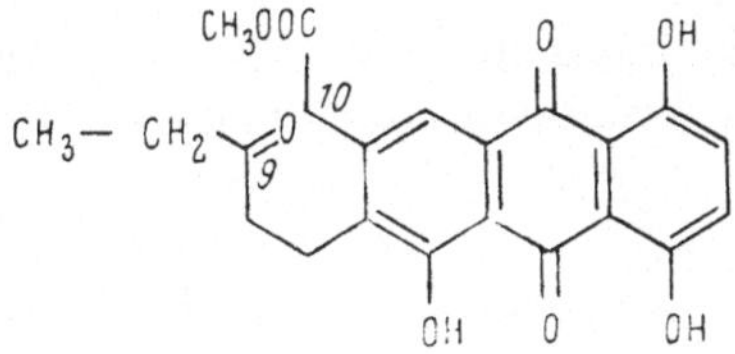

(LIII.)

Beim Erhitzen von ζ-Pyrromycinon entsteht neben dem Anhydroderivat (LII) eine kristallisierte, rote Verbindung vom Schmp. 182—183°, die mit ζ-Pyrromycinon isomer ist und außer den Carbonylbanden der Carbomethoxygruppe und des Chinonsystems noch eine weitere Carbonylbande bei 1720 cm^{-1} (5,81 μ) hat. Beim Erwärmen in Pyridin-Triäthylamin verwandelt sie sich in optisch inaktives ζ-Pyrromycinon. Auf Grund dieser Eigenschaften hat man ihr Formel (LIII) zugeschrieben. Man kann ihre Entstehung als Umkehrung der Kondensationsreaktion auffassen, die dem Biogenese-Schema der Anthracyclinone (S. 165) zufolge zur Bildung von Ring A führt.

Die beiden Anhydroderivate (LI) und (XLIX) haben im Gegensatz zum ζ-Pyrromycinon nur noch *ein* Asymmetriezentrum ($C_{(10)}$). Sie waren daher wertvolle Vergleichssubstanzen für die Auswertung von Zirkulardichroismus-Kurven der Anthracyclinone (S. 161).

D. Aklavinone.

Aklavinon.

Aklavinon $C_{22}H_{20}O_8$ (LIVa), gelbrote Kristalle vom Schmp. 170°, $p_{Ka} = 9,9$ in 90%igem Äthanol, wird durch milde Säurehydrolyse aus Aklavin (S. 179) freigesetzt [Gordon, Jackman, Ollis und Sutherland (43)].

Mit Acetanhydrid/Pyridin liefert es ein gelbes Triacetat vom Schmp. 198° mit einer OH-Bande bei 3610 cm^{-1} (2,77 μ) in Nujol, die einer tertiären Hydroxygruppe zuzuschreiben ist.

Die Anwesenheit einer Alkoxygruppe im Aklavinon ergibt sich: a) aus dem IR-Spektrum (Nujol) durch eine Carbonylbande bei 1740 cm^{-1} (5,75 μ) und b) aus der alkalischen Hydrolyse des Aklavinons, bei der eine Säure mit $p_{Ka} = 6,85$ und 10,6 (90%iges Aceton) entsteht. Analogiegründe lassen auf Vorliegen einer Carbomethoxygruppe schließen.

Im sichtbaren und UV-Gebiet zeigt Aklavinon in Methanol folgende λ_{max}- (ε_{max})-Werte: 229 mμ (43800), 258 mμ (26700), 288 mμ (11500), 430 mμ (13100); und Aklavinon-triacetat in Äthanol: 214 mμ (28000), 259 mμ (41900), 340 mμ (7100). Diesen Spektren nach ist 1,8-Dihydroxy-anthrachinon (XIV) der Chromophor des Aklavinons; ein Ergebnis, das im Einklang steht mit den IR-Spektren des Farbstoffes und seines Triacetates.

Kochen des Aklavinons in Toluol mit Toluolsulfosäure liefert gelb-rotes, kristallisiertes Bisanhydro-aklavinon, $C_{22}H_{16}O_6$, vom Schmp. 236°,

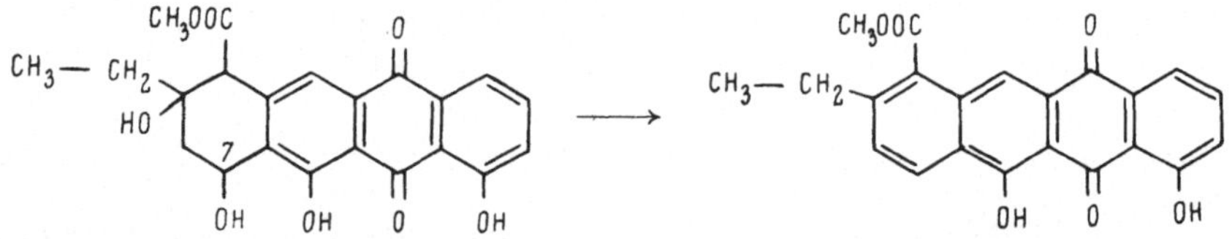

(LIVa.) Aklavinon.
(LIVb.) OH an C(7) = H, 7-Desoxy-aklavinon.

(LV.) Bisanhydro-aklavinon.

seinem Absorptions- und IR-Spektrum nach ein Derivat des 1,11-Di-hydroxy-tetracenchinons-(5,12) (IXb). Permanganat baut es zu Benzol-tetracarbonsäure-(1,2,3,4) (XLVI) ab, die man als Methylester vom Schmp. 130,5—131,5° identifizierte.

Aus diesen Befunden läßt sich für Aklavinon Formel (LIVa) und für Bisanhydro-aklavinon (LV) ableiten. (LIVa) wird durch das KMR-Spektrum des Aklavinons bestätigt (S. 156).

7-Desoxy-aklavinon.

7-Desoxy-aklavinon (LIVb) haben Ollis, Sutherland und Gordon, wie bereits erwähnt, aus der Kultur eines Aklavin produzierenden

Streptomyces-Stammes isoliert (*49, 51*). Angaben über seine Eigenschaften liegen noch nicht vor.

4. Die KMR-Spektren der Anthracyclinone.

Die KMR-Spektren von ε- und ζ-Pyrromycinon sowie von ε-Rhodomycinon und ε-Iso-rhodomycinon wurden gemessen, als die Konstitution dieser Anthracyclinone bereits durch chemische Methoden und absorptionsspektroskopischen Vergleich aufgeklärt war. Hier haben die KMR-Spektren nachträglich in allen Einzelheiten bestätigt, was mit konventionellen Verfahren erreicht worden war. Beim Aklavinon, β-Rhodomycinon, γ-Rhodomycinon, δ-Rhodomycinon und ζ-Pyrromycinon-P_2O_5-Produkt dagegen konnte Messung und Deutung ihrer KMR-Spektren rechtzeitig genug ausgenutzt werden, um die Strukturaufklärung wesentlich zu erleichtern. Wertvoll war, daß dabei die KMR-Spektren der in ihrer Struktur bekannten Anthracyclinone zum Vergleich herangezogen werden konnten.

Soweit möglich, wurden die KMR-Spektren in Deuterochloroform gemessen. War die Löslichkeit zu gering, um eine Konzentration von mindestens 2% zu erreichen, so verwendete man Trifluoressigsäure als Lösungsmittel. Sie hat gegenüber Deuterochloroform die Nachteile, daß: a) die OH-Protonen schnell mit denen des Lösungsmittels austauschen und daher nur die Signale der C—H-Protonen zu beobachten sind, und b) die Auflösung mancher Banden geringer ist.

In *Tabelle 3* (S. 156) sind die bisher vorliegenden Meßergebnisse zusammengestellt (*36*). Die dort in τ-Werten angegebenen Signale lassen sich in drei Gruppen einteilen: a) Die der Protonen von phenolischen Hydroxygruppen mit negativen τ-Werten. b) Die der aromatischen C—H-Protonen mit τ-Werten von 1,8—3,0 und c) die Signale aliphatischer C—H- und O—H-Protonen.

Für die Konstitutionsaufklärung der Anthracyclinone interessieren die Signale der aromatischen C—H-Protonen und ganz besonders die der zu Ring *A* gehörenden C—H-Protonen.

In den Pyrromycinonen (VIa, S. 129) und im δ-Rhodomycinon (XXXVIII, S. 146) ist Ring *D* symmetrisch substituiert. Die beiden chemisch gleichen Protonen an $C_{(2)}$ und $C_{(3)}$ geben sich durch ein Signal mit der relativen Intensität 2 zu erkennen, das beim ε-Pyrromycinon (XLVII) den τ-Wert 2,72 und beim δ-Rhodomycinon (XXXVIII) den τ-Wert 2,59 hat. Das Proton an $C_{(11)}$ des ε-Pyrromycinons gibt ein Signal bei $\tau = 2,31$ und das Proton an $C_{(6)}$ des δ-Rhodomycinons eines mit $\tau = 1,85$. Beide haben die relative Intensität 1.

Bei den Rhodomycinonen (IVa, S. 129) — ausgenommen δ-Rhodomycinon — sind die aromatischen C—H-Protonen chemisch verschieden und benachbart. Gegenseitige Spin-Kupplung äußert sich in einer Feinstruktur der Banden, die von L. F. Johnson analysiert worden ist (*36*).

Tabelle 3. Kernresonanzabsorptionen von Anthracyclinonen.

β-Rhodomycinon in $F_3C \cdot COOH$*			γ-Rhodomycinon in $F_3C \cdot COOH$*			δ-Rhodomycinon in $F_3C \cdot COOH$*			Zuordnung der Banden
τ	Int.	Form	τ	Int.	Form	τ	Int.	Form	
8,73	3	T	8,75	3	T	8,72	3	T	$-CH_3$
7,93	2	M	7,83	4	M	8,05	2	M	Äthyl-CH_2
7,50	2	D	7,79			7,44	2	D	Ring-CH_2, $C_{(8)}$
—	—	—	6,97	2	M	—	—	—	Ring-CH_2, $C_{(7)}$
—	—	—	—	—	—	—	—	—	$-OH$ an $C_{(7)}$
—	—	—	—	—	—	6,07	3	S	$-OCH_3$
—	—	—	—	—	—	—	—	—	$-OH$ an $C_{(9)}$
—	—	—	—	—	—	5,40	1	S	$CH_3OOC-C-H$, $C_{(10)}$
4,67	1	S	4,70	1	S	—	—	—	$HO-C-H$, $C_{(10)}$
4,48	1	M	—	—	—	4,65	1	M	$HO-C-H$, $C_{(7)}$
—	—	—	—	—	—	—	—	—	olefin. $C-H$, $C_{(8)}$
2,60	3	M	2,59	3	M	2,59	2	S	arom. $C-H$, $C_{(2)}$ und $C_{(3)}$
2,22			2,22			—	—	—	arom. $C-H$, $C_{(1)}$
—	—	—	—	—	—	—	—	—	arom. $C-H$, $C_{(11)}$
—	—	—	—	—	—	1,85	1	S	arom. $C-H$, $C_{(6)}$
—	—	—	—	—	—	—	—	—	phenol. $-OH$ an $C_{(4)}$
—	—	—	—	—	—	—	—	—	phenol. $-OH$ an $C_{(6)}$
—	—	—	—	—	—	—	—	—	phenol. $-OH$ an $C_{(1)}$
—	—	—	—	—	—	—	—	—	phenol. $-OH$ an $C_{(11)}$

ε-Rhodomycinon in $CDCl_3$**			ε-Iso-rhodomycinon in $CDCl_3$**			Aklavinon in $CDCl_3$***			Zuordnung der Banden
τ	Int.	Form	τ	Int.	Form	τ	Int.	Form	
8,85	3	T	8,84	3	T	8,90	3	?	$-CH_3$
8,32	2	Q	8,38	2	M	8,43	2	?	Äthyl-CH_2
7,72	2	D	7,70	2	D?	7,66	2	D?	Ring-CH_2, $C_{(8)}$
—	—	—	—	—	—	—	—	—	Ring-CH_2, $C_{(7)}$
—	—	—	—	—	—	—	—	—	$-OH$ an $C_{(7)}$
6,25	5	S	6,25	5	S	6,30	3	S	$-OCH_3$
—	—	—	—	—	—	—	—	—	$-OH$ an $C_{(9)}$
5,75	1	S	5,75	1	S	5,70	1	S	$CH_3OOC-C-H$, $C_{(10)}$
—	—	—	—	—	—	—	—	—	$HO-C-H$, $C_{(10)}$
4,67	1	T	4,65	1	T?	4,67	1	T?	$HO-C-H$, $C_{(7)}$
—	—	—	—	—	—	—	—	—	olefin. $C-H$, $C_{(8)}$
2,72	3	M	2,71	2	S	—	—	—	arom. $C-H$, $C_{(2)}$ und $C_{(3)}$
2,27			—	—	—	—	—	—	arom. $C-H$, $C_{(1)}$
—	—	—	—	—	—	—	—	—	arom. $C-H$, $C_{(11)}$
—	—	—	—	—	—	—	—	—	arom. $C-H$, $C_{(6)}$
$-1,9$	1	S	?†	1	S	—	—	—	phenol. $-OH$ an $C_{(4)}$
$-2,8$	1	S	?†	1	S	—	—	—	phenol. $-OH$ an $C_{(6)}$
—	—	—	?†	1	S	—	—	—	phenol. $-OH$ an $C_{(1)}$
$-3,55$	1	S	?†	1	S	—	—	—	phenol. $-OH$ an $C_{(11)}$

S = Singulett,   D = Dublett,   T = Triplett,   Q = Quadruplett,   M = Multiplett,
Int. = relative Intensität.

Literaturverzeichnis: SS. 179—182.

(Fortsetzung der Tabelle 3.)

ε-Pyrromycinon in CDCl₃**			ζ-Pyrromycinon in CDCl₃**			ζ-Pyrromycinon-P₂O₅-Produkt in F₃C·COOH*			Zuordnung der Banden
τ	Int.	Form	τ	Int.	Form	τ	Int.	Form	
8,97	3	T	8,96	3	T	8,77	3	T	—CH₃
8,47	2	Q	8,38	2	M	7,92	2	M	Äthyl-CH₂
7,48	2	M	7,84	2	M	—	—	—	Ring-CH₂, C₍₈₎
—	—	—	7,06	2	Q	6,56	2	D?	Ring-CH₂, C₍₇₎
6,67	1	S	—	—	—	—	—	—	—OH an C₍₇₎
6,48	3	S	6,27	4	S	6,08	3	S	—OCH₃
6,22	1	S	—	—	—	—	—	—	—OH an C₍₉₎
6,10	1	S	6,05	1	S	5,43	1	S	CH₃OOC—C—H, C₍₁₀₎
—	—	—	—	—	—	—	—	—	HO—C—H, C₍₁₀₎
4,75	1	T	—	—	—	—	—	—	HO—C—H, C₍₇₎
—	—	—	—	—	—	3,91	1	D?	olefin. C—H, C₍₈₎
2,72	2	S	2,72	2	S	2,75	2	S	arom. C—H, C₍₂₎ und C₍₃₎
—	—	—	—	—	—	—	—	—	arom. C—H, C₍₁₎
2,31	1	S	2,33	1	S	2,38	1	S	arom. C—H, C₍₁₁₎
—	—	—	—	—	—	—	—	—	arom. C—H, C₍₆₎
—2,0	1	S	—2,1	1	S	—	—	—	phenol. —OH an C₍₄₎
—2,7	1	S	—2,4	1	S	—	—	—	phenol. —OH an C₍₆₎
—2,85	1	S	—2,88	1	S	—	—	—	phenol. —OH an C₍₁₎
—	—	—	—	—	—	—	—	—	phenol. —OH an C₍₁₁₎

* Mit einem Varian A-60-Spektrometer von Dr. A. MELERA, Varian Research Laboratory, Zürich, gemessen. In F₃C·COOH werden nur C—H-Protonen gemessen, da die O—H-Protonen mit dem Lösungsmittel austauschen. Die τ-Werte sind um etwa 0,20 kleiner, wenn F₃C·COOH statt CDCl₃ als Lösungsmittel verwendet wird.

** Von Dr. J. SHOOLERY und Dr. LeRoy F. JOHNSON, Varian Associates, Palo Alto, California, gemessen.

*** J. J. GORDON, L. M. JACKMAN, W. D. OLLIS und I. O. SUTHERLAND (*43*).

† Trotz Verwendung einer gesättigten Lösung waren die Banden nur sehr schwach ausgebildet und wurden nicht ausgemessen.

Zusammenfassend ist festzustellen: Die Signale der aromatischen C—H-Protonen lassen ohne weiteres erkennen, ob ein Anthracyclinon vom Pyrromycinon- (VIa) oder Rhodomycinon-Typ (IVa) vorliegt, und beim ε-Pyrromycinon (XLVII, S. 149) und δ-Rhodomycinon (XXXVIII) zeigen sie, daß sich beide durch die Stellung eines C—H-Protons an Ring *B* unterscheiden.

Wertvoller sind die Aussagen, die man an Hand von Tabelle 3 über die Struktur des Ringes *A* machen kann. Signale geben in diesem Ring die C—H-Protonen und die Protonen der aliphatischen Hydroxygruppen. Die Signale der O—H-Protonen lassen sich von denen der C—H-Protonen leicht unterscheiden. Denn nach Säurezugabe kommt es zu einem schnellen Austausch zwischen den Protonen der Hydroxygruppen und denen des Lösungsmittels, der zur Folge hat, daß die Signale der Hydroxygruppen von der Stelle, die sie vor Säurezugabe einnehmen, verschwinden und in *einem* peak zusammenfallen, dessen τ-Wert von der Menge der zugesetzten Säure abhängt. Mit dieser Technik konnte man zeigen, daß die alkoholischen O—H-Protonen der Pyrromycinone bei $\tau = 6{,}67$, $6{,}22$ bzw. $6{,}27$ absorbieren und die des ε-Rhodomycinons bei $\tau = 6{,}25$. Die Signale liegen demnach ganz oder zum Teil unter dem der Methoxygruppe.

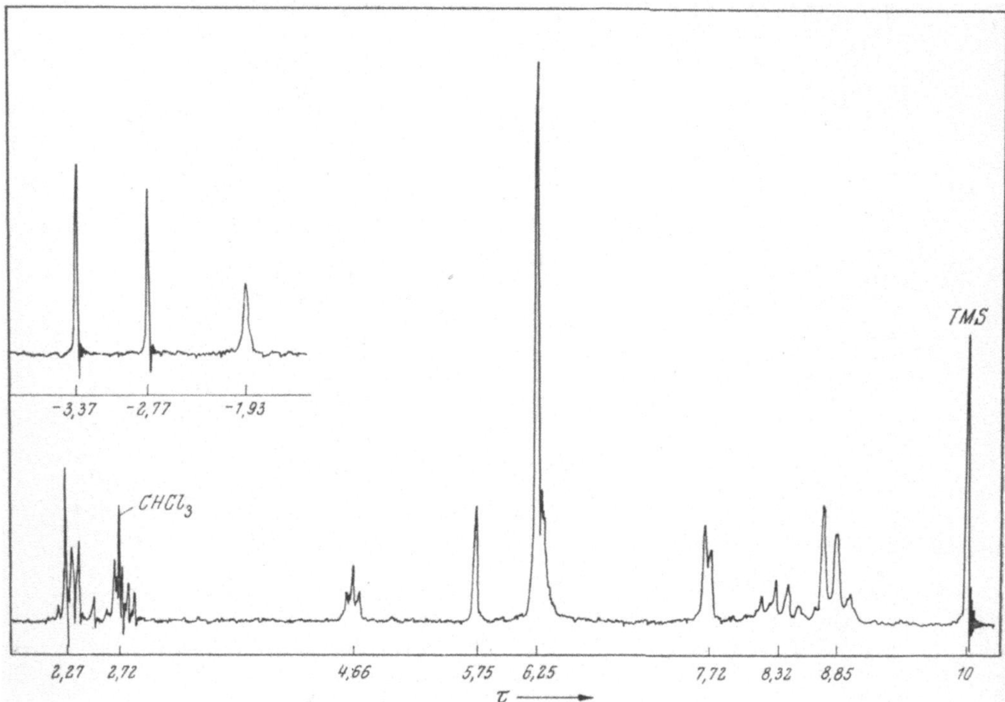

Abb. 6. KMR-Spektrum von ε-Rhodomycinon in Deuterochloroform. [Aus: Dissertation H. Brockmann jr., Göttingen 1963.]

Am Beispiel des ε-Rhodomycinon-Spektrums *(Abb. 6)* soll nun erläutert werden, welchen Protonen die Signale aus Ring *A* (LVI, S. 159) zuzuordnen sind.

Die Signale bei $\tau = 8{,}85$ (relative Intensität 3) und $\tau = 8{,}32$ (relative Intensität 2) sind charakteristisch für eine isolierte Äthylgruppe. Die beiden Methylen-Protonen verursachen eine Triplettaufspaltung der Methylbande ($\tau = 8{,}85$), und die drei Methylwasserstoffatome erzeugen ihrerseits eine Quadruplett-Aufspaltung der Methylenbande ($\tau = 8{,}32$).

Das Dublett mit $\tau = 7{,}72$ kann der zum aromatischen Ringsystem β-ständigen $C_{(8)}$-Methylengruppe zugeschrieben werden. Die Aufspaltung zum Dublett zeigt, daß von den beiden der Methylengruppe benachbarten C-Atomen nur *eines* mit *einem* Wasserstoffatom verbunden ist. Das Signal dieses Protons sollte durch die benachbarte Methylengruppe aufgespalten sein, wobei die gleiche Kopplungs-Konstante zu erwarten ist wie bei der Dublettaufspaltung der Methylenbande bei $\tau = 7{,}72$.

Ein Signal, das dieser Forderung genügt, liegt bei $\tau = 4{,}67$. Aus seiner chemischen Verschiebung kann man schließen, daß das zu ihm gehörende Proton an einem C-Atom steht, das a) einem aromatischen Ringsystem benachbart ist und von dem b) eine Bindung zu einem Sauerstoffatom ausgeht.

Ferner gibt das Signal mit $\tau = 4{,}67$ sogar Auskunft über die Konstellation des Wasserstoffatoms an $C_{(7)}$. Denn ein symmetrisches Triplett

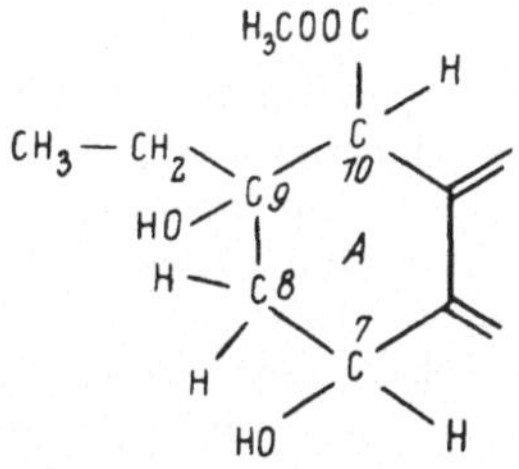

(LVI). Ring *A* von ε-Rhodomycinon.

können die benachbarten Methylenprotonen nur dann erzeugen, wenn sie vom Wasserstoffatom an $C_{(7)}$ gleich weit entfernt sind und infolgedessen auch die Kopplungs-Konstanten gleich groß sind. Gleich weit entfernt von den beiden Methylen-Protonen ist der Wasserstoff an $C_{(7)}$ aber nur, wenn er äquatorial und die Hydroxygruppe an $C_{(7)}$ dementsprechend axial steht.

Der scharfe peak bei $\tau = 6{,}25$ ist den Protonen der Carbomethoxygruppe zuzuordnen; er fehlt daher in den Spektren von β-Rhodomycinon und γ-Rhodomycinon.

Durch das Singulett bei $\tau = 5{,}75$ schließlich gibt sich das Proton an $C_{(10)}$ zu erkennen. Denn wie aus der chemischen Verschiebung hervorgeht, steht das zu diesem Signal gehörende Proton an einem C-Atom, das einem aromatischen Ring und einer Carbomethoxygruppe benachbart ist. Die Singulett-Struktur zeigt, daß keines der beiden benachbarten C-Atome Wasserstoff trägt. Dieses Signal fehlt im Spektrum von β- und γ-Rhodomycinon. An seine Stelle tritt hier ein scharfer peak bei $\tau = 4{,}67$ bzw. 4,70, der typisch ist für das Proton eines C-Atoms, a) mit einem aromatischen Ring und b) mit einem Sauerstoffatom verbunden ist.

5. Zur Stereochemie der Anthracyclinone.

Ring A der Anthracyclinone enthält zwei oder drei asymmetrische Kohlenstoffatome ($C_{(9)}$ und $C_{(10)}$ bzw. $C_{(7)}$, $C_{(9)}$ und $C_{(10)}$). Zu den stereochemischen Problemen, die sich damit ergeben, gehört die angesichts ihrer strukturellen Verwandtschaft naheliegende Frage, ob die Anthracyclinone an allen Asymmetriezentren die gleiche Konfiguration haben. Versuche, diese Frage durch Vergleich der Rotationsdispersionskurven zu beantworten, haben nicht zum Ziel geführt; die Absorption der Anthracyclinone und ihrer Acetate ist zu groß, um eine einwandfreie Messung der Rotationsdispersion zu gestatten.

Dagegen lassen sich, wie Brockmann jr. und Legrand (*37, 38*) kürzlich gezeigt haben, an Hand von Zirkulardichroismus-Kurven Aussagen über die Konfiguration der Anthracyclinone machen.

Als Zirkulardichroismus bezeichnet man die verschieden starke Absorption von links- oder rechts-zirkular polarisiertem Licht durch Verbindungen, deren Chromophor entweder selber asymmetrisch gebaut ist oder — wenn das nicht der Fall ist — in der Nähe eines Asymmetriezentrums liegt. Sind ε_l und ε_r die molaren Extinktionskoeffizienten für links- und rechtszirkular polarisiertes Licht, so ist die Größe des Zirkulardichroismus definiert als $\Delta\varepsilon = \varepsilon_l - \varepsilon_r$. Die Kurven, die man erhält, wenn man $\Delta\varepsilon$ in Abhängigkeit von der Wellenlänge aufträgt, können positiv oder negativ sein; je nachdem, ob der Absorptionskoeffizient für links-zirkulares Licht größer oder kleiner ist als für rechts-zirkulares. Die Zirkulardichroismus-Kurven bringen die Wechselwirkung zwischen polarisiertem Licht und asymmetrischen Verbindungen bis in Einzelheiten einfacher und häufig genauer zum Ausdruck als die Kurven der Rotationsdispersion (*53*). Daß dennoch der Zirkulardichroismus bis vor kurzem kaum zur Untersuchung asymmetrischer Verbindungen herangezogen worden ist, hat meßtechnische Gründe. Die dichroitische Absorption ist sehr klein, sie beträgt im günstigsten Fall einige Hundertstel der mit natürlichem Licht gemessenen Absorption. Die Anforderung, die man an eine Apparatur zur Messung von Zirkulardichroismus-Kurven stellen muß, sind daher sehr hoch. Sie konnten erst in jüngster Zeit durch ein von Grosjean und Legrand (*44*) entwickeltes Gerät erfüllt werden, mit dem man Zirkulardichroismus-Kurven im sichtbaren und UV-Gebiet kontinuierlich ebenso bequem messen kann wie eine gewöhnliche Absorptionskurve.

Abb. 7 zeigt die mit diesem Gerät gemessenen Kurven von Anthracyclinonen. Der Meßbereich 270—390 mμ wurde aus folgendem Grunde gewählt.

Aus gleichartigen Zirkulardichroismus-Kurven verschiedener Verbindungen läßt sich nur dann auf gleiche Konfiguration schließen, wenn die Verbindungen den gleichen Chromophor enthalten. Diese Forderung ist für die Pyrromycinone und Rhodomycinone erfüllt. Von ihrem Chromophor, dem 1,4,5-Trihydroxy-anthrachinon (XIII, S. 129), unterscheidet sich der Chromophor (XII) der Iso-rhodomycinone durch eine zusätzliche α-Hydroxygruppe. Sie wirkt sich sehr deutlich auf die Lage der langwelligen Absorptionsmaxima aus (Abb. 1 und 2, SS. 132, 133), nur wenig dagegen auf die Absorption zwischen 270 und 390 mμ. Denn in diesem Gebiet zeigen die Pyrromycinone, Rhodomycinone und Iso-rhodomycinone eine Bande, die bei allen praktisch die gleiche Lage und Extinktion hat und deshalb einem allen Anthracyclinonen gemeinsamen Teilchromophor zugeordnet werden kann. Die zwischen 270 und 390 mμ gemessenen Zirkulardichroismus-Kurven zeigen somit die Einflüsse der

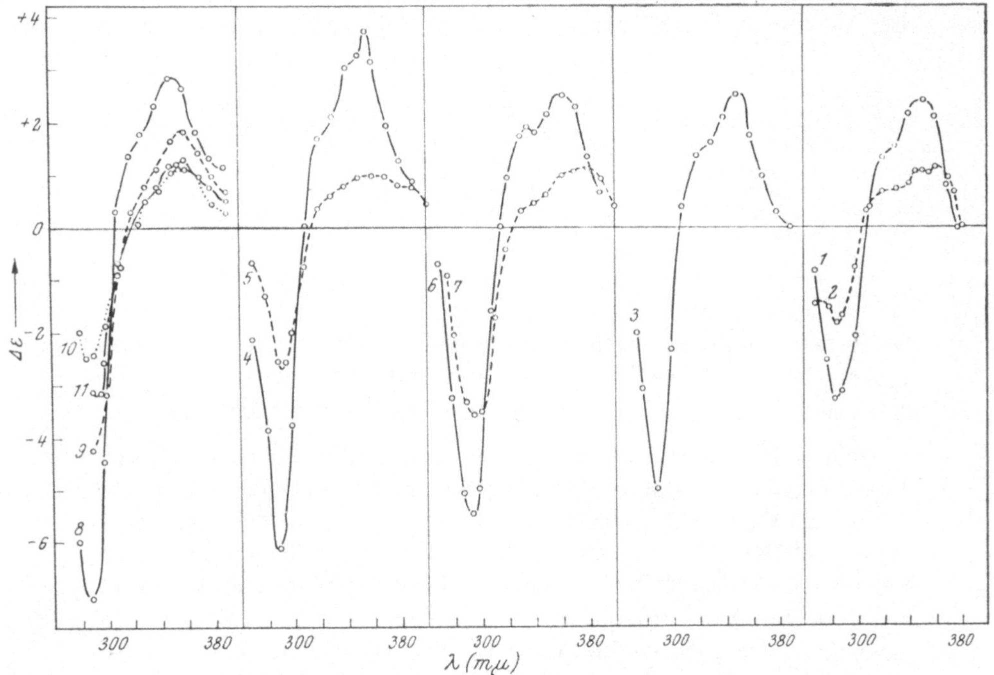

Abb. 7. Zirkulardichroismus-Kurven von Anthracyclinonen in Dioxan [BROCKMANN jr. und LEGRAND (37, 38)]. $I = \beta$-Rhodomycinon; $2 = \gamma$-Rhodomycinon; $3 = \delta$-Rhodomycinon; $4 = \varepsilon$-Rhodomycinon; $5 = \zeta$-Rhodomycinon; $6 = \varepsilon$-Iso-rhodomycinon; $7 = \zeta$-Iso-rhodomycinon; $8 = \varepsilon$-Pyrromycinon; $9 = \zeta$-Pyrromycinon; $10 = \zeta$-Pyrromycinon-P_2O_5-Produkt; $11 = \zeta$-Pyrromycinon-PTS-Produkt. [Aus: Tetrahedron 19, 395 (1963).]

einzelnen Asymmetriezentren an $C_{(9)}$ und $C_{(10)}$ bzw. $C_{(7)}$, $C_{(9)}$ und $C_{(10)}$ auf ein und denselben Chromophor und erfüllen damit die oben genannte Voraussetzung für eine Vergleichbarkeit der Kurven.

Die Zirkulardichroismus-Kurven der Anthracyclinone (Abb. 7) haben alle den gleichen charakteristischen S-förmigen Verlauf. Desgleichen die Kurven des im Abschnitt III, 3 (S. 153) beschriebenen ζ-Pyrromycinon-P_2O_5-Produktes (LI) und ζ-Pyrromycinon-PTS-Produktes (XLIX). Beide Verbindungen sind für den Vergleich wichtig, weil sie nur *ein* Asymmetrie-zentrum ($C_{(10)}$) (LVII) enthalten, das — ihrer Herkunft aus ζ-Pyrromycinon entsprechend — bei beiden die gleiche Konfiguration hat. Ihre Kurven sind daher einander gleich; bei entgegengesetzter Konfiguration wären sie spiegelbildlich.

ζ-Pyrromycinon (XLVIIIa, S. 149), ζ-Rhodomycinon (XXVIa, S. 142) und ζ-Iso-rhodomycinon (XVb, S. 139) haben außer dem Asymmetriezentrum an $C_{(10)}$ noch ein zweites an $C_{(9)}$ (LVIII), das weiter vom Chromophor entfernt ist als $C_{(10)}$ und daher sicherlich keinen großen Einfluß auf den Zirkulardichroismus hat. Da die Kurven der drei Verbindungen (XLVIIIa, XXVIa, XVb) im Verlauf und in der

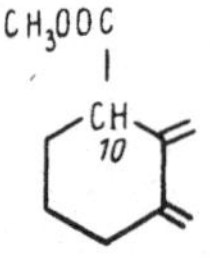

(LVII.) 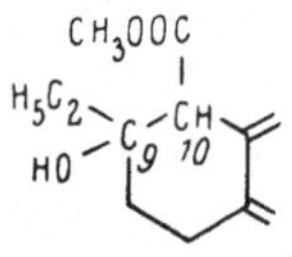   (LVIII.) 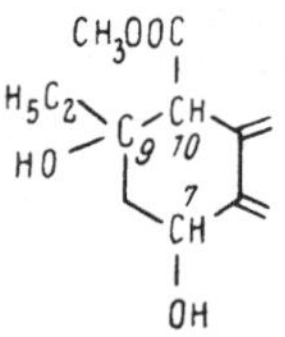   (LIX.)

Größe der Amplitude denen von (LI) und (XLIX) gleich sind, kann als sicher gelten, daß alle an $C_{(10)}$ die gleiche Konfiguration haben. Über die Konfiguration an $C_{(9)}$ dagegen sagen die Kurven nichts aus.

Beim ε-Pyrromycinon (XLVII, S. 149), ε-Rhodomycinon (XXIVa, S. 142) und ε-Iso-rhodomycinon (XVa) kommt zu den beiden Asymmetriezentren an $C_{(9)}$ und $C_{(10)}$ noch ein drittes an $C_{(7)}$, das dem Chromophor nahe ist (LIX). Die Kurven der drei Verbindungen haben den gleichen Verlauf wie die von ζ-Pyrromycinon, ζ-Rhodomycinon und ζ-Iso-rhodomycinon, ihre Amplitude ist jedoch erheblich größer, und zwar bei allen um nahezu den gleichen Betrag; ein Ergebnis, das nur verständlich ist, wenn die drei Verbindungen (XLVII, XXIVa und XVa) an $C_{(10)}$ die gleiche Konfiguration haben wie die anderen Anthracyclinone und das dritte Asymmetriezentrum an $C_{(7)}$ bei allen dreien gleich konfiguriert ist.

Zusammenfassend ergibt sich somit: a) Alle Anthracyclinone mit einer Carbomethoxygruppe an $C_{(10)}$ haben an diesem Asymmetriezentrum die gleiche Konfiguration. b) Alle Anthracyclinone mit einer Hydroxygruppe an $C_{(7)}$ haben an $C_{(7)}$ die gleiche Konfiguration.

Aussagen über die relative Konfiguration innerhalb des Moleküls kann man bei den eben genannten Anthracyclinonen nicht machen; wohl aber beim β-Rhodomycinon (XXVIIIa, S. 142) und γ-Rhodomycinon (XXXa, S. 142). Denn daß die $\Delta\varepsilon_{max}$-Werte von β-Rhodomycinon fast doppelt so groß sind wie die von γ-Rhodomycinon, läßt sich, da der Einfluß des Asymmetriezentrums an $C_{(9)}$ auf den Zirkulardichroismus gering ist, nur verstehen, wenn die beiden Hydroxygruppen des β-Rhodomycinons an $C_{(7)}$ und $C_{(10)}$ in bezug auf den Chromophor sterisch gleich angeordnet sind, oder anders gesagt, wenn die eine oberhalb und die andere unterhalb der Ringebene liegt.

Nimmt man an, daß die Hydroxyle an $C_{(9)}$ und $C_{(10)}$ des γ-Rhodomycinons deshalb so träge mit Perjodsäure reagieren, weil sie *trans*-ständig sind, so kommt man für Ring A des β- und γ-Rhodomycinons zu den Stereoformeln (LX a) und (LX b), die noch nichts über die absolute Konfiguration aussagen.

Entsprechende Formeln (LXI a und LXI b) kann man auf Grund folgender Befunde und Überlegungen für ε- und ζ-Pyrromycinon, δ-, ε- und ζ-Rhodomycinon sowie ε- und ζ-Iso-rhodomycinon ableiten. Wie auf S. 153 gezeigt, entsteht aus ζ-Pyrromycinon beim Erhitzen eine Anhydroverbindung, in der die durch Eliminierung entstandene Doppel-

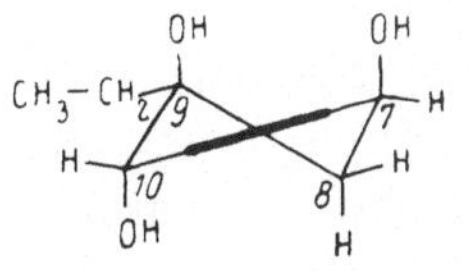

(LX a.)
(LX b.) OH an C(7) = H.

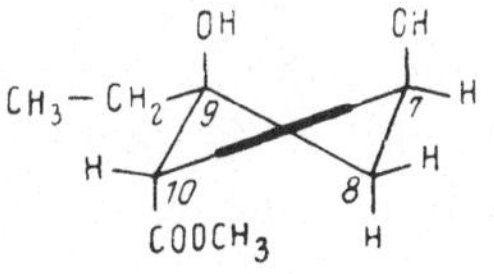

(LXI a.)
(LXI b.) OH an C(7) = H.

bindung zwischen $C_{(9)}$ und $C_{(10)}$ liegt (LII). Da Wasser bei Pyrolysen meistens *cis*-eliminiert wird, läßt die Bildung der Anhydroverbindung (LII) darauf schließen, daß die Hydroxygruppe an $C_{(9)}$ des ζ-Pyrromycinons in *cis*-Stellung zum Wasserstoff an $C_{(10)}$ steht. Da man ferner aus ε-Iso-rhodomycinon-tetramethyläther und 2,2-Dimethoxy-propan ein cyclisches Isopropyliden-derivat gewinnen konnte [BROCKMANN jr. (*36*)], müssen die Hydroxygruppen an $C_{(7)}$ und $C_{(9)}$ in ε-Iso-rhodomycinon — und nach dem oben Gesagten demnach auch in den anderen Anthracyclinonen, die eine Carbomethoxygruppe tragen — *cis*-ständig sein.·

Die in (LX) und (LXI) wiedergegebenen Konstellationen sind die stabilsten, weil sich in cyclischen *cis*-1,3-Diolen zwischen axialen Hydroxygruppen eine Wasserstoffbrücke bilden kann und weil die Äthylgruppe als größter Substituent äquatoriale Lage hat.

6. Zur Biogenese der Anthracyclinone.

Wie BIRCH und Mitarb. (*1*) gezeigt haben, steht die Konstitution vieler in höheren Pflanzen und Mikroorganismen vorkommender Phenol- und Chinonderivate mit einer zuerst von J. N. COLLIE ausgesprochenen Hypothese in Einklang, nach der solche Verbindungen in der Zelle durch eine „Kopf-Schwanz-Verknüpfung" von Acetat-Einheiten aufgebaut werden. Insbesondere bei Derivaten von Hydroxy-chinonen, z. B. Hydroxy-anthrachinonen, stimmt die gefundene Konstitution mit der hypothetischen so häufig überein, daß in Fällen, in denen für ein natürlich vorkommendes Hydroxy-chinonderivat mehrere Formeln zur Diskussion stehen, diejenige den Vorzug verdient, die mit der Acetathypothese in Einklang steht.

Werden die Anthracyclinone in der Zelle der Acetathypothese entsprechend aufgebaut, so könnte durch „Kopf-Schwanz-Verknüpfung" von zehn Essigsäure-Einheiten zunächst ein Polyketon (LXIIb) und daraus — durch Kondensation der Carbonyle mit den reaktionsfähigen Methylengruppen — das tetracyclische Zwischenprodukt (LXIIIb) entstehen. In diesem Fall müßte die Äthylgruppe der Anthracyclinone im Verlauf der Biosynthese formal gesehen durch Methylierung aus der Methylgruppe von (LXIIb), (LXIIIb) oder einem der Folgeprodukte hervorgehen: etwa so, daß ein aus elf Acetat-Einheiten entstandenes Polyketon (LXIIc) oder eines seiner Folgeprodukte, z. B. (LXIIIc), im Sinne der Formel (LXIV) durch eine als „Einkohlenstoffquelle" fungierende Verbindung, z. B. Methionin, methyliert und anschließend der Rest CH_3CO durch eine Ketonspaltung entfernt wird.

Außer dieser Möglichkeit, die Anthracyclinone mit einer Äthylgruppe auszurüsten, ließ sich noch eine zweite voraussehen: Einbau der fertigen Äthylgruppe dadurch, daß sich eine Propionat-Einheit mit neun Acetat-Einheiten zum Polyketon (LXIIa) kondensiert, nach dessen Cyclisierung zu (LXIIIa) dann bereits das Kohlenstoffgerüst der Anthracyclinone vorliegt. Daß die *Streptomyces*-Zelle diesen Weg einschlägt, zeigten Versuch von OLLIS und SUTHERLAND (*49*). Ein zur ε-Pyrromycinon-Synthese befähigter *Streptomyces*-Stamm bildete, wenn seine Kulturlösung das Natriumsalz der ^{14}C-Propionsäure (LXX) enthielt, ein ε-Pyrromycinon (LXXI), das beim oxydativen Abbau Essigsäure ohne ^{14}C und Propionsäure mit ^{14}C lieferte.

In anderen Ansätzen diente ^{14}C-markiertes Natriumacetat als Substrat. $Ba(OH)_2$-Abbau des entstandenen ε-Pyrromycinons (LXXII) und oxydativer Abbau der aus ihm gewonnenen Bisanhydroverbindung (LXXIII) zu Benzol-tetracarbonsäure-(1,2,3,4) (LXXV) zeigten, daß neun Acetat-Einheiten in das ε-Pyrromycinon-Molekül eingebaut worden waren. Aus dem geringen ^{14}C-Gehalt der beim Kuhn-Roth-Abbau aus (LXXII) erhaltenen Propionsäure und Essigsäure ergab sich, daß $C_{(9)}$ und die Äthylgruppe von (LXXII) anderen Ursprungs waren als der Rest des Moleküls.

Kommt es beim Zwischenprodukt (LXIIIa) zu einer Reduktion der $C_{(2)}$- und $C_{(7)}$-Carbonylgruppe und anschließender Eliminierung der $C_{(2)}$-

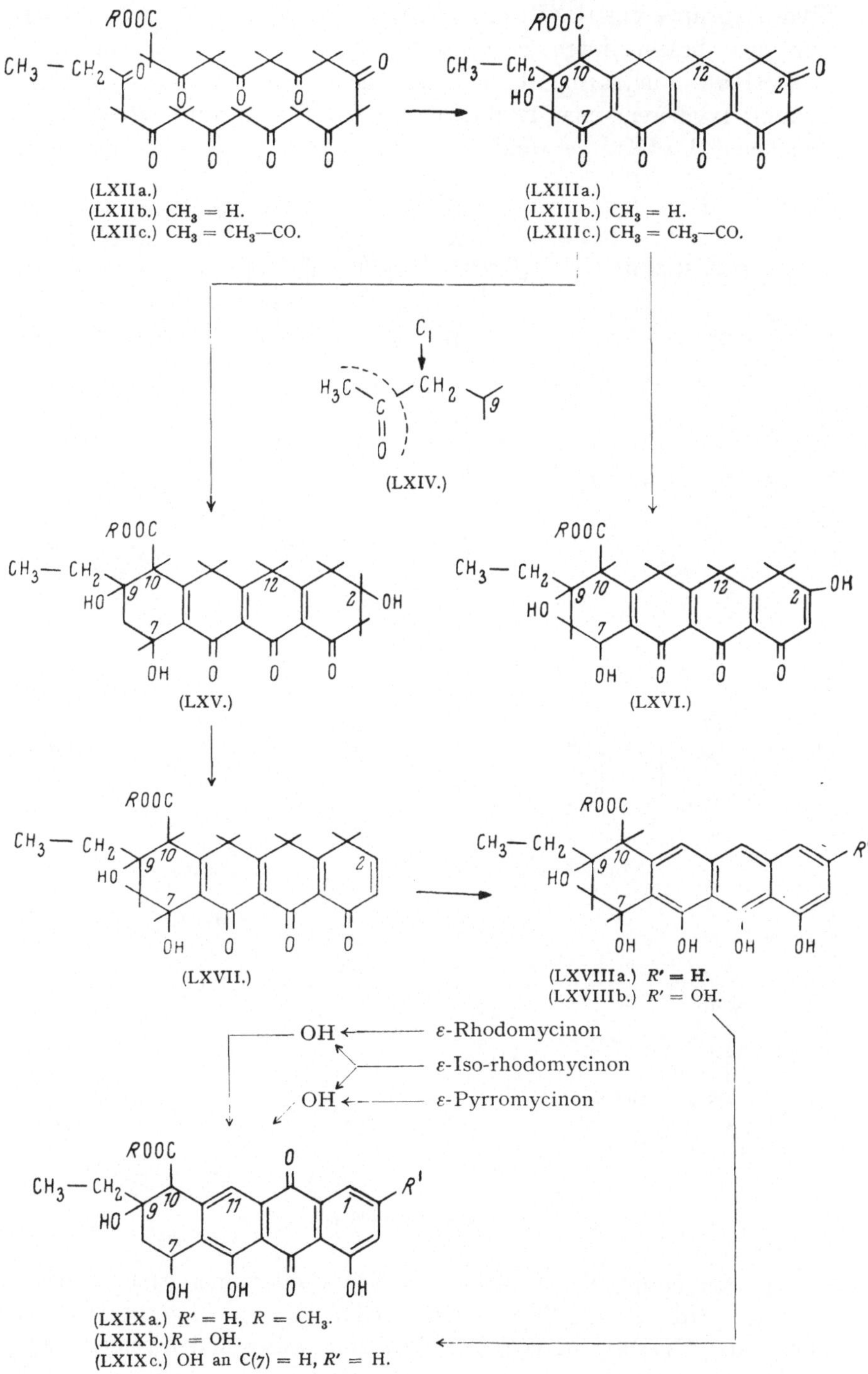

(LXIIa.)
(LXIIb.) $CH_3 = H.$
(LXIIc.) $CH_3 = CH_3—CO.$

(LXIIIa.)
(LXIIIb.) $CH_3 = H.$
(LXIIIc.) $CH_3 = CH_3—CO.$

(LXIV.)

(LXV.)

(LXVI.)

(LXVII.)

(LXVIIIa.) $R' = H.$
(LXVIIIb.) $R' = OH.$

OH ←——— ε-Rhodomycinon
       ——— ε-Iso-rhodomycinon
OH ←——— ε-Pyrromycinon

(LXIXa.) $R' = H, R = CH_3.$
(LXIXb.) $R = OH.$
(LXIXc.) OH an C(7) $= H, R' = H.$

Formelübersicht 6.

Hydroxygruppe aus (LXV) in Form von Wasser, so bildet sich (LXVII) und aus diesem durch Enolisierung das Chrysazinanthranol-Derivat (LXVIIIa). Chrysazinanthranol läßt sich leicht zu Chrysazin (XIV, S. 129) oxydieren, eine Reaktion, die (LXVIIIa) (mit $R = CH_3$) in Aklavinon (LXIXa) verwandeln würde.

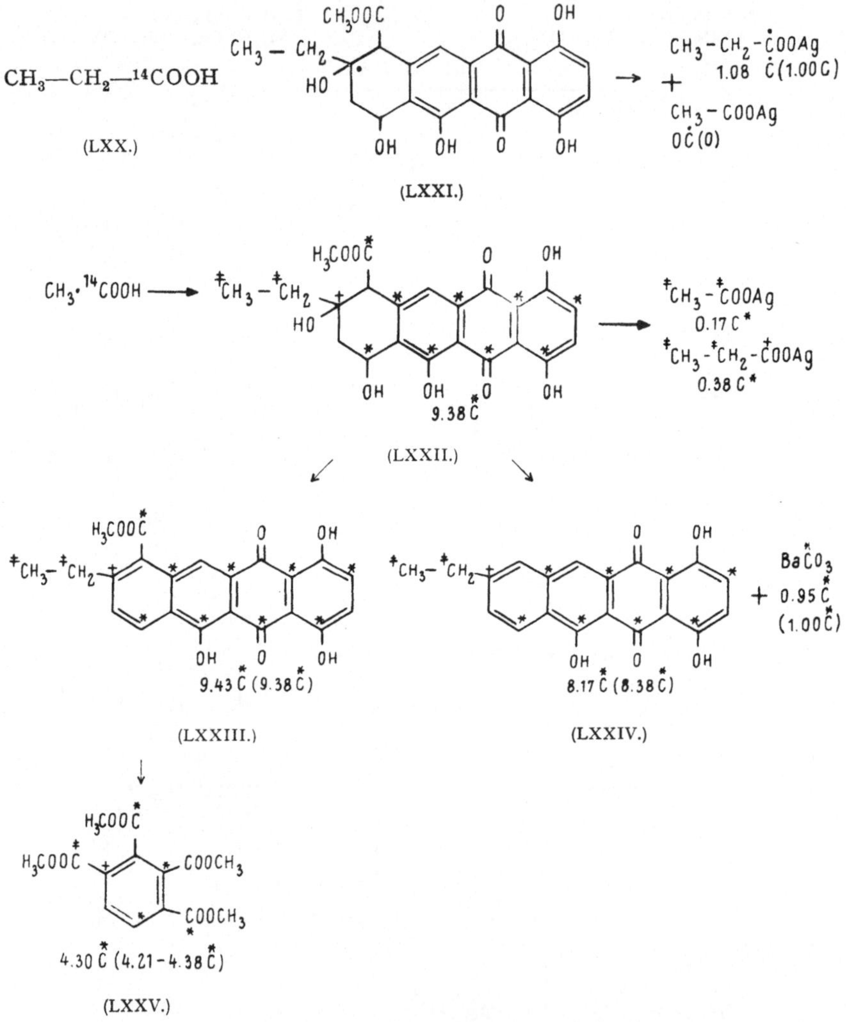

(LXX.)

(LXXI.)

(LXXII.)

(LXXIII.)

(LXXIV.)

(LXXV.)

Formelübersicht 7.

Hydroxylierung des Aklavinons (LXIXa) oder einer seiner Vorstufen an $C_{(1)}$ würde zum ε-Pyrromycinon (XLVII), an $C_{(11)}$ zum ε-Rhodomycinon (XXIVa), an $C_{(1)}$ *und* $C_{(11)}$ zum ε-Iso-rhodomycinon (XVa) führen.

Literaturverzeichnis: SS. 179—182.

Analoge Hydroxylierungen nimmt man bei der Biogenese von Hydroxy-anthrachinonen an. Helminthosporin, Islandicin und Cynodontin, drei von *Penicillium*-Arten produzierte Hydroxy-anthrachinone, entstehen der Acetat-Hypothese nach dadurch, daß aus acht Acetat-Einheiten hervorgegangene Chrysophansäure (LXXVI) an $C_{(1)}$ oder $C_{(8)}$ bzw. $C_{(1)}$ *und* $C_{(8)}$ hydroxyliert wird.

(LXXVI.)

Jedem der drei Anthracyclinone ε-Pyrromycinon (XLVII), ε-Rhodomycinon (XXIVa) und ε-Iso-rhodomycinon (XVa) entspricht eines, dem bei sonst gleicher Konstitution die Hydroxygruppe an $C_{(7)}$ fehlt [ζ-Pyrromycinon (XLVIIIa), ζ-Rhodomycinon (XXVIa), ζ-Iso-rhodomycinon (XVb)]. Sie könnten im Laufe der Biogenese aus Aklavinon (LXIXa) oder einer seiner Vorstufen durch reduktive Entfernung des $C_{(7)}$-Carbonylsauerstoffatoms hervorgehen; z. B. aus (LXVII) durch H_2O-Eliminierung von $C_{(7)}$ und $C_{(8)}$ und anschließende Hydrierung der entstandenen Doppelbindung. Das so zu erwartende Desoxy-aklavinon (LXIXc) haben OLLIS und SUTHERLAND (*51*), wie oben erwähnt, tatsächlich gefunden. Seine Hydroxylierung an $C_{(1)}$, $C_{(11)}$ oder $C_{(1)}$ *und* $C_{(11)}$ würde zum ζ-Pyrromycinon (XLVIIIa), ζ-Rhodomycinon (XXVIa) bzw. ζ-Iso-rhodomycinon (XVb) führen.

Nimmt man an, daß die Zelle die Carboxy- bzw. Carbomethoxygruppe an $C_{(10)}$ durch eine Hydroxygruppe ersetzen kann (wobei die Hydroxylierung vor oder nach der Decarboxylierung stattfinden könnte), so würde sich die Bildung von β- und γ-Rhodomycinon sowie β- und γ-Iso-rhodomycinon zwanglos in das Biogenese-Schema einfügen. β- und γ-Rhodomycinon würden dann aus ε-Rhodomycinon bzw. ζ-Rhodomycinon oder einer ihrer Vorstufen hervorgehen, β- und γ-Iso-rhodomycinon aus ε-Iso-rhodomycinon bzw. ζ-Iso-rhodomycinon oder einer ihrer Vorstufen.

Die praktische Bedeutung des vorstehenden Biogenese-Schemas liegt darin, daß es: a) gewisse Kriterien für die Richtigkeit einer Anthracyclinonformel liefert und b) Voraussagen darüber gestattet, welche Anthracyclinone außer den bereits bekannten sonst noch von *Streptomyces*-Arten synthetisiert werden könnten.

Die Kriterien betreffen die Stellung der Carbomethoxy- und Äthylgruppe sowie die Anordnung der Hydroxygruppen. Aus einem Polyketon (LXIIa) können theoretisch zwei Verbindungen mit gleichem Ringsystem entstehen, nämlich (LXIIIa) und (LXXVII). Voraussage der Hypothese ist demnach: a) Die Äthylgruppe aller Anthracyclinone steht an Ring A in β-Stellung zum Anthrachinon-Ringsystem. b) Die

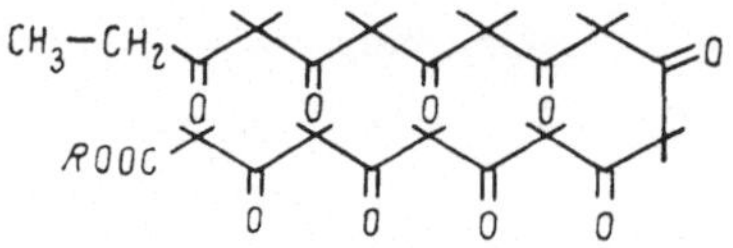

Carbomethoxygruppe ist in allen Anthracyclinonen der Äthylgruppe benachbart; ob an $C_{(8)}$ oder $C_{(10)}$, bleibt offen. Die zunächst angenommene ε-Pyrromycinon-Formel (22) mit der Carbomethoxygruppe an $C_{(8)}$ war somit von der Acetathypothese aus gesehen ebenso wahrscheinlich wie die richtige Formel (XLVII, S. 149) mit der Carbomethoxygruppe an $C_{(10)}$.

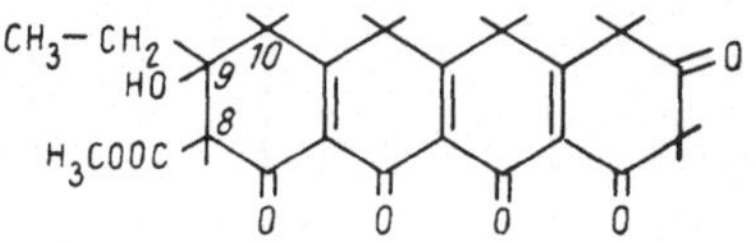

Für die Stellung der Hydroxygruppen ergibt sich aus der Acetathypothese folgendes: a) Bei allen Anthracyclinonen *mit* einer Carbomethoxygruppe sind die Hydroxyle des Ringes A mit $C_{(7)}$ und (oder) $C_{(9)}$ verknüpft. Bei den Anthracyclinonen *ohne* Carbomethoxygruppe ist wahrscheinlich eine Hydroxygruppe mit $C_{(9)}$ verknüpft. b) Die zum Anthrachinon-Ringsystem gehörenden Hydroxygruppen stehen, wenn, wie beim Aklavinon und Desoxy-aklavinon, nur zwei vorhanden sind, an $C_{(4)}$ und $C_{(6)}$. Ist wie bei den Pyrromycinonen und Rhodomycinonen das Anthrachinon-Ringsystem mit drei α-Hydroxygruppen besetzt, so nehmen diese entweder 1,4,6- oder 4,6,11-Stellungen ein. Diese Vorhersage ist, wie oben gezeigt, beim ε-Pyrromycinon bestätigt worden und macht für ε-Rhodomycinon Formel (XXIVa, S. 142) wahrscheinlicher als (XXIVb).

Die Kriterien für die Stellung der α-Hydroxygruppen gelten selbstverständlich nur unter den oben genannten Voraussetzungen, d. h. wenn die Zelle Sauerstoff von $C_{(4)}$ oder $C_{(6)}$ nicht entfernen kann. Wie kürzlich die Konstitutionsaufklärung des δ-Rhodomycinons gezeigt hat (13, 14), muß bei dessen Biogenese jedoch Sauerstoff von $C_{(6)}$ abgespalten werden. In solchen Fällen kann man aus der Acetathypothese keine Kriterien mehr für die Stellung der α-Hydroxygruppen ableiten.

Von den Asymmetriezentren der Anthracyclinone bilden sich der Hypothese nach die an $C_{(9)}$ und $C_{(10)}$ bei der Kondensation der dem

Literaturverzeichnis: SS. 179—182.

Äthylrest benachbarten Carbonylgruppe mit der zur Carboxy- bzw. Carbomethoxygruppe benachbarten Methylengruppe (LXIIa → LXIIIa); einem Syntheseschritt, der *vor* der weiteren „Ausdifferenzierung" der Zwischenprodukte liegt und daher bei allen Anthracyclinonen der gleiche sein sollte. Zu erwarten ist demnach, daß alle Anthracyclinone mit einer Carbomethoxygruppe die gleiche Konfiguration an $C_{(9)}$ und $C_{(10)}$ haben. Für $C_{(10)}$ wird diese Erwartung durch die Zirkulardichroismus-Kurven (S. 161) bestätigt. Danach ist kaum zu bezweifeln, daß alle auch an $C_{(9)}$ gleich konfiguriert sind, denn daß bei einer Reaktion, die zur Bildung von zwei Asymmetriezentren führt, das eine immer die gleiche Konfiguration annimmt, das andere aber nicht, ist so gut wie ausgeschlossen.

Falls Anthracyclinone *ohne* Carbomethoxygruppe aus solchen *mit* Carbomethoxygruppe entstehen, müssen beiden Typen an $C_{(9)}$ gleiche Konfiguration haben. Formal wird bei dieser Umwandlung die Carbomethoxygruppe durch eine Hydroxygruppe ersetzt. Dabei bleibt nach den auf S. 163 angeführten Befunden die Konfiguration an $C_{(10)}$ erhalten.

Das Asymmetriezentrum an $C_{(7)}$ entsteht durch Reduktion einer Carbonylgruppe. Daß sich die dabei beteiligten Enzyme bei den verschiedenen *Streptomyces*-Arten und -Stämmen in ihrer Stereospezifität unterscheiden, ist ebenso unwahrscheinlich wie die Annahme, daß derartiges bei der Ausbildung der Asymmetriezentren an $C_{(9)}$ und $C_{(10)}$ der Fall sein könnte. Wie oben gezeigt, ist diese Überlegung richtig, denn wie aus den Zirkulardichroismus-Kurven hervorgeht, sind alle Anthracyclinone mit einer Hydroxygruppe an $C_{(7)}$ an diesem C-Atom gleich konfiguriert.

Fragt man, welche bisher unbekannten Anthracyclinone noch gefunden werden könnten, so ergibt sich aus dem Biogenese-Schema, daß zu rechnen ist mit:

a) Zwei Vertretern, die aus Aklavinon und Desoxy-aklavinon dadurch entstehen, daß deren Carbomethoxygruppe durch eine Hydroxygruppe ersetzt wird.

b) Zwei Vertretern, die sich in gleicher Weise von ε-Pyrromycinon (XLVII) und ζ-Pyrromycinon (XLVIIIa) ableiten.

c) Drei Anthracyclinonen, die sich von δ-Rhodomycinon dadurch ableiten, daß 1.) die 7-Hydroxygruppe durch Wasserstoff und 2.) in dem so entstandenen Derivat bzw. im δ-Rhodomycinon die Carbomethoxy- durch eine Hydroxygruppe ersetzt wird.

Diesen Überlegungen liegt die Annahme zugrunde, daß bei allen Synthesen die $C_{(2)}$-Carbonylgruppe des Zwischenproduktes (LXIIIa,

S. 165) entfernt wird. Wäre das nicht der Fall, würde sie enolisieren, sich dadurch (LXVI) bilden und dessen Enolisierungsprodukt (LXVIII b) so weiterreagieren, wie für (LXVIII a) angenommen, so entstünden Anthracyclinone, die sich von den bisher bekannten durch eine Hydroxygruppe an $C_{(2)}$ unterscheiden. Anhaltspunkte für die Existenz solcher Vertreter haben sich bisher nicht gefunden, wobei allerdings zu bemerken ist, daß man noch nicht systematisch nach ihnen gesucht hat.

IV. Die Anthracycline.

Die bisher bekannten Anthracycline sind Glykoside von β-Rhodomycinon, β-Iso-rhodomycinon, γ-Rhodomycinon, ε-Pyrromycinon oder Aklavinon. Rhodomycin A und B sowie Iso-rhodomycin A, die als erste isoliert worden sind, bestehen aus einem Mol Aglykon und einem oder zwei Molen der Tridesoxy-dimethylamino-hexose Rhodosamin (LXXVIII, S. 171). Später hat man in den Cinerubinen A und B sowie im γ-Rhodomycin III und IV Anthracycline gefunden, die neben Rhodosamin noch andere, stickstoff-freie Zucker enthalten. Da diese viel leichter hydrolytisch abgespalten werden als Rhodosamin und dies — enzymatisch katalysiert — eventuell schon vor Aufarbeitung der Kulturen passieren kann, läßt sich nicht immer entscheiden, ob ein Anthracyclin mit 1 oder 2 Molen Rhodosamin „nativ" oder ein Sekundärprodukt ist.

Produziert ein Stamm mehrere Anthracyclinone, so muß man damit rechnen, jedes von ihnen im Mycel oder Kulturmedium in Gestalt mehrerer, durch Zahl und Art ihrer Zuckerreste unterschiedener Glykoside anzutreffen, d. h. wenn ein Stamm z. B. vier Anthracyclinone synthetisieren und sie mit maximal vier Zuckerresten verknüpfen kann, so ist mit dem Vorkommen von mindestens 16 verschiedenen Anthracyclinen zu rechnen.

Nach den bisherigen Erfahrungen kann man verallgemeinernd voraussagen, daß sich Anthracycline mit gleichem Aglykon und verschiedener Anzahl Zuckerreste chromatographisch trennen lassen. Anders liegen die Verhältnisse, wenn Gemische aus Glykosiden verschiedener Anthracyclinone vorliegen. Denn, da bereits die Chromatographie von Anthracyclinon-Gemischen mühsam sein kann — besonders, wenn es um die Trennung von Rhodomycinon/Iso-rhodomycinon-Paaren geht —, wird verständlich, daß der Chromatographie Grenzen gesetzt sind, wenn ein Stamm mehrere Anthracyclinone und daraus Anthracycline mit verschiedener Anzahl von Zuckerresten aufbaut.

Die Gewähr, ohne allzu großen Aufwand zu einheitlichen Anthracyclinen zu kommen, bieten daher nur Stämme, die bevorzugt oder

Literaturverzeichnis: SS. 179—182.

ausschließlich Glykoside *eines* Anthracyclinons bilden. Dementsprechend sind die bisher bekannten Anthracycline aus Kulturen isoliert worden, die als Aglykon bevorzugt β-Rhodomycinon, γ-Rhodomycinon, β-Iso-rhodomycinon, ε-Pyrromycinon oder Aklavinon produzierten.

Die Arbeiten zur Konstitutionsaufklärung der Anthracycline stecken noch in den Anfängen. Sie sind über eine Baustein-Analyse nicht wesentlich hinausgekommen und galten vor allem der Isolierung und Aufklärung der Zucker. Über sie soll zunächst berichtet werden.

1. Die Zucker der Anthracycline.

Rhodosamin.

Rhodosamin, $C_8H_{17}NO_3$ (LXXVIII), wurde zuerst als Salz der REINECKE-Säure sowie als Hydrochlorid vom Schmp. 152—153°; $[\alpha]_D^{20}: -48,2°$ (Wasser) aus Hydrolysaten von Rhodomycin A und B isoliert [BROCKMANN und SPOHLER (*28*)]. Es reduziert Fehlingsche Lösung, gibt eine positive Ninhydrin- und Jodoform-Reaktion, hat im IR-Spektrum keine Carbonylbande, enthält zwei acetylierbare Hydroxyle und liefert bei der Kuhn-Roth-Oxydation 0,9 Mol Essigsäure.

Bei kurzem Erwärmen mit Bariumhydroxidlösung entstehen unter Abspaltung von Dimethylamin 2-Desoxy-*L*-fucose (LXXIX) und

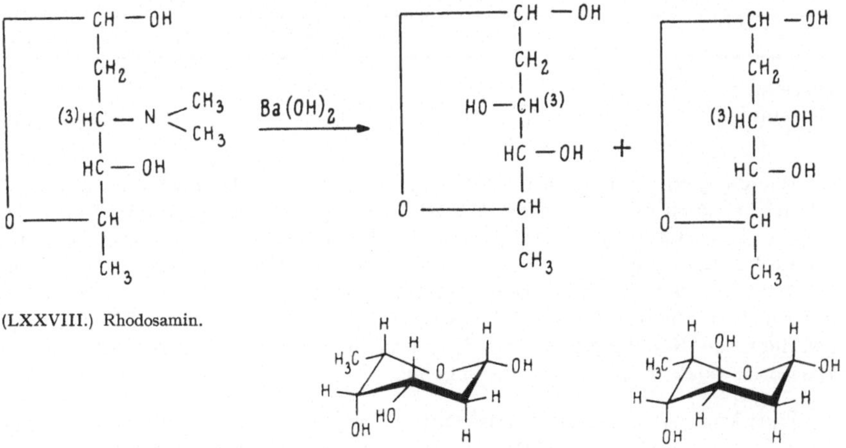

(LXXVIII.) Rhodosamin.

(LXXIX.) 2-Desoxy-*L*-fucose.

(LXXX.) *L*-Boivinose.

L-Boivinose (LXXX) [BROCKMANN und WAEHNELDT (*33*)]. Demnach kommt dem Rhodosamin Formel (LXXVIII) zu.

Die Aufklärung der Konfiguration gelang auf folgendem Wege [BROCKMANN und WAEHNELDT (*33*), BROCKMANN, SPOHLER und WAEH-

Neldt, (*30*)]: Rhodosamin bildet mit Acetanhydrid/Pyridin zwei kristallisierte Diacetate, $C_{12}H_{21}O_5N$, die sich chromatographisch an Kieselgel G aus Chloroform/Aceton trennen lassen. Das schneller wandernde Diacetat I hat in Aceton $[\alpha]_D^{20}$: — 20° ± 1°, das langsamer wandernde Diacetat II hat $[\alpha]_D^{20}$: 109° ± 3°.

Das KMR-Spektrum beider Diacetate, in Deuterochloroform mit Tetramethylsilan als internem Standard aufgenommen, zeigte Signale, deren τ-Werte und Kopplungskonstanten in *Tabelle 4* angegeben sind.

Tabelle 4. KMR-Spektren von Rhodosamin-acetaten.

τ-Werte.

	C—CH_3	N(CH_3)$_2$	O—*Ac*	O—*Ac*
Diacetat I	8,82	7,72	7,87	7,85
Diacetat II	8,88	7,71	7,92	7,86

	Proton an				
	$C_{(1)}$	$C_{(2)}$	$C_{(3)}$	$C_{(4)}$	$C_{(5)}$
Diacetat I...............	4,23	8,3—7,9	7,57	4,79	6,24
Diacetat II...............	3,67	8,2—7,9	7,35	4,69	5,94

Kopplungskonstanten (in Hz).

	$J_{1,2e}$	$J_{1,2a}$	$J_{2a,3}$	$J_{2e,3}$	$J_{3,4}$	$J_{4,5}$	$J_{5,6}$
Diacetat I..................	4,0	8,0	12,0	5,0	2,0	2,0	6,5
Diacetat II..................	2,3	2,3	10,5	6,0	2,0	2,0	6,5

Wie die große Kopplungskonstante $J_{2a,3}$ zeigt, steht das Proton an $C_{(3)}$ axial, womit die äquatoriale Lage der Dimethylaminogruppe bewiesen ist. Die kleine Kopplungskonstante $J_{3,4}$ läßt sich dann nur durch eine Axial-Äquatorial-Kopplung der Protonen an $C_{(3)}$ und $C_{(4)}$ erklären; d. h. die Acetoxygruppe an $C_{(4)}$ steht axial und damit in *cis*-Stellung zur Dimethylaminogruppe an $C_{(3)}$. Aus der Umwandlung des Rhodosamins in 2-Desoxy-*L*-fucose (LXXIX) und *L*-Boivinose (LXXX) folgt, daß die kleine Kopplungskonstante $J_{4,5}$ nicht durch äquatorial-äquatoriale, sondern äquatorial-axiale Konstellation bedingt ist.

Die Analyse des Protonensignals von $C_{(1)}$ zeigt, daß die beiden Acetate Anomere sind. Dem symmetrischen Triplett ($\tau = 3,67$) des Diacetates II zufolge sind die benachbarten Methylen-Protonen gleich weit vom $C_{(1)}$-Proton entfernt, was nur bei axialer Stellung der Acetoxygruppen möglich ist. Diacetat II ist daher das α-Anomere (LXXXII) (wofür auch seine größere Linksdrehung spricht) und Diacetat I das β-Anomere (LXXXIa), bestätigt durch die unterschiedlichen Kopplungskonstanten $J_{1,2a}$ und $J_{1,2e}$.

Literaturverzeichnis: SS. 179—182.

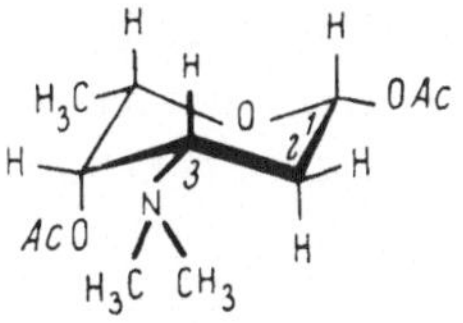

$$\text{(LXXXIa.)}$$
$$\text{(LXXXIb.)} \quad Ac = H. \qquad \text{(LXXXII.)}$$

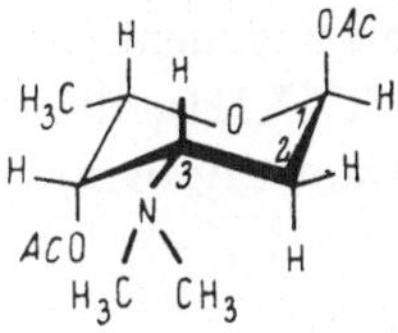

Rhodosamin ist somit die 2,3,6-Tridesoxy-3-dimethylamino-L-*lyxo*-hexose (LXXXIb) [willkürlich wie (LXXIX) und (LXXX) als β-Anomeres formuliert].

2-Desoxy-L-fucose.

2-Desoxy-L-fucose, $C_6H_{12}O_4$ (LXXIX), Schmp. 94—97°; $[\alpha]_D^{20}$: — 136° (Aceton), zuerst isoliert von ISELIN und REICHSTEIN (45), wurde von BROCKMANN und WAEHNELDT (31) aus Hydrolysaten von γ-Rhodomycin III und γ-Rhodomycin IV abgetrennt.

Rhodinose.

Rhodinose, $C_6H_{12}O_3$, eine Tridesoxyhexose, gut löslich in Chloroform, $[\alpha]_D^{20}$: — 11° ± 1,6° (Aceton), wurde von BROCKMANN und WAEHNELDT (32) amorph aus Hydrolysaten von γ-Rhodomycin IV isoliert und in ihrer

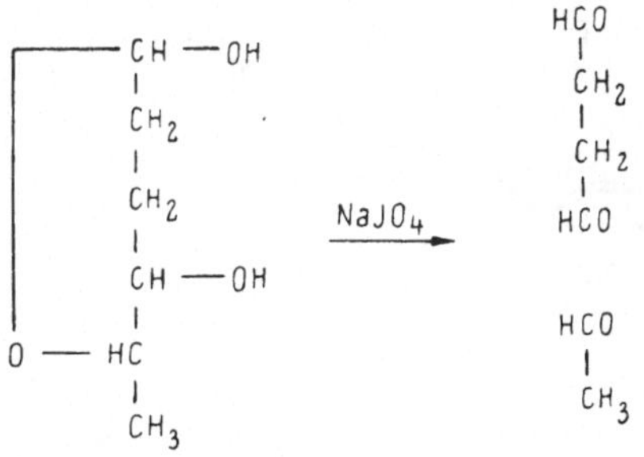

(LXXXIII.) Rhodinose.

Konstitution aufgeklärt. Sie reduziert Fehlingsche Lösung, ammoniakalische Silbernitratlösung und Triphenyl-tetrazoliumchlorid langsamer als Glucose und hat im Papierchromatogramm (Butanol/Eisessig/Wasser 4 : 1 : 1) den auffallend hohen R_F-Wert 0,71 (2-Desoxy-L-fucose: $R_F = 0,49$). Die KELLER-KILIANI-Reaktion und die WEBB-Reaktion auf 2-Desoxyzucker sind negativ.

Oxydation der Rhodinose mit Natriumperjodat lieferte Acetaldehyd (LXXXV) und Succindialdehyd (LXXXIV), die als 2,4-Dinitrophenylhydrazon isoliert und identifiziert wurden. Damit ergibt sich für Rhodinose die Formel (LXXXIII), die durch das in Deuterochloroform (interner Standard: Tetramethylsilan) gemessene KMR-Spektrum bestätigt wird.

2. Anthracycline des ε-Pyrromycinons.

Pyrromycin.

Pyrromycin wurde als kristallisiertes, gelbrotes, wasserlösliches Hydrochlorid $C_{30}H_{35}NO_{11} \cdot HCl$ vom Schmp. 162—163° (Zers.), $[\alpha]_D^{20}$: $+ 132 \pm 27°$ (Methanol), aus *Streptomyces*-Kulturen isoliert [Brockmann und Lenk (23)]. Es hemmt das Wachstum von *St. aureus* und *B. subtilis* bis zu Verdünnungen von $1 : 3 \times 10^5$. Durch milde Säurehydrolyse (0,1 n-HCl, 3—4 Stdn. bei 65—70°) wird es in 1 Mol ε-Pyrromycinon (XLVII, S. 149) und 1 Mol Rhodosamin (LXXVIII) gespalten.

Da die Absorptionskurve des Pyrromycins in Gestalt und Lage der Maxima der des ε-Pyrromycinons gleicht, kann im Pyrromycin das ε-Pyrromycinon nicht über eine seiner phenolischen Hydroxyle mit Rhodosamin verknüpft sein.

Mit Acetanhydrid/Pyridin liefert Pyrromycin ein hellgelbes, kristallisiertes Tetraacetat, $[\alpha]_D^{20}$: $+ 52° \pm 2,5°$ (Chloroform). Da ε-Pyrro-

LXXXVI.) Pyrromycin.

mycinon fünf und Rhodosamin zwei Hydroxygruppen enthält und zur glykosidischen Verknüpfung jeder der beiden Partner eine Hydroxygruppe beisteuert, müssen im Pyrromycin noch fünf Hydroxyle vorhanden sein. Vier davon werden mit Acetanhydrid/Pyridin acetyliert. Da die fünfte, schwer acetylierbare, zweifellos die tertiäre Hydroxygruppe des ε-Pyrromycinons ist, kann dieses nur über seine $C_{(7)}$-Hydroxygruppe mit dem Rhodosamin verbunden sein. Und da Pyrromycin im Gegensatz zum Rhodosamin *nicht* mit Fehlingscher Lösung reagiert, ist das Rhodosamin über sein $C_{(1)}$-Hydroxyl mit dem Aglykon verknüpft. Damit ergibt sich für Pyrromycin die Formel (LXXXVI). Zu klären bleibt, ob eine α- oder β-glykosidische Bindung vorliegt.

Literaturverzeichnis: SS. 179—182.

Cinerubine.

Cinerubin A und B wurden von ETTLINGER, GÄUMANN, HÜTTER, KELLER-SCHIERLEIN, KRADOLFER, NEIPP, PRELOG, REUSSER und ZÄHNER (40) aufgefunden, isoliert und eingehender untersucht. Beide sind kristallisierte, rote, isomere Basen, deren Analysenzahlen auf $C_{44}H_{59}NO_{18} \pm CH_2$ passen. Auf diese Formeln bezogen, enthalten sie ein Methoxyl, zwei N-Methyl- und mindestens vier C-Methylgruppen.

Die Absorptionsspektren der beiden Cinerubine sind im sichtbaren und UV-Gebiet deckungsgleich und dem des 1,4,5-Trihydroxy-anthrachinons (XIII, S. 129) ähnlich. Die in Kaliumbromid aufgenommenen IR-Spektren zeigen nur geringe Unterschiede im „finger print"-Gebiet.

Bei Hydrolyse mit 0,5 n-Schwefelsäure (20 Min., 100°) liefern beide Cinerubine ε-Pyrromycinon (XLVII, S. 149) und ein Gemisch aus drei Zuckern, die papierchromatographisch im System Butanol/Eisessig/ Wasser (4 : 1 : 1) getrennt wurden.

Nach Entwicklung mit Anilinphthalat in n-Butanol erschienen die Zucker als gelbe, im UV-Licht hell fluoreszierende Flecke, beim Cinerubin A mit R_F: 0,03, 0,47 und 0,75, beim Cinerubin B mit R_F: 0,03, 0,47 und 0,62. Glucose zeigte unter diesen Bedingungen R_F: 0,09. Mit Anilin/Diphenylamin/Phosphorsäure ließen sich nur die Flecken der beiden rascher wandernden Zucker nachweisen. Mit ammoniakalischer Silbernitratlösung kam der am schnellsten wandernde Zucker kaum zum Vorschein.

Die Vermutung, eines der Abbauprodukte sei ein Dimethylaminozucker vom Typ des Rhodosamins oder Desosamins, hat sich bestätigt. In Hydrolysaten von Cinerubin A und B wurde später Rhodosamin nachgewiesen; ferner ließ sich a) einer der stickstoff-freien Zucker als 2-Desoxy-L-fucose identifizieren und b) bei sehr milder Säurehydrolyse der Cinerubine als Zwischenprodukt Pyrromycin nachweisen [BROCKMANN und WAEHNELDT (31)].

Cinerubin A, ein dunkelrotes, mikrokristallines Pulver, schmilzt bei 155—158°, erstarrt bei 160—180° wieder zu nadelförmigen Kristallen, die ab 249° schmelzen. Wegen Schwerlöslichkeit in anderen Solvenzien wurde die Dissoziationskonstante in Dimethylsulfoxid bestimmt. Gef. $_{PK}$ DSO = 7,95. Mit Pikrinsäure bildet Cinerubin A ein rotes, kristallisiertes Pikrat $C_{43}H_{57}NO_{18} \cdot C_6H_3N_3O_7$ vom Schmp. 156—158°.
Cinerubin B kristallisiert aus Benzol/Methanol in feinen, orangeroten Stäbchen, die zwischen 168 und 178° schmelzen, dann zu langen, dünnen Nadeln erstarren, um schließlich bei 240—243° zu zerfließen. $_{PK}$ DSO = 7,18.

Für antibiotische Wirksamkeit s. *Tabelle 5* (S. 176).

Das Wachstum von *Pasteurella pestis*, *Escherichia coli* und *Salmonella typhosa* wurde bei Verdünnungen von $1 : 10^4$ noch nicht gehemmt, während solche Lösungen Protozoen wie *Entamoeba histolytica* abtöteten.

Tabelle 5. Antibiotische Wirksamkeit der Cinerubine.

Test-Mikroorganismus	Wirksame Grenzkonzentration	
	Cinerubin A	Cinerubin B
Bacillus megatherium	$1 : 10^7$	$1 : 10^6$
Corynebacterium diphtheriae	$1 : 10^9$	$1 : 10^7$
Micrococcus pyogenes aureus	$1 : 10^5$	$1 : 10^5$
Streptococcus faecalis	$1 : 10^6$	$1 : 10^5$
Streptococcus mitis	$1 : 10^7$	$1 : 10^7$
Streptococcus pyogenes	$1 : 10^7$	$1 : 10^6$
Endomyces albicans	$1 : 10^7$	$1 : 10^6$
Candida vulgaris	$1 : 10^7$	$1 : 10^6$
Mycobacterium Tbc	$1 : 10^6$	$1 : 10^5$

Die Cinerubine wirken auch cytostatisch. Bei Hühner-Fibroblasten-Kulturen veränderten sie in Verdünnungen von $1 : 10^6$—$1 : 10^8$ das morphologische Aussehen der Nucleoli und den Zellteilungsvorgang. Ferner wurde das Wachstum von Crocker-Sarkom 180, Ehrlich-Carcinom (solide Form), Adeno-carcinom EO-771, Walker-Carcino-sarkom 256 gehemmt. Cinerubin A ist wirksamer als Cinerubin B und hemmt auch das Wachstum des Uterus Epitheliom T-8, Guerin- und das Flexner-Jobling-Carcinom. Beide Cinerubine sind für Ratte und Maus sehr giftig. Nach Ansicht von Prelog und Mitarb. kann die Wirkung auf Tiertumoren kaum als spezifisch angesehen werden.

Rutilantine.

In einer Mitteilung über Rutilantinon [Ollis und Sutherland (*50*)] ist eine Gruppe tiefroter, basischer Antibiotica erwähnt, von denen verschiedene Salze dargestellt wurden und die bei milder Säurehydrolyse ε-Pyrromycinon (identisch mit Rutilantinon) liefern. Nähere Angaben fehlen. Es ist möglich, daß der eine oder andere Vertreter dieser Gruppe mit Cinerubin A bzw. B identisch ist.

3. Anthracycline der Rhodomycinone.

Rhodomycin A.

Rhodomycin A, das am längsten bekannte Anthracyclin [Brockmann, Bauer und Borchers (*3, 4, 5, 10*)], wurde als kristallisiertes Perchlorat und Hydrochlorid aus *Str. purpurascens*-Kulturen isoliert und anfangs durch fraktionierte Gegenstromverteilung, später durch präparative Ring-Papierchromatographie von beigemengtem Rhodomycin B und Iso-rhodomycin A getrennt.

Rhodomycin A-hydrochlorid, $C_{36}H_{48}N_2O_{12} \cdot 2\,HCl$, rote Prismen vom Schmp. $205°$, leicht löslich in Wasser, $[\alpha]^{18}_{606-760}: +178° \pm 10°$

Literaturverzeichnis: SS. 179—182.

(Methanol), liefert bei milder Säurehydrolyse 1 Mol β-Rhodomycinon (XXVIIIa) und 2 Mol Rhodosamin (LXXVIII, S. 171) [BROCKMANN und SPOHLER (29)].

Die Absorptionskurve von Rhodomycin A hat den gleichen Verlauf und die gleichen λ_{max}-Werte wie die von β-Rhodomycinon; d. h. von dessen phenolischen Hydroxylen ist im Rhodomycin A keines an der Bindung des Rhodosamins beteiligt. Dieses muß demnach mit Ring A des β-Rhodomycinons verbunden sein. Ob die beiden Rhodosamin-Reste einzeln mit je einem Hydroxyl des Ringes A glykosidisch verknüpft sind oder als Disaccharid am Aglykon (XXVIIIa) hängen, ist unbekannt. Der Verlauf der sauren Partialhydrolyse, bei der Rhodomycin B als Zwischenprodukt auftritt, spricht für eine Disaccharid-Struktur.

Tabelle 6. Antibiotische Wirksamkeit von Rhodomycin A* (*10*).

Test-Mikroorganismus	Antibiotisch wirksame Grenzkonzentration
St. aureus	$1 : 2 \times 10^7$
Corynebacterium diphtheriae	$1 : 10^7$
B. subtilis	$1 : 2 \times 10^6$
E. coli	$1 : 2 \times 10^4$
Enterococcen	$1 : 2 \times 10^5$
Streptococcen	$1 : 2 \times 10^7$

* Getestet in Hißscher serumhaltiger Nährlösung.

Rhodomycin B.

Rhodomycin B wurde durch präparative Ring-Papierchromatographie im System Butanol/Formamid/Wasser (1 : 1 : 1) gereinigt und als kristallisiertes, rotes Hydrochlorid vom Schmp. 180°, $[\alpha]_{606-700}^{18}$: $+ 174° \pm 10°$ (Methanol) isoliert [BROCKMANN und PATT (27)]. Es hemmt das Wachstum von St. aureus-Stämmen bis zur Verdünnung $1 : 5 \times 10^6$.

Die freie Base ist beim isoelektrischen Punkt in Wasser viel weniger löslich als freies Rhodomycin A. Bei milder Säurehydrolyse liefert sie 1 Mol β-Rhodomycinon (XXVIIIa, S. 142) und 1 Mol Rhodosamin (LXXVIII) [BROCKMANN und SPOHLER (29)]. Danach hat Rhodomycin B-hydrochlorid die Formel $C_{28}H_{33}NO_{12} \cdot HCl$.

Die Analysen der ersten kristallisierten Hydrochlorid-Präparate (*27*) stimmten nicht mit dieser Formel überein, weil die Kristalle Lösungsmittel (Chloroform) enthielten.

γ-Rhodomycine.

Die erst in jüngster Zeit aufgefundenen γ-Rhodomycine wurden bisher als amorphe, chromatographisch einheitliche Basen isoliert. Ihre Zu-

sammensetzung zeigt die folgende Tabelle [Brockmann und Waehneldt (*31*)].

Tabelle 7. Baustein-Analyse der γ-Rhodomycine.

	Gehalt in Mol			
	γ-Rhodomycinon	Rhodosamin	2-Desoxy-L-fucose	Rhodinose
γ-Rhodomycin I.....	1	1	—	—
γ-Rhodomycin II	1	2	—	—
γ-Rhodomycin III ...	1	2	1	—
γ-Rhodomycin IV ...	1	2	1	1

Da die Absorptionskurven der γ-Rhodomycine (Benzol) in Gestalt
und λ_{max}-Werten der Kurve des γ-Rhodomycinons gleichen, kann keines
von dessen phenolischen Hydroxylen in den γ-Rhodomycinen Zuckerreste
tragen; vielmehr müssen diese Reste an Ring A des γ-Rhodomycinons
stehen. Wären an dieser Bindung beide Hydroxyle von Ring A beteiligt,
so könnten die vier Zuckerreste des γ-Rhodomycins IV entweder in Gestalt
von zwei Disacchariden oder als Trisaccharid und Monosaccharid mit
dem Aglykon verknüpft sein. Gibt es an Ring A dagegen nur *eine*
Verknüpfungsstelle — in Analogie zu vielen anderen Glykosiden das
wahrscheinlichere —, so ist γ-Rhodomycin IV ein Chromo-tetrasaccharid.

γ-Rhodomycin IV wird durch 0,1 n-Salzsäure bei 20—70° über
γ-Rhodomycin III, γ-Rhodomycin II und γ-Rhodomycin I zu γ-Rhodo-
mycinon abgebaut. Demnach wäre, wenn es ein Chromo-tetrasaccharid
ist, seine „Baustein-Sequenz": γ-Rhodomycinon—Rhodosamin—Rhodosamin—2-Desoxy-L-fucose—Rhodinose.

Die antibiotische Wirksamkeit der γ-Rhodomycine ist gering. Die gegen *Bac.*
subtilis wirksame Grenzkonzentration lag für γ-Rhodomycin IV bei etwa $1:3 \times 10^5$,
für die anderen bei etwa $1:4 \times 10^4$.

Iso-rhodomycin A.

Iso-rhodomycin A-hydrochlorid, dunkelrote Kristalle vom Schmp. 220°,
$[\alpha]^{20}_{606-760}$: $+268° \pm 30°$ (Methanol), wurde durch präparative Ring-
Papierchromatographie von Rhodomycin A abgetrennt [Brockmann und
Patt (*27*)]. Bei milder Säurehydrolyse liefert es β-Iso-rhodomycinon
(XXIIa). Die Absorptionskurve des Iso-rhodomycins A gleicht in
Gestalt und λ_{max}-Werten der Kurve des β-Iso-rhodomycinons. Über
die Natur der Zuckerreste ist nichts bekannt. Nimmt man in Analogie
zu Rhodomycin A an, daß Iso-rhodomycin A aus einem Mol β-Iso-
mycinon und zwei Molen Rhodosamin besteht, so wäre die Summen-

formel des Hydrochlorides $C_{36}H_{48}N_2O_{13} \cdot 2\,HCl$, mit der die Analysenzahlen (*27*) gut übereinstimmen.

Iso-rhodomycin A-hydrochlorid hemmte das Wachstum eines *St. aureus*-Stammes bis zur Verdünnung $1:4 \times 10^7$. Ein dem Rhodomycin B entsprechendes Iso-rhodomycin B wurde in kleiner Menge aus *Str. purpurascens*-Stämmen abgetrennt.

Antibiotica der Mycetin-Violarin-Gruppe.

Eine Gruppe als Mycetine bzw. Violarine bezeichneter roter Antibiotica wurde zuerst von KRASSILNIKOV und KORENIAKO (*46*) beschrieben und später von YACUBOV, KHOKHLOV und BLINOV (*55*) durch Papierchromatographie in mehrere Komponenten aufgetrennt. Diese geben bei saurer Hydrolyse ein Gemisch aus Rhodomycinonen und Iso-rhodomycinonen, von denen einige im R_F-Wert mit β- und γ-Rhodomycinon bzw. β- und γ-Iso-rhodomycinon übereinstimmen [KHOKHLOV, BLINOV und BROCKMANN jr., unveröffentlicht].

4. Anthracycline des Aklavinons.

Aklavin.

Aklavin wurde von STRELITZ, FLON, WEISS und ASHESHOV (*52*) aus Mycel und Kulturlösung eines *Streptomyces*-Stammes (A 1165) abgetrennt, durch Gegenstromverteilung (Äthylacetat/0,2 m Phosphatpuffer pH 6,3) oder Verteilungschromatographie an Celite (Wasser/Methanol/Heptan/Benzol, $1:2:3:0,6$) gereinigt und als kristallisiertes, bräunlichgelbes Pikrat bzw. orangegelbes Helianthat oder Hydrochlorid isoliert. Aus der Analyse des Pikrates ergibt sich für die freie Base die Formel $C_{30}H_{37}NO_{11}$.

Die Autoren zweifeln an der Homogenität ihrer Präparate und halten es für wahrscheinlich, daß Gemische mehrerer ähnlicher Verbindungen vorliegen. Bei milder Säurehydrolyse des Aklavins entsteht Aklavinon (*43*).

Aklavin ist gegen verschiedene Phagen wirksam. Im Verdünnungstest wurden bei Mikroorganismen u. a. folgende antibiotisch wirksame Grenzkonzentrationen gefunden: *St. aureus:* $1:2,5—5 \times 10^5$; *Bac. mycoides:* $1:5 \times 10^5$; *Bac. subtilis:* $1:2,5 \times 10^5$; *Escherichia coli:* $1:2—4 \times 10^3$. Ausführliche Angaben über die Wirksamkeit bei anderen Mikroorganismen finden sich in der Originalarbeit. Von der Maus wurden intravenös 150 mg Aklavin/kg vertragen.

Literaturverzeichnis.

1. BIRCH, A. J.: Biosynthetic Relations of Some Natural Phenolic and Enolic Compounds. Fortschr. Chem. organ. Naturstoffe **14**, 186 (1957).

2. BOLDT, P.: Dissertation, Göttingen, 1958.

3. Brockmann, H.: Pyrromycines and Pyrromycinones, Rhodomycines and Rhodomycinones. Sostanze Naturali. Roma Academia Nazionale dei Lincei, p. 33, 51 (1961).

4. Brockmann, H. und K. Bauer: Rhodomycin, ein rotes Antibioticum aus Actinomyceten. Naturwiss. 37, 492 (1950).

5. Brockmann, H., K. Bauer und I. Borchers: Rhodomycin, ein rotes Antibioticum. Chem. Ber. 84, 700 (1951).

6. Brockmann, H. und P. Boldt: Zur Kenntnis des β-Rhodomycinons. Naturwiss. 44, 616 (1957).

7. — — Die Konstitution des ε-Iso-rhodomycinons. Naturwiss. 47, 134 (1960).

8. — — ε-Iso-rhodomycinon. Chem. Ber. 94, 2174 (1961).

9. Brockmann, H., P. Boldt und J. Niemeyer: β-Rhodomycinon und γ-Rhodomycinon. Chem. Ber. 96, 1356 (1963).

10. Brockmann, H. und I. Borchers: Rhodomycin, II. Mitteil. Chem. Ber. 86, 261 (1953).

11. Brockmann, H. und H. Brockmann jr.: Die Konstitution des ε-Rhodomycinons. Naturwiss. 48, 161 (1961).

12. — — ε-Rhodomycinon. Chem. Ber. 94, 2681 (1961).

13. — — Die Konstitution des δ-Rhodomycinons. Naturwiss. 50, 20 (1963).

14. — — δ-Rhodomycinon. Chem. Ber. 96, 1771 (1963).

15. — — Zur Konstitution der Pyrromycinone. Naturwiss. 47, 135 (1960).

16. — — 1,11-Dihydroxy-tetracenchinon. Naturwiss. 50, 519 (1963).

17. Brockmann, H., H. Brockmann jr., J. J. Gordon, W. Keller-Schierlein, W. Lenk, W. D. Ollis, V. Prelog und I. O. Sutherland: Über die Identität von Rutilantinon mit Pyrromycinon und über die Lage der Carbomethoxygruppe in Pyrromycinonen. Tetrahedron Letters 1960, Nr. 8, 25.

18. Brockmann, H. und G. Budde: Zur spektroskopischen Identifizierung des Stammkohlenwasserstoffes mehrkerniger Oxy-chinone. Chem. Ber. 86, 432 (1953).

19. Brockmann, H., L. Costa Plá und W. Lenk: Pyrromycinone, eine neue Gruppe roter Actinomyceten-Farbstoffe. Naturwiss. 69, 477 (1957).

20. Brockmann, H. und B. Franck: Rhodomycinone und Iso-rhodomycinone. Chem. Ber. 88, 1792 (1955).

21. Brockmann, H. und W. Lenk: ζ-Pyrromycinon. Angew. Chem. 69, 477 (1957).

22. — — Pyrromycinone. Chem. Ber. 92, 1880 (1959).

23. — — Pyrromycin. Chem. Ber. 92, 1904 (1959).

24. — — ζ-Pyrromycinon. Naturwiss. 47, 135 (1960).

25. Brockmann, H. und W. Müller: Hydroxy-tetracenchinone. Chem. Ber. 92, 1164 (1959).

26. Brockmann, H. und J. Niemeyer: Zur Konstitution des β- und γ-Rhodomycinons. Naturwiss. 48, 570 (1961).

27. Brockmann, H. und P. Patt: Iso-rhodomycin A, ein neues Antibioticum aus *Streptomyces purpurascens*. Chem. Ber. 88, 1455 (1955).

28. Brockmann, H. und E. Spohler: Rhodosamin, eine neue Dimethylamino-desoxy-aldohexose. Naturwiss. 42, 154 (1955).

29. — — Zur Konstitution des Rhodomycins A. Naturwiss. 48, 716 (1961).

30. BROCKMANN, H., E. SPOHLER und TH. WAEHNELDT: Isolierung, Konstitution und Konfiguration des Rhodosamins. Chem. Ber. **96** (1963) (im Druck).

31. BROCKMANN, H. und TH. WAEHNELDT: Eine neue Gruppe von Rhodomycinen. Naturwiss. **48**, 717 (1961).

32. — — Rhodinose, eine Tridesoxyhexose. Naturwiss. **50**, 43 (1963).

33. — — Konstitution und Konfiguration des Rhodosamins. Naturwiss. **50**, 92 (1963).

34. BROCKMANN, H. und E. WIMMER: Synthese des Descarbomethoxy-η-iso-pyrromycinons. Naturwiss. **48**, 162 (1961).

35. — — Synthese von Descarbomethoxy-η-iso-pyrromycinon und anderen Hydroxy-tetracenchinonen. Chem. Ber. **96**, 2399 (1963).

36. BROCKMANN jr., H.: Dissertation, Göttingen, 1963.

37. BROCKMANN jr., H. und M. LEGRAND: Zirkulardichroismus von Pyrromycinonen, Rhodomycinonen und Iso-rhodomycinonen. Naturwiss. **49**, 374 (1962).

38. — — Zirkulardichroismus von Pyrromycinonen, Rhodomycinonen und Iso-rhodomycinonen. Tetrahedron **19**, 395 (1963).

39. ENTSCHEL, R., C. H. EUGSTER und P. KARRER: Über die Anwendung des modifizierten Chromsäure-Abbaus zur Konstitutionsaufklärung eines Carotinoid-farbstoffs (Capsanthin). Helv. Chim. Acta **39**, 1263 (1956).

40. ETTLINGER, L., E. GÄUMANN, R. HÜTTER, W. KELLER-SCHIERLEIN, F. KRA-DOLFER, L. NEIPP, V. PRELOG, P. REUSSER und H. ZÄHNER: Cinerubine. Chem. Ber. **92**, 1867 (1959).

41. FROMMER, W.: Variationsbreite und Kulturkonstanz einer Rhodomycin bilden-den Actinomycetenart. Arch. Mikrobiol. **23**, 105 (1955–56).

42. — Zur Physiologie der Rhodomycinbildung. Arch. Mikrobiol. **23**, 385 (1955–56).

43. GORDON, J. J., L. M. JACKMAN, W. D. OLLIS and I. O. SUTHERLAND: Aklavinone. Tetrahedron Letters **1960**, Nr. 8, 28.

44. GROSJEAN, M. et M. LEGRAND: Appareil de mesure de dichroisme circulaire dans le visible et l'ultraviolet. C. R. hebd. Séances Acad. Sci. **251**, 2150 (1960).

45. ISELIN, B. und T. REICHSTEIN: 2-Desoxy-*L*-fucose. Helv. Chim. Acta **27**, 1200 (1944).

46. KRASSILNIKOV, N. A. und A. J. KORENIAKO: Mikrobiologiya (USSR) **8**, 673 (1939).

47. LINDENBEIN, W.: Über einige chemisch interessante Actinomycetenstämme und ihre Klassifizierung. Arch. Mikrobiol. **17**, 361 (1952).

48. OLLIS, W. D. and I. O. SUTHERLAND: A new Family of Antibiotics. In: W. D. OLLIS, Recent Developments in the Chemistry of Natural Phenolic Compounds. New York: Pergamon Press. 1961.

49. OLLIS, W. D., I. O. SUTHERLAND, R. C. CODNER, J. J. GORDON and G. A. MILLER: The Incorporation of Propionate in the Biosynthesis of ε-Pyrro-mycinone (Rutilantinone). Proc. Chem. Soc. (London) **1960**, 347.

50. OLLIS, W. D., I. O. SUTHERLAND and J. J. GORDON: Rutilantinone. Tetrahedron Letters **1959**, Nr. 16, 17.

51. OLLIS, W. D., I. O. SUTHERLAND and P. L. VEAL: Synthetical Studies regarding ε-Pyrromycinone. Proc. Chem. Soc. (London) **1960**, 349.

52. STRELITZ, F., H. FLON, U. WEISS and I. N. ASHESHOV: Aklavin, an Antibiotic Substance with Antiphage Activity. J. Bacteriol. **72**, 90 (1956).

53. Velluz, L. und M. Legrand: Zirkulardichroismus und Raumstruktur. Angew. Chem. **73**, 603 (1961).

54. Waehneldt, Th.: Diplomarbeit, Göttingen, 1960.

55. Yacubov, G. Z., J. M. Khokhlov and N. O. Blinov: A Study of the Conditions for Partitioning Antibiotics of the Mycetin-Violarin Group by Paper Chromatography. (Original russisch.) Mikrobiologiya (USSR) **31**, 526 (1962).

(Eingegangen am 15. April 1963.)

Folsäure und Folat-Enzyme.

Von **L. Jaenicke** und **C. Kutzbach**, Köln.

Mit 3 Abbildungen.

Inhaltsübersicht.

Seite

I. Einleitung 184

II. Das Vitamin Folsäure 187

  1. Entdeckung der Folsäure und ihrer Konjugate 187

  2. Konjugat-spaltende Enzyme 190

  3. Vorkommen, Bedarf und Ausscheidung 190

III. Auf- und Abbau der Folsäure-Cofaktoren 192

  1. Biogenese der Folsäure 192

  2. Biologischer Abbau der Folsäure 195

  3. Enzymatische Reduktion der Folsäure zum Cofaktor 197

IV. Chemie der Folat-Verbindungen 199

  1. Folsäure 199

    a. Isolierung 199

    b. Konstitution und physikalische Eigenschaften 200

    c. Chemische Eigenschaften 201

    d. Folsäure-Synthesen 203

  2. Reduktion von Folsäure 204

    a. Dihydrofolsäure und das Problem der Dihydrofolat-Isomerie 204

    b. 5,6,7,8-Tetrahydro-folsäure 207

  3. Mit Einkohlenstoff-Körpern substituierte Folsäuren 209

    a. 10-Formyl-folsäure 209

    b. 10-Formyl-tetrahydrofolsäure 209

    c. 5-Formyl-tetrahydrofolsäure 211

    d. 5,10-Methinyl-tetrahydrofolsäure 212

    e. 5-Formimino-tetrahydrofolsäure 215

    f. 5,10-Methylen-tetrahydrofolsäure 216

    g. 5-Methyl-tetrahydrofolsäure 219

  4. Folsäure-Analoge 220

  5. Spektren von Folat-Verbindungen 222

  6. Analyse und Trennung von Folsäure-Verbindungen 225

    a. Chemische Verfahren 225

    b. Polarographie 225

    c. Mikrobiologische Methoden 226

    d. Chromatographische Trennung 227

Seite

V. Das Einkohlenstoff-Reservoir 229

   1. Herkunft der Ameisensäure 229

   2. Glycin als Quelle von Einkohlenstoff-Körpern 231

   3. Der Einkohlenstoff-Donator Serin 232

   4. Herkunft der Methylgruppe...................................... 233

VI. Folat-katalysierte Enzym-Reaktionen 236

   1. Der Transhydroxymethylierungs-Cyclus........................... 236

     a. Serin-Aldolase... 236

     b. Transhydroxymethylierungs-Reaktionen....................... 239

     c. Methylen-tetrahydrofolat-Dehydrogenase 239

   2. Methylgruppen-Genese... 240

     a. Thymidylat-Bildung 240

     b. Methylen-tetrahydrofolat-Reduktase: 5-Methyl-tetrahydrofolsäure. 242

     c. Methionin-Bildung... 243

       α. Die Gesamt-Reaktion 243

       β. Zusammenhänge zwischen Folsäure und Vitamin B_{12} 244

       γ. Der Acceptor der Methylgruppe.......................... 245

   3. Transformylierungs-Cyclen 246

     a. Abbau von Histidin 246

     b. Deacylase und Glutamyl-Transferase 247

     c. Aktivierte Ameisensäure im Purin-Stoffwechsel.............. 248

       α. Vergärung von Purinen 248

       β. Tetrahydrofolat-Formylase............................. 249

       γ. Transformylierungen 251

VII. Zusammenfassung 253

Literaturverzeichnis 254

Abkürzungen. Folgende Abkürzungen werden verwendet: ADP und ATP = Adenosin-di- bzw. -triphosphat; CMP = Cytosinmonophosphat; UMP = Uridin-monophosphat; TMP = Thymidinmonophosphat; FMN = Flavinmononucleotid; FAD = Flavin-Adenin-dinucleotid; DPN und TPN = Diphospho- und Triphospho-pyridin-nucleotid (= NAD bzw. NADP); pAB = p-Aminobenzoesäure; Glu = Glutaminsäure; P_a = Orthophosphat (HPO_4^{--}).

Die Folsäure-Verbindungen erhalten die nachstehenden Symbole: F = Folsäure; FH_2 = Dihydrofolsäure; FH_4 = Tetrahydrofolsäure; CF = Citrovorumfaktor.

Die Kohlenstoff-Substituenten werden bezeichnet als: Formyl = CHO—; Methinyl = $CH^+<$; Formimino = CHNH—. Die vorgestellten Ziffern bedeuten das jeweils substituierte Stickstoff-Atom (vgl. Formelübersicht 4, S. 201).

I. Einleitung.

Seit die letzte, von Albert (*1*) verfaßte, zusammenfassende Darstellung der Chemie und Biochemie der Folsäure und ihrer abgeleiteten Coenzyme in dieser Reihe erschienen ist, sind acht Jahre intensiver Forschung vergangen. In dieser Zeit ist unser Wissen, besonders um die physiologischen Funktionen des Vitamins, erheblich erweitert worden,

Literaturverzeichnis: SS. 254—274.

so daß eine erneute Übersicht über die erzielten Fortschritte gerechtfertigt erscheint.

Während man vor einigen Jahren zwar überzeugt war, daß Folsäure in den Stoffwechsel der Einkohlenstoffkörper aktivierend und übertragend eingreift, waren die Hinweise doch stets mehr indirekter Natur und stammten aus in vivo-Versuchen, aus mikrobiologischen Wuchsstoff-Studien und aus Ergebnissen der Chemotherapie. Unmittelbar in vitro hatte man gerade erst die ersten Anhaltspunkte, daß es Kohlenstoff-Fragment-übertragende Enzymsysteme gibt, die Folsäurederivate benötigen und die damit den Aufbau zellwichtiger Bausteine katalysieren. Es tauchte damals auch bereits die Vermutung auf, daß Folsäure selbst nicht der eigentliche Cofaktor ist, sondern daß das ,,Coenzym F'' die labile 5,6,7,8-Tetrahydro-folsäure darstellt, was sich dann in späteren Untersuchungen bestätigt hat.

Aus diesem Grund begann man sich auch eingehender mit der Chemie der tetrahydrierten Pterine zu beschäftigen, aber der im folgenden gegebene Überblick wird erkennen lassen, daß wir auf dem Gebiet der biochemischen Phänomene und des physiologischen Zusammenwirkens sehr viel weiter gelangt sind als auf dem der Chemie der Cofaktor-Aktivität, die noch immer im Stadium der Vermutungen und der indirekten Beweisführungen steht und über wenig fruchtbare Anwendung naheliegender mechanistischer Routine aus analogen Vorgängen nicht hinausgekommen ist.

In gleicher Weise wie die anderen von essentiellen Vitaminen abgeleiteten Cofaktoren, erfüllen auch die vom Vitamin Folsäure stammenden Transportmetaboliten eine recht engumgrenzte biologische Aufgabe, nämlich die Bindung, Bereitstellung und Übertragung der kleinsten Zellbausteine, die sich von Ameisensäure, Formaldehyd und Methanol ableiten. Ihre Herkunft und funktionelle Form sind in der *Tabelle 1* zusammengefaßt. Die Umsetzungen der Kohlenstoff-Fragmente lassen sich

Tabelle 1. Von Folsäure-Coenzymen abhängige Stoffwechsel-Systeme.

System	Umgesetztes chemisches Radikal	Coenzymatische Wirkform
Methionin-Biosynthese Thymin-Biosynthese	$-CH_3$	Methyl-
Serin-Biosynthese 5-Hydroxymethylcytosin- Biosynthese	$-CH_2OH$	Hydroxymethyl- oder Methylen- (Formaldehyd)
Purin-Biosynthese Valin-Biosynthese	$-CHO$	Formyl- oder Methinyl- (Ameisensäure)
Purin-Gärung Histidin-Abbau	$-CHNH$	Formimino- (Formamid)

demnach in zwei Gruppen einteilen: a) solche, in denen die Einkohlenstoff-Einheit unverändert am Cofaktor aktiviert und übertragen wird (Transformylierung, Transhydroxymethylierung); b) solche, in denen die Oxydationsstufe der Kohlenstoff-Einheit am Cofaktor geändert wird und dann erst die Übertragung erfolgt (Formiat/Formaldehyd- und Formaldehyd/Methyl-Umwandlung).

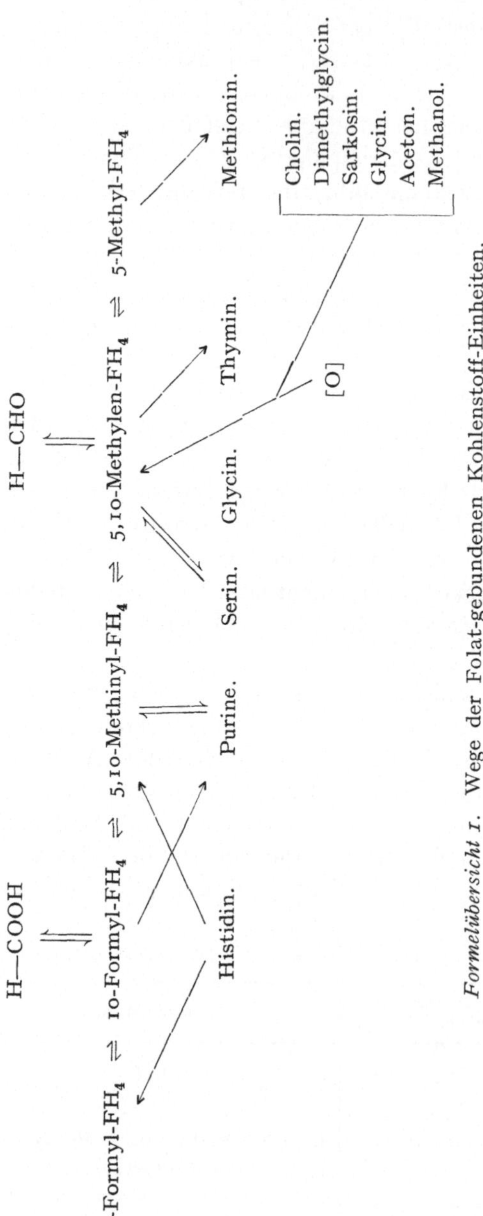

Formelübersicht 1. Wege der Folat-gebundenen Kohlenstoff-Einheiten.

Die Teilnahme der FH_4 an diesen Reaktionen ist klar erwiesen und in späteren Abschnitten eingehender diskutiert. Die verschiedenen enzymatischen Reaktionen lassen sich zu einer Reaktionsfolge zusammenfassen, in der die Oxydationsstufe des Cofaktor-gebundenen Einkohlenstoffrestes schrittweise verändert werden kann, so daß sich die schon lange bekannten biologischen Zusammenhänge aus *Formelübersicht 1* ergeben.

Bereits aus der beistehend angegebenen Reaktionskette kann entnommen werden, daß die Transportmetaboliten selbst, die bisher aus biologischem Material isoliert wurden, sämtlich Tetrahydrofolsäure als Grundkörper haben und daß die funktionellen Gruppen an einem engumgrenzten Teil der Moleküle, nämlich an die Stickstoffatome 5 und 10, gebunden sind (Numerierung: S. 201), so daß man die von diesen eingeschlossene Äthylengruppierung für die Wirkgruppe des Coenzyms halten möchte (*141*).

Literaturverzeichnis: SS. 254—274.

Feststeht, daß an den Folat-Cofaktoren biologisch gebundener potentieller Formaldehyd reversibel zu Formiat oxydiert oder zu Methyl reduziert werden und schließlich in allen diesen Formen auf spezifische Acceptoren übertragen werden kann und daß diese Vorgänge in der Zelle mehr oder weniger frei reversibel sind, so daß die gesamten Kohlenstoff-Bausteine ineinander übergehen und im Zellgeschehen wechselweise genutzt werden können (*252*). Damit aber weisen gerade die Folsäure-Cofaktoren eine Vielseitigkeit der Funktion auf, die die gemeinsame Deutung der Mechanismen außerordentlich erschwert und sie auch in Gegensatz zu den anderen Cofaktoren stellen, die viel einheitlichere Vorgänge katalysieren. Deren Mechanismus läßt sich daher auf einen allgemeinen Nenner bringen und auf spezielle Wirkgruppen zurückführen, wenn auch die Substrate selbst eine lange Reihe homologer Verbindungen sein können, die aber gleichartig und einsinnig umgesetzt werden. So ist die Wasserstoff-übertragende Rolle der Flavin- und Nicotinamid-Coenzyme klar umrissen und als Eigenschaft der heterocyclischen Ringe erkannt, ebenso wie die Acylaktivierung an der Thiolgruppe von Coenzym A, die Aldehyd-Aktivierung am Thiazol-Kohlenstoff-(2) der Cocarboxylase oder die verschiedenen Bindungs-Lockerungen der Pyridoxal-katalysierten Reaktionen von Aminosäuren durch die Bildung der Schiffschen Base mit dem heterocyclischen Aldehyd.

Im folgenden sollen Wirkformen, Chemismen und Funktionen der Folat-Coenzyme zusammenfassend abgehandelt werden, um einen Einblick in die komplexen Verzweigungen dieses Teilgebietes des Stoffwechsels zu geben, das sich noch immer im fließenden Wechsel befindet.

Die enzymatischen Reaktionen sind in einer Zusammenstellung von HUENNEKENS (*124*), die Chemismen des Moleküls in der Übersicht von RABINOWITZ (*244*) besprochen. Mit der Wirkungsweise der Folat-abhängigen Reaktionen beschäftigen sich vornehmlich die Arbeiten von BUCHANAN (*53*), HUENNEKENS (*125*) und JAENICKE (*136, 138—140*). Die biologischen Aspekte sind bei HUTNER (*133*) und WRIGHT (*359*) herausgehoben.

II. Das Vitamin Folsäure.

1. Entdeckung der Folsäure und ihrer Konjugate.

Als man zu Beginn der zwanziger Jahre erkannte, daß viele Erkrankungen auf einem Mangel bestimmter oligodynamischer Nahrungsbestandteile beruhten, setzte eine intensive Suche nach derartigen Vitaminen ein, und in kurzer Zeit waren diesen Bemühungen viele glänzende Erfolge beschieden. Die Erforschung der Rolle der Folsäure als Vitamin (vgl. *302*) begann 1931, als WILLS Remissionen bei Sprue durch Verfütterung von Hefe erreichte. Spätere Untersuchungen haben gezeigt, daß man bei Rhesusaffen experimentell eine megaloblastische Anämie mit Schleimhautschädigungen usw. erzeugen kann, die sich durch ein in Leber und Hefe enthaltenes Vitamin (Vitamin M) heilen läßt. Schließlich brauchen Kücken zum normalen Gedeihen und zur normalen Blutbildung essentielle Nahrungsstoffe (Vitamin R, S, U), die aus Hefe extrahiert wurden. Alle diese Substanzen erwiesen sich schließlich als Folat-Verbindungen (*302*), wie *Tabelle 2* (S. 188) zeigt. Ihre Reinigung

Tabelle 2. Aus natürlichem Material isolierte Folsäure-aktive Substanzen.

Name	Quelle	Wuchsstoff für	Konstitution	Literatur
Vitamin M	Hefe	Rhesusaffe	Pteroylglutaminsäure	Day (vgl. *302*)
Norit-Eluat-Faktor	Hefe	*Lactobacillus casei*, Huhn	Pteroylglutaminsäure	Snell (*295*)
Folsäure	Spinat	*L. casei, Streptococcus faecalis* R	Pteroylglutaminsäure	Mitchell (*199*)
Leber-*L. casei*-Faktor	Leber	*L. casei*	Pteroylglutaminsäure	Stokstad (*304*)
Vitamin B$_c$	Leber, Hefe	Huhn	Pteroylglutaminsäure	Pfiffner (*236*)
Vitamin B$_c$-Konjugat	Hefe	Huhn; *St. faecalis* R (nach Spaltung)	Pteroyl-heptaglutaminsäure	Pfiffner (*238*)
Fermentations-*L. casei*-Faktor	*Corynebacterium* sp.	*L. casei*; *St. faecalis* R (nach Spaltung)	Pteroyl-triglutaminsäure	Stokstad (*130*)
Citrovorum-Faktor	Pferdeleber	*Pediococcus cerevisiae (Leuconostoc citrovorum)*	5-Formyl-tetrahydropteroyl-glutaminsäure	Sauberlich (*269*)

Literaturverzeichnis: SS. 254—274.

und Isolierung aus natürlichem Material, in dem sie meist nur in kleinsten Mengen enthalten ist, konnte erst dann erfolgreich werden, als das besonders von SNELL (*293, 294*) aufgenommene intensive Studium der Ernährung der Mikroorganismen zeigte, daß die gleiche unbekannte Substanz auch ein Wachstumsfaktor bestimmter Milchsäurebakterien ist. Binnen kurzem wurde von SNELL und PETERSON (*295*) der Norit-Eluat-Faktor in Leber und Hefe entdeckt, der notwendig ist für das Wachstum des Bakteriums *Lactobacillus casei* und sich später auch als der für die Blutbildung im Hühnchen verantwortliche Cofaktor erwies. Zur gleichen Zeit gelang es MITCHELL (*199*), eine saure Substanz nahezu rein aus Spinatblättern darzustellen, die das Wachstum von *L. casei* und *Streptococcus faecalis* R fördert. Sie wurde nach ihrer Herkunft „*Folsäure*" genannt, ein Name, der sich dann zur Bezeichnung einer ganzen Stoffgruppe eingeführt hat, zu der auch noch komplexere Derivate gehören. Während PFIFFNER (*235, 236*) aus Leber den Anämie-Faktor isolierte (Vitamin B_c), teilte STOKSTAD (*130*) mit, daß ein Corynebacterium eine Folat-Substanz besonders am Ende der Wachstumsperiode in das Medium abgibt, die das Wachstum von *L. casei* stärker förderte als Folsäure, aber im Gegensatz zu dieser unwirksam gegenüber *St. faecalis* war. Dies war um so überraschender, als Folsäure selbst bei beiden Mikroorganismen etwa gleich wirksam ist. Bald wurde die Ursache dieser Diskrepanz manifest. Durch eine in tierischem Gewebe verbreitete Peptidase (vgl. S. 190) wird der Fermentation *L. casei*-factor in Vitamin B_c und zwei Moleküle Glutaminsäure gespalten. Man bezeichnet Folsäuren mit zusätzlichen Glutamyl-Resten als „*Konjugate*". In Hefe wurde dann ein weiteres Konjugat mit einer unverzweigten Kette von sieben Glutaminsäure-Resten in γ-Verknüpfung (Heptaglutamat) gefunden, und schließlich haben WITTENBERG, NORONHA und SILVERMAN (*217a, 351*) wahrscheinlich gemacht, daß in verschiedenen natürlichen Vorkommen, wie dem Gasdrüsen-Gewebe der Qualle *Physalia physalis* L., der Hühnerleber usw., eine ganze Skala von formylierten Konjugaten vorliegen. Über gemischte Konjugate wurde hin und wieder berichtet. So hat B. E. WRIGHT (*358, 358 a*) eine Substanz-Klasse aus Extrakten von *Clostridium cylindrosporum* isoliert, deren Pteroylglutamat-Anteil an $N_{(5)}$ formyliert ist und in deren Seitenkette neben Glutamat die miteinander verwandten Aminosäuren Glycin, Serin und Alanin in unbestimmter Sequenz vorkommen, die darüber hinaus verschiedene Mengen von Pentose und Phosphat enthalten. Andererseits haben RABINOWITZ und HIMES (*245*) in Extrakten des gleichen Bakteriums, die etwa 200mal mehr mikrobiologisch nachweisbare Folat-Aktivität enthalten als tierische Leber, durch Austauscherchromatographische Verfahren eindeutig nachgewiesen, daß der dort im Energie-Stoffwechsel eine Schlüsselfunktion erfüllende Folat-Cofaktor (vgl. S. 248) zu mehr als 80% das Triglutamat der FH_4 darstellt. Ebenso wurde aus *Escherichia coli*-Zellen ein 5-Formyl-pteroyltriglutamat als Cofaktor der Methioninbildung isoliert (*106, 108*). Das Konjugat in *B. subtilis* (*109*) ist vermutlich ebenfalls das Triglutamat.

Viel diskutiert wird zur Zeit, daß die natürlichen Folat-Cofaktoren stets Konjugate der FH_4 seien, jedoch kommt unstreitig auch das Monoglutamat, besonders in tierischem Gewebe, in reichlichen Konzentrationen vor. Es ist aber bereits früher (*103*) darauf hingewiesen worden, daß die synthetische FH_4 vielleicht nur ein bequemer zugängliches Modell des eigentlichen Cofaktors darstellt.

Eine eingehende Analyse verschiedener Vorkommen durch die heute einfachen und hochempfindlichen Kombinations-Verfahren von DEAE-Chromatographie und differentieller Wuchsstoff-Analyse (S. 226) wird sicher zeigen, daß die Konjugate häufiger angetroffen werden als bisher angenommen. Sie haben auch sehr spezifische Funktionen. So haben Guest und Woods (*104, 108*) in eingehenden Untersuchungen an Methionin-bildenden Mutanten von *E. coli* gezeigt, daß in einem Fall Monoglutamate der Pteroinsäure als Cofaktor genügen, in einem anderen aber nur das Triglutamat den Mangel an einem B_{12}-Enzym ausgleichen kann (S. 244).

2. Konjugat-spaltende Enzyme.

Eine die Konjugate spaltende Carboxypeptidase, die auch als Konjugase bezeichnet wird, ist in Schweineniere (*23 a*) und Hühnerpankreas (*198*) enthalten. Letztere wurde aus Pankreas-Extrakten hoch angereichert, hat ein pH-Optimum bei pH 7 bis 8 und erweist sich als sehr spezifische γ-Peptidase, so daß als Endprodukte Glutaminsäure und Pteroyldiglutamat erhalten werden (*66*). Ein ähnliches Enzym fand Torii (*313*) in Hundeleber und Volcani (*320*) in *Flavobacterium polyglutamicum*. Das Schweinenieren-Enzym dagegen mit dem pH-Optimum zwischen pH 4 und 5 ist weniger spezifisch, spaltet auch die letzte Glutamyl-glutamat-Bindung bis zum Pteroylmonoglutamat (*6*), arbeitet aber langsam. Zum vollständigen Abbau der höheren Konjugate der FH_4 zieht man daher die Zusammenwirkung beider Konjugasen vor.

Man erhält dann die gleichen Folat-Konzentrationen im Bakterienwachstumstest mit *St. faecalis* und dem nur auf tetrahydrierte Verbindungen ansprechenden *Pediococcus cerevisiae* wie mit dem wesentlich umständlicheren Hühnchen-Wachstumstest, der sowohl Folsäure, seine reduzierten Formen, wie auch alle Konjugate erfaßt. Es liegt nahe, daraus zu entnehmen, daß biologisch die Folsäure stets in der Tetrahydroform vorkommt, die die gleiche relative Aktivität für alle drei Testorganismen zeigt.

3. Vorkommen, Bedarf und Ausscheidung.

Die ungefähren Konzentrationen in mg/g Trockengewicht, die man in tierischen Geweben findet, liegen zwischen 0,5 und 15: Leber 2 bis 20; Niere 0,5 bis 15; Herz 0,5 bis 1; in pflanzlichen Quellen zwischen 0,5 bis 2,0: grüne Blätter 0,5 bis 1,2; Gurken 1,0 bis 2,0.

Literaturverzeichnis: SS. 254—274.

Der Gehalt mancher Bakterien und der Hefe ist wesentlich höher. Es ist nach den vorhandenen Angaben schwer zu entscheiden, ob die Folat-Aktivitäten als unhydrierte Folsäure oder aber als hydrierte Folat-Cofaktoren, als unkonjugierte Monoglutamate oder als Konjugate vorliegen.

Zahlreiche Mikroorganismen sind zur Bildung von Folsäure befähigt, und man kennt auch die Bedingungen, unter denen das geschieht (S. 192). Aus diesem Grund ist es aber auch nicht möglich, eine tägliche Mindestmenge für den Menschen anzugeben. Sein Bedarf kann weitgehend außer durch die Nahrung, in der das Vitamin weit verbreitet ist, auch durch symbiontische Synthese durch Darmbakterien gedeckt werden. Ein Folsäuremangel stellt sich daher nur unter pathologisch verschlechterten Resorptionsbedingungen oder aber bei extremem Bedarf ein, wie er dann und wann in der Schwangerschaft vorkommen mag.

Die Symptome der megaloblastischen Folsäuremangel-Anämien, auch derer, die in der Schwangerschaft auftreten, sind: Veränderungen im roten Blutbild, Schleimhautschäden und Infektionsbereitschaft, aber auch schwere Degenerationen im Nervensystem. Sie sprechen wohl prompt auf Gaben von Folsäure an, am besten die atypischen Formen der perniciösen Anämie, d. h. solche, die nicht mit Achlorhydrie und Nervenschädigungen einhergehen, und die tropische Sprue. Bei der vollständigen typischen perniciösen Anämie jedoch vermag Folsäure die gefürchteten neurologischen Symptome auf die Dauer nicht zu verhindern. Der eigentliche Antiperniciosa-Faktor ist das Vitamin B_{12}; Folsäure kann nur einen Teil der B_{12}-Mangelschäden verhindern, da die Funktionen dieser beiden Vitamine in bestimmten Stoffwechselketten, nämlich den zur Bildung von Methionin und Thymidylsäure führenden, eng verwoben sind.

Überschüssig aufgenommene Folsäure wird im Urin als $N_{(10)}$-Formyl-FH_4 ausgeschieden (*283*, *284*), wie überhaupt diese Verbindung die biologische Transportform des Cofaktors zu sein scheint, denn auch im Serum normaler Personen liegt ein großer Teil der Folat-Aktivität als $N_{(10)}$-Formyl-FH_4 vor. Anders ist das bei bestimmten Anämien (*204*, *325*, S. 246). Hier findet man hohe Konzentrationen an 5-Methyl-tetrahydrofolat, die bis zu 80% der gesamten Folat-Aktivität ausmachen kann. Da diese Verbindung sich als Wuchsstoff für *L. casei* und *St. faecalis* wie ein Konjugat verhält, hat man früher angenommen, daß die Serumfolsäure und damit auch die aktiven Folatverbindungen der Enzymsysteme derartige Konjugate wären. Sicher würde sich bei Nachprüfung mancher Literaturangaben herausstellen, daß es sich auch hier um 5-Methyl-FH_4 gehandelt hat. Interessant ist, daß die erhöhten Konzentrationen an Methylfolat und die relative Verteilung der Folatverbindungen sich wieder der Norm nähert, wenn man bei Vitamin B_{12}-Mangel Methionin oder Vitamin B_{12} gibt (*217*).

III. Auf- und Abbau der Folsäure-Cofaktoren.

1. Biogenese der Folsäure.

Durch Hemmungsanalyse an Mikroorganismen hatte Shive (*280*) erkannt, daß Folsäure nacheinander in den Stoffwechsel von Purinen, von Thymin, von Serin und von Methionin eingreift, und daß die Wirkungen der Folsäure durch Sulfonamide blockiert werden können. Diese Erscheinung findet ihre Erklärung in der Struktur und Biogenese der Folate, die im Molekülverband *p*-Aminobenzoesäure enthalten. Deren biologischer Einbau wird durch Sulfonamide unterbunden und damit die Funktion der Folsäure-katalysierten Reaktionen verhindert.

Pteroylglutaminsäure entsteht biologisch auf der Stufe der Dihydro-verbindung *(Formelübersicht 2)* und wird erst durch anschließende

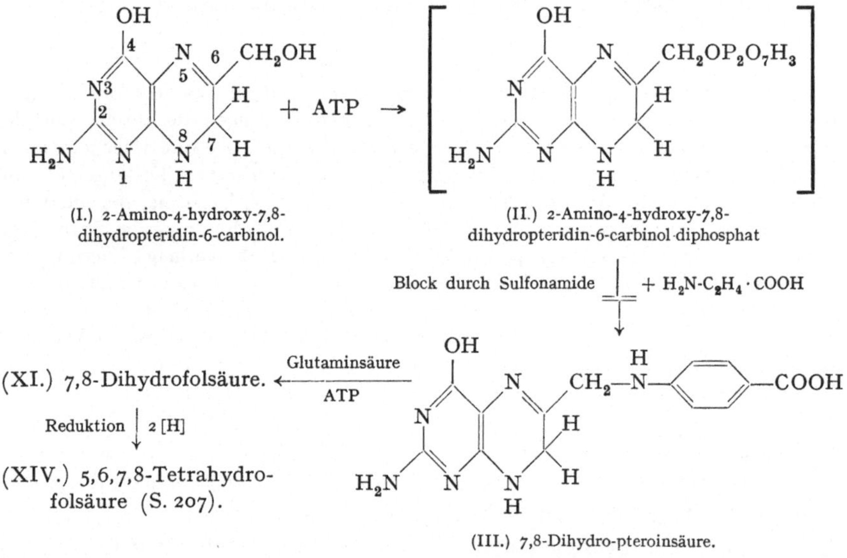

(I.) 2-Amino-4-hydroxy-7,8-dihydropteridin-6-carbinol.

(II.) 2-Amino-4-hydroxy-7,8-dihydropteridin-6-carbinol·diphosphat

Block durch Sulfonamide | + H_2N-C_2H_4·COOH

(XI.) 7,8-Dihydrofolsäure. ← Glutaminsäure / ATP

Reduktion | 2 [H]

(XIV.) 5,6,7,8-Tetrahydro-folsäure (S. 207).

(III.) 7,8-Dihydro-pteroinsäure.

Formelübersicht 2. Biogenese der Folsäure.

Reduktion in die Cofaktorform übergeführt. Die Untersuchungen über die Biogenese der Folsäure (*357*) haben aber noch zu keinem abschließenden Bild geführt. Es herrscht jedoch allgemeine Übereinstimmung, daß das Dihydropterin-6-carbinol-Derivat (I) das Vorläufer-Pterin sein wird.

Woods (*215*) wies auf die Möglichkeit hin, daß ruhende Bakterien das komplexe Molekül der Pteroylglutaminsäure aus den konstituierenden Bestandteilen aufbauen können. Er fand auch, daß 2-Amino-4-hydroxy-6-formyl-pteridin (Pterin-6-aldehyd) die Synthese stimuliert (*172*). Korte und Mitarb. (*170*) verfolgten diese Beobachtungen weiter. Sie zeigten die Aufnahme von ^{14}C-markiertem Xanthopterin in Folsäure bei verschiedenen Bakterien, erhielten jedoch keinen Einbau von Guanin in das Vitamin. Weygand und Mitarb. (*333*) fanden fast zur gleichen Zeit, daß sich bestimmte Bakterienstämme adaptieren lassen, 2-Amino-4-hydroxy-6-carboxy-pteridin (Pterin-6-carbonsäure) zur Folatbildung zu verwenden, die ihrerseits durch Sulfonamide gehemmt wird *(314)*. Auch über die Genese des Pterinanteils selbst können definitive Angaben noch nicht gemacht werden. Durch die glänzenden Untersuchungen von WEYGAND *(338)* weiß man aber, daß die Pterine der Schmetterlingsflügel aus Purinen gebildet werden oder mindestens eine gemeinsame Zwischenstufe haben, denn das nach Gabe von ^{14}C-Glycin und ^{14}C-Formiat beobachtete Markierungsmuster läßt sehr weitgehend parallellaufende Wege für die Bildung beider Ringsysteme erkennen. Zum gleichen Ergebnis führen die Untersuchungen von VIEIRA und SHAW *(318)* bei der Folat-Synthese, da $C_{(2)}$-markiertes Adenin unverdünnt in den von *Corynebacterium* sp. ausgeschiedenen Folsäure-Konjugaten gefunden wurde, während der Kohlenstoff 8 verlorenging. Einen indirekten Hinweis für den Übergang von Guanin in Pterine haben ESPOSITO und FLETCHER *(82)* gegeben: Das als Fungistaticum wirksame Kupferoxinat wird durch Pterine oder mögliche Pterin-Vorstufen, wie Guanin, Xanthin, 2,4,5-Triamino-6-hydroxy-pyrimidin, entgiftet, nicht aber durch Adenin oder 2,4-Diamino-5,6-dihydroxypyrimidin. Sie schließen daraus, daß ein spezifisches, Cu^{++}-enthaltendes Enzymsystem in der Pterin-Biosynthese die Umwandlung von Pterinvorläufern (Purinen) in 4,5-Diaminopyrimidine katalysiert und das Oxinat ein Hemmstoff dieses Enzyms ist.

Eine bedeutsame neue Beobachtung ist die Möglichkeit, die Purin-Pterin-Umwandlung, die bisher nur im in vivo-System zugänglich war, auch in vitro ausführen zu können. REYNOLDS und BROWN *(255)* teilten kürzlich mit, daß es ihnen gelungen ist, in Kohle-behandelten Coli-Extrakten Guanosin oder Guanosin-phosphate in Folsäure umzuwandeln, wenn gleichzeitig *p*-Aminobenzoylglutamin-säure, ATP und Phosphat zugegen sind. Das System kann durch Cu^{++}-Ionen stimuliert werden. Guanin ist nur in Gegenwart von Ribose-5-phosphat oder Phosphoribosyl-pyrophosphat brauchbar; Xanthin- und Inosin-Derivate sind unwirksam. Durch Chromatographie wurde nachgewiesen, daß als Produkt tatsächlich Dihydrofolsäure entsteht, deren spezifische Radioaktivität nicht wesentlich geringer als diejenige des eingesetzten Guanins ist. Auch hier wird eine Markierung in $C_{(8)}$ nicht im Folsäure-Molekül wiedergefunden.

Aus verschiedenen Arbeitsgruppen ist weiter berichtet worden, daß Ribose in die Dreikohlenstoff-Folge $C_{(7)}$, $C_{(6)}$, $C_{(9)}$ der Folsäure eingeht und dabei vielleicht zunächst ein dem Biopterin nahestehendes Seitenketten-Pterin auftritt *(45, 140, 338)*. Tatsächlich hat auch ZIEGLER *(368)* gefunden, daß Kohle-behandelte Hefe-Präparate das Sepiapterin, ein natürlich vorkommendes Tetrahydrobiopterin-Derivat, in Folsäure umwandeln.

Die Verkürzung der Seitenkette eines solchen Pterin-Derivats scheint demnach nicht ausgeschlossen, falls der Bildung von Folsäure-aktiven Substanzen hierbei nicht eine vollständige Abspaltung der Seitenkette und Neukondensation mit einer Kohlenstoff-Einheit zugrunde liegt. Denn das $C_{(3)}$ der Ribose geht nur unter Verdünnung in $C_{(9)}$ der Folsäure über. Es ist deshalb möglich, daß die Seitenkette erst aus entsprechenden Vorläufern an den Pterin-Ring ankondensiert wird. Beobachtungen in dieser Richtung liegen bereits vor: H. S. FORREST konnte aus einem Gemisch von Dihydropterin, α-Ketobutyrat und Thiaminpyrophosphat eine Biopterin-aktive Substanz synthetisieren (Privatmitteilung). Im Folsäure-synthetisierenden System konnte allerdings Folsäure nicht aus Formaldehyd, Dihydropteridin und *p*-Aminobenzoylglutaminsäure erhalten werden (JAENICKE, unveröff.).

Die ersten Versuche zur Folsäure-Synthese mit zellfreien Extrakten hat KATSUNUMA *(150, 151, 151 a)* unternommen. Verwendet wurden

Extrakte von *Mycobacterium avium*. Diese können sowohl Xanthopterin wie auch Pterin-6-carbonsäure mit *p*-Aminobenzoesäure und Glutaminsäure kondensieren, vorausgesetzt, daß Thiaminpyrophosphat und Biotin zugegen sind. Es wird daher möglicherweise eine Kohlenstoffeinheit aus CO_2 oder Formaldehyd an das Pterin angefügt und dieses dann mit *p*-Aminobenzoesäure verbunden (*150*). Darüber hinaus katalysiert der Extrakt auch die Bildung von *p*-Aminobenzoylglutaminsäure aus *p*-Aminobenzoesäure und *L*-Glutaminsäure, wobei *p*-Aminobenzoyl-adenylat als Zwischenprodukt angegeben wird und Sulfonamide eine spezifische Hemmwirkung haben.

Das Folat-synthetisierende Mycobacterien-System scheint sich aber von den Systemen anderer Bakterien grundsätzlich zu unterscheiden. Shiota (*277*) fand nämlich, daß Extrakte von *L. arabinosus* Folat-aktive Substanzen in Gegenwart von ATP und Mg^{++}-Ionen sowohl aus 2-Amino-4-hydroxy-6-hydroxymethyl-pterin (Pterin-6-carbinol) wie der entsprechenden 6-Carbonsäure synthetisieren. Brown (*47*) beschrieb ein analoges Enzym aus Extrakten von *E. coli*. Dies kann aber als Substrate nur Pterin-6-aldehyd und -6-carbinol — oder reduzierte Formen — verwenden. Da *p*-Aminobenzoesäure viel wirksamer ist als *p*-Aminobenzoylglutaminsäure, wurde postuliert, daß als Zwischenstufe Pteroinsäure — oder eine reduzierte Form — gebildet wird. Brown (*48*) wies auch nach, daß der Angriff der Sulfonamide in der Kondensation des Pterins mit *p*-Aminobenzoesäure stattfindet. Bisher ist aber der Einbau eines Sulfonamids in ein Folsäure-Analoges im in vitro-System noch nicht erwiesen worden, wenn auch die Ergebnisse von Brown sich als Konkurrenzreaktion des Sulfonamids mit *p*-Aminobenzoesäure um das Pterin deuten lassen und die Beobachtungen von Wacker (*320 a*) mit markierten Verbindungen ebenfalls nahelegen, daß Bakterienzellen die *p*-Aminobenzoesäure-Antagonisten tatsächlich in Folsäure-analoge Moleküle aufnehmen. Zellfreie Extrakte von sulfonamid-resistenten Mutanten von *Pneumococcus* zeigen in der Folat-Synthese die gleichen Resistenzcharakteristika wie intakte Zellen (*354*); es ist also tatsächlich eine unmittelbare enzymatische Reaktion, die auf die Blockierung anspricht.

Jaenicke und Chan (*142*) wiesen schließlich mit Extrakten aus Hefe und aus Coli-Bakterien nach, daß der eigentliche Pterin-Donator in der Folsäuresynthese ein Dihydropterin-6-carbinol (I, S. 192) ist, daß dieses mit ATP in ein intermediäres Pyrophosphat (II) umgewandelt wird und mit *p*-Aminobenzoyl-Verbindungen zu Dihydro-pteroat-Verbindungen (III) kondensiert, die chromatographisch identifiziert und, im Fall der Dihydrofolate, auch nach weiterer enzymatischer Reduktion zum Tetrahydrofolat-Cofaktor, an ihrer katalytischen Aktivität in Einkohlenstoff-Reaktionen nachgewiesen werden konnten.

Literaturverzeichnis: SS. 254—274.

Das postulierte Dihydropterin-diphosphat (II) wurde von SHIOTA (*279*) synthetisiert und kann tatsächlich in *L. arabinosus*-Extrakten Dihydropterin-carbinol und ATP ersetzen. Ähnliche Beobachtungen liegen für das Coli-System vor (JAENICKE, unveröff.).

In beiden Fällen ist jedoch zu bedenken, daß die noch ungereinigten Bakterien-Extrakte festgebundene Adenin-Nucleotide enthalten können, die zunächst durch das aktive Pyrophosphat zu ATP aufphosphoryliert werden und erst dann — eventuell in einer Simultan-Reaktion — mit passend angelagertem Dihydro-pterincarbinol und *p*-Aminobenzoylglutaminsäure kondensieren. Auch die erhaltenen Austauschversuche geben keine sichere Grundlage für die Diskussion eines Bildungsmechanismus des Intermediärprodukts. Erst die Reinigung des Enzyms würde erkennen lassen, ob das Diphosphat in einer einstufigen Pyrophosphorylierung oder in zwei Kinaseschritten (mit gebundenem Enzym-Pterinphosphat-Komplex) entsteht.

Da bei den Versuchen von JAENICKE und CHAN (*142*) *p*-Aminobenzoe-säure weniger wirksam war als *p*-Aminobenzoylglutaminsäure, schließen sie auf eine direkte Bildung des kompletten Folsäure-Moleküls. SHIOTA (*278*) bestätigte diese Beobachtung. Dagegen stehen die Beobachtungen von BROWN, WEISMAN und MOLNAR (*49*), die die Eigenschaften des Folat-synthetisierenden Systems in *E. coli* eingehend untersuchten. Danach ist es wahrscheinlicher, daß das biogene Primär-Produkt Dihydropteroinsäure ist, das erst in weiteren Schritten zur Folsäure vollendet wird; hohe *p*-Aminobenzoesäure-Konzentrationen hemmen die Synthese.

Die Bildung der Pteroyl-Konjugate ist erst unvollkommen untersucht, da die geringen Umsätze und die komplizierte Analytik die Bearbeitung erschweren. Nach eigenen vorläufigen Ergebnissen (mit M. SILVERMAN) sind Extrakte von einigen Mikroorganismen in der Lage, die Seitenkette der Folate in Gegenwart von ATP und Glutamat um ein bis zwei Einheiten zu verlängern. Bevorzugt ist Dihydrofolsäure als Acceptor. Der Bilanz scheint eine Spaltung von ATP in ADP und Phosphat, also möglicherweise eine Reaktion in Art der Glutathion-Bildung, zugrunde zu liegen.

2. Biologischer Abbau der Folsäure.

Der biologische Abbau des Folsäuremoleküls in Bakterienzellen (*300*), Gewebshomogenaten (*363*) und menschlichen Erythrocyten (*40*) führt zu Pterin-6-aldehyd und *p*-Aminobenzoylglutaminsäure (*93, 156*). Da dieser Vorgang ATP und ein reduzierendes System braucht, verläuft er wohl über FH_4, die dann spontan in Gegenwart von O_2 oder H_2O_2 in der nachstehenden Weise zerfällt (*257 a*) (Gleichung [1]):

$$FH_4 + O_2 \rightarrow p\text{-Aminobenzoylglutaminsäure} + \text{Pterin-6-aldehyd.} \quad [1]$$

Damit stimmt auch überein, daß der enzymatische Abbau durch Aminopterin, das die Folatreduktase blockiert (SS. 197, 221), oder in reduzierendem Milieu gehemmt wird (*71, 91*). Auch 5-Formyl-FH_4 wird in gleicher

Weise, offenbar nach vorausgehender Deformylierung, abgebaut (*71*). Dagegen erfordert diese Reaktion in Schweineleber Glutaminsäure, und bei eingehender Untersuchung konnte von Silverman (*286*) gezeigt werden, daß tatsächlich eine Transformylierung im Sinn der Gleichung [2]:

$$5\text{-CHO-FH}_4 + \text{Glutamat} \rightleftharpoons \text{N-Formyl-Glutamat} + \text{FH}_4 \qquad [2]$$

$$(K = 0{,}05{-}0{,}13)$$

erfolgt, an die sich unter aeroben Bedingungen Reaktion Gl. [1] anschließt. Unter anaeroben Bedingungen wird die reversible N-5-Formyl-tetrahydrofolat-Glutamat-Transformylase-Reaktion Gl. [2] vornehmlich von rechts nach links arbeiten.

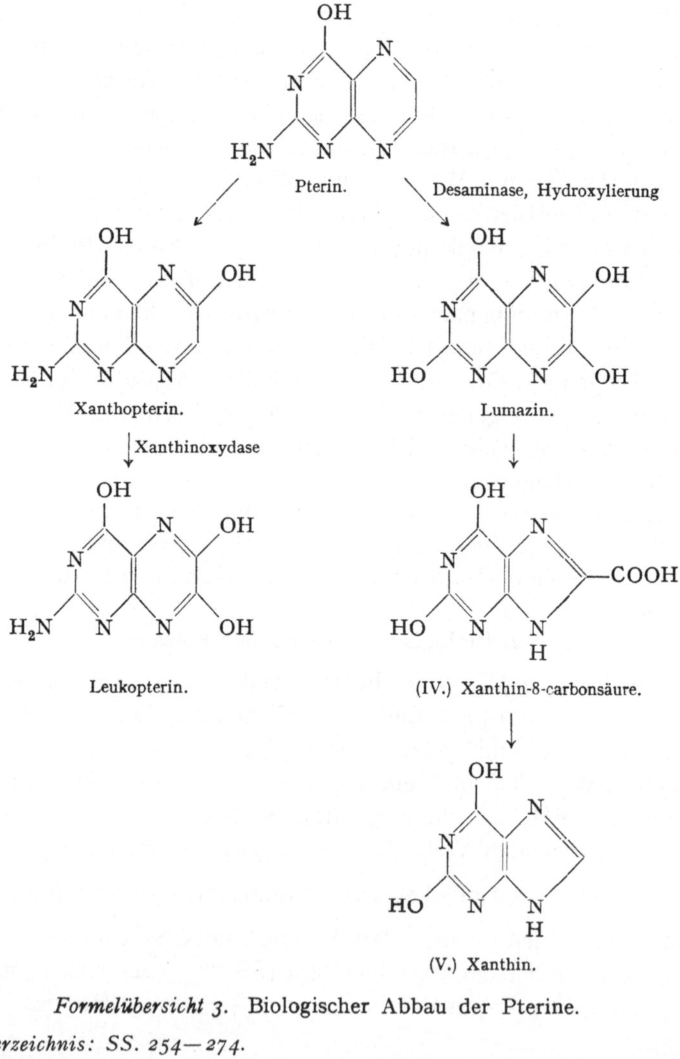

Formelübersicht 3. Biologischer Abbau der Pterine.

Literaturverzeichnis: SS. 254—274.

Bei der „aeroben N-10-Formylierung" der Folsäure (*251*) in Mitochondrien-freien Schweineleber-Homogenaten mag es sich um einen weiteren primären oder sekundären Abbaumechanismus handeln. Er führt nach fluorometrischen Messungen zu je 1 Mol 10-Formylfolsäure, Pterin-6-aldehyd und *p*-Aminobenzoylglutaminsäure je Mol umgesetzten Einkohlenstoff-Donator. Mitochondrien dagegen enthalten eine Oxydase, deren Produkte Pterin-6-aldehyd und *p*-Aminobenzoylglutamin-säure sind.

Außer durch Spaltung des Moleküls kann Folsäure auch durch Desaminierung inaktiviert und abgebaut werden. Enzyme, die diese Reaktion ausführen, wurden in *Alcaligenes metalcaligenes* gefunden (*174*). Es entsteht 2,4-Dihydroxy-pteroylglutaminsäure. Das Pteringerüst selbst wird ebenfalls von Bakterien nach vorausgehender Desaminierung abgebaut. Man erhält zunächst Lumazin-Derivate. Während deren weiteres Schicksal aber bei einem aus Faulschlamm isolierten Bakterium unklar blieb (E. RÜHL, unveröff.), konnte von McNUTT (*175*) ein Abbau unter Ringverengerung zu Xanthin-8-carbonsäure (IV, S. 196) und dann zu Xanthin (V) beobachtet werden *(Formelübersicht 3)*. Die Oxydation der Pterine, und zwar nur an $C_{(7)}$, erfolgt durch die Xanthinoxydase (*68*) nach entsprechender Wasseranlagerung an die 6,7-Doppelbindung. Das Enzym ist äußerst spezifisch; es greift Pterine nicht an, wenn sie in Stellung 6 hydroxyliert sind (*20*). 6-Formyl-pterin wird in geringen Konzentrationen zur 6-Carbonsäure oxydiert, in höheren ist es der wirksamste Hemmstoff der Purin- und Pterin-Oxydation (*148*).

3. Enzymatische Reduktion der Folsäure zum Cofaktor.

Bevor die in der Zelle gebildete Dihydrofolsäure biologisch verwendet werden kann, muß sie in Tetrahydrofolsäure umgewandelt werden (vgl. z. B. *73*, *115*). Dazu ist eine enzymatische Reduktion erforderlich. Das katalysierende Enzym, Folat- bzw. Dihydrofolat-Reduktase (*92*), hat daher eine wichtige regulierende Funktion. Untersuchungen von PETERS und GREENBERG (*232*), von ZAKRZEWSKI und NICHOL (*364*, *367*), von WERKHEISER (*332*) und von MATHEWS und HUENNEKENS (*188 a*) haben sich eingehend mit diesem Enzym befaßt, das im allgemeinen sowohl die Reduktion von Folat zu Dihydrofolat (Gl. [3]) wie die von Dihydrofolat zu Tetrahydrofolat (Gl. [4]) mit Hilfe von reduzierten Pyridinnucleotiden ausführen kann:

$$\text{F} + \text{TPNH (DPNH)} + \text{H}^+ \rightleftharpoons \text{FH}_2 + \text{TPN}^+ \text{(DPN}^+\text{)}, \qquad [3]$$

$$\text{FH}_2 + \text{TPNH (DPNH)} + \text{H}^+ \rightleftharpoons \text{FH}_4 + \text{TPN}^+ \text{(DPN}^+\text{)} \qquad [4]$$

$$(\text{K} = 5,6 \times 10^{11}) \ (\text{pH} = 0).$$

Die Reduktase wurde in Leber verschiedener Tiere, in Leukozyten (*281*) und in Mikroorganismen (*34*) gefunden und ist aus Hühnerleber (*223*) durch Alkohol-Chloroform-Fällung, Ammonsulfat-Fraktionierung und

Hydroxylapatit-Adsorption 600fach angereichert worden. Zum Nachweis lassen sich die spektralen Veränderungen verwenden (*34*), aber auch die Fluoreszenz-Zunahme beim Übergang von FH_2 zu FH_4 (*281*), die vermutlich auf einem anschließenden spontanen Zerfall in fluoreszierende Produkte beruht. Das Enzym hat mehrere überaus bemerkenswerte Eigenschaften (*122a*): Es katalysiert die TPNH-spezifische Reduktion von $7,8$-FH_2 zu $5,6,7,8$-FH_4. Es hat zwei pH-Optima, bei pH 4,5 und 7,5. In beiden Fällen entsteht Tetrahydrofolsäure als Reaktionsprodukt. Folat und DPNH können ebenfalls als Substrate verwendet werden, aber mit viel geringeren Umsätzen. Nur mit sehr großen Mengen TPN oder FH_4 ist eine Umkehr der Reaktion möglich (*188 a*). Das Geschwindigkeitsverhältnis von Folat zu Dihydrofolat beträgt 1 : 20 und bleibt über den ganzen Reinigungsprozeß konstant (*367*). Ebenso ist die Hemmbarkeit beider Stufen für Aminopterin gleich. Daher wird auf das gleiche Enzym für beide Reduktions-Schritte geschlossen. Folat-Reduktase reduziert außer Folsäure und seinen Konjugaten auch an $C_{(9)}$ und/oder $N_{(10)}$ substituierte Folate sowie Pteroinsäure und Pterin-6-aldehyd, nicht aber $N_{(10)}$-Formylfolsäure (*365*), 6-carboxylierte oder 6-methylierte Pterine und Xanthopterin. Alles in allem ist sie aber eine nicht sehr spezifische Pterin-Reduktase.

Nath und Greenberg (*212*) haben demgegenüber in Thymus eine DPNH-abhängige spezifische Dihydrofolat-Reduktase gefunden, die also nur die zweite der beiden Reduktionen bewirken kann. Auch die von Blakley (*34*) aus *St. faecalis* R 300fach angereichert TPNH-benötigende Folat-Reduktase reduziert nur FH_2 und seine Konjugate.

Die Hemmung der Folat-Reduktase durch Folsäure-Antagonisten (Aminopterin, Amethopterin) hat in jüngster Zeit besonderes Interesse gewonnen. Der Effekt selbst ist bereits seit längerer Zeit bekannt (*72, 330*), ebenso wie seine praktische Irreversibilität (vgl. *233*). Die benötigten Inhibitor-Konzentrationen sind ungewöhnlich niedrig: Blakley (*34*) erhielt bei dem *St. faecalis*-Enzym vollständige Hemmung mit $2,4 \times 10^{-9}$ M Amethopterin. In der gleichen Größenordnung liegen die von Slavíková (*290*) angegebenen Hemmwerte. Werkheiser (*332*) hat den Mechanismus der Hemmung eingehend untersucht und markiertes Aminopterin zu 80% im löslichen Überstand von Zellhomogenaten und dort an Protein fixiert gefunden. Die Folat-Antagonisten binden sich mit der Folat-Reduktase in stöchiometrischem Verhältnis und nahezu irreversibel (pseudo-reversibel), denn nur lange Dialyse in Gegenwart von Folat kann das inaktivierte Enzym reaktivieren. Die Reaktivierung geht exakt der Menge freigewordenen Aminopterins parallel: $3,3 \times 10^{-7}$ M Aminopterin, die vollständige Hemmung geben, werden durch 2×10^{-3} M Folat zu 85% vom Enzym verdrängt, wobei 85% der Enzymaktivität wiederhergestellt werden. Das Phänomen ließe sich nutzen, um die die Folsäure bindenden Gruppen im Enzym zu ermitteln.

Bei der Entwicklung von Resistenz gegen Folsäure-Antagonisten in Zellkulturen steigt die Menge der Folat-Reduktase beträchtlich an und kompensiert wohl dadurch für den Ausfall des zum Abbinden des Hemmstoffs nötigen Enzymproteins (*110, 198 a*). Auch in leukämischen Leukozyten werden erhöhte Reduktase-Konzentrationen angetroffen (*21, 347 a*). Dagegen beruht das Bedürfnis von *Pediococcus cerevisiae* für hydrierte Folat-Derivate nicht auf einem Mangel an Reduktase, sondern einer Permease, die Folsäure nicht transportieren kann, wohl aber Tetrahydrofolsäure (*356*).

Die reduzierte Pyridin-Nucleotide benötigende Folat-Reduktase ist aber nicht das einzige Enzym, das Folsäure reduzieren kann. WRIGHT (*360, 361*) beschreibt eine Pterin-Reduktase in *Clostridium sticklandii*, die entsprechend Gl. [5] Folsäure und unhydrierte konjugierte Formen reduziert und nicht durch Aminopterin gehemmt wird.

$$F + \text{Pyruvat} + CoA \rightarrow FH_2 + \text{Acetyl-CoA} + CO_2. \qquad [5]$$

In Schafsleber wurde neben einer TPNH-spezifischen Reduktase (*232, 233*) ein System beschrieben, das zwei verschiedene Enzyme für die beiden Reduktionsschritte enthält (*41, 153 a*). Der Folsäure zu Dihydrofolsäure reduzierende erste Schritt benötigt ein Flavin zur vollen Aktivität und wird durch Atebrin oder Riboflavin gehemmt. Der zweite Schritt ist vermutlich mit pyridinnucleotid-abhängiger Dihydrofolat-Reduktase identisch.

Zusätzlich ist ein Mitochondrien-Enzym in Rattenleber nachgewiesen worden, das Folsäure in 5-Formyl-FH_4 überführt (*218*). Es konnte löslich gemacht werden, wobei der zu FH_2 reduzierende Schritt leicht in Lösung geht, während die zweite Reduktionsstufe fester gebunden ist. Das gesamte komplexe System benötigt außer anaeroben Bedingungen und Serin als C_1-Donator auch Pyridoxalphosphat, TPNH, ATP, Mg^{++}-Ionen und Homocystein.

Die Umwandlung von Folsäure in den Citrovorumfaktor, die im intakten Organismus vielfach verfolgt wurde, wird bei Ratten durch Östradiol (*73*) stimuliert. Auf einen Zusammenhang zwischen Östrogenen und TPN- oder Folat-katalysierten Reaktionen ist man schon öfters aufmerksam geworden. Es ist möglich, daß die Beobachtungen von HEISLER und SCHWEIGERT (*115*) damit in Zusammenhang stehen. Extrakte von *L. casei* brauchen für diese Umwandlung außer dem Reduktionsmittel und dem Einkohlenstoffdonator einen noch unbekannten Cofaktor.

IV. Chemie der Folat-Verbindungen.

1. Folsäure.

a. Isolierung.

In den Jahren 1940—1948 gelang zwei Arbeitsgruppen unabhängig die Gewinnung kristallisierter Folsäure aus Wuchsstoff-Konzentraten (Zusammenfassung, *302*). PFIFFNER und Mitarb. (*235*) isolierten die Verbindung, die sie Vitamin B_c nannten, aus Leber, die zuvor zur Hydrolyse der Konjugate einer Autolyse unterworfen wurde (*236*). Bei der später vom gleichen Arbeitskreis berichteten Reindarstellung aus

Hefe (*236*) wurde die Freisetzung der Folsäure durch die Konjugase aus Schweineniere bewirkt (S. 190). Stokstad (*301*) behandelte ein aus Leber angereichertes Präparat mit Methanol und Säure und erhielt daraus reinen kristallisierten Methylester der Folsäure. Nach Verseifung wurde die freie Säure durch Umkristallisieren aus verdünnter Essigsäure in schönen Blättchen erhalten (*304*). Reine, aber nicht kristallisierte „Folsäure" wurde auch von Mitchell und Mitarb. (*200, 201*) aus Spinatblättern gewonnen. Die isolierten Verbindungen, die zuerst mit verschiedenen Namen belegt worden waren (S. 187), erwiesen sich in ihren chemischen Eigenschaften und im Wachstumstest mit *L. casei* und *St. faecalis* (*144*) als identisch.

In allen Fällen wurden ähnliche Methoden der Anreicherung und Reinigung benutzt (*302*): Adsorption an und Elution von Aktivkohle, Super-Filtrol oder Amberlite 4 R, Chromatographie an Aluminiumoxydgel, Extraktion mit Butanol aus schwach saurer Lösung, Fällung als Zink- oder Blei-Salz und schließlich Kristallisation der fast reinen Verbindung aus verdünnter Säure.

Zur gleichen Zeit wurde auch ein Pteroyltriglutamat (Fermentation *L. casei*-factor) aus dem Kulturmedium eines Corynebacteriums (*130*) rein dargestellt. Bei der anaeroben alkalischen Hydrolyse entsteht aus dieser Verbindung unter Racemisierung *d,l*-Folsäure mit halber mikrobiologischer Aktivität des natürlichen Vitamins und zwei Mol Glutaminsäure (*303*).

Binkley (*23*) sowie Pfiffner (*238*) isolierten aus Hefe ein mikrobiologisch unwirksames Folsäurekonjugat (B_c-Konjugat), dessen Struktur (*237*) als Pteroyl-heptaglutaminsäure aufgeklärt wurde.

Da die Polyglutamate durch die γ-Glutamyl-spezifischen Konjugasen aus Hühnerpankreas (*198*) und Schweine-Niere (*23a*) zu mikrobiologisch voll wirksamer Folsäure abgebaut werden, ergibt sich für die Folsäurekonjugate die Struktur von Pteroyl-γ-glutamyl-glutamaten (*66*), die auch durch Synthese erhärtet ist.

b. *Konstitution und physikalische Eigenschaften.*

Formelübersicht 4 zeigt die Konstitution der Folsäure (VI)*. Das Molekül ist aus einem Pteridin, *p*-Aminobenzoesäure und *L*-(+)-Glutaminsäure aufgebaut, die in der dargestellten Art miteinander verknüpft sind. Wie fast alle natürlich vorkommenden Pteridine enthält auch die Folsäure ein 2-Amino-4-oxy-pteridin als Baustein. Der systematische

---

* Die Numerierung der Atome entspricht dem „Ring-Index" (*226*). In älteren deutschen und englischen Arbeiten findet sich noch die nebenstehende Bezeichnung:

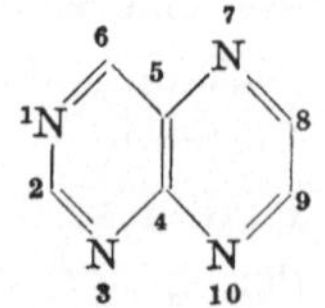

Name der Verbindung ist demnach [(2-Amino-4-oxy-6-pteridyl)-methyl]-*p*-aminobenzoyl-*L*-glutaminsäure. Die Verbindung aus dem Pteridin und

R_1 = 2-Amino-4-oxy-6-methyl-pteridin.
R_2 = *p*-Aminobenzoesäure.
R_3 = *L*-Glutaminsäure.

Formelübersicht 4. Konstitution von Folsäure.

der *p*-Aminobenzoesäure wird Pteroinsäure genannt. Folsäure sollte demnach korrekter als Pteroylglutaminsäure bezeichnet werden, und diese Bezeichnung findet sich auch häufig in der Literatur.

Es ist vorgeschlagen worden, die Bezeichnung „Folsäure" nur noch als Sammelnamen für sämtliche Pteroylglutamate zu gebrauchen, die Namen für die einzelnen Verbindungen aber systematisch mit dem Stamm „Pteroyl-" zu bilden. Bis jedoch verbindliche Richtlinien festliegen, möchten wir an unserer bisherigen Gewohnheit, den Namen „Folsäure" auch für die spezielle Verbindung zu benutzen, festhalten; ohne besonderen Zusatz ist stets das natürliche L-($+$)-Isomere gemeint. Die Bezeichnungen für die durch Substitution oder Reduktion aus der Grundsubstanz hervorgehenden Derivate leiten wir von diesem Namen ab: z. B. 10-Formyl-tetrahydro-folsäure. Für diese Derivate haben sich zum Teil auch noch Trivial- oder halbsystematische Namen gehalten, die im folgenden jeweils erwähnt werden.

Folsäure (Mol.-Gew. 441,4) kristallisiert aus wäßrigem Medium in gelben, speerförmigen Blättchen (Zers.-P. 250°) als Dihydrat, das oberhalb 140° 2 Mol Wasser abgibt. In den meisten organischen Lösungsmitteln mit Ausnahme von Eisessig ist sie nahezu unlöslich. In Wasser lösen sich 10 mg/l bei 0° und 500 mg/l bei 100°; vom Dinatriumsalz 15 g/l bei 0°. Unterhalb pH 5 ist Folsäure unlöslich, so daß die genaue Bestimmung der pKa-Werte der Carboxylgruppen nicht möglich ist. pKa_3 = 8,2 (enolische Hydroxylgruppe). Mit Zink-, Silber-, Blei- und Kobalt-Ionen bilden sich schwerlösliche Salze oder Komplexe (*236*). Für optische und spektrale Eigenschaften der Folate siehe S. 222.

Das IR-Spektrum der Folsäure (*323*) und die magnetische Suszeptibilität (*353*) können als Reinheitskriterien dienen.

c. Chemische Eigenschaften.

Folsäure kann oxidierend und reduzierend an der $C_{(9)}$-$N_{(10)}$-Bindung gespalten werden. Bei der aeroben alkalischen Hydrolyse (*303*) entsteht unter Verlust der biologischen Aktivität eine fluoreszierende Verbindung und ein primäres aromatisches Amin, das sich nach der Bratton-Marshall-Methode (*42*) diazotieren und kuppeln läßt. Durch 2 n-Schwefelsäure wird das Amin weiter zu *p*-Aminobenzoesäure und Glutaminsäure hydrolysiert; die Peptidbindung zwischen diesen beiden Bestandteilen muß die Carboxylgruppe der *p*-Aminobenzoesäure einschließen, da die aromatische Aminogruppe schon vor der hydrolytischen Spaltung frei war. Die fluoreszierende Verbindung ist mit der von Mowat (*205*) synthetisierten Pterin-6-carbonsäure (VII) identisch. Die Stellung der Carboxylgruppe

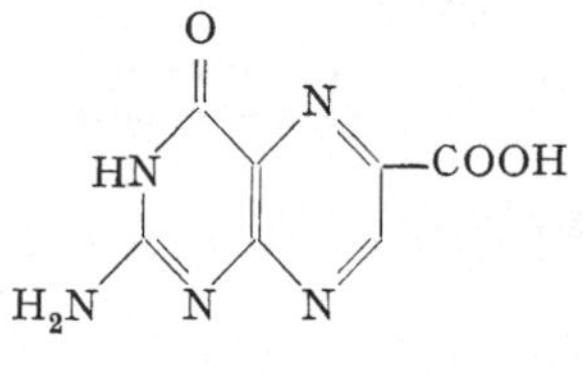

(VII.) Pterin-6-carbonsäure. 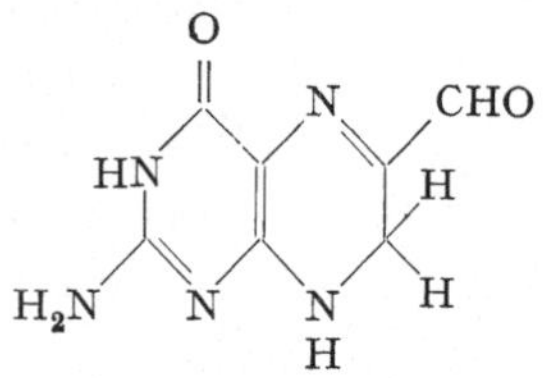   (VIII.) Dihydropterin-6-aldehyd.

in 6 des Pteridinringes geht daraus hervor, daß die Verbindung durch Oxydation von 6-Methyl-pterin dargestellt werden kann, dessen Struktur eine eindeutige Synthese von Boon und Leigh (*36*) erhärtet. Pterin-6-carbonsäure wurde auch von Wittle und Mitarb. (*352*) bei der Oxydation von Folsäure mit Permanganat oder Chlorat erhalten.

Die Hydrolyse der Folsäure mit 0,5 n-schwefliger Säure bei Raumtemperatur (*132*) spaltet das Molekül an der gleichen Stelle in *p*-Aminobenzoylglutaminsäure und fluoreszierenden Dihydropterin-6-aldehyd (VIII) (*322, 340*), der unter anaeroben alkalischen Bedingungen in einer Cannizzaro-Reaktion in gleiche Mengen Dihydropterin-6-carbonsäure und -6-carbinol dismutiert; die Carbonsäure unterliegt der Autoxydation, wogegen sich das Carbinol unter Wasserabspaltung und Umlagerung zu 6-Methyl-pterin aromatisiert. Bei den Spaltungsreaktionen der Folsäure werden stets Pterine mit einer Einkohlenstoff-Seitenkette isoliert, die aus einer Methylenbrücke stammt; andernfalls müßte die Hydrolyse bereits ohne zusätzliche oxydierende oder reduzierende Bedingungen erfolgen. Ein einfaches Modell, N-Benzyl-*p*-amino-benzoesäure, wird durch Alkali ebenfalls nur bei Luftzutritt gespalten (*205*). [Ausführlichere Darstellung des Strukturbeweises vgl. (*302*)].

Auch durch Licht wird Folsäure in ähnlicher Weise zersetzt. Es bildet sich zuerst Pterin-6-aldehyd, der nach Autoxydation zur Carbonsäure zu 2-Amino-4-oxy-pteridin decarboxyliert wird (*176, 254*). Diese Zersetzungsreaktion wird durch Riboflavin als Sensibilisator schon in kleinsten Mengen beschleunigt (*270*).

Die Nitrierung oder Halogenierung der Folsäure führt einen oder zwei Substituenten in den Benzolring ein (*61*). Durch Einwirkung von salpetriger Säure in kalter Salzsäure wird 10-Nitroso-folsäure gebildet (*63*); unter weniger milden Bedingungen erfolgt Desaminierung der 2-Aminogruppe zu einer Hydroxylgruppe (*355, 9 a*).

Literaturverzeichnis: SS. 254—274.

d. Folsäure-Synthesen.

Die Konstitution der Folsäure (VI) wurde unmittelbar nach der Isolierung durch eine Anzahl unabhängiger Synthesen bewiesen. Diese sind in früheren Zusammenfassungen ausführlich dargestellt (*302, 1*), so daß hier nur die wichtigsten Synthese-Prinzipien kurz beschrieben und einige neuere Arbeiten hinzugefügt werden. Fast alle bisher veröffentlichten Folsäuresynthesen benutzten die Kondensation von 2,4,5-Triamino-6-oxy-pyrimidin (TAOP, IX) mit p-Aminobenzoyl-L-glutaminsäure (X) und einem Dreikohlenstoffkörper nach dem allgemeinen Prinzip:

$$\text{(IX.) TAOP} \quad + \quad \begin{array}{c} \text{YC} - \text{CZ} \\ | \\ \text{XC} \end{array} \quad + \quad \text{H}_2\text{N}-\!\langle\ \rangle\!-\text{CONH}-\underset{\underset{\text{CH}_2-\text{CH}_2-\text{COOH}}{|}}{\overset{\overset{\text{COOH}}{|}}{\text{CH}}}$$

(IX.) TAOP.   (X.) p-Aminobenzoylglutaminsäure.

(VI.) Folsäure. (S. 201.)

Als Dreikohlenstoffkörper eignen sich viele Verbindungen, die Carbonyl- und/oder Halogenfunktionen als reaktive Gruppen (X-Y-Z) tragen. Für die Kondensationsreaktion können grundsätzlich drei Wege beschritten werden:

(1.) Die gleichzeitige Kondensation aller drei Komponenten in einem „Eintopfverfahren“. (2.) Die Synthese des Pterins aus Pyrimidin und Dreikohlenstoffkörper und anschließende Kondensation mit der p-Aminobenzoylglutaminsäure. (3.) Die Bildung einer Zwischenverbindung aus p-Aminobenzoylglutaminsäure und Dreikohlenstoffkörper, die dann mit dem Pyrimidin kondensiert wird. Diese Möglichkeiten wurden sämtlich zuerst von der „Lederle-Gruppe“ experimentell untersucht (*9*):

(1.) WALLER (*323*) kondensierte TAOP (IX) mit α,β-Dibrompropionaldehyd und p-Aminobenzoylglutaminsäure (X) in wäßrigem Medium. Auf dem gleichen Wege erhielt er bei Verwendung von p-Aminobenzoesäure Pteroinsäure (Formelübersicht 4, S. 201). Zahlreiche Variationen dieses Prinzips ergeben sich durch die Verwendung anderer Dreikohlenstoffkörper (*335, 158, 149*). Darüber hinaus ist die Darstellung spezifisch ^{14}C-markierter Folsäure möglich. WEYGAND (*334*) gelang die Synthese von 9-^{14}C-Folsäure durch Verwendung von 3-^{14}C-Dibrompropionaldehyd, wogegen mit Tribromaceton-3-^{14}C ein Gemisch von 63% 7-^{14}C- und 37% 9-^{14}C-Folsäure erhalten wurde (*339*). [Wegen der Synthese von 2-^{14}C-Folsäure siehe (*337*).]

(2.) Den zweiten Weg beschritt HULTQUIST (*126*) mit der Kondensation von TAOP, α,β-Dibrompropionaldehyd, Kaliumjodid und Pyridin zu N-[(2-Amino-4-oxy-6-pteridyl)-methyl]-pyridiniumjodid, das mit p-Aminobenzoylglutaminsäure in Äthylenglykol zu Folsäure reagiert. Weitere Synthesen dieser Art gehen vom

Pterinaldehyd (*336*), dem 6-Brommethyl- (*39*) oder dem 6-Hydroxymethyl-pterin aus. Der Pterinaldehyd kann unter anderem auch durch Kondensation von TAOP mit einer Ketohexose und anschließende Glykol-Spaltung des 6-Tetrahydroxybutyl-pterins gewonnen werden (*85, 336, 340a*). Dieser Weg entspricht etwa der heutigen Vorstellung von der Biosynthese des Pterinteils der Folsäure (S. 193). SLETZINGER (*291*) veröffentlichte neuerdings eine Methode, die im Vergleich zu den älteren Verfahren wesentlich reinere Folsäure liefert. Aus dem Produkt der Umsetzung von TAOP und 2-Brom-3,3-diäthoxypropionaldehyd wird durch Acetylierung und Hydrolyse des Acetals 2-Acetamido-4-oxy-pteridin-6-aldehyd gewonnen, der unter Einwirkung von Thiokresol mit *p*-Aminobenzoylglutaminsäure in guter Ausbeute kondensiert werden kann; das 85% reine Produkt läßt sich gut umkristallisieren und gibt dann bei der alkalischen Hydrolyse reine Folsäure.

(3.) Die dritte theoretische Möglichkeit wurde aus praktischen Erwägungen nur wenig genutzt. ANGIER (*10*) kondensierte *p*-Aminobenzoylglutaminsäure-diäthylester mit Redukton und setzte das Produkt mit TAOP zu Folsäure um. Nach diesem Verfahren wurden auch andere Pteroinsäurederivate, wie Ester, Amid und Glycinat, dargestellt.

Konjugate der Folsäure werden nach den gleichen Methoden gewonnen, indem man die entsprechenden *p*-Aminobenzoyl-γ-glutamyl-glutaminsäureester einsetzt (*37, 38, 206*).

Im Handel erhältliche Folsäure ist im allgemeinen für biochemische Untersuchungen zu unrein. Erprobte Reinigungsverfahren sind das mehrmalige Ausfällen aus konzentrierter Säure durch Verdünnen, das Umkristallisieren aus verdünnten Säuren unter Zusatz von Aktivkohle (*102*), die Extraktion der Verunreinigungen mit Butanol (*26*) und die Chromatographie an Cellulose (*267*).

2. Reduktion von Folsäure.

Der Pyrazinring der Folsäure läßt sich bereits unter milden Bedingungen hydrieren, unter denen aromatische Verbindungen und Pyrimidine noch nicht reduziert werden. Mit Wasserstoff am Platinkatalysator oder mit Dithionit wird je nach den Bedingungen Dihydrofolsäure (FH_2) oder Tetrahydrofolsäure (FH_4) gebildet. Die Stellung der neu-eingeführten Wasserstoffatome im Pyrazinring ergibt sich besonders klar aus polarographischen Untersuchungen (*5*) an der 5-Formyl-tetrahydrofolsäure (S. 225).

Reduziert man Folsäure im sauren Medium mit Zink oder mit Wasserstoff am Palladium-Kontakt (*132*), so wird das Molekül in ähnlicher Weise gespalten, wie es bei der Tetrahydrofolsäure beschrieben ist (S. 208).

a. Dihydrofolsäure und das Problem der Dihydrofolat-Isomerie.

Bei der Hydrierung über Platinoxyd in 0,1 n-NaOH erhielt O'DELL (*219*) unter Aufnahme von einem Mol H_2 Dihydrofolsäure. Eine spätere Nachprüfung zeigte aber, daß die Reduktion in einer zweiten, langsameren Stufe auch bis zur Tetrahydrofolsäure weitergeht (*26, 197*). Die Darstellung reiner Dihydrofolsäure erfolgt nach FUTTERMAN (*92*) durch Reduktion mit Dithionit und Ascorbat bei Zimmertemperatur; nach einer Verbesserung durch BLAKLEY (*33*) kann auf diesem Wege kristalline FH_2 erhalten werden, die in allen Eigenschaften dem durch Hydrierung gewonnenen Produkt gleicht. Kristallisierte Dihydrofolsäure ist in der Kälte für längere Zeit stabil. In alkalischer Lösung wird sie bei Luftzutritt teilweise zu Folsäure reoxydiert, zum Teil zerfällt sie aber auch schneller als Folsäure in ein diazotierbares Amin und mehrere nicht näher identifizierte Pterine (*26, 366*). In saurer Lösung bildet sich rasch ein gelbes Abbauprodukt mit einem Absorptionsmaximum bei 420 mμ, das zur quantitativen Bestimmung der Dihydrofolsäure herangezogen wurde (*361*). Für Dihydrofolsäure lassen sich drei isomere Strukturen schreiben: (XI), (XII) und (XIII).

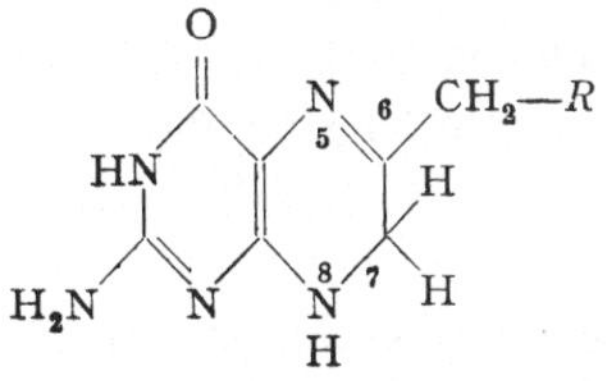

(XI.) 7,8-FH_2. 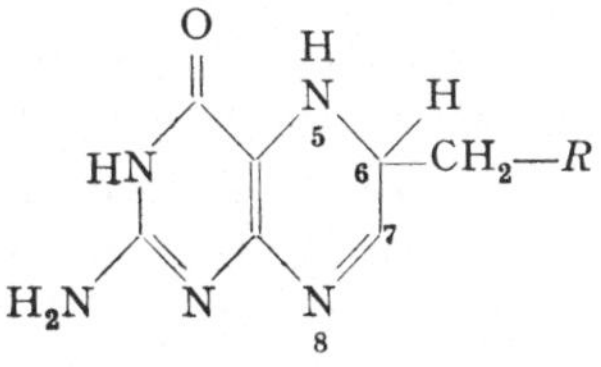   (XII.) 5,6-FH_2.

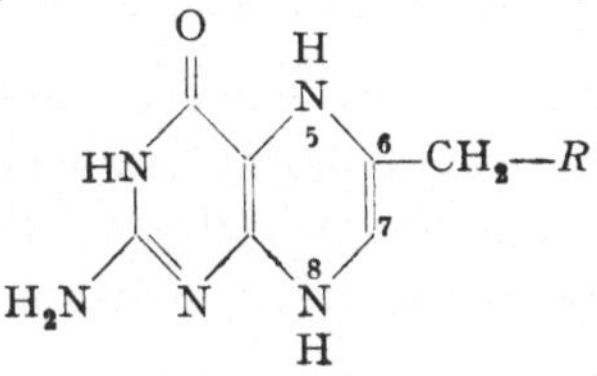

(XIII.) 5,8-FH_2.

$R = p$-Aminobenzoyl-glutaminsäure (X).

O'DELL (*219*) beobachtete, daß Xanthopterin katalytisch leicht zu 7,8-Dihydroxanthopterin reduziert wird, dessen Struktur durch eindeutige Synthesen festliegt (*36, 79*). Dagegen konnte Isoxanthopterin unter den gleichen Bedingungen nicht reduziert werden. Dies galt als Beweis, daß die Reduktion der Pterine an der 7,8-Doppelbindung einsetzen muß, und daher wurde der Dihydrofolsäure die Struktur einer 7,8-FH_2 (XI) zugewiesen. Die enolische Doppelbindung wurde bei der katalytischen Reduktion nicht angegriffen. Kürzlich berichteten aber ALBERT und MATSUURA (*3*), daß 4,7-Dihydroxy-pteridin leicht zu 4,7-Dihydroxy-5,6-dihydro-pterin reduziert wird, so daß beim Isoxanthopterin anscheinend ein Sonderfall vorliegt, der für die Struktur der FH_2 nicht beweisend sein kann. Ein anderes Argument für die 7,8-Dihydrostruktur ist die bemerkenswerte Stabilität der 5-Formyl-FH_4 (XVII, S. 210) gegen Oxydation. Aber es ist ebensogut möglich, daß dies auf einem Einfluß der Formylgruppe auf den ganzen hydrierten Ring beruht als auf einem Schutz der 5-Stellung, an der die Oxydation in Umkehr der Reduktion zuerst angreifen sollte. Zudem konnte Kaufman (*152*) wahrscheinlich machen, daß bei der Oxydation von Tetrahydrofolsäure oder Tetrahydro-6,7-dimethyl-pterin mit Dichlorphenol-indophenol und anderen Oxydationsmitteln zuerst 5,6-Dihydroverbindungen entstehen, die sich in einer spontanen Reaktion in Isomere umlagern, denen vermutlich die 7,8-Dihydrostruktur zukommt, und Donaldson und Keresztesy (*78a*) berichteten über die nichtenzymatische Oxydation von 5-Methyl-FH_4 zu 5-Methyl-5,6-FH_2 durch verschiedene Oxydationsmittel.

Einen direkten Beweis für die 7,8-Dihydrostruktur der Dihydrofolsäure gaben Osborn und Huennekens (*223*); sie zeigten, daß chemisch synthetisierte Dihydrofolsäure bei der enzymatischen Reduktion durch Dihydrofolsäure-Reduktase (S. 197) zu 100% umgesetzt wird. Im Unterschied zu 7,8-FH_2 ist bei 5,6-FH_2 das Kohlenstoffatom 6 asymmetrisch. Chemisch synthetisierte 5,6-FH_2 (XII) wäre ein Diastereomeren-Gemisch und könnte daher nicht voll aktiv sein, wie es auch bei synthetischer Tetrahydrofolsäure der Fall ist (*113, 32*).

Diese Versuche lassen allerdings auch eine 5,8-Dihydrostruktur (XIII) zu, die aber deshalb als unwahrscheinlich gilt, weil derartige Pterine bisher nur mit einem Substituenten an $N_{(8)}$ dargestellt werden konnten (*239*).

Perault und Pullman (*228*) schließen aus Strukturberechnungen nach der LCAO-Methode, daß Folsäure am leichtesten zur 7,8-Dihydrostufe reduziert wird, während sich die 5,8-Dihydroform nach ihren Berechnungen durch besonders schwere Reduzierbarkeit zur Tetrahydro-Stufe auszeichnet.

Unter diesen Voraussetzungen hat somit die 7,8-Dihydrostruktur die größte Wahrscheinlichkeit.

Enzymatisch wird eine Dihydrofolsäure bei der Reduktion von Folsäure mit einem Enzym aus *Cl. sticklandii* (*360*) und durch Oxydation von Tetrahydro-folsäure in der Thymidylatsynthetase-Reaktion (S. 240) gebildet. Diesen Präparaten wurde auf Grund des gleichen Spektrums und chromatographischen Verhaltens (*321*) ebenfalls die 7,8-Dihydrostruktur zugeschrieben. Blakley (*35*) zeigte darüber hinaus, daß Dihydrofolsäure aus der Thymidylatsynthetase-Reaktion für Dihydro-folat-Reduktase aus *St. faecalis* ein ebenso gutes Substrat ist wie chemisch synthetisierte, und ähnliche Ergebnisse erhielten auch Nath und Greenberg (*212*), die auch die enzymatisch mit dem Enzym aus *Cl. sticklandii* gebildete Dihydrofolsäure in ihre Untersuchung einbezogen, mit einem Thymusenzym. Mathews (*187*) und Huennekens (*122a*) synthetisierten Dihydrofolsäure sowohl chemisch, einmal durch Reduktion von Folsäure mit Dithionit, dann durch Oxydation von FH_4 mit Indophenol nach Kaufman (*152*), als auch auf den beiden enzymatischen Wegen, und reinigten die Produkte an DEAE-Cellulosesäulen. Alle Präparate hatten gleiches Spektrum und zeigten identische R_f-Werte in mehreren Solventien. Durch Dihydrofolat-Reduktase aus Hühnerleber, Schafleber und Kalbsthymus wurden sie sämtlich gleich schnell und vollständig reduziert. Dagegen berichteten Nath und Greenberg (*212*), daß die von ihnen aus Schafleber angereicherte Reduktase die enzymatisch gewonnenen Präparate erheblich langsamer reduziert als das chemisch synthetisierte Produkt.

Literaturverzeichnis: SS. 254—274.

Wenn diese Beobachtung tatsächlich darauf zurückzuführen ist, daß verschiedene Isomere der Dihydrofolsäure vorliegen, dann sind diese mit den anderen Enzymen deshalb nicht entdeckt worden, weil sie sich spontan oder durch Einwirkung einer „Tautomerase" in $7,8$-FH_2 (XI, S. 205) umlagern könnten. Diese Form muß man nach dem Gesagten für das spezifische Substrat der Reduktasen halten, zumal auch die in ihrer Struktur gesicherte (S. 206) 5-Methyl-$5,6$-FH_2 (*78 a*) nicht reduziert wird. Doch sollte auch diese Annahme noch mittels eindeutig synthetisierter $7,8$-Dihydropterine (*2, 36*) gesichert werden. Für die enzymatisch erzeugten Dihydrofolsäuren käme dann in erster Linie die $5,6$-Dihydro-Struktur (XII) in Betracht. Reaktionsmechanistische Überlegungen machen es aber unwahrscheinlich, daß das Primärprodukt der Thymidylatsynthetase-Reaktion eine $5,6$-FH_2 ist; vielmehr wäre danach die $7,8$-Dihydro-Struktur leichter einzusehen (vgl. S. 242).

In jedem Fall bedarf dieses Problem der Struktur der Dihydrofolsäuren noch weiterer experimenteller Bemühungen. Z. B. hat jüngst ZAKRZEWSKI (*365 a*) gewichtige Argumente für die $5,8$-Dihydro-Struktur der chemisch synthetisierten FH_2 beigebracht. Sie beruhen darauf, daß bei der Reduktion mit Dithionit (*288*) in T_2O Tritium aus dem Medium erst im Reduktions-Schritt $FH_2 \rightarrow FH_4$ in $C_{(7)}$ eingebaut wird. Daraus wird es bei der Reinigung — im Gegensatz zu den anderen drei möglichen Stellungen — nicht mehr ausgetauscht.

Alle Dihydrofolsäure-Tautomeren (wenn es tatsächlich welche gibt) scheinen sehr ähnliche Spektren zu haben. Ein Strukturbeweis auf Grund des Vergleichs dieses Spektrums mit den Spektren anderer Dihydropterine ist unzuverlässig, da deren Absorptionsmaxima in hohem Maße von der Substitution abhängen.

b. 5,6,7,8-Tetrahydro-folsäure.

Führt man die katalytische Hydrierung der Folsäure in Eisessig durch (*219*), so erhält man $5,6,7,8$-Tetrahydro-folsäure (XIV). Die Folsäure geht dabei in dem Maße, wie sie hydriert wird, in Lösung. Das Produkt kann durch Ausfällen mit Äther (*102*) oder durch Lyophilisieren des Lösungsmittels (*248, 114*) in fester Substanz gewonnen werden. In neutraler wäßriger Lösung (*26, 250*) und in konz. Ameisensäure (*189*) ist Folsäure löslich und daher geht die Hydrierung rascher, doch bildet sich in Ameisensäure infolge teilweiser Formylierung kein sauberes Produkt.

(XIV.) 5,6,7,8-Tetrahydro-folsäure (FH_4).

Die Darstellung von Tetrahydro-folsäure wurde ferner durch Reduktion von Folsäure mit Natrium-borhydrid (*103*) und kürzlich mit Dithionit und Ascorbat bei $75°$ beschrieben (*288*). Es ist infolgedessen zweifelhaft, ob die mit Dithionit und Ascorbat bei Zimmertemperatur hergestellte Dihydrofolsäure in allen Fällen ganz frei von Tetrahydro-folsäure ist.

Die Zahl der bei der Reduktion eingeführten Wasserstoffatome kann titrimetrisch mit Jodlösung (*141 a, 366*), durch Messung der O_2-Aufnahme beim Schütteln mit Platin im Warburg-Apparat und kolorimetrisch durch Verfolgen der Oxydation mit Dichlorphenol-indophenol (*152*) bestimmt werden.

Wie bereits erwähnt, wird bei der Hydrierung ein zweites Asymmetrie-Zentrum am Kohlenstoffatom 6 gebildet, so daß ein Diastereomeren-gemisch — d,l-L-$(+)$-Tetrahydro-folsäure — entsteht. Dabei wird vorausgesetzt, daß das bereits vorhandene asymmetrische Zentrum in der Glutaminsäure die Reduktion nicht merklich in einer Richtung lenkt. In Übereinstimmung mit dieser Auffassung wird das synthetische Produkt in enzymatischen Reaktionen nur zu 50% umgesetzt (113). Die Trennung der optischen Isomeren gelang bisher noch nicht. Mathews und Huennekens (188) gewannen das natürliche Isomere — l,L-FH_4 — durch Reduktion von Dihydrofolsäure mit Reduktase aus Hühnerleber und chromatographische Reinigung.

Tetrahydro-folsäure ist in fester Substanz und in Lösungen sehr oxydationsempfindlich. Das frische, nahezu weiße Produkt färbt sich an Luft rasch braun, kann aber im zugeschmolzenen Röhrchen bei — 10° längere Zeit unzersetzt aufbewahrt werden. In Lösung läßt sich Tetrahydro-folsäure durch Zusatz hoher Konzentrationen an Sulfhydryl-verbindungen oder anderen Reduktionsmitteln stabilisieren (103), worunter sich Dimercaptopropanol (BAL) (31) und Ascorbinsäure (15) anscheinend besonders wirksam erwiesen haben. Äthylendiamintetraacetat (EDTA, Versene) hat ebenfalls einen konservierenden Effekt, da Schwermetall-ionen die Selbstzersetzung der Tetrahydro-folsäure merklich katalysieren.

Kaufman (152) zeigte, daß die Wirkung der SH-Verbindungen darin besteht, durch Luftoxydation zunächst entstandenes 5,6-Dihydropterin wieder zu reduzieren. Diese Reduktion kann auch durch TPNH nichtenzymatisch bewirkt werden, läßt sich aber nicht auf 7,8-Dihydropterine ausdehnen. Ähnliche Verhältnisse finden sich, wie Kaufman anführt, auch bei den Flavinen mit Isoalloxazinstruktur, die ebenfalls durch TPNH nichtenzymatisch reduziert werden, im Gegensatz zu den isomeren Alloxazinen. Folate und Flavine haben eng verwandte Strukturelemente, und es ist daher möglich, daß auch bei der Oxydation der Tetrahydro-folsäure die Elektronen in zwei Stufen statt paarweise eliminiert werden. Kaufman konnte aber bei Messungen der paramagnetischen Resonanz frisch oxydierter Tetrahydro-folsäure keinerlei Hinweis für das Auftreten von freien Radikalen erhalten.

Das Redoxpotential des Systems FH_2/FH_4 wurde bisher nicht direkt gemessen, doch ist bekannt, daß das Gleichgewicht der enzymatischen Reduktion der Dihydro-folsäure mittels TPNH weit auf Seiten der Tetrahydro-folsäure liegt (223). Mathews und Huennekens ($122a$, $188a$) bestimmten die Gleichgewichts-Konstante zu $5,6 \times 10^4$ bei pH $= 7$ und errechneten daraus unter Verwendung des Potentials des TPN/TPNH-Systems ($E_0' = -0,32$ V) einen Wert von —0,19 V für FH_2/FH_4.

Frisch hergestellte Lösungen von Tetrahydro-folsäure zeigen ein Absorptionsmaximum bei 298 mμ; ohne Zusatz von Reduktionsmitteln verschiebt sich dieses rasch zum Maximum der Dihydrofolsäure bei 283 mμ, die durch Reduktion mit Dihydrofolat-Reduktase nachgewiesen werden konnte (223). An Luft zerfällt Tetrahydro-folsäure weiter zu stöchiometrischen Mengen von p-Aminobenzoylglutaminsäure (26, 366) und Formaldehyd aus $C_{(9)}$ (Jaenicke und Brode, unveröff.); vom Pterinteil des Moleküls wird nur eine kleinere Menge als Xanthopterin wiedergefunden (*26*). Die Bildung des diazotierbaren Amins wurde zur Bestimmung der Tetrahydro-folsäure herangezogen (*92, 367*). Mehrere Autoren zeigten, daß das Amin nur zu etwa 50% in einer raschen Reaktion gebildet wird. BLAKLEY (*26*) schüttelte Lösungen von Tetrahydro-folsäure an Luft. Nach einer Stunde gab die Analyse 40% diazotierbares Amin, jedoch konnte die enzymatische Aktivität der Probe noch zu 90% durch Hydrieren wiederhergestellt werden. Es muß daher angenommen werden, daß in einer raschen Reaktion eine zu FH_4 rehydrierbare Verbindung (5,6-FH_2?) entsteht, die unter den Bedingungen des Bratton-Marshall-Tests — vielleicht durch die Einwirkung des Nitrits — p-Aminobenzoylglutaminsäure abspaltet. In einer langsamen Reaktion erfolgt der weitere Zerfall unter Spaltung an der $C_{(6)}$-$C_{(9)}$-Bindung.

Weshalb dieser Zerfall nur zu etwa 50% stattfindet, während die biologische Aktivität vollständig verlorengeht, konnte bisher nicht befriedigend erklärt werden. Die Möglichkeit, daß die beiden Diastereomeren sich in ihrer Stabilität so sehr unterscheiden, hat wenig für sich, da ein derartiger Einfluß des zweiten optischen Zentrums sich andernfalls schon bei der Hydrierung durch einseitige Bevorzugung der Bildung eines Diastereomeren bemerkbar machen sollte. Ferner müßte sich dann zeigen lassen, daß das natürliche Isomere der Tetrahydro-folsäure beim Spontanzerfall 100% diazotierbares Amin ergibt.

3. Mit Einkohlenstoff-Körpern substituierte Folsäuren.

(Formelübersicht 5.)

a. 10-Formyl-folsäure.

Durch Formylierung von Folsäure mit starker Ameisensäure bei 100° (*98*) oder mit Ameisensäure/Essigsäureanhydrid (*103*) erhält man 10-Formyl-folsäure (XV), und aus Pteroinsäure in gleicher Weise das Rhizopterin (10-Formyl-pteroinsäure) (*355*). Diese Verbindung war als erstes formyliertes Pteroyl-derivat von RICKES (*256*) aus dem Kulturmedium von *Rhizopus nigricans* isoliert worden. Dadurch wurde erstmals die Aufmerksamkeit auf die Beteiligung der Folate am Ameisensäure-Stoffwechsel gelenkt (*98*). 10-Formyl-folsäure wurde von SILVERMAN (*285*) aus Pferdeleber und von RAUEN (*253*) aus inkubierten Schweineleber-Homogenaten isoliert. Durch 0,1 n-NaOH bei Raumtemperatur oder bei pH 10—12 in der Hitze (*60*) wird die Formylgruppe abgespalten. Die Stellung dieser Gruppe ergab sich durch Vergleich mit dem Rhizopterin, dessen Struktur durch Abbau und Synthese aufgeklärt war (*355*).

b. 10-Formyl-tetrahydro-folsäure.

Die Hydrierung von Formyl-folsäure in Eisessig oder die Formylierung von Tetrahydro-folsäure führt nach Neutralisierung des Ansatzes zu 10-Formyl-tetrahydro-folsäure (XVI). Infolge der großen Zersetzlichkeit

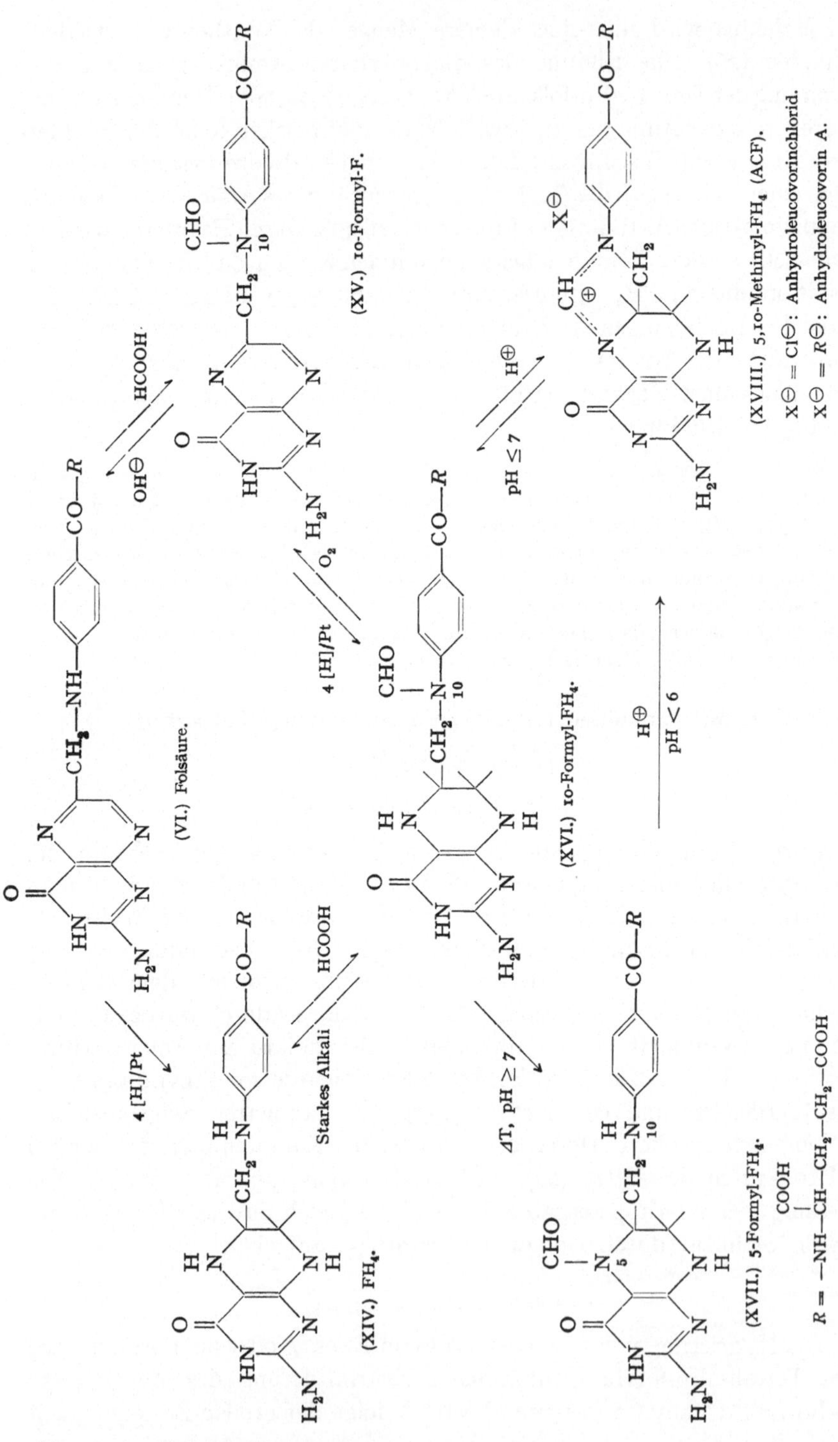

Formelübersicht 5. Bildung und Umwandlung der Formyl-tetrahydrofolate.

Literaturverzeichnis: SS. 254—274.

dieser Verbindung sind die Präparate jedoch so unrein, daß die Verbindung auf diesem Wege nicht in definierter Form gefaßt werden kann. JAENICKE und GREENBERG (*134*, *103*) konnten zeigen, daß 10-Formyl-FH$_4$ das Endprodukt der biologischen Aktivierung der Ameisensäure darstellt und somit als die „aktivierte Ameisensäure" (*252*) betrachtet werden kann (vgl. S. 248). Erst in jüngster Zeit gelang mit schonenden Methoden die Isolierung der Substanz aus Geweben und Körperflüssigkeiten (*283*, *287*, *351*), während früher stets nur das Oxydationsprodukt 10-Formyl-folsäure oder die Umlagerungsform, der „Citrovorumfactor", isoliert werden konnte.

c. 5-Formyl-tetrahydro-folsäure.

Beim längeren Stehenlassen oder besser beim Erhitzen schwach alkalischer Lösungen von 10-Formyl-FH$_4$ unter anaeroben Bedingungen wandert die Formylgruppe an das Stickstoffatom 5; es entsteht 5-Formyl-tetrahydro-folsäure (XVII) (Leucovorin, Folinic acid-SF) (*261*, *189*) (Mol.-Gew. 473), eine bemerkenswert stabile Verbindung, die sich gut reinigen und kristallisieren läßt. Anlaß für die Syntheseversuche, die zu dieser Verbindung führten, war die Entdeckung des „Citrovorumfactors" (Folinic acid). Dieser stellt das *l,L*-Diastereomere der synthetischen Verbindung dar und wurde bald nach der Synthese auch aus Leber isoliert (*269*, *155*). Das synthetische Produkt wurde als Calcium-Salz und als freie Säure (pKa-Werte: 3,1; 4,8; 10,4) rein dargestellt. Über das Calcium-Salz konnte das natürliche Isomere auf Grund geringerer Löslichkeit in reiner Form erhalten werden (*64*). Diese *l,L*-5-Formyl-tetrahydro-folsäure hat die gleiche mikrobiologische Aktivität wie der natürliche Citrovorum-Faktor. Als stabilste der Cofaktor-wirksamen Folatverbindungen wird diese Substanz meist als Vergleichsstandard für mikrobiologische Tests verwendet. Das IR-Spektrum wurde von WRIGHT (*359*) gemessen.

Die Summen-Formel, $C_{20}H_{23}O_7N_7$, und der analytische Nachweis einer Formylgruppe, die nur unter stark sauren oder alkalischen Bedingungen abgespalten wird, bestätigen die Struktur eines Formyltetrahydrofolats, wie es schon der Syntheseweg annehmen läßt. Das Problem der Strukturaufklärung (*60*, *242*) betraf daher vor allem die Frage nach der Stellung der Formylgruppe. Eine Betrachtung der sterischen Verhältnisse des Moleküls und Untersuchungen an 2,4-Diamino-5-formamido-pyrimidin und 2-Amino-4-oxy-5-formyl-6-methyl-tetrahydro-pteridin als Modellen (*60*) machten von vornherein wahrscheinlich, daß die Formylgruppe bei der Umlagerung an das N$_{(5)}$ gewandert ist.

Einen schönen Beweis für die Verknüpfung der Formylgruppe mit dem Tetrahydro-pyrazinring gaben polarographische Untersuchungen von ALLEN und Mitarb. (*5*). Bei pH 9 fanden sie für Pterine, Dihydropterine und Tetrahydropterine jeweils charakteristische Halbwellenpotentiale, während Pyrimidine und *p*-Amino-benzoylglutaminsäure nicht reduziert werden. 5-Formyl-FH$_4$ gibt ebenfalls keine Welle im Polarographen, doch nach Behandlung mit verdünnter Säure, die zum Verlust der mikrobiologischen Aktivität für *P. cerevisiae* führt, wird das charakteristische Potential eines Tetrahydropterins gefunden.

Mit salpetriger Säure in kalter Mineralsäure erhält man aus 5-Formyl-FH$_4$ ein 10-Nitroso-Derivat (*60*), woraus wiederum hervorgeht, daß N$_{(10)}$ unsubstituiert ist. Nach kurzem Stehenlassen der 5-Formyl-FH$_4$ mit verdünnter Säure ist aber keine Nitrosierung mehr möglich, da ein neuer Ringschluß dann N$_{(5)}$ und N$_{(10)}$ verbindet.

d. *5,10-Methinyl-tetrahydro-folsäure.*

Das Produkt dieses Ringschlusses unter Einwirkung von Protonen ist 5,10-Methinyl-tetrahydro-folsäure (XVIII, S. 210) (Anhydro-citrovorumfaktor, ACF, Anhydro-leucovorin) mit einer positiven Ladung, die in dem neugeschlossenen Imidazolinium-Ring nach Art einer Amidinium-Mesomerie verschiedenen Positionen zugeordnet werden kann. Die gleiche Verbindung entsteht auch bei Säurebehandlung von 10-Formyl-FH$_4$ und ist daher das Primärprodukt bei der Hydrierung der Formylfolsäure in Eisessig. Beim Neutralisieren einer Lösung von 5,10-Methinyl-FH$_4$ bildet sich zunächst hauptsächlich 10-Formyl-FH$_4$ (XVI), die sich unter den oben angegebenen Bedingungen wieder in das 5-Isomere (XVII)

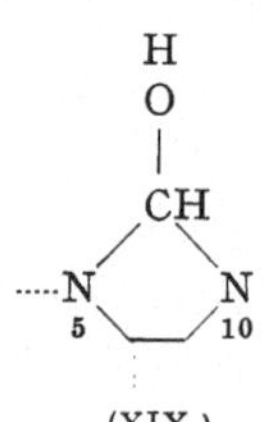

(XIX.)

umlagern kann, womit der Kreis geschlossen ist (siehe Formelübersicht 5, S. 210). Diese Umlagerung kann entweder über die cyclische Methinyl-Form oder über eine Methylol-Form (XIX) verlaufen (*292*). Bei der enzymatischen Umwandlung des 5-Isomeren in das 10-Isomere konnten Kay und Mitarb. (*153*) bei Verwendung eines Hühnerleberenzyms die Methinyl-Zwischenstufe ausschließen. — Bei Luftzutritt wird 10-Formyl-FH$_4$ zu 10-Formylfolsäure (XV) oxydiert, und das ist der Grund für die Inaktivierung der 5-Formyl-FH$_4$ und die Oxydierbarkeit des Tetrahydropyrazinringes nach Säurebehandlung, wie oben beschrieben.

5,10-Methinyl-FH$_4$ wurde als Chlorid isoliert (*59*). Die Verbindung zeichnet sich durch ein charakteristisches UV-Spektrum mit einem Maximum bei 355 mμ aus, das gegenüber der 5-Formyl-FH$_4$ um 70 mμ zum sichtbaren Bereich hin verschoben ist, so daß die Substanz gelb gefärbt ist. Diese Verschiebung wird auf die Ausbildung einer durchgehenden Konjugation zwischen dem Pterin- und dem *p*-Aminobenzoyl-Kern im Molekül zurückgeführt, die über die Methin-Brücke zwischen N$_{(5)}$ und N$_{(10)}$ verbunden sind.

Diese Auffassung konnte von May, Shive et al. (*189*) an einem einfachen Modell bestätigt werden. Beim Ansäuern einer Lösung von N-Formyl-N,N′-diphenyl-äthylendiamin (XX) entsteht nämlich in einer analogen Reaktion unter Verschiebung des Absorptionsmaximums von 245 nach 315 mμ die N,N'-Diphenylimidazolinium-Verbindung (XXI). Die coplanare Lage der π-Elektronensysteme ist wohl auch der Grund

(XX.) N-Formyl-N,N'-diphenyl-äthylendiamin.

(XXI.)

für die starke Fluoreszenz der 10-Methinyl-FH$_4$ und des Modells. Die Struktur der Anhydroformyl-FH$_4$ ist ein weiterer Beweis für die Bindung der Formylgruppe des Citrovorum-Faktors an das N$_{(5)}$ der Tetrahydrofolsäure.

Verschiedene isomere Formen der 5,10-Methinyl-FH$_4$ sind: Anhydroleucovorinchlorid mit einem Mol ionogenem Chlor und Anhydroleucovorin A, wahrscheinlich eine betainartige Verbindung, in der eine Carboxylgruppe der Glutaminsäure das Gegenion für das Imidazolinium-Kation stellt (*59*, *60*).

Die Geschwindigkeit der Cyclisierungsreaktion der beiden isomeren Formyl-tetrahydrofolate zu 5,10-Methinyl-FH$_4$ ist der H$^+$-Ionen-Konzentration proportional (*189*); dabei cyclisiert das N$_{(10)}$-Isomere zirka 5- bis 6mal schneller (*244*).

In schwach alkalischer Lösung unter anaeroben Bedingungen geht die Anhydroformyl-Verbindung sehr rasch in 10-Formyl-FH$_4$ über (*189*). Bei pH 7 ist diese Reaktion wesentlich langsamer und in hohem Grade von den anwesenden Anionen abhängig, in dem Sinne, daß 0,1 M-Phosphat, Pyrophosphat und Arsenat die „Cyclohydrolyse" im Vergleich zu Maleat-Puffer um etwa das 7fache beschleunigen (*244*, *310*, *111*). Dieser Effekt konnte ebenfalls an den N-Formyldiaryläthylendiaminen (vgl. XX) als Modellen beobachtet werden (KUTZBACH, unveröff.), doch war eine stichhaltige Begründung noch nicht möglich. Die Diaryläthylendiamine lassen sich als einfachste Analoge der Wirkgruppe der Tetrahydro-folsäure betrachten (*141*). Ihre Formyl-Derivate und Formaldehyd-Kondensationsprodukte (S. 216) haben zur Strukturaufklärung und auch zum Verständnis der Wirksamkeit der entsprechenden Cofaktor-Formen beigetragen.

Die Messung des Cyclisierungs-Gleichgewichts von 5- und 10-Formyl-FH$_4$ zwischen pH 2 und 7 führt zu Aussagen über die relative Hydrolyse-

Energie dieser beiden Verbindungen und der Anhydro-Form (*124, 125, 153*), wie in *Tabelle 3* dargestellt ist.

Tabelle 3. **Werte zur Berechnung der relativen Hydrolysen-Energie der Formyl-tetrahydrofolate (*124*).**

Gleichgewicht	K	$\Delta G^{\circ}_{25^{\circ}}$ [cal] bei pH 7
1. $H^+ + 5\text{-}CHO\text{—}FH_4 \rightleftharpoons 5,10\text{-}CH{=}FH_4{}^+ + H_2O$	$7 \times 10^2\,M^{-1}$	5620
2. $H^+ + 10\text{-}CHO\text{—}FH_4 \rightleftharpoons 5,10\text{-}CH{=}FH_4{}^+ + H_2O$	$9,1 \times 10^5\,M^{-1}$	1400
3. $5\text{-}CHO\text{—}FH_4 \rightleftharpoons 10\text{-}CHO\text{—}FH_4$	$7,7 \times 10^{-4}$	4220*
Zum Vergleich:		
4. $H^+ + (XX) \rightleftharpoons (XXI) + H_2O$	$2,3 \times 10^4\,M^{-1}$	3580**

* Berechnet aus 1. und 2.   ** Kutzbach (unveröff.).

Die freie Energie der Hydrolyse der Formylgruppe nimmt also in der Reihe: 5-Formyl < 10-Formyl < 5,10-Methinyl-FH_4 zu. Das Gleichgewicht der Formyl-gruppenübertragung von 5-Formyl-FH_4 auf Glutaminsäure (*282*) (S. 248) liegt mit einer Konstanten von circa 0,1 auf seiten der Formyl-tetrahydro-folsäure, so daß wir für diese etwa die freie Hydrolysenenergie einer normalen Formamid-Bindung annehmen dürfen, die etwa 2,0 kcal beträgt. Dann errechnet sich für die 10-Formyl-Verbindung ein Wert von etwa 6,5 kcal und 7,9 kcal für die cyclische Form. Diese stehen in guter Übereinstimmung mit der Beobachtung, daß die Bildung von 10-Formyl-FH_4 in der Formylase-Reaktion (S. 249) in reversiblem Gleichgewicht mit der Spaltung einer ATP-Pyrophosphat-Bindung steht (*103 a*) und diese Reaktion von einem Purin-abbauenden Mikroorganismus zur Synthese von ATP ausgenutzt wird (*247*) (S. 249). Für die Hydrolysen-Energie einer energiereichen Bindung des ATP wurden 7,8 kcal errechnet (*257*). Eine Erklärung für eine so große Hydrolysen-Energie eines Säureamids bietet eine Betrachtung der mesomeren Grenzstrukturen (XXII). Dabei zeigt sich, daß das freie Elektronenpaar am $N_{(10)}$ sowohl von der Mesomerie des aromatischen Systems wie der Formamidgruppe beansprucht wird, so daß beide sich nicht richtig ausbilden können (hindered resonance). Durch die hydrolytische Spaltung wird diese Behinderung aufgehoben und unter Energiegewinn entstehen zwei freie, mesomeriefähige Moleküle. Es handelt sich also grundsätzlich um die gleiche Ursache wie bei anderen biologisch wichtigen „energiereichen" Bindungen, z. B. der Nucleotid-pyrophosphate oder auch der CoA-Thiolester.

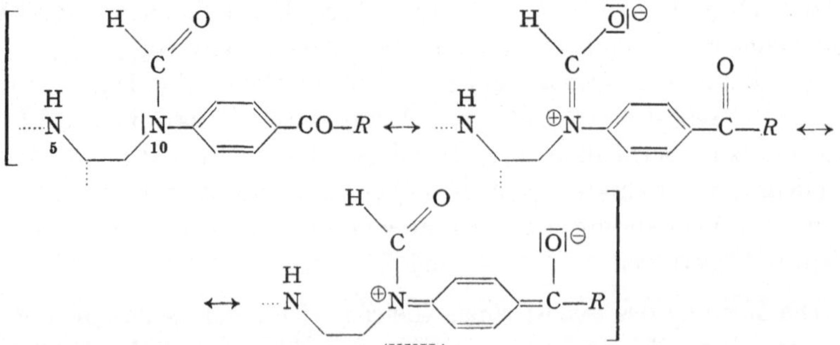

(XXII.)

Die Reihenfolge der Hydrolysenenergie der Formyl-tetrahydrofolate wird auch durch die LCAO-Berechnungen von PERAULT und PULLMAN (*228*) bestätigt und erklärt die Beobachtung, daß nur 10-Formyl-FH_4 und die Anhydro-Form als Formyl-donatoren in biologischen Transformylierungs-Reaktionen wirksam sind (*53, 111*). Es scheint vernünftig, anzunehmen, daß auch die $N_{(10)}$-Verbindung, wenn sie transformylierend wirkt, am Enzym in Methinyl-tetrahydro-folsäure oder eine ähnlich polarisierte Form übergeführt wird. Diese Verbindung, mit einer partiellen positiven Ladung auf dem Kohlenstoff, wird dann von einem nucleophilen Amino-stickstoff angegriffen, wobei ein sehr aktives Orthoamid *(Formelübersicht 6)* ent-steht (*58*), das in Richtung des größten Energiegewinnes hydrolysiert wird. Eine ähnliche Reaktion in umgekehrter Richtung spielt sich in der Cyclodeaminase-Reaktion der 5-Formimino-tetrahydro-folsäure (XXIII) ab.

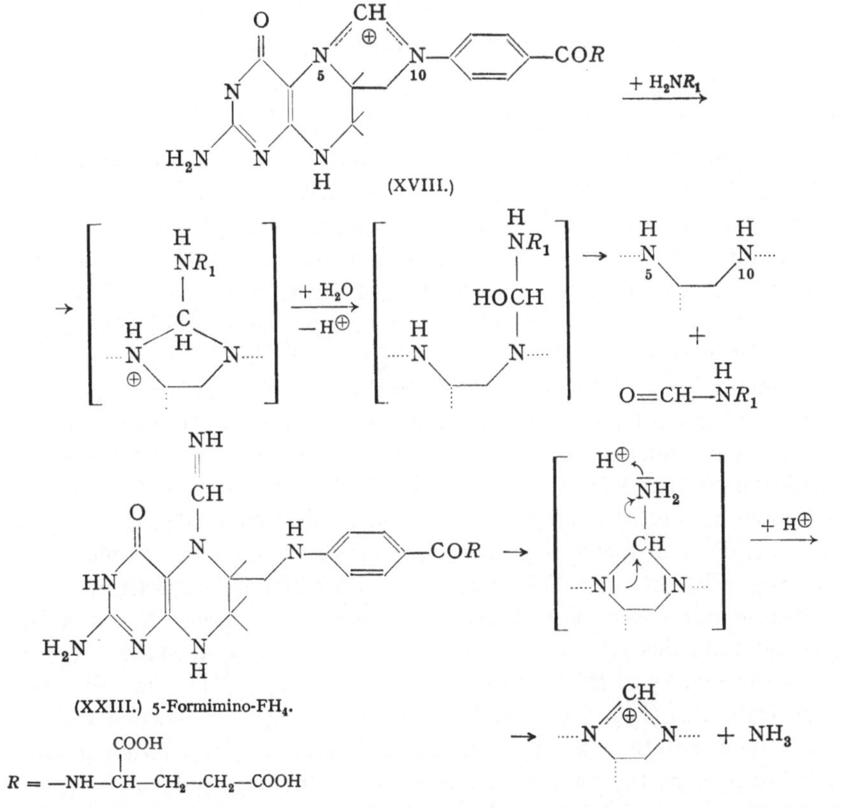

Formelübersicht 6. Orthoamide als Zwischenstufe in Reaktionen von 5,10-Methinyl-tetrahydro-folsäure.

e. 5-Formimino-tetrahydro-folsäure.

Diese Verbindung wurde zuerst von RABINOWITZ und PRICER (*247*) bei Untersuchungen des enzymatischen Abbaues der Purine und dann von TABOR und WYNGAARDEN (*310*) bei ihren Arbeiten über den Histidin-Abbau entdeckt. Formimino-glycin bzw. Formimino-glutaminsäure, die

Endprodukte dieser Abbauwege, übertragen die Formiminogruppe auf Tetrahydro-folsäure (Gln. [17, 24], SS. 247, 249). Mit angereicherten Enzymen wurde (XXIII) auf beiden Wegen synthetisiert (*249, 310*) und chromatographisch gereinigt, wobei es mit 0,2 N-Essigsäure als einziges Folat von Dowex-1-acetat eluiert wird. Nach dem Lyophilisieren bleibt ein gelbes Pulver, das gegen Luftsauerstoff recht stabil und auch im Spektrum 5-Formyl-FH_4 ähnlich ist. In saurer Lösung wandelt sich die Verbindung langsam unter Abspaltung von Ammoniak in 5,10-Methinyl-FH_4 (XVIII) um, wie am Ansteigen der Absorption bei 350 mμ verfolgt werden kann. Die gleiche Umwandlung wird biologisch durch Cyclodeaminase bewirkt (*307*) (Gl. [18], S. 247).

f. 5,10-Methylen-tetrahydro-folsäure.

Beobachtungen über die Beteiligung von Tetrahydro-folsäure an der biologischen Umwandlung des β-C-Atoms von Serin in Methylgruppen führten zum Postulat einer 5-Hydroxymethyl-tetrahydro-folsäure (*330*). Jaenicke und Greenberg (*135, 103*) erhielten durch Umsetzung von Tetrahydro-folsäure und Formaldehyd bei neutralem pH eine neue Folatverbindung, die in enzymatischen Reaktionen als „aktiver Formaldehyd" reagierte. Eine identische Verbindung konnten sie aus Ansätzen, die die bereits länger bekannte Umwandlung von Serin zu Glycin in Anwesenheit von Tetrahydro-folsäure (*163, 25, 4*) durchführten, isolieren.

Kisliuk (*159, 160*) zeigte, daß Formaldehyd zusammen mit dem Folat auf einer kleinen Dowex-1-chlorid-Säule festgehalten wird, während freier Formaldehyd durchläuft: in einer äquimolekularen Mischung der beiden Reaktionspartner waren nach 45 Sek. Reaktionszeit 77% des Aldehyds gebunden und nicht mehr eluierbar. Eine große Reihe reduzierter Pterine und Folate, bei denen $N_{(5)}$ oder $N_{(10)}$ substituiert waren, zeigen keinen derartigen Effekt. Der Folat-gebundene Formaldehyd tauscht in rascher Reaktion mit freiem Formaldehyd-^{14}C aus. Da freies $N_{(5)}$ und $N_{(10)}$ zur Bindung des Aldehyds erforderlich sind, schlug Kisliuk (*160*) die Struktur einer 5,10-Methylen-tetrahydro-folsäure (XXIV) vor, die schon von Jaenicke (*135*) und von Blakley (*25*) postuliert worden war, der auch beobachtete, daß die Modellverbindung Diphenyläthylendiamin mit Formaldehyd unter Ringschluß zu N,N'-Diphenylimidazolidin (XXV) reagiert (*26*). Er bestimmte ferner die Dissoziationskonstante des FH_4-Formaldehyd-Kondensationsproduktes, einmal durch Adsorption an einer Säule (*29, 30*) und zum anderen unter Ausnutzung der Veränderungen des UV-Maximums bei der Reaktion von Tetrahydro-folsäure und von Pterinen mit Formaldehyd (*31*) und erhielt einen Wert von etwa 10^{-4} M bei pH 7,2. Dagegen sind die Dissoziationskonstanten der Komplexe von Formaldehyd und Folatverbindungen, an denen $N_{(5)}$ oder $N_{(10)}$ blockiert ist, um zirka 10^2 größer. Die spektralen Veränderungen beim Übergang von Tetrahydro-folsäure zu Methylen-tetrahydro-folsäure zeigen zwei isosbestische Punkte, ein wichtiger Anhalt, daß nur ein Produkt entsteht.

(XXIV.) 5,10-Methylen-FH$_4$.

(XXV.) N,N'-Diphenylimidazolidin.

OSBORN et al. (*224*) reinigten das Produkt aus Tetrahydro-folsäure und Formaldehyd an einer Cellulose-Säure und erhielten eine Substanz mit 85—100% enzymatischer Aktivität (bezogen auf $^1/_2$ Pterin) in der enzymatischen Oxydation zu 5,10-Methinyl-FH$_4$. Wenn chemisch synthetisierte Tetrahydro-folsäure eingesetzt wird, erhält man natürlich auch die Methylen-tetrahydro-folsäure als Diasteromeren-Paar.

Die Funktion der Bildungsgeschwindigkeit, deren Maximum bei pH 4,5 gefunden wurde, in Abhängigkeit vom pH (*224*) hat zwei deutliche Wendepunkte, die darauf hinweisen, daß zwei dissoziierende Gruppen an der Bindung des Formaldehyds beteiligt sind. Die Wendepunkte liegen bei 3,2 und 5,2 pH und entsprechen damit den pKa-Werten von N-Methyl-*p*-aminobenzoylglutaminsäure (2,9) und dem N$_{(5)}$ hydrierter Pterine (5,1—5,6).

Die Modellverbindung Diphenyläthylendiamin reagiert auch mit anderen Aldehyden zu stabilen Produkten mit einem Imidazolidinring (*324*), und das gleiche muß auch für Tetrahydro-folsäure angenommen werden. Ho und Mitarb. (*121*) beschrieben ein derartiges Kondensationsprodukt der Tetrahydro-folsäure mit der Carbonylgruppe von Glyoxylsäure (XXVI), dessen biologische Bedeutung noch unklar ist. Die Verbindung geht durch Autoxydation in ein Oxalyl-Derivat der Tetrahydro-folsäure über, das an sich seiner Struktur nach leicht zu 5,10-Methinyl-FH$_4$ decarboxylieren sollte.

(XXVI.)

Methylen-tetrahydro-folsäure (XXIV) kann auch durch Reduktion von Anhydroformyl-tetrahydro-folsäure mit etwa äquimolaren Mengen Natriumborhydrid gewonnen werden (*224, 103*), wogegen die umgekehrte Reaktion nicht gelang. Biologisch stehen diese beiden aktivierten Einkohlenstoffkörper in einer mit TPN gekoppelten Reaktion im reversiblen Gleichgewicht, dessen Konstante zu $1,7 \times 10^{-2}$ bestimmt wurde (*124*) (S. 239, Gl. [13]). Die Bindung an die Tetrahydro-folsäure verändert dabei das Redoxpotential des Systems Ameisensäure/Formaldehyd ($E'_0 = -0,7$ V), so daß es in der Nähe des TPN$^+$/TPNH-Systems ($E'_0 = -0,32$ V) zu liegen kommt.

Diese Verschiebung ins Positive findet eine Erklärung durch die Ladung der oxydierten Form, in der die Onium-Energie des TPN$^+$ erhalten wird. JAENICKE und Brode (*141*) bestimmten an di-*p*-substituierten Modellverbindungen mit weitgehend variierten Substituenten Redoxpotentiale für das System Diaryl-imid-

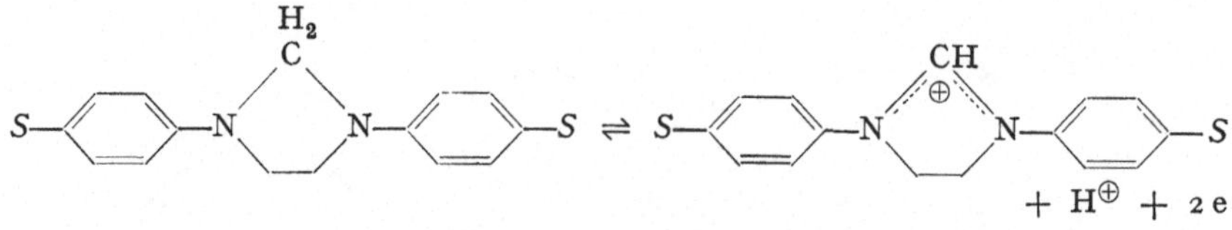

$$S=: \quad -H, \quad -Br, \quad -CH_3, \quad -OC_2H_5, \quad -NO_2, \quad -COOH, \quad -\beta\text{-Naphthyl}.$$

Formelübersicht 7. Modell der Methylen-FH_4-dehydrogenase-Reaktion (*141*).

azolinium/Diaryl-imidazolidin *(Formelübersicht 7)* und fanden Werte zwischen $+ 0,2$ und $+ 0,6$ V, die sogar höher liegen, als wir es für die natürliche Verbindung erwarten müssen. Die Ursache für diese Differenz kann nur ein wichtiges Strukturelement sein, das die groben Modelle nicht enthalten und das daher wohl im Tetrahydropyrazin-Ring zu suchen ist.

In Analogie zur „Aktivierten Ameisensäure" wird 5,10-Methylen-FH_4 (XXIV) häufig als „Aktiver Formaldehyd" bezeichnet, doch besteht ein Unterschied insofern, als die biologische Bildung der 10-Formyl-FH_4 der Zufuhr von ATP-Energie bedarf (S. 249), während die Bildung der Methylen-FH_4 aus Tetrahydro-folsäure und Formaldehyd exergon verläuft. Dennoch ist der Formaldehyd in der Verbindung mit dem Cofaktor in der Lage, Reaktionen einzugehen, zu denen freier Formaldehyd nicht befähigt ist, da die Kondensation mit der Aminogruppe wie in einer Mannich-Verbindung (vgl. *Formelübersicht 8*) die Ausbildung eines Übergangszustandes mit einem resonanz-stabilisierten Carbonium-Ion (XXVII) ermöglicht, das mit nucleophilen Partnern, wie z. B. dem Pyridoxal-aktivierten Glycin, ohne weitere Energiezufuhr reagieren kann (*136, 138, 135*). Von den beiden möglichen isomeren Stellungen des Carbonium-Ions sollte wohl bei Transhydroxymethylierungs-Reaktionen die

a) Mannich-Reaktion:

$$R_2NH + CH_2O \rightarrow R_2N-CH_2OH \xrightarrow[-H_2O]{+ H^\oplus} \left[R_2N-CH_2^\oplus \leftrightarrow R_2\overset{\oplus}{N}=CH_2 \right] \xrightarrow{+ X^\ominus}$$
$$\rightarrow R_2N-CH_2-X$$

b) Transhydroxymethylierung:

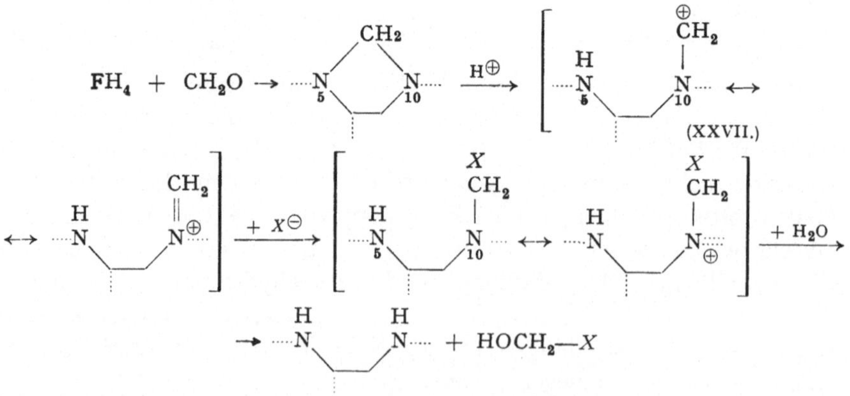

Formelübersicht 8. Vorgeschlagener Mechanismus der Transhydroxymethylierungs-Reaktion.

Literaturverzeichnis: SS. 254—274.

am $N_{(10)}$ bevorzugt sein, da der hydroxymethylierte Acceptor von dem aromatischen Amin-Stickstoff leichter abgelöst werden kann. Die Bindung des Substrates an den Cofaktor bewirkt also eine kinetische Beschleunigung der Reaktion, und nicht eine Verschiebung des Gleichgewichtes.

Allgemein lassen sich Cofaktoren unterteilen in solche, die in ,,energiereicher Bindung" aktivierte Gruppen zur Reaktion bringen, und solche, die durch geeignete Polarisation der Gruppe die Reaktion beschleunigen, indem sie die Wirkungsweise des Enzyms unterstützen. Typische Vertreter der ersten Art sind Coenzym A, Biotin und die Nucleotidphosphate, während Pyridoxalphosphat und Thiaminpyrophosphat dem zweiten Typ angehören. Tetrahydro-folsäure wirkt einmal als Träger der Ameisensäure in einer energiereichen Bindung im Sinne des ersten, zum anderen als Aktivator des Formaldehyds im Sinne des zweiten Typs. Im Hydroxylmethyl-thiaminpyrophosphat (J. Koch, unveröff.) ist der Formaldehyd im entgegengesetzten Sinne wie in der Methylen-tetrahydro-folsäure polarisiert, so daß die Kondensation mit elektrophilen Partnern möglich wird.

g. 5-Methyl-tetrahydro-folsäure.

Bei Untersuchungen über die Biosynthese der Methylgruppe des Methionins aus ,,aktivem Formaldehyd" mit einem reduzierenden System beobachteten JAENICKE (*137*) und BUCHANAN (*171, 171 a*) die Anreicherung einer neuen, bisher unbekannten Folatverbindung, als sie den Acceptor der Methylgruppe — Homoystein oder eine nahe verwandte Verbindung — aus dem System wegließen. Die isolierte Verbindung bewirkt bei der erneuten Inkubation mit dem Acceptor die Bildung von Methionin. Eine im Spektrum völlig gleichartige Verbindung mit halber Aktivität im enzymatischen System und im mikrobiologischen Test synthetisierten SAKAMI (*268*) und JAENICKE (*137*) durch Reduktion von Methylen-FH_4 mit Natriumborhydrid, wobei offenbar das 5-Isomere der Carboniumgrenzform des ,,aktiven Formaldehyds" (XXVII) ein Hydridion anlagert. Durch Abspaltung mit HJ wurde in der natürlichen (*171, 171 a*) und der synthetischen (*137*) Verbindung eine Methylgruppe nachgewiesen, die

$$\text{(XXVIII.)} \quad \text{5-Methyl-FH}_4.$$

(XXVIII.) 5-Methyl-FH₄.

höchstwahrscheinlich an das Stickstoffatom 5 gebunden ist (XXVIII), da die Hydrierung von 10-Methyl-folsäure ein völlig anderes Produkt liefert.

Kürzlich wurde berichtet, daß dieses neue 5-Methyl-tetrahydrofolat (XXVIII) mit einer natürlich vorkommenden Verbindung identisch ist, die schon 1959 von DONALDSON und KERESZTESY (*75*) aus Pferdeleber isoliert und später als Barium-Salz kristallisiert wurde (*76*). Sie nannten ihre ungeklärte Verbindung Prefolic A, weil sie durch ein enzymatisches System in Citrovorum-Faktor umgewandelt werden kann. Der erste Schritt dieser Umwandlung besteht in der Oxydation zu Methylen-tetrahydro-folsäure (XXIV, S. 217), die in Tetrahydro-folsäure und Formaldehyd dissoziiert (77, 78). Umgekehrt ließ sich aus Methylen-FH_4 enzymatisch und chemisch durch Reduktion mit Borhydrid aktive Prefolic A erhalten (154) (S. 242). Nichtenzymatisch wird 5-Methyl-FH_4 durch verschiedene Oxydationsmittel zu 5-Methyl-FH_2 oxydiert, die ihrerseits durch Ascorbat oder Sulfhydryl-Verbindungen wieder zum Ausgangsmaterial reduziert werden kann ($78a$). Da dieser Oxydations-Reduktions-Cyclus keine Racemisierung der Verbindung am $C_{(6)}$ bewirkt, kommt dem Oxydationsprodukt die Struktur einer 5-Methyl-5,6-FH_2 zu.

5-Methyl-FH_4 ist Wuchsstoff für *L. casei*, aber inaktiv für *St. faecalis* und *P. cerevisiae* und verhält sich damit ebenso wie das Triglutamat der Tetra-hydro-folsäure. Man muß daher annehmen, daß viele ältere Angaben über das Vorkommen von Konjugaten der Folsäure in Geweben und Körperflüssigkeiten (118), die auf dem Test mit den drei Organismen basieren, zu berichtigen sind.

4. Folsäure-Analoge.

Folsäure und ihre biologischen Derivate sind im Stoffwechsel an vielen Reaktionen beteiligt, denen eine Schlüsselstellung in der Bio-synthese von Nucleinsäuren und Proteinen zukommt. Eine Folge davon sind die schweren Symptome, die wir beim Folsäuremangel an höheren Lebewesen beobachten können. Auf der anderen Seite aber bietet eine künstliche, gesteuerte Erzeugung von Folsäuremangel Aussicht auf eine chemotherapeutische Beeinflussung übermäßiger Zellteilungsprozesse, insbesondere in der Krebs- und Leukämie-Therapie. Eine große Zahl mehr oder weniger abgewandelter Analoge der Folsäure wurden daher synthetisiert und auf ihre Wirksamkeit als Antagonisten im Wuchsstoff-test und in der Erzeugung künstlichen Folsäuremangels untersucht. Für eine eingehendere Darstellung siehe (302, 147, 122, 346, 184, 94).

Die wichtigste Entdeckung machten Seeger und Mitarb., als sie Folat-Verbindungen darstellten, die statt der 4-Hydroxy- eine Amino-gruppe am Pterinkern tragen (272, 273). Diese Verbindungen zeichnen

Aminopterin:	$R_1 = -NH_2$;	$R_2 = -H$;	$R_3 = $ *L*-Glutaminsäure.
Amethopterin:	$R_1 = -NH_2$;	$R_2 = -CH_3$;	$R_3 = $ *L*-Glutaminsäure.
10-Methyl-folsäure:	$R_1 = -OH$;	$R_2 = -CH_3$;	$R_3 = $ *L*-Glutaminsäure.
Pteroyl-asparaginsäure:	$R_1 = -OH$;	$R_2 = -H$;	$R_3 = $ *L*-Asparaginsäure.

Formelübersicht 9. Konstitution von Folsäure-Analogen.

Literaturverzeichnis: SS. 254—274.

sich durch besonders hohe Wirksamkeit aus, und Aminopterin und Amethopterin (Methotrexate) *(Formelübersicht 9)* sind bis heute die wichtigsten Folat-Antagonisten geblieben. Es entspricht daher auch praktischen Gesichtspunkten, wenn man die Vielzahl der beschriebenen Folsäure-Analogen in die folgenden beiden Gruppen einteilt (*147*):

(1) Verbindungen, die einen 2,4-Diaminopyrimidin-Ring oder 2,4-Diaminotriazin-Ring enthalten. Diese Verbindungen hemmen schon in kleinen Konzentrationen sehr stark das Wachstum von *St. faecalis*, der meist als Testorganismus verwendet wird, und von verschiedenen Versuchstieren. Die Hemmung kann nur durch Citrovorum-Faktor, nicht aber durch Folsäure vollständig aufgehoben werden. Neben den genannten gehören zu dieser Gruppe insbesondere 4-Amino-pteroyl-asparaginsäure, Derivate des 2,4-Diaminopyrimidins, wie das Pyrimethamin (2,4-Diamino-5-*p*-chlorphenyl-6-äthyl-pyrimidin), Derivate des Aminopterins mit Halogensubstituenten im Benzolkern, ferner 2,4-Diamino-pterine, -chinazoline und -triazine und neuerdings das 6-Phenyl-2,4,7-triamino-pterin.

(2) Derivate der Folsäure mit unveränderter Substitution am Pterinkern, insbesondere methylierte Folsäure mit einer Methylgruppe in 7-, 9- oder 10-Stellung (*62, 87, 127, 185*) und Pteroyl-Verbindungen mit anderen Aminosäuren als Glutaminsäure, z. B. Pteroylasparaginsäure (*129*). Vertreter dieser Gruppe hemmen wesentlich schwächer, und die Hemmung kann bereits durch Folsäure aufgehoben werden. Einige haben sogar geringe Wuchsstoffwirksamkeit, wie Pteroyladipin- und -pimelinsäure. 10-Methylfolsäure ist ein stärkerer Inhibitor als 9-Methylfolsäure. Für den medizinischen Gebrauch sind sie kaum von Bedeutung.

Die Wirksamkeit der Antagonisten der Gruppe (1) beruht im wesentlichen auf einer Blockierung der Folatreduktase-Reaktion (S. 198). Da Folsäure nur in der tetrahydrierten Form Cofaktor ist, wird durch die Hemmung der Reduktase praktisch vollständiger Folsäuremangel erzeugt. Die Antagonisten reagieren mit dem Enzym in stöchiometrischer Weise; der gebildete Komplex hat eine 10^5mal kleinere Dissoziationskonstante als der Enzym-Substratkomplex (*332*) und nimmt damit unter allen bisher bekannten Hemmstoffen enzymatischer Reaktionen eine Ausnahmestellung ein. K_I wurde zu etwa 10^{-9} bestimmt (*34, 290*). Im Vergleich dazu beträgt die K_M für FH_2 etwa 10^{-6} (*34*).

Nach Berechnungen von PERAULT und PULLMAN (*229*) sind in den 4-Aminoanalogen $N_{(1)}$ und die Aminogruppe am $C_{(2)}$ basischer als in den Folaten; dies könnte für die festere Bindung verantwortlich sein. Eine andere Möglichkeit ist die Bildung eines Donor-Acceptorkomplexes mit einem Tryptophan des Enzymproteins. Derartige Komplexe wurden von FUJIMORI (*90*) beobachtet, wobei das Folat der Acceptor ist. Die Antagonisten werden durch Folatreduktase nicht reduziert (*233*).

Die Antimetaboliten lassen sich aber chemisch ebenso wie die Folate zur Dihydro- und Tetrahydro-Stufe reduzieren. Dihydro-aminopterin (*365b*) hemmt die Folat-Reduktasen in gleichem Maße wie die unhydrierte Verbindung (*122a, 365b*), dagegen hat Aminopterin-H$_4$ nur noch eine K$_I$ von etwa 10^{-6} (*122a, 290*). Die verstärkte Inhibitorwirkung der tetrahydrierten Verbindung für Kulturen von *St. faecalis* und *P. cerevisiae* (*161*) scheint auf einem Gehalt an Dihydroverbindung zu beruhen (*365b*). Die Hemmung anderer Folatreaktionen, allerdings in wesentlich geringerem Maße als bei der Reduktase, wurde von Tabor und Wyngaarden (*310*) sowie von Slavík (*289, 290*) beschrieben. Wahba und Friedkin (*321*) beobachteten bei der Thymidylatsynthetase-Reaktion in Anwesenheit von Substratmengen FH$_4$ eine Hemmung durch die Tetrahydro-Form von Aminopterin und Amethopterin, während die unhydrierten Formen wirkungslos waren. Eigene Beobachtungen betreffen Versuche mit der FH$_4$-Formylase (E. Brode, Dissertation). Diese Hemmeffekte sind zum Teil kompetitiver Natur.

5. Spektren von Folat-Verbindungen.

Die Spektren der wichtigsten Folsäureverbindungen sind in den Abb. 1—3 dargestellt. Die *Tabelle 4* enthält die Absorptionsmaxima und die zugehörigen Extinktionskoeffizienten.

Folsäure zeigt in alkalischer Lösung drei Maxima: bei 256, 282 und 365 mμ; im Neutralen verbinden sich die beiden kurzwelligeren zu einem Maximum bei

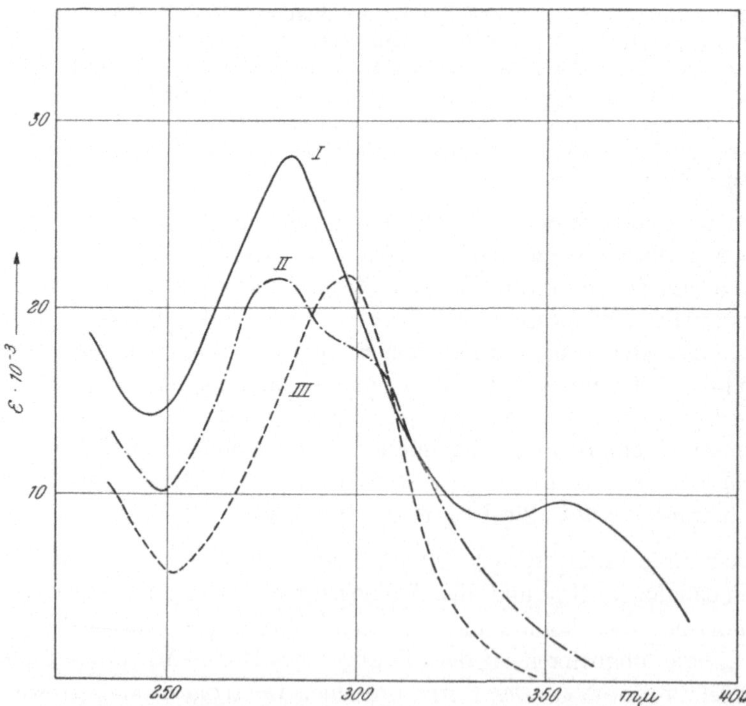

Abb. 1. Spektren von Folat-Verbindungen (bei pH 8): I, Folsäure; II, Dihydro-folsäure; III, Tetrahydro-folsäure.

Literaturverzeichnis: SS. 254—274.

Tabelle 4. Spektrale Daten wichtiger Folatverbindungen.

Verbindung	pH	λ_{max} (mμ)	λ_{min} (mμ)	$\varepsilon_{max} \times 10^{-3}$
Pteroinsäure............	13	256; 275; 356	320	45 (256 mμ)
Folsäure	13	256; 282; 356	330	30; 26; 9,8
Folsäure	7	282; 350	330	27; 7
FH$_2$.................	7,5	283	250	21
FH$_4$.................	7,5	298	245	22
10-Formyl-FH$_4$	7,5	260	240	17
5-Formyl-FH$_4$	13	282	245	32,6
5,10-Methinyl-FH$_4$	1	355	305	24,9
5,10-Methylen-FH$_4$......	7,2	294	245	32
5-Methyl-FH$_4$	7	287	246	28

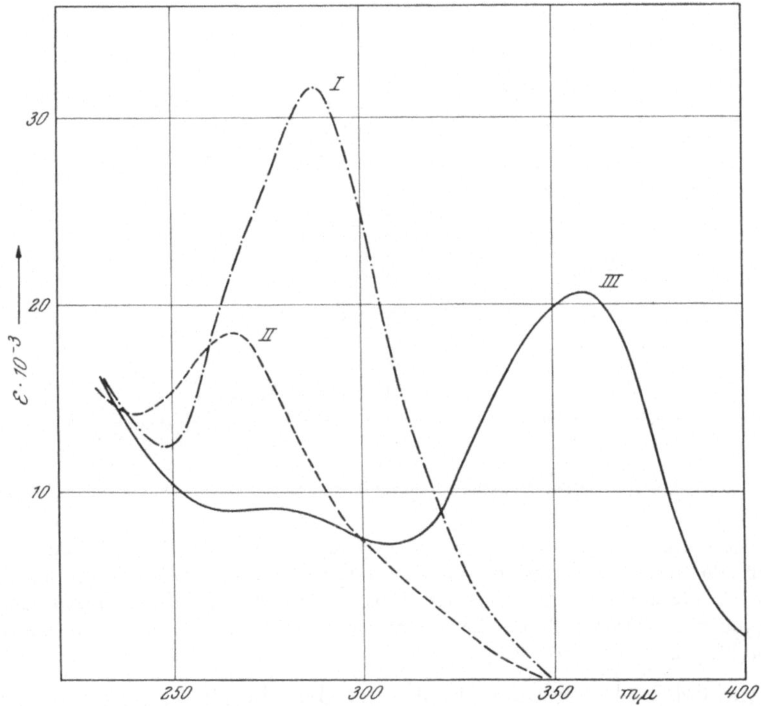

Abb. 2. Spektren von Folat-Verbindungen (I und II bei pH 8; III bei pH 2): I, 5-Formyl-FH$_4$; II, 10-Formyl-FH$_4$; III, 5,10-Methinyl-FH$_4$.

290 mμ und im Sauren wird schließlich nur ein Maximum bei 300 mμ gefunden. Das Maximum bei 280 mμ kann man durch Vergleich dem pAB-Teil, die beiden anderen dem Pterin-Teil des Moleküls zuschreiben. Das UV-Spektrum der Pteroinsäure gleicht dem der Folsäure, und auch die Anfügung weiterer Glutaminsäure in den Konjugaten beeinflußt das Spektrum kaum noch. Das Spektrum der 5-Formyl-

FH$_4$ deckt sich nahezu mit dem eines äquimolekularen Gemisches des Formimino-pyrimidins und der N-Methyl-p-ABGlu. Dagegen weichen das Spektrum der Folsäure und eines Gemisches von 6-Methylpterin und N-Methyl-p-ABGlu voneinander ab. Diese Abweichung wird auf eine hyperkonjugative Verbindung der beiden π-Systeme in der Folsäure über die C$_{(9)}$-Methylenbrücke zurückgeführt (*186*).

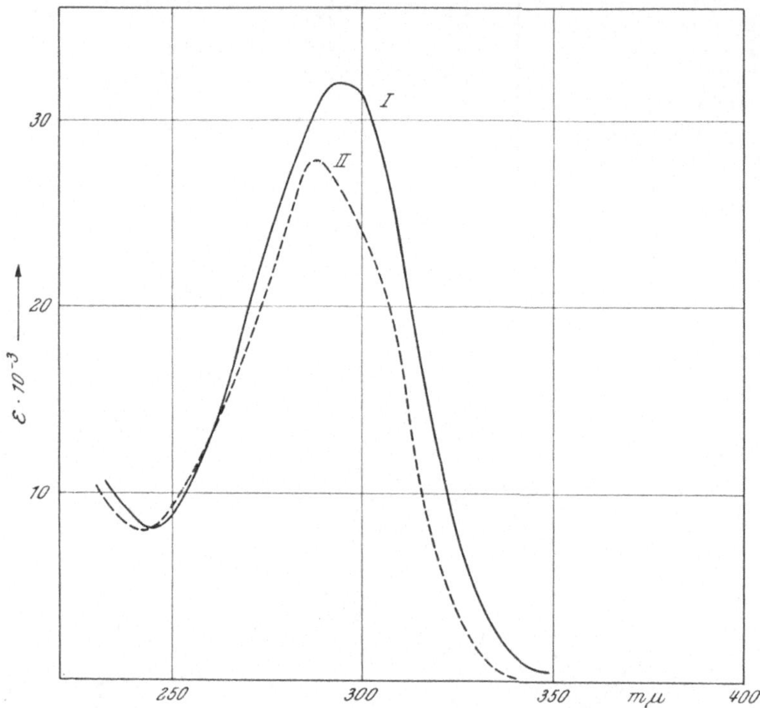

Abb. 3. Spektren von Folat-Verbindungen (bei pH 7): I, 5,10-Methylen-FH$_4$; II, 5-Methyl-FH$_4$.

Fluoreszenz zeigen wie die meisten Pterine so auch viele Folat-Verbindungen. Nicht alle sind aber bisher so weit gereinigt worden, daß sich entscheiden ließe, ob die beobachtete Fluoreszenz der Substanz selbst oder Zersetzungsprodukten zu eigen ist. Folsäure fluoresziert in reiner Form nicht, dagegen sind sämtliche Formylderivate Fluoreszenzträger.

Alle Folsäureverbindungen enthalten die Glutaminsäure in der natürlichen L-(+)-Form und sind daher optisch aktiv. Bei der Hydrierung zur Tetrahydrostufe wird ein neues optisches Zentrum geschaffen: es entsteht ein Diastereomeren-Paar, dessen Trennung nur im Fall der 5-Formyl-FH$_4$ (Leucovorin) gelungen ist. Die absolute Konfiguration des natürlichen Isomeren ist nicht bekannt.

Die spezifischen Drehwinkel der Diastereomeren-Gemische verschiedener Folatverbindungen (F, FH$_2$, FH$_4$, 5-Formyl-FH$_4$) betragen $+ 15°$ bis $+ 20°$; die der l,L-Formen von FH$_4$ und 5-Formyl-FH$_4$ $- 16,9°$ bzw. $- 15,1°$ (*340 a, 188, 64*).

Literaturverzeichnis: SS. 254—274.

6. Analyse und Trennung von Folsäure-Verbindungen.

Zahlreiche Veröffentlichungen befassen sich mit der quantitativen Bestimmung der Folsäure in Arzneimitteln und in biologischem Material, so daß hier nur ein knapper Überblick gegeben werden kann. Arbeitsvorschriften sind in den „Methods of Biochemical Analysis" erschienen (*146*). Für die quantitative Bestimmung der Folsäure in biologischem Ausgangsmaterial kommt dem überaus empfindlichen mikrobiologischen Test die größte Bedeutung zu. Die vorhergehende Trennung der Folat-Gemische, mit denen man es dabei im allgemeinen zu tun hat, an der Säule oder im Papierchromatogramm gestattet die genaue Identifizierung der Bestandteile. Die chemischen und polarographischen Methoden erlauben dagegen nur die Bestimmung des gesamten Folatgehalts einer Probe.

a. Chemische Verfahren.

Die Spaltungsreaktion der Folsäure an der $C_{(9)}$-$N_{(10)}$-Bindung bei Reduktion mit Zn/HCl oder, besser, $TiCl_3$ (*95*) wurde zu einer analytischen Methode ausgebaut; das entstehende primäre aromatische Amin (*p*-Aminobenzoylglutaminsäure) wird nach BRATTON und MARSHALL (*42*) diazotiert und gekuppelt und der entwickelte Farbstoff im Photometer bei 550 mμ gemessen (*131*). In ähnlicher Weise kann FH_4 bestimmt werden, die schon ohne Einwirkung eines Reduktionsmittels unter Freisetzung des diazotierbaren Amins zerfällt.

Eine wesentlich empfindlichere Methode besteht in der Oxydation der Folatverbindungen mit Permanganat zu Pterincarbonsäure, die dann fluorometrisch bestimmt wird (*7*). Die Methode gestattet auch die Bestimmung konjugierter Folate ohne vorherige Behandlung mit Konjugasen. Das könnte der Grund sein, daß etwas höhere Werte als bei der mikrobiologischen Analyse erhalten werden. Folsäuremengen von etwa 0,25 μg an können bestimmt werden.

b. Polarographie.

Die polarographische Reduktion der Folsäure wurde von mehreren Autoren beschrieben (*256, 181, 55*). ALLEN (*5*) benutzte das Verfahren bei Untersuchungen mit dem Ziel der Strukturaufklärung des Leucovorins (S. 211). Leucovorin selbst gab keine Welle, aber nach Behandlung der Verbindung mit verd. Mineralsäure fand ALLEN bei pH 9,0 drei deutliche Halbwellenpotentiale, die durch Vergleich mit den Polarogrammen einfacher und hydrierter Pterine zugeordnet werden konnten: — 0,88 V für Folsäure, — 1,4 V für Dihydrofolsäure und ein anodisches Potential bei 0,20 (?) für die tetrahydrierte Verbindung. Das Auftreten der Potentiale für F und FH_2 ist auf eine teilweise Oxydation der Probe bei der Vorbehandlung zurückzuführen. Die Menge CF kann aus der Summe aller drei Potentialströme errechnet werden und befindet sich in guter Übereinstimmung mit dem mikrobiologischen Test. Neuere

Arbeiten über die polarographische Bestimmung der Folsäure in Gemischen mit anderen Vitaminen sind von Asahi (13) und Heyrovský (120) erschienen. Die Methode wurde auf biologisches Material bisher nicht angewendet.

c. Mikrobiologische Methoden.

Die Unfähigkeit einiger Hefen und Bakterien, Folsäure oder ihre Cofaktor-Formen zu synthetisieren, ermöglicht den mikrobiologischen Wachstumstest auf Folsäureverbindungen. Ihre spezifischen Ansprüche, die sie an die Zugabe von Folsäure-Derivaten im Medium stellen, erlauben eine Differenzierung der verschiedenen Verbindungen entsprechend ihrer Oxydationsstufe und der Länge der Konjugat-Seitenkette. Für diese Differenzierungs-Analyse verwendet werden vor allem die drei Mikroorganismen *L. casei* (ATCC 7469), *St. faecalis* R (ATCC 8034) und *P. cerevisiae* (ATCC 8081) (146). Die beiden ersteren können hydrierte und unhydrierte Pteroyl-Verbindungen verwerten, jedoch nur die kurzkettig-konjugierten Formen, nämlich *St. faecalis* die mit 0 bis 2, *L. casei* die mit 1 bis 3 Glutamylresten. Man kann demnach mit *L. casei* die Summe der niedrigen Folat-Konjugate, mit *St. faecalis* die einfachen Folsäuren bestimmen. Die differierenden Angaben über die Wuchsstoff-Aktivität hydrierter Folat-Derivate und Konjugate sind durch die unreinen Präparate oder die oxydierende Vorbehandlung verursacht. Allgemein kann man sagen, daß unter gleichmäßigen Bedingungen kein Unterschied zwischen hydrierten und unhydrierten Folaten bezüglich der Wuchsstoffwirkung besteht, daß aber die verschiedenen Mikroorganismen auf unterschiedliche Konzentrationen ansprechen. Halbmaximales Wachstum erhält man bei *St. faecalis* mit 0,3 mμg/ml, bei *L. casei* aber bereits mit 0,05 mμg/ml. *P. cerevisiae*, dem anscheinend die Folat-Permease fehlt (356), so daß er auf die Zufuhr hydrierter Folate angewiesen ist, zeigt halbmaximales Wachstum bei 0,15 mμg/ml. Diese

Tabelle 5. Wuchsstoff-Aktivität von Folat-Verbindungen (287).

Verbindung	Wachstums-Aktivität* im Test mit		
	L. casei	*St. faecalis*	*P. cerevisiae*
Pteroinsäure, Rhizopterin	—	+	—
Folsäure, 10-Formyl-folsäure, FH_2	+	+	—
FH_4, 10-Formyl-FH_4, 5-Formyl-FH_4, 5,10-Methinyl-FH_4, 5,10-Methylen-FH_4	+	+	+
Prefolic A (5-Methyl-FH_4)	+	—	—
Pteroyltriglutamat und Pteroyltriglutamat-H_4	+	—	—
Pteroylheptaglutamat	—	—	—

* + = 60—100%, — = < 5% Wachstum, bezogen auf Citrovorum-Faktor als Standard.

Literaturverzeichnis: SS. 254—274.

Werte sind auf *d,l*-CF als Standard bezogen. Bei *P. cerevisiae* sind alle tetrahydrierten Folsäuren gleich aktiv, auch die an $N_{(5)}$ oder $N_{(10)}$ formylierten, mit Ausnahme der 5-Methyl-FH, die nur Wuchsstoff für *L. casei* ist. Das Wachstum wird im allgemeinen nach 24 Stunden durch optische Trübungsmessung gegen einen Vergleichswert bestimmt. Nach 72 Stunden kann auch die gebildete Säure titriert werden.

Die biologisch gebildeten Tetrahydrofolat-Verbindungen weisen durch die Asymmetrie an $C_{(6)}$ als einheitliche diastereomere Formen die doppelte Wachstumsaktivität auf wie die synthetischen Racemate. Keiner der Mikroorganismen spricht auf die höheren Konjugate an, so daß sich die in *Tabelle 5* wiedergegebenen Verhältnisse ergeben (*287*).

Die Ergebnisse der Tabelle wurden durch die Anwendung der von SILVERMAN und Mitarb. (*283, 284, 286*) entwickelten und von BAKERMAN (*15*) ausführlich dargestellten „aseptischen Ascorbat-Technik" erhalten, die darin besteht, daß die hydrierten Folate durch Zusatz von Ascorbat vor der Oxydation beim Autoklavieren geschützt werden, oder daß diese empfindlichen Verbindungen erst dem fertig sterilisierten Medium aseptisch zugesetzt werden. In Ascorbat-haltiger Lösung unterhalb 75° bleiben die hydrierten Folate stabil und wird auch 10-Formyl-FH_4 nicht in CF umgelagert (*351*).

Nach den älteren Methoden, in denen das Ausgangsmaterial einem längeren Autoklavieren ohne Zusatz von schützenden Reduktionsmitteln ausgesetzt wurde, konnte nur der gesamte Folatgehalt der Probe mit *L. casei* oder *St. faecalis* bestimmt werden, und der Gehalt an Citrovorum-Faktor, der als einzige hydrierte Verbindung diese Bedingungen übersteht, mit *P. cerevisiae*. Die Methoden der Ausführung dieser Tests und des vorherigen Abbaus der Konjugate mit spezifischen Enzymen sind in einer Reihe von Handbüchern und Übersichtsartikeln zusammengefaßt (*146, 214, 293, 302, 346*).

Außer den meistgebrauchten Wachstumstests mit den genannten drei Bakterienarten wurden Methoden der Folsäurebestimmung mit dem Protozoon *Tetrahymena geleii* W (*69, 157*) und mit dem thermophilen Bakterium *B. coagulans* (*14*) beschrieben. *T. geleii* wächst auf Konjugaten ebenso gut wie auf freien Folsäureverbindungen, möglicherweise deshalb, weil das Protozoon eine Konjugase enthält. Das langsame Wachstum ist ein Nachteil des Verfahrens. 0,01—1,0 mμg werden sicher bestimmt. *T. geleii* wird durch Antagonisten im Wachstum nicht gehemmt. *B. coagulans* (ATCC 3084, nicht mehr im Katalog) gestattet die Bestimmung von 0,1—10 mμg/ml Folat in einem Wachstumstest von 24 Stunden bei 55°. Das Bakterium gibt gleiches Wachstum mit *p*AB, F, CF und 10-Formyl-F, wächst aber nicht mit *p*ABGlu und Pteroinsäure. Die Verwendung der Mikroorganismen zur Lokalisierung von Folaten auf Papierchromatogrammen wird unten beschrieben.

d. Chromatographische Trennung.

RAUEN und Mitarb. (*253*) bedienten sich bei der Isolierung der 10-Formyl-folsäure mit Vorteil der Gegenstromverteilung im System n/50 HCl/*n*-Butanol, die zur Trennung von Pterinen weitere Anwendung gefunden hat (*254*).

Die ersten Versuche zur Trennung von Folat-Verbindungen durch Chromatographie an Säulen mit Cellulose oder Solka-Floc (*224, 225, 267, 283*) brachten kaum mehr als eine Abtrennung von Verunreinigungen. In speziellen Fällen brachte die Verwendung von Harz-Ionenaustauschern gute Erfolge, doch bleibt diesem Verfahren eine breite Anwendung versagt, da die Elution der adsorbierten Verbindungen nur unter stark sauren Bedingungen möglich ist, wo die empfindlicheren Folate schon zersetzt werden. Mittels eines Essigsäuregradienten an Dowex-1-acetat isolierten Rabinowitz und Pricer (249) sowie Tabor und Wyngaarden (310) Formimino-FH$_4$ aus enzymatischen Ansätzen. Die Verwendung von Dowex-1-formiat (283) und von Dowex-1-chlorid wurde ebenfalls beschrieben (119).

Ein entscheidender Fortschritt wurde erst durch die Verwendung modifizierter Cellulosen erreicht (296). Diese Austauscher binden die Folate nicht so fest wie die Harze und gestatten die Elution schon mit Phosphat-Puffern bei pH 6,0. Das Verfahren wurde zuerst von Usdin und Porath (316) zur Trennung von Folsäurekonjugaten an TEAE-Cellulose beschrieben. Die gleichen Autoren beschreiben auch ein Verfahren zur säulen-elektrophoretischen Trennung von Folaten. Die Chromatographie an TEAE wurde erfolgreich auf die Trennung und Identifizierung der Folsäureverbindungen im Blut angewendet (315). Mehrere Arbeiten aus jüngerer Zeit beschreiben die Verwendung von DEAE-Säulen (220, 245, 288). Mathews und Huennekens (188) trennten FH$_4$ und FH$_2$ an DEAE mit einem Tris-Gradienten (0,005—0,2 M pH 7). Die Chromatographie an DEAE wurde von Silverman und Mitarb. (287) zu einem empfindlichen analytischen Verfahren vervollkommnet.

Die Folate werden dabei auf einer zirka 20 cm langen Säule aus einer 50/50-Mischung von DEAE-Cellulose und Hyflo-Supercel mit einem 0,6% ascorbathaltigen 0—0,5 M Phosphatgradienten pH 6,0 getrennt. Die 5-ml-Fraktionen werden im mikrobiologischen Test mit den drei üblichen Organismen identifiziert; dieser empfindliche Nachweis gestattet unter den standardisierten Bedingungen die Trennung von 0,5—1 μg jeder Folatverbindung. Dieses schonende und empfindliche Verfahren erlaubte den Nachweis, daß die Folsäure in den Geweben und Körperflüssigkeiten überwiegend in der Cofaktor-Form der 10-Formyl-FH$_4$ (287, 351) oder der Prefolic A vorliegt (287), während man auf Grund der älteren Testverfahren den CF für die meistverbreitete Cofaktor-Form hielt.

Methoden der Papierchromatographie wurden vielfach zur analytischen und mikropräparativen Trennung von Folaten verwendet. Die Autoxydation der hydrierten Folate kann dabei bestenfalls teilweise durch Zusatz von Thioglykol oder EDTA zu den Solventien oder durch Chromatographie unter Stickstoff (26) vermieden werden. Aus der Fülle der einschlägigen Arbeiten seien einige zitiert, in denen eine größere Zahl von Rf-Werten für wichtige Folsäureverbindungen enthalten sind (26, 103, 170, 224, 289, 315, 345).

Die meistverwendeten Solventien sind: wäßriger Phosphatpuffer verschiedener Konzentration; Formiat/Ameisensäure; 3% Ammoniumchlorid oder -acetat; Äthanol/Wasser/(Ammoniak); Butanol/Eisessig.

Zweiphasige Systeme wurden von Wieland angegeben (345). Beim Vergleich der in der Literatur angegebenen Rf-Werte findet man häufig große Differenzen, deren Ursache kleine Unterschiede der Methode (Papier, Temperatur), aber auch Zersetzung der untersuchten Substanzen während der Chromatographie sein können.

Literaturverzeichnis: SS. 254—274.

Die Lokalisierung der Verbindungen auf den Chromatogrammen erfolgt am einfachsten auf Grund ihrer Absorption oder Fluoreszenz bei Betrachtung unter UV-Licht. Ein Spray-Reagens, das auf der Reduktion des Folats mit Titantrichlorid und darauffolgender Kupplung der p-Aminobenzoylglutaminsäure mit einem Diazo-Reagens beruht, wurde von KOMENDH (*168*) angegeben und von HATEFI (*114*) modifiziert. Eine spezifische Identifizierung ermöglicht die Bioautographie-Technik, bei der die entwickelten Papierstreifen auf eine mit dem Testorganismus inokulierte Agarplatte gedrückt werden, in der der Wachstumsfaktor fehlt. Dort, wo die aktiven Verbindungen mit der Platte in Berührung kommen, wird Wachstum beobachtet (*345, 350*). Die Methode kann in ihrer Empfindlichkeit gesteigert werden, wenn man den Nährböden Tetrazoliumsalze zusetzt, die durch wachsende Mikroorganismen zu stark gefärbten Formazanen reduziert werden. Diese „Tetrazolium-Bioautographie" (*317*) erreicht nahezu die Nachweisempfindlichkeit des Wachstumstests in flüssiger Kultur: 0,01 γ können sicher gefunden werden.

V. Das Einkohlenstoff-Reservoir.

Auf die engen Beziehungen zwischen Ameisensäure oder Formaldehyd und den verschiedenen Donatoren und Acceptoren der Formyl- und Hydroxymethyl-Gruppe wurde man zunächst durch Isotopenversuche am lebenden Organismus aufmerksam. ^{14}C-Formiat ist hierbei ein Vorläufer sowohl des β-Hydroxymethyl-Kohlenstoffs von Serin (*266*), wie auch der Amidin-Kohlenstoffe der Purine ($C_{(2)}$, $C_{(8)}$) (*50—54, 96, 97, 103*), oder von Histidin ($C_{(2)}$) (*203*), die überdies alle miteinander im Stoffwechsel-Gleichgewicht stehen (*80a, 297*). Formaldehyd wird in diese Verbindungen in gleicher Weise eingebaut wie Formiat; darüber hinaus tauschen die beiden Kohlenstoff-Einheiten nicht nur miteinander, sondern auch mit der Methylgruppe in Cholin, Methionin oder Thymin aus (*81*). Die Zu- und Abflüsse der Einkohlenstoff-Reservoirs sind in *Formelübersicht 10* dargestellt (S. 230).

1. Herkunft der Ameisensäure.

Ameisensäure ist ein gewöhnliches Nebenprodukt bakterieller Gärungen. Die erste Erkenntnis, daß sie an biologisch wichtigen Reaktionen aktiv teilnimmt, kam von BUCHANANS Untersuchungen über die Biogenese der Purine. Es wurde gefunden, daß Formiat in die Ureidkohlenstoffe (2 und 8) der Harnsäure eingebaut wird (*54*). Diese Entdeckung erweckte das Interesse an möglichen anderen Rollen der Ameisensäure im Stoffwechsel und stimulierte die weiteren Untersuchungen, die dann zu dem Komplex der Einkohlenstoff-Reaktionen geführt haben, bei denen von Ameisensäure hergeleitete Kohlenstoff-Reste in Stoffwechselreaktionen eingesetzt werden.

Da exogene Ameisensäure in die Einkohlenstoff-Reaktionen eintreten kann (*264*), können auch solche biologische Reaktionen, bei denen Ameisensäure entsteht, zum Einkohlenstoff-Pool beitragen (*252*). Dazu gehört z. B.

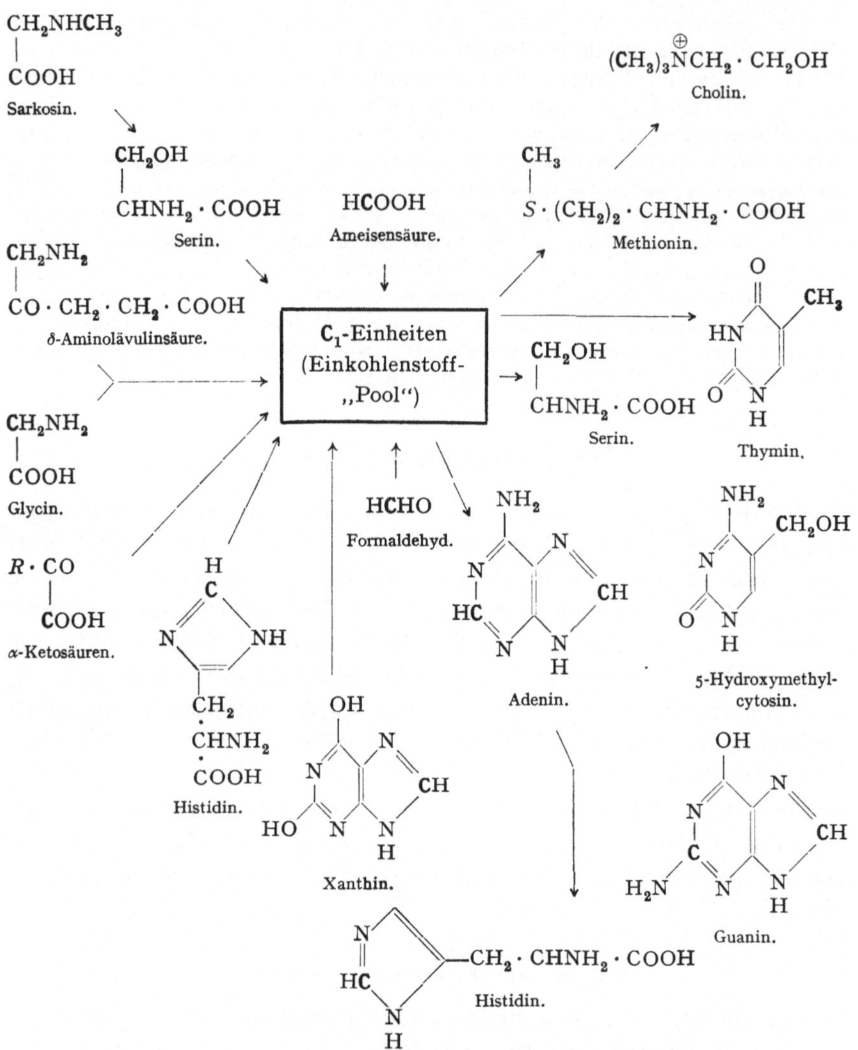

Formelübersicht 10. Zu- und Abflüsse des Kohlenstoff-Reservoirs.

die phosphoroklastische Spaltung von Pyruvat (*305*) und von α-Keto-valeriansäure, der Abbau von Glycin (*265, 266*) und Tryptophan (*193*) oder die noch unbekannte Bildung von Ameisensäure bei Insekten.

Die phosphoroklastische Spaltung der Brenztraubensäure, in der nach Gl. [6] Formiat gebildet wird, ist im Mechanismus noch weitgehend unbekannt.

$$CH_3COCOOH + H_3PO_4 \rightarrow CH_3COOPO_3H_2 + HCOOH. \qquad [6]$$

Ein dieser Reaktion ähnlicher Vorgang, der möglicherweise zum Formiat-Pool beisteuert, ist die Spaltung der α-Ketovaleriansäure, die aus Valin entsteht. Nach

Literaturverzeichnis: SS. 254—274.

den Untersuchungen von WEBB (*326, 327*) an Amethopterin-gehemmten Kulturen von *A. aerogenes* wird α-Ketovaleriansäure in Isobutyryl CoA und aktives Formiat gespalten:

$$(CH_3)_2CH \cdot CHNH_2 \cdot COOH \rightarrow (CH_3)_2CH \cdot CO \cdot COOH \xrightarrow[\text{(Co-Faktoren)}]{\text{CoASH}}$$

$$\rightarrow (CH_3)_2CH \cdot CO \cdot SCoA + \text{,,HCOOH``}. \qquad [7]$$

Diese Reaktionsfolge kann auch die Beobachtung von WINKLER und DE HAAN (*349*) erklären, nach der eine Sulfonamidhemmung bei *E. coli* vollständig überwunden wird, wenn außer den bereits genannten Verbindungen Xanthin, Thymin, Serin und Methionin (vgl. S. 192) auch Valin zugefügt wurde. Das Cofaktor-Bedürfnis ist unbekannt.

Ähnlich liegt das bei der Formiatbildung (*243*) aus Glyoxylsäure (*194*) oder aus Oxalsäure (*276*) durch das Enzym Oxalat-decarboxylase (*243 a*). Über die Bildung von Formiat beim Abbau heterocyclischer Moleküle wird im Zusammenhang mit den daran beteiligten Folat-Enzymen berichtet (S. 246).

2. Glycin als Quelle von Einkohlenstoff-Körpern.

Obwohl Glycin zu den am längsten bekannten Aminosäuren gehört, gelang es doch erst in neuerer Zeit, bessere Einsicht in seine Umwandlungen zu gewinnen. Wichtig war dabei die Erkenntnis, daß der Glycin-Stoffwechsel mit dem von Serin eng verbunden ist. Der reversible Übergang von Glycin zu Serin (*266, 275*) hat also eine große biochemische Bedeutung.

Während der Abbau beider Aminosäuren je nach der Stoffwechsellage verschiedene Wege einschlagen kann (*80*), dürfte die Neusynthese von Glycin fast ausschließlich über Serin erfolgen (*11*). Glycin wird durch Transaminierung oder oxydative Desaminierung in Glyoxylat umgewandelt (*328*). Dies kann dann oxydativ durch ein Mitochondrien-Enzym in Formiat und CO_2 gespalten werden (*208*). Dazu sind als Cofaktoren DPN, TPP, Mg^{++} und Glutaminsäure erforderlich. Als Produkte wurden CO_2 und N-Formylglutamat isoliert, das nun direkt oder indirekt gespalten wird (*208*). Auch δ-Aminolävulinsäure, die ja ein Abkömmling von Glycin ist, soll mit dem Formiat-Reservoir der Zelle im Gleichgewicht stehen. Die Spaltung wird über α-Ketoglutaraldehyd angegeben, der sogar — reversibel (?) — zu Succinat und Formyl gespalten wird (*331*). In Kaninchen-Reticulocyten verdünnt δ-Aminolävulinat im Gegensatz zu Formiat oder Oxalat das aus dem $C_{(1)}$ von markiertem Glycin entstehende $*CO_2$ (*314 a*). Meerschweinchen-Leberschnitte dagegen markieren Glutamin aus Glycin-2-*C in der Weise, wie sie für die Bildung eines markierten Kohlenstoff-Fragments zu erwarten ist, das sich mit Glycin zu Serin wieder kondensiert und so 2,3-markiertes Pyruvat entstehen läßt (*143*).

Die Umwandlung der Methylengruppe von Glycin in Einkohlenstoff-Einheiten auf dem Weg Glycin $\rightarrow$ Glyoxylat $\rightarrow$ Formiat ist zuerst von NAKADA (*209*) und von WEINHOUSE (*328*) aus Isotopen-Verdünnungsversuchen erschlossen worden. FLEMING und CROSBIE (*84*) weisen jedoch darauf hin, daß der Beweis nicht schlüssig ist, denn bereits nichtenzymatisch kann es in Gegenwart z. B. von Cu^{++}-Ionen zu einer Desaminierung von Glycin zu Glyoxylat kommen. Gegen den postulierten Weg spricht auch, daß eine Transaminierung von Glutaminsäure auf Glyoxylat (*241*) oder der Übergang des α-C in die Ureidpositionen der Purine nicht nachgewiesen werden kann (*240*). Die Einführung des α-C von Glycin in den Stoffwechsel erfolgt hier also nicht über Serin (*65*). Dagegen wurde in Extrakten der Glycin-Serin-Mangelmutanten *E. coli* PA/15 eine Reaktion gefunden, durch die, FH_4- und Pyridoxal-abhängig, Glycin in CO_2 (aus $C_{(1)}$) und Formaldehyd (aus $C_{(2)}$) gespalten wird (*240*), der in die Reaktionen der Methylen-FH_4 eingeht. Eine gleichartige Reaktion in löslichen Enzymen aus den auf Glycin wachsenden Mikroorganismen *Cl. acidi-urici* und *Diplococcus glycinophilus*, durch die Glycin zu Ammoniak, CO_2 und Acetat abgebaut wird, beschrieben Sagers und Gunsalus (*263*). Auch hier entsteht unter Umgehung von Glyoxylsäure aus dem α-C in DPN- und Pyridoxal-phosphat-abhängiger Reaktion zunächst FH_4-gebundener Formaldehyd, daraus in den bekannten Wegen über aktive Ameisensäure Formiat, entsprechend *Schema 1*.

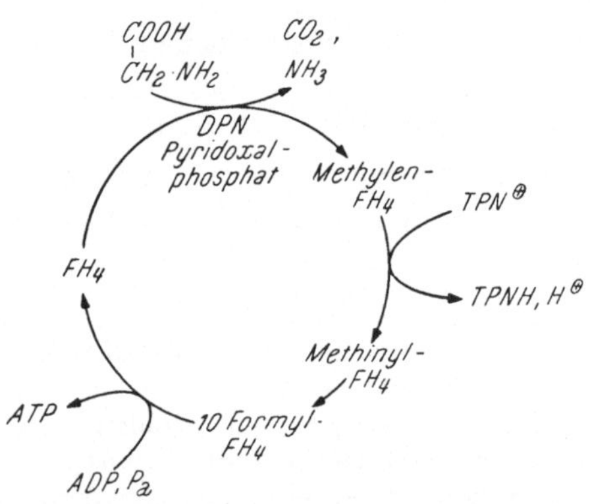

Schema 1. Glycin-Abbau nach Sagers und Gunsalus (*263*).

3. Der Einkohlenstoff-Donator Serin.

Zweifellos die wichtigste physiologische Quelle der Einkohlenstoff-Einheiten ist das β-Kohlenstoff-Atom von Serin. Seine leichte Zugänglichkeit und leichte Nutzung machen es zur Hauptquelle der biologischen C_1-Einheiten. Man kann diese Gruppe geradezu als einen gebundenen und entgifteten Formaldehyd auffassen. Sie stammt letzten Endes aus den Kohlenstoff-Atomen $C_{(1)}$ und $C_{(6)}$ von Glucose, denn das Dreikohlenstoff-Gerüst wird von einer Verbindung geliefert, die sich von den Triosen der Glykolyse ableitet (*11*). Wahrscheinlich ist Hydroxypyruvat oder 3-Phospho-hydroxypyruvat direkter Vorläufer (*133 a, 166 a*).

Zuerst zeigte Shemin (*275*), daß bei der Spaltung von Serin HCHO und Glycin entstehen, deren spezifische Markierung in $C_{(1)}$ und $C_{(2)}$ unverändert war. Lascelles und Woods (*173*) fanden die Teilnahme von Pyridoxalphosphat und Folsäure. Schließlich wurde in verschiedenen

Arbeitskreisen der Mechanismus und die Bindung des entstehenden Formaldehyds an FH_4 als Transportmetaboliten aufgeklärt.

Ein weiterer Weg, der zu dieser Verbindung führt, ist die oxydative Umlagerung von Sarkosin in Serin und anschließende Aldol-Spaltung (*178*, *179*). Die erste Reaktion erfordert eine partikuläre Oxydase, die durch Digitonin in Lösung gebracht werden konnte, und ist besonders auch als Modell der für die Pharmakologie so bedeutsamen oxydativen Demethylierung von heterocyclisch gebundenen Methylgruppen von Bedeutung. Man nimmt für den Abbau von Sarkosin den in *Formel-*

$$\begin{array}{c} COOH \\ | \\ CH_3HN \cdot CH_2 \end{array}$$

Sarkosin.

$$\begin{array}{c} COOH \\ | \\ H_2N \cdot CH_2 \end{array} \qquad [HCHO] \rightarrow CH_2O$$

Glycin. Formaldehyd.

$$\begin{array}{c} COOH \\ | \\ H_2N \cdot CH \\ | \\ CH_2OH \end{array}$$

Serin.

Formelübersicht 11. Abbau von Sarkosin nach MACKENZIE und ABELES (*179*).

übersicht 11 skizzierten Mechanismus an, kennt aber weder die Natur des Formaldehyd-Zwischenprodukts noch des Elektronenacceptors bei diesem Oxydationsvorgang. Das Reaktionsschema erklärt jedoch die beobachteten Austausch- und Verdünnungs-Reaktionen recht gut. Ameisensäure tritt hier intermediär nicht auf.

In gleicher Weise wie Sarkosin wird nach MACKENZIE (*180*) auch Dimethylglycin zunächst zu Formaldehyd und Sarkosin oxydiert und dann auf dem gezeigten Weg weiter umgewandelt, so daß die beiden Methylgruppen in das Einkohlenstoff-Reservoir gelangen.

4. Herkunft der Methylgruppe.

Die labilen Methylgruppen werden aus Vorläufern der „aktiven" C_1-Einheiten gebildet (*44*). So können Extrakte aus Leber (*19*, *210*, *211*) oder Coli-Bakterien (*116*) den Kohlenstoff aus Formiat, Formaldehyd oder dem β-C von Serin zur Überführung von Homocystein in Methionin verwenden. Gerade diese letzte Reaktion, in der die CH-Bindungen der Hydroxymethylgruppe intakt bleiben (*12*, *177*), hat gezeigt, daß dabei FH_4 als Cofaktor dient (*106*, *145*), aber auch Vitamin B_{12} beteiligt ist (*298*). Bei Pflanzen dagegen scheinen Formaldehyd und der Methylen-Kohlenstoff von Glycin im Stoffwechsel zur Methylbildung bevorzugt zu werden (*43*).

Tabelle 6. Folat-Cofaktor-abhängige

Nr.	Enzym	System-Namen*	System-Nr.*	Literatur
1.	Glycin-Formimino-transferase	N-Formiminoglycin: FH_4-5-Iminotransferase	2.1.2.4	(*247 a, 262*)
2.	Glutamat-Formimino-transferase	N-Formimino-*L*-glutamat: FH_4-5-Iminotransferase	2.1.2.5	(*307*)
3.	Cyclodesaminase (Formimino-FH_4-Cyclodesaminase)	5-Formimino-FH_4-Ammoniaklyase (cyclisierend)	4.3.1.4	(*307*)
4.	Glycinamid-ribotid-(GAR)-Transformy-lase (5-Phospho-ribosylglycinamid-Formyltransferase)	5-Phosphoribosyl-N-formylglycinamid: FH_4-Formyltransferase	2.1.2.2	(*96, 111*)
5.	Cyclohydrolase			(*307*)
6.	Tetrahydrofolat-Formy-lase (Formyl-tetra-hydrofolat-Synthetase)	Formyl-FH_4-Ligase (ADP)	6.3.4.3	(*103, 249a, 341*)
7.	Aminoimidazol-carboxamid-**ribotid** (AICAR)-Trans-formylase (5'-Phosphoribosyl-aminoimidazol-carboxamid-Transformylase)	5'-Phosphoribosyl-5-formamido-4-imidazol-carboxamid: FH_4-10-Formyltrans-ferase	2.1.2.3	(*96, 111*)
8.	Formyl-tetrahydro-folat-Deformylase (Formyl-FH_4-Deacylase)	10-Formyl-FH_4-Amido-hydrolase	3.5.1.10	(*221*)
9.	Formyl-tetrahydro-folat-Reduktase (Methylen-FH_4-Dehydrogenase)	5,10-Methylen-FH_4: NADP-Oxydoreduktase	1.5.1.5	(*113, 135, 250*)
10.	Thymidylat-Synthetase			(*89, 128, 192*)
11.	Dihydrofolat-Reduktase	FH_4: NAD(P)-Oxydoreduktase	1.5.1.3	(*34, 92, 223, 365*)
12.	Folat-Reduktase	FH_2: NAD(P)-Oxydoreduktase	1.5.1.4	(*365*)
13.	Desoxycytidylat-Trans-hydroxymethylase			(*83 a*)
14.	Serin-Aldolase (Serin-Desmolase, Serin-Trans-hydroxymethylase)	*L*-Serin: FH_4-5,10-Hydroxymethyl-trans-ferase	2.1.2.1	(*4, 27, 135, 164*)

* System-Namen und -Nummern nach: Report of the Commission on Enzymes of the International Union of Biochemistry. London: Pergamon Press. 1961.

Literaturverzeichnis: SS. 254—274.

Enzym-Reaktionen.

Reaktions-Gleichung	Gleichungs-Nr.	Seite
$FH_4 + R{-}NH \cdot CH({=}NH) \rightleftharpoons 5\text{-}CHNH\text{-}FH_4 + RNH_2$ $R = CH_2{-}COOH$	[24]	249
$R = HOOC{-}(CH_2)_2{-}CH{-}COOH$ $\mid$	[17]	247
$5\text{-}CHNH\text{-}FH_4 + H^+ \rightarrow 5,10\text{-}CH^+{=}FH_4 + NH_3$	[18]	247
$5,10\text{-}CH^+{=}FH_4 + GAR \rightarrow FH_4 + CHO\text{-}GAR$	[32]	252
$5,10\text{-}CH^+ = FH_4 + H_2O \rightleftharpoons 10\text{-}CHO\text{-}FH_4 + H^+$	[19]	247
$10\text{-}CHO\text{-}FH_4 + ADP + H_3PO_4 \rightleftharpoons FH_4 + HCOOH + ATP$	[25]	249
$10\text{-}CHO\text{-}FH_4 + AICAR \rightleftharpoons FH_4 + CHO\text{-}AICAR$	[30]	251
$10\text{-}CHO\text{-}FH_4 + H_2O \rightarrow FH_4 + HCOOH$	[20]	248
$10\text{-}CHO\text{-}FH_4 + TPNH + H^+ \rightleftharpoons 5,10\text{-}CH_2\text{-}FH_4 + TPN^+ + H_2O$ $(5,10\text{-}CH^+{=}FH_4 + TPNH \rightleftharpoons 5,10\text{-}CH_2\text{-}FH_4 + TPN^+)$	[13]	239
$5,10\text{-}CH_2\text{-}FH_4 + \text{Desoxy-UMP} \rightarrow TMP + FH_2$	[14]	241
$FH_2 + TPNH + H^+ \rightleftharpoons FH_4 + TPN^+$	[4]	197
$F + TPNH + H^+ \rightleftharpoons FH_2 + TPN^+$	[3]	197
$5,10\text{-}CH_2\text{-}FH_4 + \text{Desoxy-CMP} + H_2O \rightleftharpoons$ $\rightleftharpoons 5\text{-}CH_2OH\text{-}desoxy\text{-}CMP + FH_4$	[11]	239
$5,10\text{-}CH_2\text{-}FH_4 + \text{Glycin} + H_2O \overset{(PyrP)}{\rightleftharpoons}$ $\rightleftharpoons \text{Serin} + FH_4$	[10]	239

(Fortsetzung S 236)

(Fortsetzung der Tabelle 6)

Nr.	Enzym	System-Namen*	System-Nr.*	Literatur
15.	Methylen-tetrahydro-folat-Reduktase			(78, 137, 171)
16.	Methionin-Synthetase-System			(104, 112, 137, 268)
17.	Formyl-Transferase			(286, 282)
18.	Cyclodehydrase			(231)
19.	Folinsäure-Isomerase			(101)

VI. Folat-katalysierte Enzym-Reaktionen.

Die allgemeinen Stoffwechsel-Funktionen der Folsäure hatte man zuerst bei Wuchsstoff-Untersuchungen an Mikroorganismen erkannt. Besonders bedeutungsvoll war das Auffinden formylierter Folsäure-Derivate [Gordon et al. (98)] und die Anhäufung von Stoffwechsel-Zwischenprodukten, deren Verwandtschaft mit um ein Kohlenstoff-Atom reicheren kompletten Stoffwechsel-Bausteinen augenfällig ist, wie z. B. die Beziehung des bei Sulfonamidblockierung entstehenden Amino-imidazolcarboxamids zu den Purinen (346). Die spezifische Rolle der Folsäure-Cofaktoren, nämlich der Derivate der Tetrahydro-folsäure, klärten dann in vitro-Untersuchungen (252) und die Reinigung der katalysierenden Folat-Enzyme. Im folgenden Abschnitt soll das Zusammenspiel der verschiedenen Folat-abhängigen Enzyme im Stoffwechsel behandelt werden. Wir wollen bei dieser Betrachtung davon ausgehen, daß Tetrahydro-folsäure in der lebendigen Substanz ebenso wie die übrigen Coenzyme katalytisch wirkt. Es müssen also im Leben alle Vorgänge, die zu einer „Beladung" des freien Cofaktors durch einen Einkohlenstoff-Rest führen, durch Prozesse kompensiert werden, in denen freie Tetrahydro-folsäure wieder zurückgebildet wird.

Tabelle 6 (S. 234) gibt zunächst eine Übersicht über die verschiedenen Folat-Enzyme und die von ihnen katalysierten Gesamt-Reaktionen.

1. Der Transhydroxymethylierungs-Cyclus.

a. *Serin-Aldolase.*

Die enzymatische Umwandlung von Serin in Glycin ist wegen der Bedeutung sowohl im Ameisensäure-Stoffwechsel wie der Einkohlenstoff-Einheiten eingehend untersucht worden (202, 275, 297). Isotopenversuche, in denen die Deutero-markierte β-Gruppe von Serin ohne Deuterium-verlust in Methylgruppen eingebaut wurde, und Verdünnungsversuche bewiesen klar, daß der unmittelbare Vorläufer nicht freier Formaldehyd

Reaktions-Gleichung	Gleichungs-Nr.	Seite
$5,10\text{-}CH_2\text{-}FH_4 + FADH_2 \underset{\text{(Menadion)}}{\overset{\text{(DPNH, H}^+\text{)}}{\rightleftarrows}} 5\text{-}CH_3\text{-}FH_4 + FAD$	[15]	242
$5\text{-}CH_3\text{-}FH_4 + \text{Homocystein} \xrightarrow{\text{(Vit. B}_{12}\text{)}} \text{Methionin} + FH_4$	[16]	243
$5\text{-}CHO\text{-}FH_4 + \text{Glutamat} \rightleftharpoons FH_4 + N\text{-Formyl-glutamat}$	[22]	248
$5\text{-}CHO\text{-}FH_4 + H^+ \overset{\text{ATP}}{\rightleftharpoons} 5,10\text{-}CH^+ = FH_4 + H_2O$		
$5\text{-}CHO\text{-}FH_4 + \text{ATP} \rightleftharpoons 10\text{-}CHO\text{-}FH_4 + \text{ADP} + H_3PO_4$	[21]	248

oder freie Ameisensäure, sondern ein gebundener C_1-Körper ist, der aus beiden gebildet werden kann (*297*). Die Teilnahme der Folsäure, bereits aus Fütterungsversuchen an Tieren (*11*) und aus Wachstumsversuchen an Mikroorganismen geschlossen, wurde zuerst von WELCH und NICHOL (*330*) postuliert und fast gleichzeitig von verschiedenen Seiten in Zellextrakten nachgewiesen (*25*, *135*, *159*, *163*, *164*). Als Zwischenprodukt wurde ein Kondensationsprodukt aus Formaldehyd und FH_4 isoliert (*67*, *135*), dessen Struktur jetzt allgemein als Methylen-FH_4 angesehen wird. Die Verbindung wurde synthetisch dargestellt (*28*, *29*, *135*, *160*) (Gl. [8]) und erwies sich als aktiv in der Übertragungs-Reaktion auf Glycin nach Gl. [9].

$$CH_2O + FH_4 \rightleftharpoons CH_2 \cdot FH_4 \qquad (K = 1,3 \times 10^4). \qquad [8]$$

$$CH_2 \cdot FH_4 + \text{Glycin} \underset{\text{phosphat}}{\overset{\text{Pyridoxal-}}{\rightleftarrows}} \text{Serin} + FH_4 \qquad (K = 2,76 \times 10^3). \qquad [9]$$

$$CH_2O + \text{Glycin} \underset{\substack{\text{Pyridoxal-}\\\text{phosphat}}}{\overset{FH_4}{\rightleftarrows}} \text{Serin}. \qquad [10]$$

Über ein, Reaktion [8] auch bei großer Verdünnung katalysierendes „Formaldehyd-aktivierendes" Enzym aus Taubenleber-Acetonpulver berichten OSBORN und Mitarb. (*225*). Seine physiologische Rolle ist nicht durchaus einzusehen.

Die die Reaktion [10] katalysierende Serinaldolase (Serindesmolase oder Serin-Transhydroxymethylase) ist aus Leber angereichert worden (*4*, *135*, *164*). Sie ist spezifisch für *L*-Serin und das biologische Diastereomere der FH_4 und besonders dadurch ausgezeichnet, daß sie mit zwei Cofaktoren arbeitet: Serinaldolase ist ein Pyridoxal-Enzym (*4*, *29*, *135*, *164*). Dieses Cofaktor-Bedürfnis war auf Grund der Vorstellung über die Spaltungs-Reaktionen von Aminosäuren vorauszusehen (*195*). Die α,β-Bindung wird durch die Bildung der Schiffschen Base mit dem Pyridoxal-Enzym gelockert, und gleichzeitig greift das Elektronenpaar eines Stickstoffs von FH_4 am β-C an, so daß bei diesem push-pull-Prozeß*, wie ihn *Formel-übersicht 12* darstellt, schließlich Methylen-FH_4 und Pyridoxal-gebundenes Glycin entstehen.

Formelübersicht 12. Reaktionsmechanismus der Serinhydroxymethylase.

Dieser Mechanismus läßt sich im Modell reproduzieren (*46*): Während Serin bei pH 5,5 in Gegenwart von Pyridoxal oder dem FH_4-Modell Diphenyl-äthylen-diamin und Al^{+++}-Ionen praktisch stabil ist, zerfällt es in Gegenwart beider Co-faktoren in beträchtlichem Ausmaß unter Freisetzung von Glycin. Der α-Wasser-stoff bleibt dabei unangetastet, da er an der Reaktion nicht teilnimmt, denn diese erfolgt, wie aus der obigen Formulierung zu erkennen ist, als push-pull-Prozeß lediglich am β-Kohlenstoff, was auch die Entdeckung einer spezifischen α-Methyl-serin-Hydroxymethylase durch Wilson und Snell (*348a*) bestätigt.

Die Kinetik des Enzyms wird spektrophotometrisch (*32*) oder mano-metrisch nach Perjodatoxydation (*27*) oder nach Koppeln mit Methylen-FH_4-Dehydrogenase und Formiatdehydrogenase (*123*) gemessen. Die in dem Bakterium *Cl. sticklandii* HF vorkommende Serinaldolase (*362*) ist insofern besonders zu erwähnen, als sie an Stelle der Pteroyl-mono-glutamate für die katalytische Funktion neben anorganischem Phosphat ein Pteroyl-polyglutamat braucht. Diese Klasse von Cofaktoren, CoC genannt (*358, 359*), sind teils Pteroylglutamate mit 1, 2 oder 5 Mol Glutaminsäure und anderen Aminosäuren, teils Derivate der $N_{(5)}$-Formyl-FH_4. Hier wurde das erstemal erkannt, daß Polyglutamate dort eine kata-lytische Funktion ausüben können, wo Monoglutamate in Substrat-mengen erforderlich sind.

---

* Als push-pull-Reaktion (Stoß-Zug-Reaktion) bezeichnet man eine Substitution, bei der der Eintritt des Substituenten durch gleichzeitigen Angriff eines dritten Reaktionspartners (Lösungsmittel, Gruppe des Enzyms usw.) auf die austretende Gruppe erleichtert wird. Im allgemeinen handelt es sich um nucleophile Substitutions-mechanismen; in diesem Fall hat der ziehende Reaktionspartner die Eigenschaften einer Lewis-Säure.

b. Transhydroxymethylierungs-Reaktionen.

Die Bildung von Serin durch Transhydroxymethylierung auf Glycin hat im Stoffwechsel nur geringe Bedeutung. Eine unentbehrliche Transhydroxymethylierungs-Reaktion ist hingegen die Wirkung der Desoxy-CMP-Hydroxymethylase (*234*), die in *E. coli*-Zellen erst nach Infektion mit $T6_r^+$-Phagen induziert wird (*169*). Das in sehr guter Ausbeute aus solchen Zellen 40fach angereicherte Enzym (*83 a*) bewirkt die Reaktion Gl. [11] (d = desoxy-):

$$\text{Methylen-FH}_4 + \text{dCMP} + \text{H}_2\text{O} \rightarrow \text{5-CH}_2\text{OH-dCMP} + \text{FH}_4. \quad [11]$$

Auch Wasser kann der Acceptor sein, und es entsteht dann freier Formaldehyd. Dieser kann nach Untersuchungen von Batt (*18*) auf aktivierten Glykolaldehyd übertragen werden, wobei Dihydroxyaceton erhalten wird. Es wäre dies ein Weg, auf dem die C_1-Körper in das C_2 der Essigsäure nach den Gl. [10] und [12] eingehen könnten:

$$\text{CH}_2\text{OH} \cdot \text{CHNH}_2 \cdot \text{COOH} \xrightarrow[\text{Pyr P}]{\text{FH}_4} \text{CH}_2\text{NH}_2 \cdot \text{COOH} + \text{CH}_2\text{O}. \quad [10]$$

$$\text{CH}_2\text{O} + \text{CH}_2\text{OH} \cdot \text{CO} \cdot \text{COOH} \rightarrow \text{CH}_2\text{OH} \cdot \text{CO} \cdot \text{CH}_2\text{OH} + \text{CO}_2. \quad [12]$$

c. Methylen-tetrahydrofolat-Dehydrogenase.

Die Isotopenuntersuchungen am Gesamtorganismus hatten erkennen lassen, daß die β-Gruppe von Serin mit Formiat im Gleichgewicht steht (*275*). Im zellfreien System haben dann Blakley (*25*), Sakami (*164*) und Jaenicke (*134, 103*) die Bildung von Serin aus Formyl-FH_4 und Glycin in Gegenwart von reduzierten Pyridinnucleotiden beobachtet und gefolgert, daß die enzymatische Reduktion von Folat-gebundenem Formiat zu Folat-gebundenem Formaldehyd der Stoffwechselweg ist, durch den sich die früheren Beobachtungen erklären lassen, in denen der Formiat-Kohlenstoff in zahlreichen Substanzen gefunden wurde, die die Kohlenstoffeinheit auf einer weiter reduzierten Stufe enthalten, oder in denen umgekehrt Serin-Hydroxymethyl in Formylgruppen vorkommt. Das von Jaenicke (*135*) aufgefundene und von Huennekens (*113, 222*) und Blakley (*250*) weit gereinigte, für die Dehydrierung verantwortliche Enzym Hydroxymethyl-tetrahydrofolat-Dehydrogenase (Formylfolat-Reduktase) katalysiert die reversible Reaktion (Gl. [13]):

$$\text{Methylen-FH}_4 + \text{TPNH}^+ \rightleftharpoons \text{10-Formyl-FH}_4 + \text{TPNH} + \text{H}^+$$
$$(\text{K} = 1,7 \times 10^{-2}). \quad [13]$$

Das Enzym aus normalen Zellen ist stets TPN^+-spezifisch; in Ehrlich-Asciteszellen dagegen findet sich ein entsprechendes DPN^+-Enzym (*271*). Bei dieser Oxydoreduktion wird die Oniumstruktur des Pyridinnucleotids vermutlich zunächst im Imidazolinium-Ring von Anhydroformyl-FH_4 bewahrt, denn die Oxydation von entsprechenden Modell-Imidazolidinen zu Imidazolinium-Verbindungen verläuft mit einem scharfen isosbestischen Punkt, also ohne Zwischenstufe (*141*). Entsprechend fand Huennekens (*222*), daß Cyclohydrolase-freie Dehydrogenase 10-Formyl-FH_4 sehr viel langsamer als Anhydro-Formyl-FH_4 reduzieren kann. Der Nachweis dieser Zwischenstufe gelingt spektrophotometrisch in Maleat-Puffer.

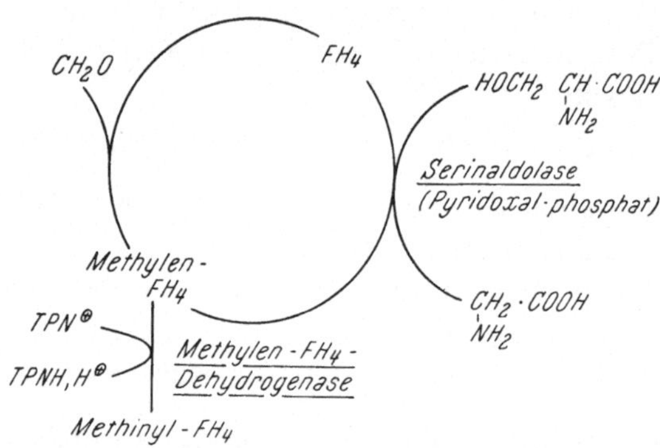

Schema 2. Enzymatische Reaktionen der Methylen-tetrahydro-folsäure (Transhydroxymethylierungs-Cyclus).

„Aktives" Formiat und, nach Isomerisierung, auch Citrovorumfaktor (*230*) können auf diese Weise zum Hydroxymethylgruppen-Reservoir beisteuern. *Schema 2* zeigt die besprochenen Reaktionen in einem cyclischen System.

2. Methylgruppen-Genese.

a. Thymidylat-Bildung.

Aus dem Hydroxymethylgruppen-„Pool" stammt aber auch die Methylgruppe von Thymin und Methionin. Auch diesen Weg haben bereits die ernährungsphysiologischen Untersuchungen gewiesen, während aus Isotopenverdünnungs-Versuchen zugleich erkannt wurde, daß beide Methylgruppen nicht unmittelbar ineinander umgewandelt werden können (*99*), sondern offenbar auf unabhängigen Wegen entstehen.

Die Methylgruppe von Thymin stammt bei *E. coli* in einer Tetrahydrofolat und TPNH benötigenden Reaktion aus dem β-Kohlenstoff von Serin oder dem α-Kohlenstoff von Glycin, die in den Versuchen von Crosbie (*24, 65*) im Stoffwechsel nichts miteinander zu tun haben. Formiat ist nur ein schlechter, das Methionin-Methyl überhaupt kein Vorläufer dieser Gruppe. Jedoch findet sich bei *P. cerevisiae* die Formylgruppe von aktivierter Ameisensäure im Thymin-Methyl (*329*).

Lösliche Extrakte, die die Umwandlung von Desoxyuridin-nucleotid in Thymidylsäure katalysieren, finden sich in Thymusdrüse (*128*), Leukozyten (*281*), *E. coli* (*89*) und *St. faecalis* (*192*). Nach Infektion von Thymin-mangel-Mutanten von *E. coli* mit T_2- oder T_5-Bakteriophagen steigt dieses Enzym ganz außerordentlich an (*17, 83*). Die Thymidylat-synthetisierenden Extrakte benötigen außer Serin als Donator und Desoxy-uridylsäure als Acceptor der Kohlenstoffeinheit FH_4, ATP, Mg^{++} und DPNH (*88, 89*). Aus ihnen wurde die Thymidylat-Synthetase angereichert (*35, 88a, 100, 321a*). Die FH_4 und der Einkohlenstoff-Donator können auch durch synthetische Methylen-FH_4 ersetzt werden. Schließlich hat MALEY (*182*) auch Uridylsäure als Acceptor verwenden können.

Die häufiger untersuchte Frage nach der Beteiligung von Vitamin B_{12} an der Thymidylsäure-Bildung (*70, 117*) ist nun dahin entschieden, daß es die Umwandlung der Ribose in die Desoxyribose des Acceptors katalysiert (vgl. *257b*).

Bereits HUMPHREYS und GREENBERG (*128*) haben aus der stöchiometrischen Bilanz zwischen Thymidin-Bildung und freigesetzter FH_2 sowie dem Ersatz katalytischer Mengen DPNH durch Substratmengen FH_4 und umgekehrt geschlossen, daß in der Thymidylat-Synthetase-Reaktion Tetrahydrofolat gleichzeitig als Träger der Kohlenstoffeinheit und als Wasserstoff-Donator für die Reduktion der Hydroxymethyl-gruppe dient (*190*). Tatsächlich gelang es WAHBA und FRIEDKIN (*321*), die Bilanz der Gesamtreaktion Gl. [14] (d = desoxy-):

$$\text{dUMP} + \text{Methylen-}FH_4 \rightarrow \text{TMP} + FH_2. \qquad [14]$$

zu beweisen. Aus ^{32}P-markierter dUMP wurde je Mol ^{32}P-TMP ein Mol FH_2 spektrophotometrisch gemessen. Diese wurde auch chromatographisch isoliert und durch Spektrum und angeschlossene enzymatische Reduktion identifiziert. Reduzierte Folat-Antagonisten hemmen die Dihydrofolat-Bildung (*191*). Durch praktisch gleichartige Ergebnisse mit *St. faecalis* und Thymus-Extrakten bestätigen BLAKLEY und Mit-arb. (*35, 191, 344*) diese Ergebnisse. Sie konnten dabei auch die Thymidylat-Synthetase besonders aus *St. faecalis* R-Extrakten 100- bis 600fach anreichern (*35*).

Im Mechanismus der enzymatischen Reduktion und Übertragung haben FLAKS und COHEN (*83*) ein 5-Hydroxymethyl-dihydrodesoxy-uridylat ausgeschlossen. Kinetische Untersuchungen von McDOUGALL und BLAKLEY (*190*) zeigten weiter, daß ein Zwischenstoff, etwa der Konstitution (XXIX) (S. 242) (*83*), sich nicht anhäuft, denn es fehlt die dabei zu erwartende Änderung der Reaktionsgeschwindigkeit. Die spektralen Änderungen zeigen zwei gut definierte isosbestische Punkte. Markiert man dagegen die FH_4 biologisch in Stellung 6 und 7 mit Tritium, so findet sich am Ende der Thymin-Bildung das Tritium in der Thymin-Methyl-gruppe wieder (*225a*). Man muß daher annehmen, daß in dem Reaktions-knäuel der Hydridwasserstoff aus dem Tetrahydropyrazin-Ring in die polarisierte Methylengruppe wandert und sich die Elektronen dann etwa in der in *Formelübersicht 13* gezeigten Weise neu ordnen (*138, 139*).

$$+ \text{5,10-Methylen-FH}_4 \longrightarrow$$

dUMP.

$$\rightarrow \quad \text{(XXIX.)} \quad \rightarrow \quad \text{CH}_3 \quad + \text{7,8-FH}_2 \quad \text{(XI.)}$$

Formelübersicht 13. Thymidylatsynthese:
Methylgruppen-Bildung durch Hydridverschiebung.

b. Methylen-tetrahydrofolat-Reduktase: 5-Methyl-tetrahydro-folsäure.

Ein wirkliches Zwischenprodukt dagegen tritt bei der Folat-katalysierten Bildung der Methylgruppe von Methionin aus Hydroxymethylgruppen auf. Anzeichen dafür ergaben sich aus Präinkubationsversuchen im Methionin-synthetisierenden System aus Schweineleber-Homogenaten. Diese bilden aus Methylen-FH$_4$ in Gegenwart von einem reduzierenden System eine Verbindung, die isoliert (*137*) und zunächst als Methyl-dihydrofolsäure angesprochen wurde (*348*). Auch Larrabee und Buchanan (*171, 171a*) fanden ein solches Zwischenprodukt mit gleichen Eigenschaften in Methionin-bildenden Coli-Extrakten. Sie erkannten sie als Methyl-FH$_4$. Ihre Wuchsstoffeigenschaften erwiesen sich identisch mit der Präfolsäure A von Donaldson und Keresztesy (*75—77*). Die gleiche, aber racemische Verbindung erhält man durch Boranat-Reduktion der Methylen-FH$_4$ (S. 219).

Die Methylenfolat-Reduktase, die zuerst von Keresztesy (*78*) beschrieben wurde und vermutlich mit dem von Hatch (*112*) aus der Coli-Mutanten 205-2 erhaltenen „205-2"-Enzym identisch ist, benötigt für die Reduktion der Methylen-FH$_4$ FAD nach Gl. [15].

$$\text{Methylen-FH}_4 + \text{FADH}_2 \underset{\text{Menadion}}{\overset{\text{DPNH, H}^+}{\rightleftharpoons}} \text{5-Methyl-FH}_4 + \text{FAD}. \quad [15]$$

Dementsprechend kommt das eintretende H-Atom als Proton aus dem Medium: Kisliuk (*162a*) fand nach enzymatischer Synthese von Methionin in D$_2$O *ein* Atom Deuterium in der Methylgruppe. 5-Methyl-

FH_2 (S. 220) ist kein Substrat der Reaktion Gl. [15] (*78a*); in Übereinstimmung damit geht Deuterium aus 5,6,7,8-deuterierter FH_4 nicht in Methionin über (*162a*), wie eine Umlagerung von Methylen-FH_4 in Methyl-FH_2 hätte erwarten lassen. Zu gleichen Ergebnissen kam auch BUCHANAN bei Untersuchung der Reaktion Gl. [15] in T_2O (*55a*). Die Verwertung von Methyl-FH_2 als Methionin-Methyl-Donator (*258*) durchläuft vermutlich ihre nichtenzymatische Reduktion zu Methyl-FH_4 durch die SH-Gruppe des Acceptors Homocystein (*78a*).

Aus $^{14}C,^{3}H$-methylmarkierter 5-Methyl-FH_4 geht die Methylgruppe unverändert in die weiteren Stufen der Methionin-Synthese ein (*162a*) und wird an die Sulfhydrylgruppe von Homocystein gebunden. In Umkehr der Reaktion Gl. [15] wird Methyl-FH_4 durch eine FAD-abhängige Oxydase zu FH_4 und Formaldehyd oxydiert, wobei Menadion oder ein anderes Chinon Wasserstoffacceptor sein können. Die Oxydation wird durch DPNH gehemmt, während umgekehrt die Bildung von Methyl-FH_4 durch Menadion vermindert wird (*78*), was sich aus Gl. [15] ergibt und den unterschiedlichen Redoxpotentialen der für Hin- und Rückreaktion spezifischen Wasserstoff-Überträger (DPN < FAD < Menadion).

c. Methionin-Bildung.

α. Die Gesamt-Reaktion. Wie beim Thymin stammt auch die Methylgruppe von Methionin aus dem Einkohlenstoff-Reservoir des Stoffwechsels, dem β-Kohlenstoff von Serin oder dem α-Kohlenstoff von Glycin und geht von dort durch Transmethylierungen (*57, 319*) in die Methylgruppen von Cholin und seinen Abkömmlingen, sowie verschiedene andere N- oder C-gebundene Methylgruppen ein.

Die Synthese von Methionin aus Formiat und Homocystein im zellfreien System gelang zuerst BERG (*19*). Dialysierte Extrakte, die dann auch aus mehreren anderen Quellen gewonnen wurden, sind abhängig von Tetrahydrofolat, ATP, Mg^{++} und TPNH (*74, 210, 211, 299, 348*) und werden durch Methionin stimuliert (*211, 299*). Während zuerst sehr komplexe Systeme untersucht wurden, in denen man beispielsweise Serin als mittelbaren Einkohlenstoff-Donator verwendete, vereinfachte der Ersatz von Serin und FH_4 durch synthetische Methylen-FH_4 die Ansätze (*348*). Es konnte aber gezeigt werden, daß S-Hydroxymethyl-homocystein nicht das Zwischenprodukt ist, das dann anschließend reduziert wird, wie man zunächst annahm (*19, 74*), sondern, daß an der Reaktion ein neues Zwischenprodukt, die 5-Methyl-FH_4, beteiligt ist (vgl. S. 242). Die weitere Übertragung dieser Methylgruppe nach Gl. [16]

$$\text{Homocystein (oder Derivat)} + \text{Methyl-FH}_4 \xrightarrow[\text{(113-3)}]{\text{Vit. B}_{12}}$$

$$\rightarrow \text{Methionin (oder Deriv.)} + \text{FH}_4 \qquad [16]$$

ist ein zur Zeit besonders in Fluß geratenes Gebiet des Stoffwechsels von Kohlenstoffeinheiten. Am besten sind die Verhältnisse bei der Synthese von Methionin in *E. coli* und einigen Methionin-losen Mutanten durch die Arbeiten von Guest und Woods (*108*) und von Buchanan (*112*) bekannt, obwohl auch hier die genaue Natur der katalytischen Schritte und die Verteilung der Funktionen zwischen den verschiedenen Cofaktoren noch nicht genügend definiert sind.

β. Zusammenhänge zwischen Folsäure und Vitamin B_{12}. Seit den Beobachtungen von Helleiner und Woods (*116*) war es bekannt, daß die Synthese von Methionin in *E. coli* außer Folat-Cofaktoren auch Vitamin B_{12} erfordert (*165*). Dieser Protein-Cobalamin-Faktor kann in *E. coli*-Extrakten durch Inkubation von Cobalamin, ATP, Mg^{++} und DPN gebildet (*107*) oder direkt aus auf Vitamin B_{12} gewachsenen Zellen gereinigt werden (*86*, *162*). Hatch (*112*) beschreibt ein solches System aus der Methionin oder Vitamin B_{12} bedürftigen Mangel-Mutanten 113-3, und Takeyama (*312*) die Reinigung und einige Charakteristika des Vitamin B_{12}-Enzyms. Dessen Apoenzym verbindet sich mit einem Co^{++}-Cobamid, das nicht mit dem Barkerschen Faktor identisch ist (*311*). Ganz analog sind wohl die Enzyme und der Faktor X, die Guest (*107*, *108*) aus der Mutanten 121/176 erhielt, die der Mutanten 113-3 entspricht.

Jüngst wurde von E. L. Smith (*105*) ein Methyl-cobamid dargestellt, das in der Methioninsynthese als Methyldonator wirksam ist, allerdings auch bereits nichtenzymatisch Aktivität besitzt. Abzuwarten bleibt vorerst, wie diese Methyl-Übertragung mechanistisch vor sich gehen kann. Geht das Methyl als $CH_3{}^+$-Ion vom Co^{III}-Cofaktor auf den S von Homocystein, würde intermediär ein Co^{I}-Komplex und daraus mit einem Proton ein Co^{III}-Hydrid entstehen; würde in einem Radikalmechanismus die Co—C-Bindung homolytisch gespalten, könnte das $CH_3{}^{\cdot}$-Radikal von der Merkaptogruppe abgefangen werden, wofür man Analogien hat. Ebenso unklar ist die Natur der Folat-Cofaktoren und ihre Beziehungen zum Vitamin B_{12}-Enzym. Jones (*106*, *145*) fand, daß formylierte und unformylierte Pteroyltriglutamate im Coli-System FH_4 ersetzen können, ja sogar, daß sie das Vitamin B_{12}-Enzym überflüssig machen. Sie ständen damit dem natürlichen Cofaktor näher, der in Form eines 5-Formyl-tetrahydropteroyl-konjugats isoliert wurde (*108*). Die Verhältnisse ähneln hiermit also den bei der Serinaldolase beschriebenen. Guest und Woods (*104*, *108*) schließen daher auf die Existenz zweier Wege zum Methionin aus Methyl-folaten, von denen der eine Cobalamin-unabhängig ist, aber ein Pteroylkonjugat erfordert. Er wird durch FH_4 gehemmt und ist in der Mutanten unterdrückt. Der zweite braucht sowohl Cobalamin wie irgendein einfaches oder konjugiertes Pteroylglutamat. Es ist der allgemeinere, der aber bei der Mutanten in Gegenwart von FH_4 beschritten wird, da diese den ersten Weg blockiert (s. oben) (*104*). Das Cobalamin stellt somit im ganzen mehr ein Schaltelement des Stoffwechsels als einen Cofaktor dar. Zu ähnlichen Folgerungen, die *Schema 3* zu illustrieren versucht, kommt auch Kisliuk (*162*).

Im Säugetiersystem sind die Dinge undurchsichtiger, da hier ein Bedürfnis der Reaktion für FAD oder Cobalamin-Derivate noch nicht nachgewiesen werden konnte. Das liegt möglicherweise mehr an den sehr ungünstigen relativen Mengenverhältnissen, die eine außerordentlich hohe Anreicherung des Enzyms erforderlich machen, und die noch nicht gelungen ist, als an grundsätzlichen Unterschieden. Man weiß lediglich aus den Untersuchungen von SAKAMI et al. (*258, 268*), daß Methylfolat in Gegenwart von ATP zur Methylierung von Homocystein dienen kann und echtes Zwischenprodukt ist (*259*), und daß ein weiterer Faktor, der aus dem Vitamin B_{12}-reichen *P. shermanii* extrahiert wurde, gealterte Methionin-Synthetase-Präparate wieder vollkommen aktivieren kann (*216*). Die im Extrakt enthaltene Substanz ist weder nach ihren Eigenschaften

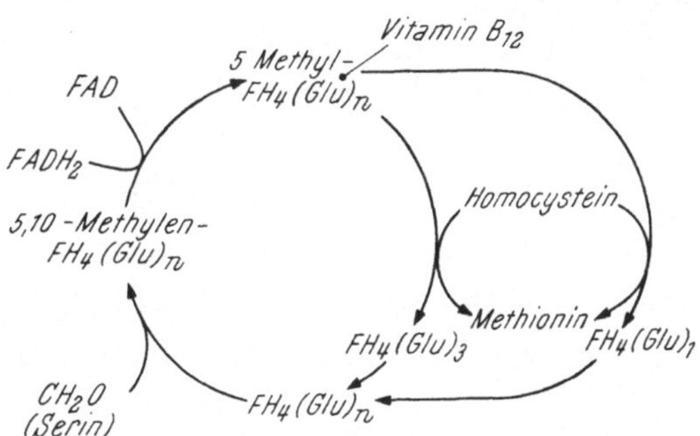

Schema 3. Methionin-Bildung in *E. coli*-Systemen.

noch nach ihrer Lichtempfindlichkeit dem Barkerschen Cofaktor verwandt, und nach dem Veraschen bleiben noch 70% der Aktivität erhalten. Andere indirekte Beobachtungen sprechen ebenfalls für die Beteiligung von einem Vitamin B_{12}-oder einem Cobalt-Komplex im Lauf der Bildung von Methionin (siehe unten).

γ. Der Acceptor der Methylgruppe. Die Frage nach dem Acceptor der Methylgruppe ist ebenfalls noch nicht befriedigend beantwortet. Die Teilnahme von Adenosylhomocystein (*348*) hat sich nicht bestätigt. Dagegen kann sowohl freies Homocystein (*299*) wie N-Acetylhomocystein (*268*) die Methylgruppe aufnehmen. Die Übertragung erfordert in Schweineleber-Extrakt (*183*) wie im Coli-System Adenosyl-methionin als Katalysator (*260*).

Über seine Rolle, wie überhaupt über den gesamten Mechanismus der Übertragung lassen sich heute nur Vermutungen anstellen. Es ist eine Quaternisierung oder zum mindesten Polarisierung der $N—CH_3$-Bindung nötig, denn vom tertiären Amin der Methyl-FH_4 ist die Methyl-Übertragung schwer vorzustellen. Ob dies durch zusätzliche Einführung eines Substituenten geschieht, wozu das immer noch benötigte ATP verwendet werden könnte, ob das Vitamin B_{12} hier seine Funktion hat oder ob eine Art Anlagerungskomplex entsteht, bei dem der Sulfoniumschwefel des Adenosylmethionins den Stickstoff polarisiert, ist zur Zeit gleich wahrscheinlich. Für die Funktion des Adenosylmethionins könnte aber auch ein allosterischer Aktivator-Ort am Enzym in Frage kommen, so daß die Rolle von Adenosylmethionin rein indirekter Art ist. Ebenso ließe sich vermuten, daß das B_{12}-Enzym als allgemeiner Elektronenüberträger dient (*227*) um das Milieu reduzierend zu halten, in dem die SH-Gruppe von Homocystein substituiert werden soll.

Für die Beteiligung von Adenosylmethionin und Vitamin B_{12} an der Methionin-Bildung läßt sich auch die Beobachtung von Noronha und Silverman (*217*) anführen, nach der die Anhäufung von 5-Methyl-FH_4 in der Leber von Vitamin B_{12}-Mangel-Ratten nach Gabe von Methionin ausbleibt und sich das Folat-Profil wieder normalisiert, indem größere Mengen $N_{(10)}$-Formyl-folat an Stelle der 5-Methylverbindung auftreten. Wie Methionin wirkt auch Vitamin B_{12} selbst. In gleiche Richtung weisen auch die Untersuchungen von Mollin (*204*); es sei hier eingefügt, daß sich der *L. casei*-Faktor, der von Mollin und Waters (*325*) gefunden wurde, als 5-Methyl-FH_4 erwiesen hat (Jaenicke, unveröff.).

3. Transformylierungs-Cyclen.

a. Abbau von Histidin.

Der Abbau von Histidin gibt in den verschiedenen untersuchten Organismen je Mol Aminosäure schließlich 1 Mol Glutaminsäure, 2 Mol

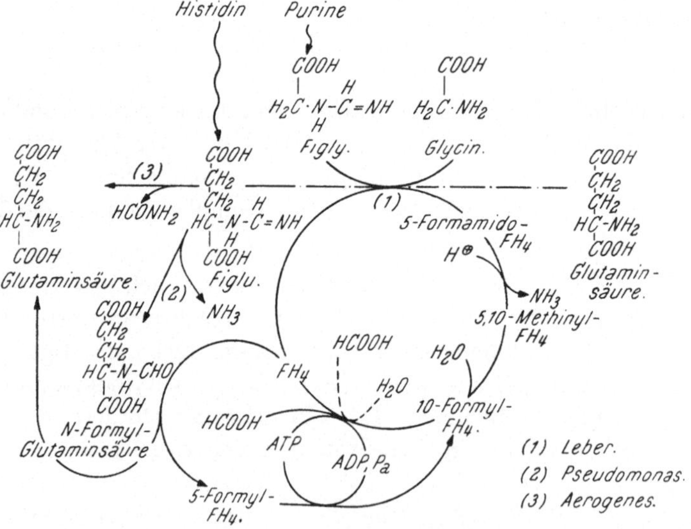

Schema 4. Endabbau von Histidin und Purinen.

Ammoniak und 1 Mol 10-Formyl-FH_4. Zunächst wird der Imidazolring der Urocaninsäure hydrolytisch aufgespalten, und es entsteht α-Form-imino-*L*-glutaminsäure (FIGLU) (*196*). Während in verschiedenen Mikroorganismen daraus Ammoniak und Formylglutaminsäure bzw. Formamid und Glutaminsäure entstehen, verläuft die Spaltung in Säugetierleber unter Zuhilfenahme von FH_4 in dem von Waelsch (*197*) und Tabor (*309*,

310) aufgefundenen und in *Schema 4* dargestellten Weg (Gl. [17]—[19]). Die Rolle der Folsäure als Formylacceptor war durch die Beobachtung gegeben, daß Leberhomogenate mit Histidin intensiv fluoreszierende 10-Formylfolsäure geben.

$$\text{FIGLU} + \text{FH}_4 \xrightleftharpoons[]{\text{Formiminotransferase}} \text{Glutamat} + 5\text{-Formimino-FH}_4$$
$$(\text{K} = 0,8 \text{ bis } 2,6). \quad [17]$$

$$5\text{-Formimino-FH}_4 \xrightarrow{\substack{\text{Formimino-FH}_4\text{-}\\ \text{cyclodesaminase}}} 5,10\text{-Methinyl-FH}_4 + \text{NH}_3. \quad [18]$$

$$5,10\text{-Methinyl-FH}_4 \xrightleftharpoons[(\text{H}_2\text{O})]{\text{Cyclohydrolase}} 10\text{-Formyl-FH}_4. \quad [19]$$

Die gesamte Formamid-Gruppe wird durch das Schweineleber-Enzym Formaminotransferase (Gl. [17]) auf $\text{N}_{(5)}$ der FH_4 übertragen, so daß 5-Formimino-FH_4 entsteht, dessen Bildung optisch gemessen werden kann (*306*). Das Enzym selbst wurde aus Schweineleber-Acetonpulver auf das 520fache angereichert, wobei gleichzeitig das Folge-Enzym, die Cyclodesaminase (Gl. [18]), durch die Formimino-FH_4 in NH_3 und Methinyl-FH_4 gespalten wird, im gleichen Maß angereichert wird. Durch Behandlung mit Chymotrypsin wird diese entfernt und eine 700- bis 1000fache Anreicherung der Transferase erzielt. Dagegen inaktiviert Behandlung mit Ammoniak-Puffer pH 10,5 nur die Transferase. Diese ist recht spezifisch für die angegebene Reaktion mit FIGLU. Formimino-glycin (FIGLY) wird nicht, Formiminoasparaginsäure (FIASP) und Formyl-glutamat werden nur mit $^1/_{1000}$ Geschwindigkeit umgesetzt. Beide Enzyme werden durch K^+- und NH_4^+-Ionen aktiviert. Die cyclische Methinylverbindung wird schließlich durch ein Enzym aufgespalten, das die im Neutralen nicht schnell genug ablaufende Spontanhydrolyse (Gl. [19]) beschleunigt. Diese 5,10-Methinyl-FH_4-Cyclohydrolase wurde aus Extrakten von *Cl. cylindrosporum* (*246*) und tierischem Gewebe (*307*) angereichert und läßt sich durch die charakteristischen Absorptions-unterschiede von Substrat und Produkt (vgl. S. 223) spektrophotometrisch bei 356 mμ nachweisen.

Wegen der in Maleatpuffer verlangsamten nichtenzymatischen Hydrolyse wird die Bestimmung in diesem Milieu ausgeführt, doch läßt sich ein pH-Optimum nicht angeben, da oberhalb pH 8 auch dann die Öffnungsgeschwindigkeiten zu groß werden. Das Enzym ist wiederum stereospezifisch für eines der Diastereomeren der FH_4. Bei Folsäure-Mangel scheiden Ratten erhebliche Mengen von FIGLU aus (*274, 308*). Das gleiche ist der Fall bei Patienten mit gewissen Anämien, bei denen Folat-Cofaktoren größtenteils als Methyl-FH_4 festgelegt und damit dem Stoff-wechsel entzogen sind (vgl. *167*).

b. *Deacylase und Glutamyl-Transferase.*

Aus 10-Formyl-FH_4 kann dann Ameisensäure freigesetzt und die FH_4 wieder in den Prozeß eingeführt werden. Die diese Reaktion katalysierende

Deacylase wurde von Huennekens (*221*) aus Rinderleber-Acetonpulver angereichert. Die stereospezifische Reaktion folgt der Gl. [20]:

$$10\text{-Formyl-FH}_4 + H_2O \rightarrow HCOOH + FH_4. \qquad [20]$$

Sie braucht als Starter Pyridinnucleotide und kann $N_{(5)}$-Formyl-FH$_4$ nicht spalten. Diese kann als Substrat nur dann verwendet werden, wenn sie durch die ATP-benötigende Isomerase gemäß Gl. [21] in 10-Formyl-FH$_4$ umgewandelt wird (*101, 231*):

$$5\text{-Formyl-FH}_4 + ATP \rightleftharpoons 10\text{-Formyl-FH}_4 + ADP + H_3PO_4. \quad [21]$$

Diese Bilanz folgt aus den Messungen von Kay und Mitarb. (*153*), die auch angeben, daß bei dieser Umlagerung die Anhydroformyl-Verbindung spektral nicht nachweisbar ist. Es ist wohl noch offen, ob die Reaktion durch ein Enzym bewirkt oder die Summe aus Spaltung und Reaktivierung durch eine aus mehreren Katalysatoren zusammengesetzte Kette ist.

Zusätzlich zur Isomerase gibt es noch ein weiteres 5-Formyl-FH$_4$ als Substrat verwendendes Enzym, die 5-Formyl-FH$_4$-Glutaminsäure-Transferase (*286*), deren Gleichgewicht allerdings weit auf der linken Seite der Gl. [22] liegt.

$$5\text{-Formyl-FH}_4 + L\text{-Glutamat} \rightarrow FH_4 + N\text{-Formyl-glutamat}$$

$$(K_{eq}\ 0{,}05\text{—}0{,}13). \quad [22]$$

Dieses Enzym wurde aus frischer Schweineleber etwa 400fach angereichert (*282*); es enthält zwar noch FIGLU-Transferase-Aktivität, ist aber spezifisch für 5-Formyl-FH$_4$ und L-Glutaminsäure.

Im unteren Teil des Schemas 4, S. 246, ist die Möglichkeit diskutiert, daß diese Isomerase und Transformylase am Abbau der Formylglutaminsäure des Histidin-Stoffwechsels teilnehmen. Die Spaltung von 10-Formyl-FH$_4$ kann aber auch mit der Bildung von ATP aus ADP in der Formylase-Reaktion gekoppelt sein, wodurch dann sogar Energie gewonnen werden kann. Diese Reaktion ist aber besonders beim Abbaustoffwechsel der Purine wichtig.

c. Aktivierte Ameisensäure im Purin-Stoffwechsel.

α. *Vergärung von Purinen.* In den Stoffwechsel der Purine greift FH$_4$ erst bei der Endspaltung und Energiegewinnung aus dem Produkt ein. Diese Reaktionsfolge hat Rabinowitz (*244*) in eleganter Weise aufgeklärt. Durch sie vermögen die Purin-vergärenden Clostridien den zunächst rein hydrolytischen Abbauvorgang auf Energiegewinn umzuschalten. *Cl. cylindrosporum* (*16*) und *Cl. acidi-urici* (*16*) verwenden als einzige Stickstoff- und Kohlenstoffquelle Purine, die zunächst sämtlich zu Xanthin umgewandelt werden. *Cl. acidi-urici* bildet dann letzten Endes 5 NH$_3$, 3 CO$_2$ und 1 Acetat, *Cl. cylindrosporum* an Stelle von

Acetat 1 Mol Formiat und je zirka $^1/_4$ Mol Glycin und Acetat. Die verschiedenen vorbereitenden Schritte sind Hydrolysen oder Decarboxylierungen und benötigen außer Schwermetallionen keine Cofaktoren. Es entsteht dabei Formiminoglycin (FIGLY) als Endprodukt (Schema 4). Dessen weiterer Stoffwechsel erfordert FH_4, ADP und Phosphat (*246, 262*), womit der folgende Gesamtabbau bewirkt wird (Gl. [23]):

$$FIGLY + ADP + H_3PO_4 \xrightleftharpoons{FH_4} ATP + Formiat + Glycin + NH_3, \quad [23]$$

der sich aus vier Reaktionsschritten zusammensetzt, von denen Gl. [18] und Gl. [19] mit den beim Histidinabbau beschriebenen identisch sind.

$$FIGLY + FH_4 \xrightleftharpoons{FIGLY\text{-}Transferase} Glycin + Formimino\text{-}FH_4$$
$$(K = 0{,}2). \quad [24]$$

$$5\text{-}Formimino\text{-}FH_4 \xrightarrow{FIFH_4\text{-}Cyclodesaminase} 5{,}10\text{-}Methinyl\text{-}FH_4 + NH_3. \quad [18]$$

$$5{,}10\text{-}Methinyl\text{-}FH_4 \xrightleftharpoons[\text{(H}_2\text{O)}]{Methinyl\text{-}FH_4\text{-}Cyclohydrolase} 10\text{-}Formyl\text{-}FH_4. \quad [19]$$

$$10\text{-}Formyl\text{-}FH_4 + ADP + H_3PO_4 \xrightleftharpoons{Formylase} FH_4 + Formiat + ATP$$
$$(K = 0{,}025). \quad [25]$$

Die einzelnen Reaktionsschritte sind eine Formiminoübertragung auf FH_4 (*246, 247, 262*), katalysiert in Analogie zu Gl. [17] durch ein spezifisches Enzym, die Formiminoglycin-FH_4-Transferase (Gl. [24]), die aus Extrakten von *Cl. cylindrosporum* isoliert wurde. Das Produkt 5-Formimino-FH_4 wurde präparativ in recht guter Ausbeute durch Ionenaustauscher-Chromatographie erhalten (*249*) und ist mit demjenigen aus der FIGLU-Transferase identisch.

Die Bildung von 10-Formyl-FH_4 aus Formimino-FH_4 durch Cyclodesaminase (*307*) und Cyclohydrolase (*247*) wurde bereits besprochen. Der letzte Schritt, der für die Bildung von ATP und damit für die Zellsynthese notwendig ist, besteht in der Umkehr der Folatformylase-Reaktion (*246*), die zunächst in einem anderen Zusammenhang entdeckt und untersucht worden war (*101, 103, 134*) und anschließend besprochen wird. Extrakte von *Cl. cylindrosporum* enthalten so große Mengen an diesem Enzym, daß es etwa 3% des Trockengewichts ausmacht (*249 a*). Auch das zeigt die Wichtigkeit und Bedeutung dieses Schrittes für die Bakterien an.

β. *Tetrahydrofolat-Formylase.* Aber auch im Verlauf der biologischen Bildung des Purinsystems wird die aktivierte Ameisensäure verwendet, und zwar zu den beiden die Ringe schließenden Amidin-Kohlenstoff-Atomen $C_{(2)}$ und $C_{(8)}$. Die Beteiligung der Folsäure an diesen besonders in Taubenlebersystemen untersuchten Reaktionsfolgen wurde von GREENBERG (*102*) und BUCHANAN (*51*) erkannt. Bald darauf konnte

Greenberg zeigen, daß der Transformylierungs-Cofaktor in diesem System durch 5-Formyl-FH_4 und ATP oder durch FH_4, Formiat und ATP ersetzt werden kann (*101*). Bei der letzten Reaktion entsteht 10-Formyl-FH_4, womit die chemische Natur der „aktiven" Ameisensäure aufgeklärt war (*103 a, 134*). Das katalysierende Enzym Tetrahydrofolat-Formylase (*103*), auch Formiat-aktivierendes Enzym genannt, bewirkt die Energie-benötigende Kondensation von Formiat mit dem $N_{(10)}$ der Tetrahydro-folsäure nach Gl. [25]:

$$FH_4 + HCOOH + ATP \rightleftharpoons \text{10-Formyl-}FH_4 + ADP + H_3PO_4$$
$$(K = 40). \quad [25]$$

Die bei entsprechender Koppelung reversible Reaktion ist die Umkehr der Energie-liefernden Formyl-FH_4-Spaltung, die beim Abbau der Purine in *Cl. cylindrosporum* besprochen wurde (S. 248). Aus diesen Bakterien wurde das Enzym nach 32facher Anreicherung einheitlich kristalli-siert (*120 a, 245, 249 a*).

Die mit diesem Enzym ausgeführten Austauschversuche zur Aufklärung des Reaktionsmechanismus gaben in keinem Fall einen Hinweis auf ein individuelles Zwischenprodukt, wohl aber auf die intermediäre Bindungsbildung zwischen Formiat und Phosphat (*120 b*). Der gesamte Vorgang läßt sich daher als push-pull-Mechanismus (XXX) behandeln, bei dem zunächst ein Enzym-ATP-Komplex

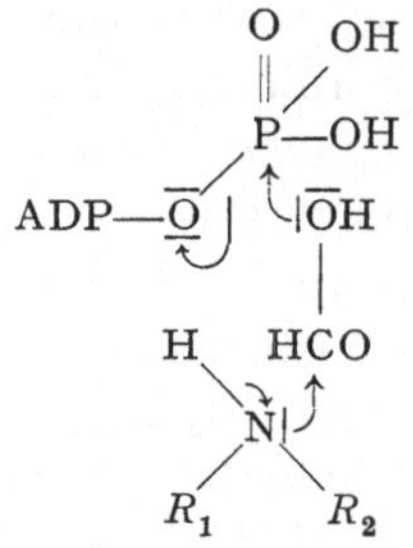

(XXX.) Push-pull-Mechanismus der Folat-Formylase-Reaktion.

entsteht, an dem die beiden anderen Reaktionspartner passend angelagert werden und schließlich der Elektronenfluß von FH_4 über Formiat zur β,γ-Bindung von ATP zur Bildung der Reaktionsprodukte 10-Formyl-FH_4, Phosphat und ADP führt. Damit ist die Reaktion analog den anderen Säureamid-Bildungen mit entsprechender ATP/ADP-Bilanz, wie der Glutamin-Synthese, der Glutathionbildung, dem P-Enzym und einigen Zwischenstufen in der Purin-Biogenese (*53*). In diesem Fall darf eine Arsenolyse von Formyl-FH_4 nur eintreten, wenn ADP zugegen ist, was auch tatsächlich gefunden wurde (*120 b*). Grundsätzlich ähnliche Ergebnisse wurden mit der Formylase aus Taubenleber erhalten, die im Grunde eine spezifische ATP-ase darstellt (*141 a*). Auch hier deuten die Resultate auf einen Mehrzentren-prozeß hin (*138, 139*), aber offenbar entsteht zunächst ein Enzym-Phosphat, und die primäre Bindung von Formiat erfolgt an $N_{(5)}$. Darauf erfolgt, stets ans Enzym gebunden, eine Umlagerung der Formylgruppe, so daß die gleichen Endprodukte erhalten werden, die durch spezifische Reaktionen identifiziert wurden (*103 a*). Auch der Mechanismus der sehr hochgereinigten Formylase

Literaturverzeichnis: SS. 254—274.

aus *Micrococcus aerogenes* (*341, 343*) verläuft, entgegen früheren Mitteilungen (*342*), über ein Enzym-gebundenes Reaktionsknäuel, so daß die verschieden Enzyme nicht nur in ihren Eigenschaften, sondern auch im Mechanismus einander gleichen. Das wird auch für die Formylase aus Erythrozyten (*22*) und Leukozyten (*347*) zutreffen.

$$ATP + E \text{ (Enzym)} \rightleftharpoons E\text{-ATP,} \qquad [26]$$

$$E\text{-ATP} + FH_4 \rightleftharpoons E\text{-}\textcircled{P}\text{-}FH_4 + ADP, \qquad [27]$$

$$E\text{-}\textcircled{P}\text{-}FH_4 + HCOOH \rightleftharpoons E\text{-10-CHO-}FH_4 + P_a \qquad [28]$$

$$E\text{-10-CHO}-FH_4 \rightleftharpoons E + 10\text{-CHO-}FH_4. \qquad [29]$$

Durch diese Reaktionsfolge (Gl. [26]—[29]) entsteht ebenso wie bei der Oxydation der 5,10-Methylen-tetrahydro-folsäure (Gl. [13], S. 239) der neuartige Typ der

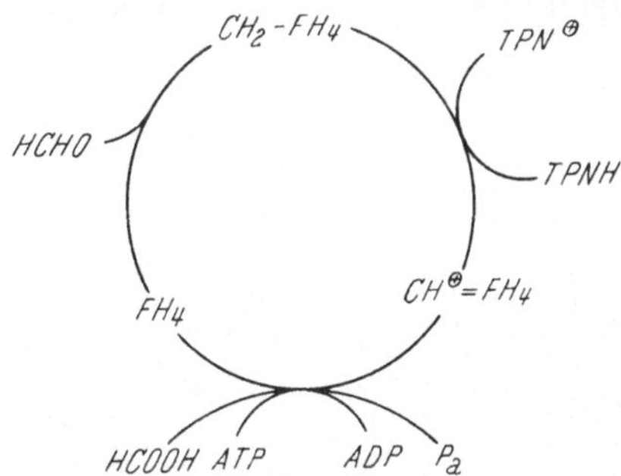

Schema 5. Energie-Koppelung der Formaldehyd-Oxydation.

energiereichen Bindung des cyclischen Carbimonium-Ions (vgl. S. 214), das als aktive Form der Ameisensäure an den weiteren Umsetzungen teilnimmt (*138*). Dadurch erhält der Folat-abhängige Stoffwechsel der C_1-Körper eine Reaktionsfolge aus Dehydrogenierung und ADP-Phosphorylierung, die, in *Schema 5* wiedergegeben, in manchen der oxydativen Phosphorylierung an die Seite zu stellen ist und dadurch gewissermaßen als ein Modell dieser Vorgänge betrachtet werden darf.

γ. Transformylierungen. Die auf diese Weise entstehende „aktivierte" Ameisensäure kann im Stoffwechsel zur Synthese einer Reihe heterocyclischer Verbindungen verwendet werden, namentlich der Purine und – indirekt – des Histidins.

Ausgangspunkt für die vollständige Aufklärung des schrittweisen enzymatischen Aufbaus der Purine war die Entdeckung, daß die Biogenese auf der Stufe der Nucleotide stattfindet und daß 4-Amino-5-imidazolcarboxamid-ribotid (AICAR), eine Verbindung, die sich in Sulfonamidblockierten Mikroorganismen und „purinlosen" Mutanten anhäuft, durch Taubenleber-Extrakte in Gegenwart von Formiat zu Inosinsäure umgewandelt wird (*96, 97*, vgl. *103*). Die für diese Reaktion verantwortlichen Enzyme (*53*) AICAR-Transformylase (Gl. [30]):

$$AICAR + Formyl\text{-}FH_4 \rightleftharpoons Formyl\text{-}AICAR + FH_4 \qquad [30]$$

und Inosinicase (Gl. [31]):

$$Formyl\text{-}AICAR \rightleftharpoons IMP + H_2O \qquad [31]$$

wurden von Buchanan (*50—53, 111*) aus Hühnerleber angereichert, aber nicht voneinander getrennt, so daß die Bildung von N-Formyl-AICAR indirekt erschlossen wurde.

Der Einbau von Formiat in die Stellung 8 der Purine durch Glycin-amid-ribotid-transformylase verläuft ähnlich. Acceptor der Formylgruppe ist Glycinamid-ribotid (GAR), das zu Formylglycinamid-ribotid formyliert wird (Gl. [32]):

$$\text{GAR} + \text{Formyl-FH}_4 \rightarrow \text{Formyl-GAR} + \text{FH}_4. \qquad [32]$$

Dies wird erst nach weiterer Einführung einer Aminogruppe auf der Stufe von Formylglycinamidin-ribotid cyclisiert (*51*). Die GAR-

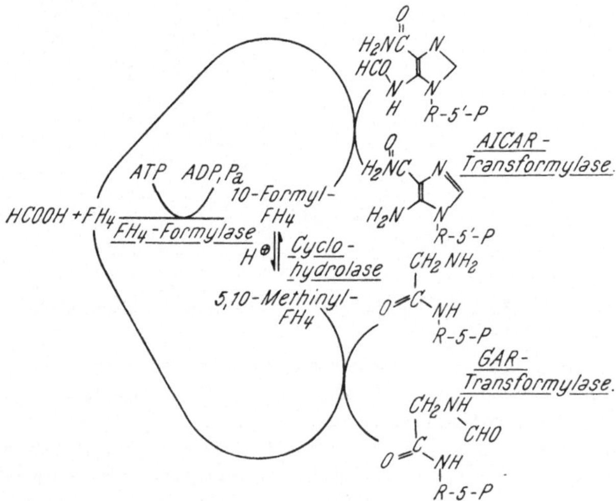

Schema 6. Transformylierungs-Cyclen.

Transformylase-Reaktion ist praktisch irreversibel. Hartman und Buchanan (*111*) fanden interessante Unterschiede bei den Transformy-lierungs-Reaktionen im Hinblick auf die Spezifität der Formyldonatoren. In Abwesenheit des Enzyms Cyclohydrolase, das die Ineinander-Um-wandlung von 10-Formyl-FH$_4$ und 5,10-Anhydroformyl-FH$_4$ katalysiert, ist die cyclische Anhydroform der spezifische Donator für die irreversible Bildung von Formylglycinamid-ribotid, während die Formylgruppe von Formyl-AICAR der 10-Formyl-FH$_4$ entstammt. In beiden Fällen wird FH$_4$ regeneriert *(Schema 6)* und kann von neuem in die Reaktionen eintreten.

AICAR-Transformylase wirkt, wenn auch nur mittelbar, bei der Biogenese von Histidin mit, das aus Pentosephosphat und einer Formamidgruppe gebildet wird, die dem $N_{(1)}$ und $C_{(2)}$ des Purinrings entstammt und zu Position 3 und 2 des Imidazolrings wird (*213*). Die Untersuchungen von Ames (*8*) und Moyed (*207*) lassen nun als ersten Schritt die Kondensation von ATP mit 5-Phosphoribose-

1-pyrophosphat erkennen, wobei Pyrophosphat und $N_{(1)}$-5'-Phosphoribosyl-ATP entstehen. Dies wird in Abwesenheit des Stickstoff-Donators in ,,Verbindung III'' (XXXI) (*207*) umgewandelt, die sich anhäuft, aber in Gegenwart von Säure wieder

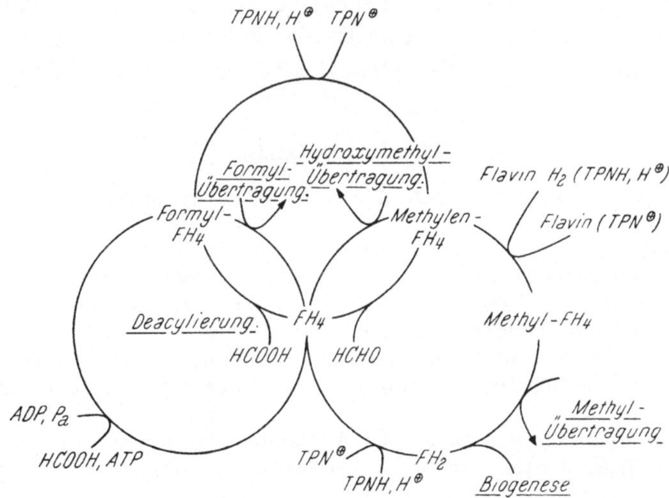

(XXXI.)

in die Ausgangssubstanzen zerfällt. In Gegenwart von Glutamin dagegen entsteht über eine weitere Zwischenstufe (*166*) Imidazolyl-glycerylphosphat und 4-Amino-imidazolcarbonsäureamid-ribotid, das durch Transformylierung, Ringschluß und Aminierung wieder in Adenylsäure übergeführt wird, während die Imidazolverbindung über mehrere anschließende Schritte zum Histidin vollendet wird.

VII. Zusammenfassung.

Betrachten wir die einzelnen besprochenen Cyclen der von Folsäure-Cofaktoren katalysierten und gesteuerten Reaktionen im Zusammenhang *(Schema 7)*, so ergibt sich ein einheitliches Bild.

Schema 7. Reaktionscyclen der Folat-Coenzyme.

Die aktivierten Kohlenstoffeinheiten, die als Tetrahydrofolat-gebundene Gruppen erkannt wurden, können durch die — miteinander durch die übertragende Tetrahydro-folsäure verknüpften — Kreisläufe hindurch und zwischen ihnen hinüber wechseln, wobei der Transport-Metabolit abwechselnd beladen und entladen wird und auf diese Weise tatsächlich als Katalysator dienen kann. Die beobachteten Umwandlungen der Einkohlenstoff-Einheiten gehen vom Einkohlenstoff-Reservoir des Serins aus, in dessen β-Kohlenstoff man eine gebundene, entgiftete, aber leicht zugängliche Form des Formaldehyds sehen darf. Die Reaktionsschritte zu seiner Mobilisierung umfassen die Aktivierung der α,β-Bindung, die Aldol-artige Abspaltung und die gleichzeitige Kondensation mit Tetrahydro-folsäure (vgl. S. 238) zu Methylen-FH_4. Indem dann die Methylengruppe durch Pyridinnucleotid oxydiert wird, läßt sich eine neuartige energiereiche Verbindung, Formyl-FH_4, gewinnen, mit deren Hilfe höhere Organismen die verschiedenen Formylierungs-Reaktionen zum Aufbau zellwichtiger Substanzen ausführen, einige spezialisierte Mikroorganismen sogar ihren Energiebedarf decken können. Wird dagegen die Methylengruppe reduziert, so entsteht Methyl-FH_4 und damit die Quelle der Methylgruppen, die sich über Methionin und Adenosylmethionin im Cholin und durch die weiteren Transmethylierungs-Reaktionen in zahlreichen Körperbausteinen finden. Aber die Hydroxymethylgruppe kann auch ohne weitere Umwandlungen übertragen werden, wobei Hydroxymethyl-Verbindungen entstehen, oder, wenn die Übertragung mit einer Reduktion verbunden ist, Methylgruppen, wie sie im Thymin vorgefunden werden. Dabei entsteht FH_2, die nun wiederum zu FH_4 reduziert und in die Cyclen zurückgeführt werden muß. Alle diese Vorgänge werden also durch Tetrahydro-folsäure katalysiert, die wir so mit Recht als eine Schlüsselsubstanz des Stoffwechsels bezeichnet haben.

Literaturverzeichnis.

1. Albert, A.: The Pteridines. Fortschr. Chem. organ. Naturstoffe 11, 350 (1954).
2. Albert, A. and S. Matsuura: Pteridine Studies. Part XV. The Reduction of 2-Hydroxypteridine. J. Chem. Soc. (London) 1961, 5131.
3. — — Pteridine Studies. Part XVII. The Reduction of Hydroxypteridines. J. Chem. Soc. (London) 1962, 2162.
4. Alexander, N. and D. M. Greenberg: Studies on the Purification and Properties of the Serine-forming Enzyme System. J. Biol. Chem. 220, 775 (1956).
5. Allen, W., R. L. Pasternak and W. Seaman: Polarographic Determination and Evidence for the Structure of Leucovorin. J. Amer. Chem. Soc. 74, 3264 (1952).
6. Allfrey, V. G. and C. G. King: An Investigation of the Folic Acid-Protein Complex in Yeast. J. Biol. Chem. 182, 367 (1950).
7. Allfrey, V., L. J. Tepley, C. Geffen and C. G. King: A Fluorometric Method for the Determination of Pteroylglutamic Acid. J. Biol. Chem. 178, 465 (1949).

8. AMES, B. N., R. G. MARTIN and B. J. GARNY: The first Step of Histidine Biosynthesis. J. Biol. Chem. **236**, 2019 (1961).

9. ANGIER, R. B., J. H. BOOTHE, B. L. HUTCHINGS, J. H. MOWAT, J. SEMB, E. L. R. STOKSTAD, Y. SUBBAROW, C. W. WALLER, D. B. COSULICH, M. J. FAHRENBACH, M. E. HULTQUIST, E. KUH, E. H. NORTHEY, D. R. SEEGER, J. P. SICKELS and J. M. SMITH, Jr.: The Structure and Synthesis of the Liver *L. casei* Factor. Science (Washington) **103**, 667 (1946).

9a. ANGIER, R. B., J. H. BOOTHE, J. H. MOWAT, C. W. WALLER and J. SEMB: Pteridine Chemistry. II. The Action of Excess Nitrous Acid upon Pteroylglutamic Acid and Derivatives. J. Amer. Chem. Soc. **74**, 408 (1952).

10. ANGIER, R. B., E. L. R. STOKSTAD, J. H. MOWAT, B. L. HUTCHINGS, J. H. BOOTHE, C. W. WALLER, J. SEMB, Y. SUBBAROW, D. B. COSULICH, M. J. FAHRENBACH, M. E. HULTQUIST, E. KUH, E. H. NORTHEY, D. R. SEEGER, J. P. SICKELS and J. M. SMITH, Jr.: Synthesis of Pteroylglutamic Acid. III. J. Amer. Chem. Soc. **70**, 25 (1948).

11. ARNSTEIN, H. R. V. and D. KEGLEVIĆ: A Comparison of Alanine and Glucose as Precursors of Serine and Glycine. Biochemic. J. **62**, 199 (1956).

12. ARNSTEIN, H. R. V. and A. NEUBERGER: The Effect of Cobalamin on the Quantitative Utilization of Serine, Glycine, and Formate for the Synthesis of Choline and Methyl Groups of Methionine. Biochemic. J. **55**, 259 (1953).

13. ASAHI, Y.: Polarographic Determination of Multivitamin Preparation. I. Ascorbic Acid, Riboflavin and Pteroylglutamic Acid. Vitamins (Kyoto) **13**, 490 (1957).

14. BAKER, H., S. H. HUTNER and H. SOBOTKA: Estimation of Folic Acid with a Thermophilic Bacillus. Proc. Soc. exp. Biol. Med. **89**, 210 (1955).

15. BAKERMAN, H. A.: A Method for Measuring the Microbiological Activity of Tetrahydrofolic Acid and Other Labile Reduced Folic Acid Derivatives. Anal. Biochem. **2**, 588 (1961).

16. BARKER, H. A. and J. V. BECK: *Clostridium acidi-urici* and *Clostridium cylindrosporum*, Organisms Fermenting Uric Acid and some Other Purines. J. Bacteriol. **43**, 291 (1942).

17. BARNER, H. D. and S. S. COHEN: Virus Induced Acquisition of Metabolic Function. IV. Thymidylate Synthetase in Thymine-requiring *Escherichia coli* Infected by T_2 and T_5 Bacteriophages. J. Biol. Chem. **234**, 2957 (1959).

18. BATT, R. D., F. DICKENS and D. H. WILLIAMSON: The Enzymic Incorporation of the β-Carbon of Serine into Dihydroxyacetone. Biochim. Biophys. Acta **45**, 571 (1961).

19. BERG, P.: A Study of Formate Utilization in Pigeon Liver Extracts. J. Biol. Chem. **205**, 145 (1953).

20. BERGMANN, F. and H. KWIETNY: Pteridines as Substrates of Mammalian Xanthine Oxidase. I. The Endproduct of the Enzymic Oxidation of Pteridines. Biochim. Biophys. Acta **28**, 613 (1958).

21. BERTINO, J. R., B. W. GABRIO and F. M. HUENNEKENS: Dihydrofolic Reductase in Human Leucemic Leucocytes. Biochem. Biophys. Res. Comm. **9**, 103 (1961).

22. BERTINO, J. R., B. SIMMONS and D. M. DONOHUE: Purification and Properties of the Formate Activating Enzyme from Erythrocytes. J. Biol. Chem. **237**, 1314 (1962).

23. BINKLEY, S. B., O. D. BIRD, E. S. BLOOM, R. A. BROWN, D. G. CALKINS, C. J. CAMPBELL, A. D. EMMETT and J. J. PFIFFNER: On the Vitamin B_C Conjugate in Yeast. Science (Washington) **100** 36 (1944).

23a. BIRD, O. D., M. ROBBINS, J. M. VANDENBELT and J. J. PFIFFNER: Observations on Vitamin Bc-Conjugase from Hog Kidney. J. Biol. Chem. **163**, 649 (1946).

24. Birnie, G. D. and G. W. Crosbie: Biosynthesis of Thymidylic Acid. Biochemic. J. **69**, 1 P (1958).

25. Blakley, R. L.: The Interconversion of Serine and Glycine: Rôle of Pteroylglutamic Acid and Other Cofactors. Biochemic. J. **58**, 448 (1954).

26. — The Interconversion of Serine and Glycine: Preparation and Properties of Catalytic Derivatives of Pteroylglutamic Acid. Biochemic. J. **65**, 331 (1957).

27. — The Interconversion of Serine and Glycine: Some Further Properties of the Enzyme System. Biochemic. J. **65**, 342 (1957).

28. — The Reactive Intermediate Formed by Formaldehyde and Tetrahydropteroylglutamate. Biochim. Biophys. Acta **23**, 654 (1957).

29. — Interaction of Formaldehyde and Tetrahydrofolic Acid and its Relation to the Enzymatic Synthesis of Serine. Nature (London) **182**, 1719 (1958).

30. — The Reaction of Tetrahydropteroylglutamic Acid and Related Hydropteridines with Formaldehyde. Biochemic. J. **72**, 707 (1959).

31. — Spectrophotometric Studies on the Combination of Formaldehyde with Tetrahydropteroylglutamic Acid and other Hydropteridines. Biochemic. J. **74**, 71 (1960).

32. — Spectrophotometric Study on the Reaction Catalyzed by Serine Transhydroxymethylase. Biochemic. J. **77**, 459 (1960).

33. — Crystalline Dihydropteroylglutamic Acid. Nature (London) **188**, 231 (1960).

34. Blakley, R. L. and B. M. McDougall: Dihydrofolic Acid Reductase from *Streptococcus faecalis* R. J. Biol. Chem. **236**, 1163 (1961).

35. — — The Biosynthesis of Thymidylic Acid. III. Purification of Thymidylate Synthetase and its Spectrophotometric Assay. J. Biol. Chem. **237**, 812 (1962).

36. Boon, W. R. and T. Leigh: Pteridines. III. Unambiguous Synthesis of Xanthopterin and 2-Amino-4-hydroxy-6-methylpteridine. J. Chem. Soc. (London) **1951**, 1497.

37. Boothe, J. H., J. H. Mowat, B. L. Hutchings, R. B. Angier, C. W. Waller, E. L. R. Stokstad, J. Semb, A. L. Gazzola and Y. SubbaRow: Pteroic Acid Derivatives. II. Pteroyl-γ-glutamylglutamic Acid and Pteroyl-γ-glutamyl-γ-glutamylglutamic Acid. J. Amer. Chem. Soc. **70**, 1095 (1948).

38. Boothe, J. H., J. Semb, C. W. Waller, R. B. Angier, J. H. Mowat, B. L. Hutchings, E. L. R. Stokstad and Y. SubbaRow: Pteroic Acid Derivatives. III. Pteroyl-γ-glutamylglutamic Acid and Pteroyl-γ-glutamyl-γ-glutamylglutamic Acid. J. Amer. Chem. Soc. **71**, 2304 (1949).

39. Boothe, J. H., C. W. Waller, E. L. R. Stokstad, B. L. Hutchings, J. H. Mowat, R. B. Angier, J. Semb, Y. SubbaRow, D. B. Cosulich, M. J. Fahrenbach, M. E. Hultquist, E. Kuh, E. H. Northey, D. R. Seeger, J. P. Sickels and J. M. Smith, Jr.: Synthesis of Pteroylglutamic Acid. IV. J. Amer. Chem. Soc. **70**, 27 (1948).

40. Braganca, B. M., I. Aravindakshan and D. S. Ghanekar: Enzymic Cleavage of Folic Acid by Extracts from Human Blood Cells. I. Preparation and Cofactor Requirement of the Enzyme System. Biochim. Biophys. Acta **25**, 623 (1957).

41. Braganca, B. M. and U. W. Kenkare: Flavins as Components of Folic Acid Reductases. Nature (London) **184**, 1488 (1959).

42. Bratton, A. C. and E. K. Marshall, Jr.: A New Coupling Compound for Sulfanilamide Determination. J. Biol. Chem. **128**, 537 (1939).

43. Bregoff, H. M. and C. C. Delwiche: The Formation of Choline and Betain in Leaf Disks of *Beta vulgaris*. J. Biol. Chem. **217**, 819 (1955).

44. BREMER, J., P. H. FIGARD and D. M. GREENBERG: The Biosynthesis of Choline and its Relation to Phospholipid Metabolism. Biochim. Biophys. Acta **43**, 477 (1960).

45. BRENNER-HOLZACH, O. und F. LEUTHARDT: Untersuchungen zur Biosynthese der Pterine bei *Drosophila melanogaster*. Helv. Chim. Acta **42**, 2254 (1959).

46. BRODE, E. und L. JAENICKE: Modelluntersuchungen zur biologischen Aktivierung der Einkohlenstoffeinheiten. II. Ein Modell der Serinhydroxymethylase-Reaktion. Biochem. Z. **332**, 259 (1960).

47. BROWN, G. M.: The Synthesis of Folic Acid by Cell-free Extracts of *Escherichia coli*. Federat. Proc. (Amer. Soc. exp. Biol.) **18**, 19 (1959).

48. — Biosynthesis of Folic Acid. II. Inhibition by Sulfonamides. J. Biol. Chem. **237**, 536 (1962).

49. BROWN, G. M., R. A. WEISMAN and D. A. MOLNAR: The Biosynthesis of Folic Acid. I. Substrate and Cofactor Requirements for Enzymatic Synthesis by Cell-free Extracts of *Escherichia coli*. J. Biol. Chem. **236**, 2534 (1961).

50. BUCHANAN, J. M.: The Enzymatic Synthesis of Inosinic Acid. Proc. Intern. Sympos. Enzyme Chem., Tokyo and Kyoto, 1957. London: Pergamon Press. 1958.

51. — Enzymic Synthesis of Purine Nucleotides. Harvey Lect. **54**, 104 (1960).

52. BUCHANAN, J. M., J. G. FLAKS, S. C. HARTMAN, B. LEVENBERG, L. LUKENS and L. WARREN: The Enzymic Synthesis of Inosinic Acid de novo. In: G. E. W. WOLSTENHOLME and C. M. O'CONNOR, Chemistry and Biology of Purines, p. 233. London: Churchill. 1957.

53. BUCHANAN, J. M. and S. C. HARTMAN: Enzymic Reactions in the Synthesis of the Purines. Adv. Enzymology **21**, 199 (1959).

53a. BUCHANAN, J. M., B. LEVENBERG, J. G. FLAKS and J. A. GLADNER: Interrelationships of Amino Acid Metabolism with Purine Biosynthesis. In: W. D. McELROY and B. GLASS, Amino Acid Metabolism, p. 743. Baltimore: Johns Hopkins Press. 1955.

54. BUCHANAN, J. M., J. C. SONNE and A. M. DELLUVA: Biological Precursors of Uric Acid Carbon. J. Biol. Chem. **166**, 395 (1946).

55. CARPENTER, K. J. and E. KODICEK: A Polarographic Study of Pteroylglutamic Acid and Related Compounds. Biochemic. J. **43**, 11 (1948).

55a. CATHOU, R. E. and J. M. BUCHANAN: Enzymatic Synthesis of the Methyl Group of Methionine. V. J. Biol. Chem. **238**, 1746 (1963).

56. CAVALIERI, L. F. and A. BENDICH: The Ultraviolet Absorption Spectra of Pyrimidines and Purines. J. Amer. Chem. Soc. **72**, 2587 (1950).

57. CHALLENGER, F.: Biological Methylation. Quart. Rev. (Chem. Soc. London) **9**, 255 (1955).

58. CLEMENS, D. H. and W. D. EMMONS: The Synthesis of Orthoamides and their Conversion to Formamidininium Salts. J. Amer. Chem. Soc. **83**, 2588 (1961).

59. COSULICH, D. B., B. ROTH, J. M. SMITH, Jr., M. E. HULTQUIST and R. P. PARKER: Acid Transformation Products of Leucovorin. J. Amer. Chem. Soc. **73**, 5006 (1951).

60. — — — — — Chemistry of Leucovorin. J. Amer. Chem. Soc. **74**, 3252 (1952).

61. COSULICH, D. B., D. R. SEEGER, M. J. FAHRENBACH, K. H. COLLINS, B. ROTH, M. E. HULTQUIST and J. M. SMITH, Jr.: Analogs of Pteroylglutamic Acid. IX. Derivatives with Substituents on the Benzene Ring. J. Amer. Chem. Soc. **75**, 4675 (1953).

62. COSULICH, D. B. and J. M. SMITH, Jr.: Analogs of Pteroylglutamic Acid. I. N^{10}-Alkyl Pteroic Acid and Derivatives. J. Amer. Chem. Soc. **70**, 1922 (1948).

63. — — N^{10}-Nitroso Pteroylglutamic Acid. J. Amer. Chem. Soc. **71**, 3574 (1949).

64. Cosulich, D. B., J. M. Smith, Jr. and H. P. Broquist: Diastereoisomers of Leucovorin. J. Amer. Chem. Soc. **74**, 4215 (1952).

65. Crosbie, G. W.: Pyrimidine Biosynthesis in *Escherichia coli*. Biochemic. J. **69**, 1 P (1958).

66. Dabrowska, W., A. Kazenko and M. Laskowski: Concerning the Specificity of Chicken-pancreas Conjugase. Science (Washington) **110**, 95 (1949).

67. Deodhar, S., W. Sakami and A. L. Stevens: Mechanism of Serine Formation. Federat. Proc. (Amer. Soc. exp. Biol.) **14**, 201 (1955).

68. DeRenzo, E. C.: Chemistry and Biochemistry of Xanthine Oxidase. Adv. Enzymology **17**, 293 (1956).

69. Dewey, V. C., R. E. Parks, Jr. and G. W. Kidder: Growth Response of *Tetrahymena geleii* to Changes in the Basal Media. Arch. Biochemistry **29**, 281 (1950).

70. Dinning, J. S., B. K. Allen, R. S. Young and P. L. Day: The Rôle of Vitamin B_{12} in Thymine Biosynthesis by *Lactobacillus leichmannii*. J. Biol. Chem. **233**, 674 (1958).

71. Dinning, J. S., J. T. Sime, P. S. Work, B. Alten and P. L. Day: The Metabolic Conversion of Folic Acid and Citrovorum Factor to a Diazotizable Amine. Arch. Biochem. Biophys. **65**, 114 (1957).

72. Doctor, V. M.: In vitro Studies on the Conversion of Folic Acid to Citrovorum Factor. J. Biol. Chem. **222**, 959 (1956).

73. — Studies in vivo on the Conversion of Folic Acid to Citrovorum Factor. J. Biol. Chem. **233**, 982 (1958).

74. Doctor, V. M., T. Patton and J. Awapara: Incorporation of Serine-3-C^{14} and Formaldehyde-C^{14} in Methionine in vitro. I. Rôle of Folic Acid. Arch. Biochem. Biophys. **67**, 404 (1957).

75. Donaldson, K. O. and J. C. Keresztesy: Naturally Occurring Forms of Folic Acid. I. "Prefolic A": Preparation of Concentrate and Enzymatic Conversion to Citrovorum Factor. J. Biol. Chem. **234**, 3235 (1959).

76. — — The Interconversion of Prefolic A and Tetrahydrofolic Acid. Federat. Proc. (Amer. Soc. exp. Biol.) **20**, 453 (1961).

77. — — Further Evidence on the Nature of Prefolic A. Biochem. Biophys. Res. Comm. **5**, 289 (1961).

78. — — Naturally Occurring Forms of Folic Acid. II. Enzymic Conversion of Methylenetetrahydrofolic Acid to Prefolic A-methyl-tetrahydrofolate. J. Biol. Chem. **237**, 1298 (1962).

78a. — — Naturally Occurring Forms of Folic Acid. III. Characterization and Properties of 5-Methyl-dihydro-folate, an Oxidation Product of 5-Methyl-tetrahydrofolate. J. Biol. Chem. **237**, 3815 (1962).

79. Elion, G. B. and G. H. Hitchings: The Identification of "β-Dihydroxantho-pterin" as 2,4-Diamino-6-hydroxy-*p*-oxazino-(2,3-d)-pyrimidine. J. Amer. Chem. Soc. **74**, 3877 (1952).

80. Elwyn, D., J. Ashmore, G. F. Cahill, Jr., S. Zottu, W. Welch and A. B. Hastings: Serine Metabolism in Rat Liver Slices. J. Biol. Chem. **226**, 735 (1957).

80a. Elwyn, D. and D. B. Sprinson: The Rôle of Serine and Acetate in Uric Acid Formation. J. Biol. Chem. **184**, 465 (1950).

81. — — The Extensive Synthesis of the Methyl Groups of Thymine in the Adult Rat. J. Amer. Chem. Soc. **72**, 3317 (1950).

82. Esposito, R. G. and A. M. Fletcher: The Relationship of Pteridine Bio-synthesis to the Action of Copper-8-hydroxyquinolinate on Fungal Spores. Arch. Biochem. Biophys. **93**, 369 (1961).

83. FLAKS, J. G. and S. S. COHEN: Virus-induced Acquisition of Metabolic Function. III. Formation and Some Properties of Thymidylate Synthetase of Bacterio-phage-infected *Escherichia coli.* J. Biol. Chem. **234**, 2981 (1959).

83a. — — Virus-induced Acquisition of Metabolic Function. I. Enzymatic Formation of 5-Hydroxymethyl-deoxycytidylate. J. Biol. Chem. **234**, 1501 (1959).

84. FLEMING, L. W. and G. W. CROSBIE: Non-enzymic Transamination between Glycine and Glyoxylate. Biochim. Biophys. Acta **43**, 139 (1960).

85. FORREST, H. S. and J. WALKER: The Effect of Hydrazine on the Condensation of Certain α-Ketols and Related Substances with 2 : 4 : 5-Triamino-6-hydroxy-pyrimidine. J. Chem. Soc. (London) **1949**, 2077.

86. FOSTER, M. A., K. M. JONES and D. D. WOODS: The Purification and Properties of a Factor Containing Vitamin B_{12} Concerned in the Synthesis of Methionine by *Escherichia coli.* Biochemic. J. **80**, 519 (1961).

87. FRANKLIN, A. L., E. L. R. STOKSTAD, M. BELT and T. H. JUKES: Biochemical Experiments with a Synthetic Preparation having an Action Antagonistic to that of Pteroylglutamic Acid. J. Biol. Chem. **169**, 427 (1947).

88. FRIEDKIN, M.: Enzymatic Conversion of Desoxyuridylic Acid to Thymidylic Acid and the Participation of Tetrahydrofolic Acid. Federat. Proc. (Amer. Soc. exp. Biol.) **16**, 183 (1957).

88a. FRIEDKIN, M., E. J. CRAWFORD, E. DONOVAN and E. J. PASTORE: Enzymatic Synthesis of Thymidylate. III. The Further Purification of Thymidylate Synthetase and its Separation from Natural Fluorescent Inhibitors. J. Biol. Chem. **237**, 3811 (1962).

89. FRIEDKIN, M. and A. KORNBERG: The Enzymatic Conversion of Deoxyuridylic Acid to Thymidylic Acid and the Participation of Tetrahydrofolic Acid. In: W. D. MCELROY and B. GLASS, Chemical Basis of Heredity, p. 609. Baltimore: Johns Hopkins Press. 1957.

90. FUJIMORI, E.: Interaction between Pteridines and Tryptophan. Proc. Nat. Acad. Sci. (USA) **45**, 133 (1959).

91. FUTTERMAN, S.: Enzymatic Inactivation of Folic Acid. Federat. Proc. (Amer. Soc. exp. Biol.) **15**, 258 (1956).

92. — Enzymatic Reduction of Folic Acid and Dihydrofolic Acid to Tetra-hydrofolic Acid. J. Biol. Chem. **228**, 1031 (1957).

93. FUTTERMAN, S. and M. SILVERMAN: The "Inactivation" of Folic Acid by Liver. J. Biol. Chem. **224**, 31 (1957).

94. GIRDWOOD, R. H.: Folic Acid, its Analogs and Antagonists. Adv. Clin. Chem. **3**, 1 (1960).

95. GLAZKO, A. J. and L. M. WOLF: Colorimetric Determination of Folic Acid and Adenine. Arch. Biochemistry **21**, 241 (1949).

96. GOLDTHWAIT, D. A. and G. R. GREENBERG: Some Methods for the Study of the de novo Synthesis of Purine Nucleotides. In: S. P. COLOWICK and N. O. KAPLAN, Methods in Enzymology, Vol. II, p. 504. New York: Academic Press. 1955.

97. GOLDTHWAIT, D. A., R. A. PEABODY and G. R. GREENBERG: The Biosynthesis of the Purine Ring. In: W. D. MCELROY and B. GLASS, Amino Acid Metabolism, p. 765. Baltimore: Johns Hopkins Press. 1955.

98. GORDON, M., J. M. RAVEL, R. E. EAKIN and W. SHIVE: Formylfolic Acid, a Functional Derivative of Folic Acid. J. Amer. Chem. Soc. **70**, 878 (1948).

99. GREEN, M. and S. S. COHEN: Studies on the Biosynthesis of Bacterial and Viral Pyrimidines. I. Tracer Studies. J. Biol. Chem. **225**, 387 (1957).

100. Greenberg, D. M., R. Nath and G. K. Humphreys: Purification and Properties of Thymidylate Synthetase from Calf Thymus. J. Biol. Chem. **236**, 2271 (1961).

101. Greenberg, G. R.: A Formylation Cofactor. J. Amer. Chem. Soc. **76**, 1458 (1954).

102. — Rôle of Folic Acid Derivatives in Purine Biosynthesis. Federat. Proc. (Amer. Soc. exp. Biol.) **13**, 745 (1954).

103. Greenberg, G. R. and L. Jaenicke: On the Activation of the One-carbon Unit for the Biosynthesis of Purine Nucleotides. In: G. E. W. Wolstenholme and C. M. O'Connor, The Chemistry and Biology of Purines, p. 204. London: Curchill. 1957.

103a. Greenberg, G. R., L. Jaenicke and M. Silverman: On the Occurrence of N^{10}-Formyl Tetrahydrofolic Acid by Enzymic Formylation of Tetrahydrofolic Acid and on the Mechanism of this Reaction. Biochim. Biophys. Acta **17**, 589 (1955).

104. Guest, J. R., S. Friedman and M. A. Foster: Alternative Pathways for the Methylation of Homocysteine by *Escherichia coli*. Biochemic. J. **84**, 93 P (1962).

105. Guest, J. R., S. Friedman, D. D. Woods and E. L. Smith: A Methyl Analogue of Cobamide Coenzyme in Relation to Methionine Synthesis by Bacteria. Nature (London) **195**, 340 (1962).

106. Guest, J. R. and K. M. Jones: Tetrahydropteroyltriglutamate as a Cofactor of Methionine Synthesis. Biochemic. J. **75**, 12 P (1960).

107. Guest, J. R. and D. D. Woods: Cobalamin and the Enzymic Formation of a Factor Concerned in the Synthesis of Methionine by *Escherichia coli*. Biochemic. J. **77**, 422 (1960).

108. — — Metabolic Interrelationships between Cobalamine and Folic Acid in the Synthesis of Methionine by *Escherichia coli*. In: H. C. Heinrich, Vitamin B_{12} and Intrinsic Factor, p. 686. Stuttgart: F. Enke. 1962.

109. Hakala, M. T. and A. D. Welch: A Polyglutamate Form of Citrovorum Factor Synthesized by *Bacillus subtilis*. J. Bacteriol. **73**, 35 (1957).

110. Hakala, M. T., S. F. Zakrzewski and C. A. Nichol: Relation of Folic Acid Reductase to Amethopterin Resistance in Cultured Mammalian Cells. J. Biol. Chem. **236**, 952 (1961).

111. Hartman, S. C. and J. M. Buchanan: Biosynthesis of the Purines. XXVI. The Identification of the Formyl Donors of the Transformylation Reactions. J. Biol. Chem. **234**, 1812 (1959).

112. Hatch, F. T., A. R. Larrabee, R. E. Cathou and J. M. Buchanan: Enzymatic Synthesis of the Methyl Group of Methionine. I. Identification of the Enzymes and Cofactors Involved in the System Isolated from *Escherichia coli*. J. Biol. Chem. **236**, 1095 (1961).

113. Hatefi, Y., M. J. Osborn, L. D. Kay and F. M. Huennekens: Hydroxymethyl Tetrahydrofolate Dehydrogenase. J. Biol. Chem. **227**, 637 (1957).

114. Hatefi, Y., P. T. Talbert, M. J. Osborn and F. M. Huennekens: Tetrahydrofolic Acid. In: H. A. Lardy, Biochemical Preparations, Vol. 7, p. 89. New York: Wiley & Sons. 1960.

115. Heisler, C. R. and B. S. Schweigert: Conversion of Pteroylglutamic Acid to the Citrovorum Factor by Preparations from *Lactobacillus casei*. J. Bacteriol. **77**, 804 (1959).

116. Helleiner, C. W. and D. D. Woods: Cobalamin and the Synthesis of Methionine by Cell-free Extracts of *Escherichia coli*. Biochemic. J. **63**, 26 P (1956).

117. HENDERSON, R. F. and J. S. DINNING: Effects of Vitamin B_{12} and its Coenzyme on Methylene Tetrahydrofolic Dehydrogenase Activity. Federat. Proc. (Amer. Soc. exp. Biol.) **21**, 471 (1962).

118. HERBERT, V.: The Assay and Nature of Folic Acid Activity in Human Serum. J. Clin. Invest. **40**, 81 (1961).

119. HERMICH, M. R., V. C. DEWEY and G. W. KIDDER: Chromatography of Pteroylglutamic Acid and Related Compounds on Ion-exchange Resins. J. Chromatography **2**, 296 (1959).

120. HEYROVSKÝ, J.: Analysis with the Electronic Polaroscope. Anal. Chim. Acta **12**, 600 (1955).

120a. HIMES, R. H. and J. C. RABINOWITZ: Formyltetrahydrofolate Synthetase. II. Characteristics of the Enzyme and the Enzymic Reaction. J. Biol. Chem. **237**, 2903 (1962).

120b. — — Formyltetrahydrofolate Synthetase. III. Studies on the Mechanism of the Reaction. J. Biol. Chem. **237**, 2915 (1962).

121. HO, P. P. K., K. G. SCRIMGEOUR and F. M. HUENNEKENS: A Novel Reaction between Glyoxylate and Tetrahydrofolate. J. Amer. Chem. Soc. **82**, 5957 (1960).

122. HOLLAND, J. F.: Folic Acid Antagonists. Clin. Pharmacology **2**, 374 (1961).

122a. HUENNEKENS, F. M.: The Role of Dihydrofolic Reductase in the Metabolism of One-Carbon Units. Biochemistry **2**, 151 (1963).

123. HUENNEKENS, F. M., Y. HATEFI and L. D. KAY: Manometric Assay and Cofactor Requirements for Serine Hydroxymethylase. J. Biol. Chem. **224**, 435 (1957).

124. HUENNEKENS, F. M. and M. J. OSBORN: Folic Acid Coenzymes and One-Carbon Metabolism. Adv. Enzymology **21**, 369 (1959).

125. HUENNEKENS, F. M., H. R. WHITELEY and M. J. OSBORN: Mechanisms of Formylation and Hydroxymethylation Reactions. J. Cellular Comp. Physiol. **54**, Suppl. 1, 109 (1959).

126. HULTQUIST, M. E., E. KUH, D. B. COSULICH, M. J. FAHRENBACH, E. H. NORTHEY, D. R. SEEGER, J. P. SICKELS, J. M. SMITH, Jr., R. B. ANGIER, J. H. BOOTHE, B. L. HUTCHINGS, J. H. MOWAT. J. SEMB, E. L. R. STOKSTAD, Y. SUBBAROW and C. W. WALLER: Synthesis of Pteroylglutamic Acid (Liver *L. casei* Factor) and Pteroic Acid. II. J. Amer. Chem. Soc. **70**, 23 (1948).

127. HULTQUIST, M. E., J. M. SMITH, Jr., D. R. SEEGER, D. B. COSULICH and E. KUH: Analogs of Pteroylglutamic Acid. II. 9-Methylpteroylglutamic Acid and Derivatives. J. Amer. Chem. Soc. **71**, 619 (1949).

128. HUMPHREYS, G. K. and D. M. GREENBERG: Conversion of Desoxyuridylic Acid to Thymidylic Acid by a Soluble Extract from Rat Thymus. Arch. Biochem. Biophys. **78**, 275 (1958).

129. HUTCHINGS, B. L., J. H. MOWAT, J. J. OLESON, E. L. R. STOKSTAD, J. H. BOOTHE, C. W. WALLER, R. B. ANGIER, J. SEMB and Y. SUBBAROW: Pteroylaspartic Acid, an Antagonist for Pteroylglutamic Acid. J. Biol. Chem. **170**, 323 (1947).

130. HUTCHINGS, B. L., E. L. R. STOKSTAD, N. BOHONOS, N. H. SLOANE and Y. SUBBAROW: The Isolation of the Fermentation *Lactobacillus casei* Factor. J. Amer. Chem. Soc. **70**, 1 (1948).

131. HUTCHINGS, B. L., E. L. R. STOKSTAD, J. H. BOOTHE, J. H. MOWAT, C. W. WALLER, R. B. ANGIER, J. SEMB and Y. SUBBAROW: A Chemical Method for the Determination of Pteroylglutamic Acid and Related Compounds. J. Biol. Chem. **168**, 705 (1947).

132. Hutchings, B. L., E. L. R. Stokstad, J. H. Mowat, J. H. Boothe, C. W. Waller, R. B. Angier, J. Semb and Y. SubbaRow: Degradation of the Fermentation *L. casei* Factor. II. J. Amer. Chem. Soc. **70**, 10 (1948).

133. Hutner, S. H., H. A. Nathan and H. Baker: Metabolism of Folic Acid and other Pterin-Pteridine Vitamins. Vitamins and Horm. **17**, 1 (1959).

133a. Ichihara, A. and D. M. Greenberg: Further Studies on the Pathway of Serine Formation from Carbohydrate. J. Biol. Chem. **224**, 331 (1957).

134. Jaenicke, L.: Occurrence of N^{10}-Formyltetrahydrofolic Acid and its General Involvement in Transformylation. Biochim. Biophys. Acta **17**, 588 (1955).

135. — Conversion of β-Carbon of Serine to N^{10}-Formyltetrahydrofolic Acid. Federat. Proc. (Amer. Soc. exp. Biol.) **15**, 281 (1956).

136. — Folsäure als Cofaktor biologischer Reaktionen. Experientia **17**, 481 (1961).

137. — Ein biologisch aktives Methylfolat. Z. physiol. Chem. (Hoppe-Seyler) **326**, 168 (1961).

138. — Die Folsäure im Stoffwechsel der Einkohlenstoff-Einheiten. Angew. Chem. **73**, 449 (1961).

139. — Mechanisms of Action of Tetrahydrofolate Cofactors in One-Carbon Transfer. In: A. V. S. de Reuck and M. O'Connor, The Mechanism of Action of Water-soluble Vitamins. Ciba Foundation Study Group No 11, p. 38. London: Churchill Ltd. 1961.

140. — Wirkformen der Folsäure, ihre Struktur und Funktion. In: H. C. Heinrich, Vitamin B_{12} and Intrinsic Factor, p. 701. Stuttgart: F. Enke. 1962.

141. Jaenicke, L. und E. Brode: Modelluntersuchungen zur biologischen Aktivierung der Einkohlenstoff-Einheiten. I. N,N'-Diaryl-äthylendiamine als Modelle der Tetrahydrofolsäure in nicht-enzymatischen Reaktionen. Liebigs Ann. Chem. **624**, 120 (1959).

141a. — — Untersuchungen über Einkohlenstoffkörper. I. Die Tetrahydrofolatformylase aus Taubenleber. Reinigung und Mechanismus. Biochem. Z. **334**, 108 (1961).

142. Jaenicke, L. und P. C. Chan: Die Biosynthese der Folsäure. Angew. Chem. **72**, 752 (1960).

143. Jenny, E. und F. Leuthardt: Über den Abbau von Glykokoll-(2-^{14}C) durch Meerschweinchenleberschnitte. Helv. Chim. Acta **44**, 78 (1961).

144. Johnson, B. C.: The Activity of "*Lactobacillus casei* Factor", "Folic Acid" and "Vitamin Bc" for *Streptococcus faecalis* and *Lactobacillus casei*. J. Biol. Chem. **163**, 255 (1946).

145. Jones, K. M., J. R. Guest and D. D. Woods: Folic Acid and the Synthesis of Methionine by Extracts of *Escherichia coli*. Biochemic. J. **79**, 566 (1961).

146. Jukes, T. H.: Compounds with Folic Acid Activity. In: D. Glick, Methods of Biochemical Analysis, Vol. II, p. 121. New York: Interscience Publ. 1955.

147. Jukes, T. H. and H. P. Broquist: Sulfonamides and Folic Acid Antagonists. (Privatmitteilung.)

148. Kalckar, H. M., N. D. Kjeldgaard and H. Klenow: 2-Amino-4-hydroxy-6-formylpteridine, an Inhibitor of Purine and Pterine Oxydases. Biochim. Biophys. Acta **5**, 516 (1950).

149. Karrer, P. und R. Schwyzer: Über die Konstitution einiger neuer Pteridine. Eine weitere Folsäuresynthese. Helv. Chim. Acta **31**, 777 (1948).

150. Katsunuma, N. und A. Shoda: Folic Acid-synthesizing System in *Mycobacterium avium*. Kôso Kagaku Shinpojiumu **12**, 124 (1957) [Chem. Abstr. **52**, 6488 (1958)].

151. KATSUNUMA, N., T. SHODA and H. NODA: Enzymic Study of Folic Acid Biosynthesis by *Mycobacterium avium*. Vitamins (Kyoto) 11, 322 (1956).

151a. — — — Folic Acid Biosynthesis of *M. avium*. J. Vitaminol. (Osaka) 3, 77 (1957).

152. KAUFMAN, S.: The Nature of the Primary Oxydation Product Formed from Tetrahydropteridines during Phenylalanine Hydroxylation. J. Biol. Chem. 236, 804 (1961).

153. KAY, L. D., M. J. OSBORN, Y. HATEFI and F. M. HUENNEKENS: The Enzymatic Conversion of N^5-Formyltetrahydrofolic Acid (Folinic Acid) to N^{10}-Formyl-tetrahydrofolic Acid. J. Biol. Chem. 235, 195 (1960).

153a. KENKARE, U. W. and B. M. BRAGANCA: Flavin Requirement and Partial Separation of Enzymes Catalysing the Reduction of Folic Acid to Tetra-hydrofolic Acid in Liver. Biochemic. J. 86, 160 (1963).

154. KERESZTESY, J. C. and K. O. DONALDSON: Synthetic Prefolic A. Biochem. Biophys. Res. Comm. 5, 286 (1961).

155. KERESZTESY, J. C. and M. SILVERMAN: Crystalline Citrovorum Factor from Liver. J. Amer. Chem. Soc. 73, 5510 (1951).

156. — — Enzymatic Cleavage of the Citrovorum Factor. J. Amer. Chem. Soc. 75, 1512 (1953).

157. KIDDER, G. W.: Studies on the Biochemistry of Tetrahymena. VI. Folic Acid as a Growth Factor for *T. geleii* W. Arch. Biochemistry 9, 51 (1946).

158. KING, F. E. and P. C. SPENSLEY: The Use of Nitro- and Halogenoketones in the Synthesis of Pteridines, including Pteroic Acid, from 2,4,5-Triamino-6-hydroxy-pyrimidine. J. Chem. Soc. (London) 1952, 2144.

159. KISLIUK, R. L.: Mechanism of Formaldehyde Incorporation into Serine. Federat. Proc. (Amer. Soc. exp. Biol.) 15, 289 (1956).

160. — Studies on the Mechanism of Formaldehyde Incorporation into Serine. J. Biol. Chem. 227, 805 (1957).

161. — Reduced Folic Acid Analogues as Antimetabolites. Nature (London) 188, 584 (1960).

162. — Further Studies on the Relationship of Vitamin B_{12} to Methionine Synthesis in Extracts of *Escherichia coli*. J. Biol. Chem. 236, 817 (1961).

162a. — The Source of Hydrogen for Methionine Methyl Formation. J. Biol. Chem. 238, 397 (1963).

163. KISLIUK, R. L. and W. SAKAMI: The Stimulation of Serine Biosynthesis in Pigeon Liver Extracts by Tetrahydrofolic Acid. J. Amer. Chem. Soc. 76, 1456 (1954).

164. — — A Study of the Mechanism of Serine Biosynthesis. J. Biol. Chem. 214, 47 (1955).

165. KISLIUK, R. L. and D. D. WOODS: Interrelationships between Folic Acid and Cobalamin in the Synthesis of Methionine by Extracts of *E. coli*. Biochemic. J. 75, 467 (1960).

166. KLOPOTOWSKI, T., M. LUZZATI and P. P. SLONIMSKI: Evidence for a new Step between ATP and 5-Amino-4-imidazolcarboxamide Ribotide in the Cyclic Process of Histidine Biosynthesis. Biochem. Biophys. Res. Comm. 3, 150 (1960).

166a. KOEPPE, R. E., M. L. MINTHORN, Jr. and R. J. HILL: Formation of Serine from Glycerol-1,3-C^{14}. Arch. Biochem. Biophys. 68, 355 (1957).

167. KOHN, J., D. L. MOLLIN and L. M. ROSENBACH: Conventional Voltage Electro-phoresis for Formiminoglutamic Acid Determination in Folic Acid Deficiency. J. Clin. Pathology 14, 345 (1961).

168. Komendh, J.: Detection of Folic Acid by Paper Chromatography. Chem. Listy **47**, 1877 (1953) [Chem. Abstr. **48**, 3850 (1954)].

169. Kornberg, A., S. B. Zimmerman, S. R. Kornberg and J. Josse: Enzymatic Synthesis of Desoxyribonucleic Acid. VI. Influence of Bacteriophage T 2 on the Synthetic Pathway in Host Cells. Proc. Nat. Acad. Sci. (USA) **45**, 772 (1959).

170. Korte, F., H. Barkemeyer und G. Synnatschke: Heterocyclen im Stoffwechsel. XI. Umwandlung von Xanthopterin-(8 a-^{14}C) und *p*-Aminobenzoesäure-(2,6-^{14}C) durch Mikroorganismen. Z. physiol. Chem. (Hoppe-Seyler) **314**, 106 (1959).

171. Larrabee, A. R. and J. M. Buchanan: A New Intermediate of Methionine Biosynthesis. Federat. Proc. (Amer. Soc. exp. Biol.) **20**, 9 (1961).

171a. Larrabee, A. R., S. Rosenthal, R. E. Cathou and J. M. Buchanan: A Methylated Derivative of Tetrahydrofolate as an Intermediate of Methionine Biosynthesis. J. Amer. Chem. Soc. **83**, 4094 (1961).

172. Lascelles, J. and D. D. Woods: The Synthesis of "Folic Acid" by *Bacterium coli* and *Staphylococcus aureus* and its Inhibition by Sulfonamides. Brit. J. Exp. Pathol. **33**, 288 (1952).

173. — — The Synthesis of Serine and *Leuconostoc citrovorum* Factor by Cell Suspensions of *Streptococcus faecalis* R. Biochemic. J. **58**, 486 (1954).

174. Levenberg, B.: Enzymatic Deamination of Pterins. Federat. Proc. (Amer. Soc. exp. Biol.) **17**, 263 (1958).

175. Levy, C. C. and W. S. McNutt: The Degradation of Xanthopterin by a Bacterium Isolated from the Soil. Federat. Proc. (Amer. Soc. exp. Biol.) **21**, 369 (1962).

176. Lowry, O. H., O. A. Bessey and E. J. Crawford: Photolytic and Enzymatic Transformations of Pteroylglutamic Acid. J. Biol. Chem. **180**, 389 (1949).

177. Lowy, D. A., G. B. Brown and J. A. Rachele: A Study of Formaldehyde-C^{14}, D$_2$ as a One-carbon Metabolite in the Rat. J. Biol. Chem. **220**, 325 (1956).

178. Mackenzie, C. G.: Conversion of N-Methyl-glycine to Active Formaldehyde and Serine. In: W. D. McElroy and B. Glass, Amino Acid Metabolism, p. 684. Baltimore: Johns Hopkins Press. 1955.

179. Mackenzie, C. G. and R. H. Abeles: Production of Active Formaldehyde in the Mitochondrial Oxidation of Sarcosine-CD$_3$. J. Biol. Chem. **222**, 145 (1956).

180. Mackenzie, C. G. and W. R. Frisell: The Metabolism of Dimethylglycine by Liver Mitochondria. J. Biol. Chem. **232**, 417 (1958).

181. Mader, W. J. and H. A. Frediani: Polarographic Determination of Folic Acid. Analyt. Chemistry **20**, 1199 (1948).

182. Maley, F.: Nucleotide Interconversions. VI. The Enzymic Formation of 5-Methyluridylic Acid. Arch. Biochem. Biophys. **96**, 550 (1962).

183. Mangum, J. H. and K. G. Scrimgeour: Cofactor Requirements and Intermediates in Methionine Biosynthesis. Federat. Proc. (Amer. Soc. exp. Biol.) **21**, 242 (1962).

184. Martin, G. J.: Biological Antagonism. New York: Blakeston & Co. 1952.

185. Martin, G. J., L. Tolman and J. Moss: *d*-(—) Methylfolic Acid, Displacing Agent for Folic Acid. Arch. Biochemistry **12**, 318 (1947).

186. Mason, S. F.: Some Aspects of the Ultraviolet Absorption Spectra of the Pteridines. In: G. E. W. Wolstenholme and M. P. Cameron, Ciba Found. Sympos. Chemistry and Biology of Pteridines. London: Churchill. 1954.

187. Mathews, C. K.: Possible Alternate Forms of Dihydrofolate. Federat. Proc. (Amer. Soc. exp. Biol.) **21**, 242 (1962).

188. Mathews, C. K. and F. M. Huennekens: Enzymic Preparation of the *l,L*-Diastereoisomer of Tetrahydrofolic Acid. J. Biol. Chem. **235**, 3304 (1960).

188a. — — Further Studies on Dihydrofolic Reductase. Federat. Proc. (Amer. Soc. exp. Biol.) **20**, 453 (1961).

189. May, M., T. J. Bardos, F. L. Barger, M. Lansford, J. M. Ravel, G. L. Sutherland and W. Shive: Synthetic and Degradative Investigations of the Structure of Folinic Acid-SF. J. Amer. Chem. Soc. **73**, 3067 (1951).

190. McDougall, B. M. and R. L. Blakley: Mechanism of the Action of Thymidylate Synthetase. Nature (London) **188**, 944 (1960).

191. — — Rôle of Reduced DPNH and Tetrahydropteroylglutamate in the Synthesis of Thymidylic Acid. Biochim. Biophys. Acta **39**, 176 (1960).

192. — — The Biosynthesis of Thymidylic Acid. I. Preliminary Studies with Extracts of *Streptococcus faecalis* R. J. Biol. Chem. **236**, 832 (1961).

193. Mehler, A. H. and W. E. Knox: The Conversion of Tryptophan to Kynurenine in Liver. II. The Enzymatic Hydrolysis of Formylkynurenine. J. Biol. Chem. **187**, 431 (1950).

194. Meister, A.: Biochemistry of the Amino Acids. New York: Academic Press. 1957.

195. Metzler, D. E., M. Ikawa and E. E. Snell: A General Mechanism for Vitamin B_6-Catalyzed Reactions. J. Amer. Chem. Soc. **76**, 648 (1954).

196. Miller, A. and H. Waelsch: The Conversion of Urocanic Acid to Formamidinoglutaric Acid. J. Biol. Chem. **228**, 365 (1957).

197. — — Formimino Transfer from Formamidinoglutaric Acid to Tetrahydrofolic Acid. J. Biol. Chem. **228**, 397 (1957).

198. Mims, V. and M. Laskowski: Studies on Vitamin Bc Conjugase from Chicken Pancreas. J. Biol. Chem. **160**, 493 (1945).

198a. Misra, D. K., S. R. Humphreys, M. Friedkin, A. Goldin and E. J. Crawford: Increased Dihydrofolate Reductase Activity as a possible Basis of Drug Resistance in Leukaemia. Nature (London) **189**, 39 (1961).

199. Mitchell, H. K., E. E. Snell and R. J. Williams: The Concentration of "Folic Acid". J. Amer. Chem. Soc. **63**, 2284 (1941).

200. — — — Folic Acid. I. Concentration from Spinach. J. Amer. Chem. Soc. **66**, 267 (1944).

201. Mitchell, H. K. and R. J. Williams: Folic Acid. III. Chemical and Physiological Properties. J. Amer. Chem. Soc. **66**, 271 (1944).

202. Mitoma, C. and D. M. Greenberg: Studies on the Mechanism of the Biosynthesis of Serine. J. Biol. Chem. **196**, 599 (1952).

203. Mitoma, C. and E. E. Snell: The Rôle of Purine Bases as Histidine Precursors in *Lactobacillus casei*. Proc. Nat. Acad. Sci. (USA) **41**, 891 (1955).

204. Mollin, D. L., A. H. Waters and E. Harriss: Clinical Aspects of the Metabolic Interrelationships between Folic Acid and Vitamin B_{12}. In: H. C. Heinrich, Vitamin B_{12} and Intrinsic Factor, p. 737. Stuttgart: F. Enke. 1962.

205. Mowat, J. H., J. H. Boothe, B. L. Hutchings, E. L. R. Stokstad, C. W. Waller, R. B. Angier, J. Semb, D. B. Cosulich and Y. SubbaRow: The Structure of the Liver *L. casei* Factor. J. Amer. Chem. Soc. **70**, 14 (1948).

206. Mowat, J. H., B. L. Hutchings, R. B. Angier, E. L. R. Stokstad, J. H. Boothe, C. W. Waller, J. Semb and Y. SubbaRow: Pteroic Acid Derivatives. I. Pteroyl-α-glutamylglutamic Acid and Pteroyl-α,γ-glutamylglutamic Acid. J. Amer. Chem. Soc. **70**, 1096 (1948).

207. Moyed, A. S. and B. Magasanik: The Biosynthesis of the Imidazole Ring of Histidine. J. Biol. Chem. **235**, 149 (1960).

208. Nakada, H. I. and I. P. Sund: Glyoxylic Acid Oxidation by Rat Liver. J. Biol. Chem. **233**, 8 (1958).

209. Nakada, H. I. and S. Weinhouse: Studies of Glycine Oxidation in Rat Tissues. Arch. Biochem. Biophys. **42**, 257 (1953).

210. Nakao, A. and D. M. Greenberg: Co-factor Requirements for the Incorporation of $H_2C^{14}O$ and Serine-3-C^{14} into Methionine. J. Amer. Chem. Soc. **77**, 6715 (1955).

211. — — Studies on the Incorporation of Isotope from Formaldehyde-C^{14} and Serine-3-C^{14} into the Methyl Group of Methionine. J. Biol. Chem. **230**, 603 (1958).

212. Nath, R. and D. M. Greenberg: Dihydrofolic Acid Reductase of Calf Thymus. Biochemistry **1**, 435 (1962).

213. Neidle, A. and H. Waelsch: The Origin of the Imidazole Ring of Histidine in *Escherichia coli*. J. Biol. Chem. **234**, 586 (1958).

214. Nichol, C. A.: Symposium on Vitamin Metabolism, No 13. New York: National Vitamin Foundation. 1956.

215. Nimmo-Smith, R. H., J. Lascelles and D. D. Woods: The Synthesis of "Folic Acid" by *Streptobacterium plantarum* and its Inhibition by Sulfonamides. Brit J. Exp. Path. **29**, 264 (1948).

216. Nollstadt, K., L. Klein and I. Ukstins: Stimulation of Methionine Methyl Synthesis in an Enzyme System from Pig Liver by *P. Shermanii* Boiled Extrakt. Federat. Proc. (Amer. Soc. exp. Biol.) **21**, 4 (1962).

217. Noronha, J. M. and M. Silverman: On Folic Acid, Vitamin B_{12}, Methionine and Formiminoglutamic Acid Metabolism. In: H. C. Heinrich, Vitamin B_{12} and Intrinsic Factor, p. 728. Stuttgart: F. Enke. 1962.

217a. — — Distribution of Folic Acid Derivatives in Natural Material. I. Chicken Liver Folates. J. Biol. Chem. **237**, 3299 (1962).

218. Noronha, J. M. and A. Sreenivasan: In vitro Conversion of Pteroylglutamic Acid to Citrovorum Factor by Rat Liver Enzymes. Biochim. Biophys. Acta **44**, 64 (1960).

219. O'Dell, B. L., J. M. Vandenbelt, E. S. Bloom and J. J. Pfiffner: Hydrogenation of Vitamin Bc (Pteroylglutamic Acid) and Related Pteridines. J. Amer. Chem. Soc. **69**, 250 (1947).

220. Oliverio, V. T.: Chromatographic Separation and Purification of Folic Acid Analogs. Analyt. Chemistry **33**, 263 (1961).

221. Osborn, M. J., Y. Hatefi, L. D. Kay and F. M. Huennekens: Evidence for the Enzymic Deacylation of N^{10}-Formyl Tetrahydrofolic Acid. Biochim. Biophys. Acta **26**, 208 (1957).

222. Osborn, M. J. and F. M. Huennekens: Participation of Anhydroleucovorin in the Hydroxymethyl Tetrahydrofolate Dehydrogenase System. Biochim. Biophys. Acta **26**, 646 (1957).

223. — — Enzymatic Reduction of Dihydrofolic Acid. J. Biol. Chem. **233**, 969 (1958).

224. Osborn, M. J., P. T. Talbert and F. M. Huennekens: The Structure of "Aktive Formaldehyde" (N^5,N^{10}-Methylene Tetrahydrofolic Acid). J. Amer. Chem. Soc. **82**, 4921 (1960).

225. Osborn, M. J., F. N. Vercamer, P. T. Talbert and F. M. Huennekens: The Enzymatic Synthesis of Hydroxymethyltetrahydrofolic Acid (Active Hydroxymethyl). J. Amer. Chem. Soc. **79**, 6565 (1957).

225a. Pastore, E. J. and M. Friedkin: The Enzymatic Synthesis of Thymidylate. II. Transfer of Tritium from Tetrahydrofolate to the Methyl Group of Thymidylate. J. Biol. Chem. **237**, 3802 (1962).

226. PATTERSON, A. M. and L. T. CAPELL: The Ring-Index. New York: Reinhold Publ. Co. 1960.

227. PEEL, J. L.: Vitamin B_{12} Derivatives and the CO_2-Pyruvate Exchange Reaction. A Reappraisal. J. Biol. Chem. **237**, PC 236 (1962).

228. PERAULT, A.-M. and B. PULLMAN: Electronic Structure and Biochemical Function of Folic Acid Coenzymes. Biochim. Biophys. Acta **44**, 251 (1960).

229. — — Structure électronique et mode d'action des antimétabolites de l'acide folique. Biochim. Biophys. Acta **52**, 266 (1961).

230. PETERS, J. M. and D. M. GREENBERG: Studies on the Conversion of Citrovorum Factor to a Serine Aldolase Cofactor. J. Biol. Chem. **226**, 329 (1957).

231. — — Citrovorum Factor Cyclodehydrase. J. Amer. Chem. Soc. **80**, 2719 (1958).

232. — — Dihydrofolic Acid Reductase. J. Amer. Chem. Soc. **80**, 6679 (1958).

233. — — Studies on Folic Acid Reduction. Biochim. Biophys. Acta **32**, 273 (1959).

234. PEZER, L. I. and S. S. COHEN: Virus Induced Acquisition of Metabolic Function. V. Purification and Properties of the Deoxycytodylate Hydroxymethylase and Studies on its Origin. J. Biol. Chem. **237**, 1251 (1962).

235. PFIFFNER, J. J., S. B. BINKLEY, E. S. BLOOM, O. D. BIRD, A. D. EMMETT, A. G. HOGAN and B. L. O'DELL: Isolation of the Antianemia Factor (Vitamin Bc) in Crystalline Form from Liver. Science (Washington) **97**, 404 (1943).

236. PFIFFNER, J. J., S. B. BINKLEY, E. S. BLOOM and B. L. O'DELL: Isolation and Characterization of Vitamin Bc from Liver and Yeast. J. Amer. Chem. Soc. **69**, 1476 (1947).

237. PFIFFNER, J. J., D. G. CALKINS, E. S. BLOOM and B. L. O'DELL: On the Peptide Nature of Vitamin Bc Conjugate from Yeast. J. Amer. Chem. Soc. **68**, 1392 (1946).

238. PFIFFNER, J. J., D. G. CALKINS, B. L. O'DELL, E. S. BLOOM, R. A. BROWN, C. J. CAMPBELL and O. D. BIRD: Isolation of an Antianemia Factor (Vitamin Bc Conjugate) in Crystalline Form from Yeast. Science (Washington) **102**, 228 (1945).

239. PFLEIDERER, W. and E. C. TAYLOR: Pteridines. XXII. 5,8-Dihydropteridines by Sodium Borohydride Reduction. J. Amer. Chem. Soc. **82**, 3765 (1960).

240. PITTS, J. D. and G. W. CROSBIE: The Conversion of Glycine into C_1-Units by *Escherichia coli*. Biochemic. J. **83**, 35 P (1962).

241. PITTS, J. D., J. A. STEWART and G. W. CROSBIE: Observations on Glycine Metabolism in *Escherichia coli*. Biochim. Biophys. Acta **50**, 361 (1961).

242. POHLAND, A., E. H. FLYNN, R. G. JONES and W. SHIVE: A Proposed Structure for Folinic Acid-SF, a Growth Factor Derived from Pteroylglutamic Acid. J. Amer. Chem. Soc. **73**, 3247 (1951).

243. QUAYLE, J. R.: Metabolism of C_1 Compounds in Autotrophic and Heterotrophic Microorganisms. Annu. Rev. Microbiol. **15**, 119 (1961).

243a. — Enzymic Decarboxylation of Oxalylcoenzyme A to Formylcoenzyme A. Biochemic. J. **87**, 10 P (1963).

244. RABINOWITZ, J. C.: Folic Acid. In: P. O. BOYER, H. A. LARDY and K. MYRBÄCK, The Enzymes, Vol. II, p. 185. New York: Academic Press. 1960.

245. RABINOWITZ, J. C. and R. H. HIMES: Folic Acid Coenzymes. Federat. Proc. (Amer. Soc. exp. Biol.) **19**, 963 (1960).

246. RABINOWITZ, J. C. and W. E. PRICER, Jr.: ATP Formation Accompanying Formiminoglycine Utilization. J. Amer. Chem. Soc. **78**, 1513 (1956).

247. — — The Enzymatic Synthesis of N^{10}-Formyltetrahydrofolic Acid and its Rôle in ATP Formation during Formiminoglycine Degradation. J. Amer. Chem. Soc. **78**, 4176 (1956).

247a. Rabinowitz, J. C. and W. E. Pricer, Jr.: Formimino-tetrahydrofolic Acid and Methenyltetrahydrofolic Acid as Intermediates in the Formation of N^{10}-Formyltetrahydrofolic Acid. J. Amer. Chem. Soc. **78**, 5702 (1956).

248. — — An Enzymatic Method for the Determination of Formic Acid. J. Biol. Chem. **229**, 321 (1957).

249. — — Formation, Isolation and Properties of 5-Formiminotetrahydrofolic Acid. Federat. Proc. (Amer. Soc. exp. Biol.) **16**, 236 (1957).

249a. — — Formyltetrahydrofolate Synthetase. I. Isolation and Crystallization of the Enzyme. J. Biol. Chem. **237**, 2898 (1962).

250. Ramasastri, B. V. and R. L. Blakley: 5,10-Methylene-tetrahydrofolate Dehydrogenase from Baker's Yeast. I. Partial Purification and some Properties. J. Biol. Chem. **237**, 1982 (1962).

251. Rauen, H. M.: Transformylierungen und Transhydroxymethylierungen. I. Allgemeine Reaktionsbedingungen und Formyldonatoren der aeroben N^{10}-Formylierung der Pteroylglutaminsäure. Biochem. Z. **328**, 562 (1957).

252. Rauen, H. M. und L. Jaenicke: Über „aktivierte Ameisensäure" und die fermentative Transformylierung. Z. physiol. Chem. (Hoppe-Seyler) **293**, 46 (1953).

253. Rauen, H. M., W. Stamm und K.-H. Kimbel: N(12)-Formylfolsäure als fermentatives Umwandlungsprodukt der Folsäure. Z. physiol. Chem. (Hoppe-Seyler) **289**, 80 (1952).

254. Rauen, H. M. und H. Waldmann: Über die Gegenstromverteilung von Pterinen. Z. physiol. Chem. (Hoppe-Seyler) **286**, 180 (1950).

255. Reynolds, J. J. and G. M. Brown: Enzymatic Formation of the Pteridine Moiety of Folic Acid from Guanosine Compounds. J. Biol. Chem. **237**, PC 2713 (1962).

256. Rickes, E. L., L. Chaiet and J. C. Keresztesy: Isolation of Rhizopterin, a new Growth Factor for *Streptococcus lactis* R. J. Amer. Chem. Soc. **69**, 2749 (1947).

257. Robbins, E. A. and P. D. Boyer: Determination of the Equilibrium of the Hexokinase Reaction and the Free Energy of Hydrolysis of Adenosine Triphosphate. J. Biol. Chem. **224**, 121 (1957).

257a. Roberts, DeWayne: Cleavage of Folic Acid by Hydrogen Peroxide. Biochim. Biophys. Acta **54**, 572 (1961).

257b. Roberts, DeWayne and C. A. Nichol: Biosynthesis of Thymine and Formyltetrahydrofolate by Vitamin B_{12}-deficient Cells of *L. leichmanii*. J. Biol. Chem. **237**, 2278 (1962).

258. Rohrbaugh, T.: Study of the Utilization of Methyl-THF for Methionine Biosynthesis. Federat. Proc. (Amer. Soc. exp. Biol.) **21**, 4 (1962).

259. Rooze, V., W. Sakami and D. Blair: The Rôle of Methyltetrahydrofolate in Methionine Biosynthesis in Pig Liver. Federat. Proc. (Amer. Soc. exp. Biol.) **21**, 4 (1962).

260. Rosenthal, S. and J. M. Buchanan: Studies on ATP Function in Methionine Biosynthesis. Federat. Proc. (Amer. Soc. exp. Biol.) **21**, 470 (1962).

261. Roth, B., M. E. Hultquist, M. J. Fahrenbach, D. B. Cosulich, H. P. Broquist, J. A. Brockman, Jr., J. M. Smith, Jr., R. P. Parker, E. L. R. Stokstad and T. H. Jukes: Synthesis of Leucovorin. J. Amer. Chem. Soc. **74**, 3247 (1952).

262. Sagers, R. D., J. V. Beck, W. Gruber and I. C. Gunsalus: A Tetrahydrofolic Acid Linked Formimino Transfer Enzyme. J. Amer. Chem. Soc. **78**, 694 (1956).

263. Sagers, R. D. and I. C. Gunsalus: Intermediary Metabolism of *Diplococcus glycinophilus*. J. Bacteriol. **81**, 541 (1961).

264. SAKAMI, W.: The Conversion of Formate and Glycine to Serine and Glycogen in the Intact Rat. J. Biol. Chem. **176**, 995 (1948).

265. — The Conversion of Glycine into Serine in the Intact Rat. J. Biol. Chem. **178**, 519 (1949).

266. — The Biochemical Relationship between Glycine and Serine. In: W. D. McELROY and B. GLASS, Amino Acid Metabolism, p. 658. Baltimore: Johns Hopkins Press. 1955.

267. SAKAMI, W. and R. KNOWLES: Purification of Folic Acid. Science (Washington) **129**, 274 (1959).

268. SAKAMI, W. and I. UKSTINS: Enzymatic Methylation of Homocysteine by a Synthetic Tetrahydrofolate Derivative. J. Biol. Chem. **236**, PC 50 (1961).

269. SAUBERLICH, H. E.: Comparative Studies with the Natural and Synthetic Citrovorum Factor. J. Biol. Chem. **195**, 337 (1952).

270. SCHEINDLIN, S., A. LEE and I. GRIFFITH: The Action of Riboflavin on Folic Acid. J. Amer. Pharm. Assoc. **41**, 420 (1952).

271. SCRIMGEOUR, K. G. and F. M. HUENNEKENS: Occurrence of a DPN-linked, N^5-N^{10}-Methylene Tetrahydrofolic Dehydrogenase in Ehrlich Ascites Tumor Cells. Biochem. Biophys. Res. Comm. **2**, 230 (1960).

272. SEEGER, D. R., D. B. COSULICH, J. M. SMITH, Jr. and M. E. HULTQUIST: Analogs of Pteroylglutamic Acid. III. 4-Amino Derivatives. J. Amer. Chem. Soc. **71**, 1753 (1949).

273. SEEGER, D. R., J. M. SMITH, Jr. and M. E. HULTQUIST: Antagonist for Pteroylglutamic Acid. J. Amer. Chem. Soc. **69**, 2567 (1947).

274. SEEGMILLER, J. E., M. SILVERMAN, H. TABOR and A. H. MEHLER: Synthesis of a Metabolic Product of Histidine. J. Amer. Chem. Soc. **76**, 6205 (1954).

275. SHEMIN, D.: The Biological Conversion of *L*-Serine to Glycine. J. Biol. Chem. **162**, 297 (1946).

276. SHIMAZONO, H. and O. HAYAISHI: Enzymatic Decarboxylation of Oxalic Acid. J. Biol. Chem. **227**, 151 (1957).

277. SHIOTA, T.: Enzymic Synthesis of Folic Acid-like Compounds by Cell-free Extracts of *Lactobacillus arabinosus*. Arch. Biochem. Biophys. **80**, 155 (1959).

278. SHIOTA, T. and H. N. DISRAELY: The Enzymic Synthesis of Dihydrofolate from 2-Amino-4-hydroxymethyl-dihydropteridine and *p*-Aminobenzoylglutamate by Extracts of *Lactobacillus plantarum*. Biochim. Biophys. Acta **52**, 457 (1961).

279. SHIOTA, T., H. N. DISRAELY and M. P. McCANN: Preparation of Dihydropteridine Diphosphate, an Intermediate in Dihydrofolate Synthesis. Biochem. Biophys. Res. Comm. **7**, 194 (1962).

280. SHIVE, W.: Utilization of Antimetabolites in the Study of Biochemical Processes in Living Organisms. Ann. New York Acad. Sci. **52**, 1212 (1950).

281. SILBER, R., F. M. HUENNEKENS and B. W. GABRIO: Dihydrofolic Reductase and Thymidylate Synthetase in the Leukocyte. Federat. Proc. (Amer. Soc. exp. Biol.) **21**, 241 (1962).

282. SILVERMAN, M.: N^5-Formyltetrahydrofolic Acid—Glutamic Acid Transformylase from Hog Liver. In: S. P. COLOWICK and N. O. KAPLAN, Methods in Enzymology, Vol. V, p. 790. New York: Academic Press. 1962.

283. SILVERMAN, M., F. G. EBAUGH, Jr. and R. C. GARDINER: The Nature of Labile Citrovorum Factor in Human Urine. J. Biol. Chem. **223**, 259 (1956).

284. SILVERMAN, M. and R. C. GARDINER: Labile Citrovorum Factor in Urine. J. Bacteriol. **71**, 433 (1956).

285. SILVERMAN, M., J. C. KERESZTESY and G. J. KOVAL: Isolation of N-10-Formylfolic Acid. J. Biol. Chem. **211**, 53 (1954).

286. Silverman, M., J. C. Keresztesy, G. J. Koval and R. C. Gardiner: Citrovorum Factor and the Synthesis of Formylglutamic Acid. J. Biol. Chem. **226**, 83 (1957).

287. Silverman, M., L. W. Law and B. Kaufman: The Distribution of Folic Acid Activities in Lines of Leucemic Cells of the Mouse. J. Biol. Chem. **236**, 2530 (1961).

288. Silverman, M. and J. M. Noronha: A Method for the Preparation of Tetrahydrofolic Acid. Biochem. Biophys. Res. Comm. **4**, 180 (1961).

289. Slavík, K., V. Slavíková and Z. Kolman: Metabolism of Folic Acid. VI. Preparation of Intermediary Antimetabolites of Folic Acid. Coll. Czech. Chem. Commun. **25**, 1929 (1960).

290. Slavíková, V. and K. Slavík: On the Mechanism of Action of some 4-Aminoanalogues of Folic Acid. Experientia **17**, 113 (1961).

291. Sletzinger, M., D. Reinhold, J. Grier, M. Beachem and M. Tishler: The Synthesis of Pteroylglutamic Acid. J. Amer. Chem. Soc. **77**, 6365 (1955).

292. Smith, C. W., R. S. Rasmussen and S. A. Ballard: Assignment of Amide Structures to the Supposed 2,3-Dihydro-2-benzimidazolols and their Acylation Products. J. Amer. Chem. Soc. **71**, 1082 (1949).

293. Snell, E. E.: Microbiological Methods in Vitamin Research. In: P. György, Vitamin Methods, Vol. I, p. 327. New York: Academic Press. 1950.

294. Snell, E. E. and H. K. Mitchell: Purine and Pyrimidine Bases as Growth Substances for Lactic Acid Bacteria. Proc. Nat. Acad. Sci. (USA) **27**, 1 (1941).

295. Snell, E. E. and W. H. Peterson: Growth Factors for Bacteria. X. Additional Factors Required by Certain Lactic Acid Bacteria. J. Bacteriol. **39**, 273 (1940).

296. Sober, H. A. and E. A. Peterson: Chromatography of Proteins on Cellulose Ion-exchangers. J. Amer. Chem. Soc. **76**, 1711 (1954).

297. Sprinson, D. B.: On the Formation of C_1 Fragments from Serine. In: W. D. McElroy and B. Glass, Amino Acid Metabolism, p. 608. Baltimore: Johns Hopkins Press. 1955.

298. Stekol, J. A., S. Weiss, E. I. Anderson, P. T. Hsu and A. Watjen: Vitamin B_{12} and Folic Acid in Relation to Methionine Synthesis from Betaine in vivo and in vitro. J. Biol. Chem. **226**, 95 (1957).

299. Stevens, A. and W. Sakami: Biosynthesis of Methionine in Liver. J. Biol. Chem. **234**, 2063 (1959).

300. Stokes, J. L. and A. Larsen: Transformation of the *Streptococcus lactis* R Factor to "Folic Acid" by Resting Cell Suspensions of Enterococci. J. Bacteriol. **50**, 219 (1945).

301. Stokstad, E. L. R.: Some Properties of a Growth Factor for *Lactobacillus casei*. J. Biol. Chem. **149**, 573 (1943).

302. — Pteroylglutamic Acid. In: W. H. Sebrell, Jr. and R. S. Harris, The Vitamins, Vol. III, p. 89. New York: Academic Press. 1954.

303. Stokstad, E. L. R., B. L. Hutchings, J. H. Mowat, J. H. Boothe, C. W. Waller, R. B. Angier, J. Semb and Y. SubbaRow: The Degradation of the Fermentation *Lactobacillus casei* Factor. I. J. Amer. Chem. Soc. **70**, 5 (1948).

304. Stokstad, E. L. R., B. L. Hutchings and Y. SubbaRow: The Isolation of the *Lactobacillus casei* Factor from Liver. J. Amer. Chem. Soc. **70**, 3 (1948).

305. Strecker, H. J.: Formate Fixation in Pyruvate by *Escherichia coli*. J. Biol. Chem. **189**, 815 (1951).

306. Tabor, H.: Histidine Degradation. In: S. P. Colowick and N. O. Kaplan, Methods in Enzymology, Vol. V, p. 784. New York: Academic Press. 1962.

307. Tabor, H. and J. C. Rabinowitz: Intermediate Steps in the Formylation of Tetrahydrofolic Acid by Formimino-glutamic Acid in Rabbit Liver. J. Amer. Chem. Soc. **78**, 5705 (1956).

308. Tabor, H., M. Silverman, A. H. Mehler, F. S. Daft and H. Bauer: L-Histidine Conversion to Urinary Glutamic Acid Derivatives in Folic-deficient Rats. J. Amer. Chem. Soc. **75**, 756 (1953).

309. Tabor, H. and C. Wyngaarden: Determination of Formimidoyl-glutamic Acid in Urine. J. Clin. Invest. **37**, 824 (1958).

310. — — The Enzymatic Formation of Formimino-tetrahydrofolic Acid, 5,10-Methinyltetrahydrofolic Acid, and 10-Formyltetrahydrofolic Acid in the Metabolism of Formiminoglutamic Acid. J. Biol. Chem. **234**, 1830 (1959).

311. Takeyama, S. and J. M. Buchanan: Enzymatic Synthesis of the Methyl Group of Methionine. III. Spectral and Electrophoretic Studies of the Prosthetic Group of the B_{12} Enzyme. J. Biochemistry (Tokyo) **49**, 578 (1961).

312. Takeyama, S., F. T. Hatch and J. M. Buchanan: Enzymatic Synthesis of the Methyl Group of Methionine. II. Involvement of Vitamin B_{12}. J. Biol. Chem. **236**, 1102 (1961).

313. Torii, M.: Enzymic Hydrolysis of a Glutamyl Peptide of *Bacillus megatherium*. J. Biochemistry (Tokyo) **46**, 513 (1959).

314. Tschesche, R.: Eine neue Deutung des antibakteriellen Wirkungsmechanismus der Sulfonamide. Z. Naturforsch. **26**, 10 (1947).

314a. Urbahn, H. und S. Rapoport: Über den Abbau von Glyzin in roten Blutzellen. Acta biol. Med. Germanica **6**, 16 (1961).

315. Usdin, E.: Blood Folic Acid Studies. VI. Chromatographic Resolution of Folic Acid Active Substances Obtained from Blood. J. Biol. Chem. **234**, 2373 (1959).

316. Usdin, E. and J. Porath: Separation of Folic Acid and Derivatives by Electrophoresis and Anion Exchange Chromatography. Ark. Kemi **11**, 41 (1957).

317. Usdin, E., G. H. Shockman and G. Toennies: Tetrazolium Bioautography. Appl. Microbiol. **2**, 29 (1954).

318. Vieira, E. and E. Shaw: The Utilization of Purines in the Biosynthesis of Folic Acid. J. Biol. Chem. **236**, 2507 (1961).

319. Vigneaud, V. du: A Trail of Research. Ithaka, N. Y.: Cornell Univ. Press. 1952.

320. Volcani, B. E. and P. Margalith: A New Species *(Flavobacterium polyglutamicum)* which Hydrolyzes the γ-Glutamyl Bond in Polypeptides. J. Bacteriol. **74**, 646 (1957).

320a. Wacker, A., H. Grisebach, A. Trebst und F. Weygand: Wirkungsmechanismus der Sulfonamide. I. Mitt. über Coenzym F. Angew. Chem. **66**, 326 (1954).

321. Wahba, A. J. and M. Friedkin: Direct Spectrophotometric Evidence for the Oxidation of Tetrahydrofolate during the Enzymatic Synthesis of Thymidylate. J. Biol. Chem. **236**, PC 11 (1961).

321a. — — The Enzymatic Synthesis of Thymidylate. I. Early Steps in the Purification of Thymidylate Synthetase of *E. coli*. J. Biol. Chem. **237**, 3794 (1962).

322. Waller, C. W., A. A. Goldman, R. B. Angier, J. H. Boothe, B. L. Hutchings, J. H. Mowat and J. Semb: 2-Amino-4-hydroxy-6-pteridine-carboxaldehyde. J. Amer. Chem. Soc. **72**, 4630 (1950).

323. Waller, C. W., B. L. Hutchings, J. H. Mowat, E. L. R. Stokstad, J. H. Boothe, R. B. Angier, J. Semb, Y. SubbaRow, D. B. Cosulich, M. J. Fahrenbach, M. E. Hultquist, E. Kuh, E. H. Northey, D. R. Seeger, J. P. Sickels and J. M. Smith, Jr.: Synthesis of Pteroylglutamic Acid (Liver *L. casei* Factor) and Pteroic Acid. I. J. Amer. Chem. Soc. **70**, 19 (1948).

324. Wanzlick, H. W. und W. Löchel: 1,2-Dianilinoäthan als Aldehyd-Reagenz. Chem. Ber. **86**, 1463 (1953).

325. Waters, A. H. and D. L. Mollin: Studies on the Folic Acid Activity of Human Serum. J. Clin. Pathology **14**, 335 (1961).

326. Webb, M.: Aminopterin Inhibition in *Aerobacter aerogenes*: Alanine and Valine Accumulation during the Inhibition and their Utilization on Recovery. Biochemic. J. **70**, 472 (1958).

327. — The Valine Carboxyl Group as a Source of Active Formate in *Aerobacter aerogenes*. J. Gen. Microbiol. **18**, xiv (1958).

328. Weinhouse, S.: The Synthesis and Degradation of Glycine. In: W. D. McElroy and B. Glass, Amino Acid Metabolism, p. 637. Baltimore: Johns Hopkins Press. 1955.

329. Weislogel, O. and T. J. Bond: Studies on Cofactor Function of Folinic Acid using Radio-labeled Compounds. Arch. Biochem. Biophys. **89**, 221 (1960).

330. Welch, A. D. and C. A. Nichol: Water-soluble Vitamins Concerned with one- and two-Carbon Intermediates. Annu. Rev. Biochem. **21**, 633 (1952).

331. Welliky, I. and D. Shemin: Metabolism of α-Amino-β-ketoadipic Acid. Federat. Proc. (Amer. Soc. exp. Biol.) **16**, 268 (1957).

332. Werkheiser, W. C.: Specific Binding of 4-Amino Folic Acid Analogues by Folic Acid Reductase. J. Biol. Chem. **236**, 888 (1961).

333. Weygand, F., E. F. Möller und A. Wacker: Mikrobiologische Synthese der Folinsäure aus 2-Amino-6-oxy-pterinaldehyd-8- und *p*-Aminobenzoyl-*L*-glutaminsäure. Z. Naturforsch. **4 b**, 269 (1949).

334. Weygand, F. und G. Schäfer: Synthese von Pteroyl-1-glutaminsäure [11-C^{14}] (Folinsäure [11-C^{14}]). Chem. Ber. **85**, 307 (1952).

335. Weygand, F. und V. Schmied-Kowarzik: Weitere Folinsäure-Synthesen. Chem. Ber. **82**, 333 (1949).

336. Weygand, F., V. Schmied-Kowarzik, A. Wacker und W. Rupp: Weitere Synthesen von Pteridinen. Chem. Ber. **83**, 460 (1950).

337. Weygand, F. und H. Simon: Herstellung isotopenhaltiger organischer Verbindungen. In: Houben-Weyl, Methoden der organischen Chemie, IV. Aufl., Bd. 4, 2. Teil. Stuttgart: Thieme. 1955.

338. Weygand, F., H. Simon, G. Dahms, M. Waldschmidt, H. J. Schliep und H. Wacker: Über die Biogenese des Leucopterins. Angew. Chem. **73**, 402 (1961).

339. Weygand, F. und O. P. Swoboda: Verlauf der Folsäure-Synthese mit 1,1,3-Tribromaceton [3-^{14}C]. Chem. Ber. **89**, 18 (1956).

340. Weygand, F., A. Wacker und V. Schmied-Kowarzik: Kondensationsprodukte von Oxyketonen und Aminoketonen mit 2,4,5-Triamino-6-oxy-pyrimidin. Experientia **4**, 427 (1948).

340a. — — — Über die Kondensationsprodukte von *p*-Tolyl-*d*-isoglucosamin und Zuckern mit 6-Oxy-2,4,5-triaminopyrimidin; eine neue Folinsäure-Synthese. Chem. Ber. **82**, 25 (1949).

341. Whiteley, H. R. and F. M. Huennekens: Mechanism of the Reaction Catalyzed by the Formate Activating Enzyme from *Micrococcus aerogenes*. J. Biol. Chem. **237**, 1290 (1962).

342. WHITELEY, H. R., M. J. OSBORN and F. M. HUENNEKENS: The Mechanism of Formate Activation. J. Amer. Chem. Soc. **80**, 757 (1958).

343. — — — Purification and Properties of the Formate Activating Enzyme from *Micrococcus aerogenes*. J. Biol. Chem. **234**, 1538 (1959).

344. WHITTAKER, V. K. and R. L. BLAKLEY: The Biosynthesis of Thymidylic Acid. II. Preliminary Studies with Calf Thymus Extracts. J. Biol. Chem. **236**, 838 (1961).

345. WIELAND, O. P., B. L. HUTCHINGS and J. H. WILLIAMS: Studies on the Natural Occurrence of Folic Acid and the Citrovorum Factor. Arch. Biochemistry **40**, 105 (1952).

346. WILLIAMS, R. J., R. E. EAKIN, E. BEERSTECHER, Jr. and W. SHIVE: The Biochemistry of the B-Vitamins. New York: Reinhold Publ. Co. 1950.

347. WILMANNS, W.: Die Tetrahydrofolsäure-abhängige Aktivierung von Ein-kohlenstoffeinheiten in normalen und pathologischen weißen Blutzellen. Klin. Wschr. **39**, 884 (1961).

347a. — Bestimmung, Eigenschaften und Bedeutung der Dihydrofolsäure-Reduktase in den weißen Blutzellen bei Leukämien. Klin. Wschr. **40**, 533 (1962).

348. WILMANNS, W., B. RÜCKER und L. JAENICKE: Zur Biogenese von Methionin. Z. physiol. Chem. (Hoppe-Seyler) **322**, 283 (1960).

348a. WILSON, E. M. and E. E. SNELL: Metabolism of α-Methylserine. I. α-Methyl-serine Hydroxymethylase. J. Biol. Chem. **237**, 3171 (1962).

349. WINKLER, K. C. and P. C. DE HAAN: On the Action of Sulfanilamide. XII. A Set of Non-competitive Sulfanilamide Antagonists for *Escherichia coli*. Arch. Biochemistry **18**, 97 (1948).

350. WINSTEIN, W. A. and E. EIGEN: Bioautographic Studies with Use of *Leuconostoc citrovorum* 8081. J. Biol. Chem. **184**, 155 (1950).

351. WITTENBERG, J. B., J. M. NORONHA and M. SILVERMAN: Folic Acid Derivatives in the Gas Gland of *Physalia physalis* L. Biochemic. J. **85**, 9 (1962).

352. WITTLE, E. L., B. L. O'DELL, J. M. VANDENBELT and J. J. PFIFFNER: Oxidative Degradation of Vitamin B_c (Pteroylglutamic Acid). J. Amer. Chem. Soc. **69**, 1786 (1947).

353. WOERNLEY, D. L.: The Magnetochemistry of Vitamins. Arch. Biochem. Biophys. **54**, 378 (1955).

354. WOLF, B. and R. D. HOTCHKISS: Genetically Modified Folic Acid Synthesizing Enzymes of Pneumococcus. Biochemistry **2**, 145 (1963).

355. WOLF, D. E., R. C. ANDERSON, E. A. KACZKA, S. A. HARRIS, G. E. ARTH, P. L. SOUTHWICK, R. MOZINGO and K. FOLKERS: The Structure of Rhizopterin. J. Amer. Chem. Soc. **69**, 2753 (1947).

356. WOOD, R. C. and G. H. HITCHINGS: Nature of the Citrovorum Factor Requirement of *Pediococcus cerevisiae*. J. Bacteriol. **79**, 524 (1960).

357. WOODS, D. D.: Biosynthesis and Breakdown of Folic Acid. Proc. 4th Internat. Congr. Biochemistry, Vol. XI, p. 97. London: Pergamon Press. 1958.

358. WRIGHT, B. E.: Poly-glutamyl Pteridine Coenzymes. J. Amer. Chem. Soc. **77**, 3930 (1955).

358a. — The Rôle of Polyglutamyl Pteridine Coenzymes in Serine Metabolism. II. A Comparison of Various Pteridine Derivatives. J. Biol. Chem. **219**, 873 (1956).

359. — Folic Acid Coenzyme Forms and Function. Proc. 4th Internat. Congr. Biochemistry, Vol. XI, p. 266. London: Pergamon Press. 1958.

360. WRIGHT, B. E. and M. L. ANDERSON: Pteridine Reductase. Biochim. Biophys. Acta **28**, 370 (1958).

361. Wright, B. E., M. L. Anderson and E. C. Herman: The Rôle of Polyglutamyl Pteridine Coenzymes in Serine Metabolism. III. The Enzymatic Formation of Dihydrofolic Acid and Dihydropterin. J. Biol. Chem. **230**, 271 (1958).

362. Wright, B. E. and T. C. Stadtman: The Rôle of Polyglutamyl Pteridine Coenzymes in Serine Metabolism. I. Cofactor Requirements in the Conversion of Serine to Glycine. J. Biol. Chem. **219**, 863 (1956).

363. Wright, L. D., H. R. Skeggs and A. D. Welch: Observations on the Occurrence of "Folic Acid" in Liver and Muscle. Arch. Biochemistry **6**, 15 (1945).

364. Zakrzewski, S. F.: Purification and Properties of Folic Acid Reductase from Chicken Liver. J. Biol. Chem. **235**, 1776 (1960).

365. — Studies on the Substrate Specificity of Folic Acid Reductase. J. Biol. Chem. **235**, 1780 (1960).

365a. — On the Structure of Dihydrofolate. Federat. Proc. (Amer. Soc. exp. Biol.) **22**, 231 (1963).

365b. Zakrzewski, S. F., M. T. Hakala and C. A. Nichol: Preparation and Biological Activity of Dihydroaminopterin. Biochemistry **1**, 842 (1962).

366. Zakrzewski, S. F. and C. A. Nichol: Labile Reduced Derivatives of Pteroylglutamic Acid. Federat. Proc. (Amer. Soc. exp. Biol.) **15**, 390 (1956).

367. — — Evidence for a single Enzyme Reducing Folate and Dihydrofolate. J. Biol. Chem. **235**, 2984 (1960).

368. Ziegler, I.: Biosynthese der Folsäure aus natürlich vorkommenden hydrierten Biopterin-Derivaten. Naturwiss. **48**, 458 (1961).

(Eingelaufen am 5. November 1962.)

Chemistry of the Natural Rotenoids.

By **L. Crombie**, London.

Contents.

	Page
I. Introduction	275
II. General Remarks on Rotenone and the Rotenoids	276
1. Isolation	278
2. Colour Tests	278
3. Nomenclature	279
III. Stereochemistry of Rotenone	279
IV. Chemistry of Rotenone	284
V. The Rotenolones and Isorotenolones	290
1. The A and B Series	290
2. The C and D Series	294
VI. The Rotenoids	295
1. Stereochemistry	295
2. Deguelin	296
3. Elliptone	297
4. Munduserone	298
5. α-Toxicarol	299
6. Sumatrol	301
7. Malaccol	302
8. Pachyrrhizone	303
9. Erosone	304
10. Dolineone	304
VII. Biogenesis and Biogenetic Connections of the Rotenoids	305
VIII. Synthesis in the Rotenoid Group	309
Addendum	316
References	316

I. Introduction.

In many warm countries of the world the natives fish by throwing preparations of crushed plant material into ponds and streams. These preparations cause stupefaction of the fish, which rise to the surface and are collected. Plants which are frequently employed in South America, tropical Africa, the East Indies and Malaya belong to the Papilionatae sub-family of the Leguminosae, especially species of the genera *Derris, Lonchocarpus, Mundulea, Millettia, Neorautanenia* and

Tephrosia. A number of these are also noted locally as insecticides and the preparation "Derris" has been available in commerce as an insecticide for half a century. Such *Derris* preparations (tuba), which originate mainly in East Asia, are now less important in Western markets than those derived from South American *Lonchocarpus* species (the latter have a variety of names, e. g. cube, timbo, barbasco and haiari) (*92*).

The most important insecticide and fish-poison in *Derris* and *Lonchocarpus* species is rotenone and it occurs together with a group of related compounds classed as rotenoids. Lists of some hundreds of plants suspected of containing rotenoids have been drawn up by Roark and others (*92, 156—159*). As the toxicity of rotenone to man is low when the administration is oral, fishing with it carries little risk.

It has been estimated that the lethal dose is about 200 g. For rats, the oral LD_{50} is 133 mg./kg., but by intraperitoneal injection the substance is considerably more toxic (LD_{50} 5 mg./kg.) (*123*).

For a long time the toxic action of rotenone has been associated with interference in the oxygen utilization processes. A relationship between the insecticidal activity of rotenoids and their ability to inhibit an insect *L*-glutamic dehydrogenase enzyme system has been claimed (*62*). Recently, using rat-liver mitochondrial preparations, it has been demonstrated that rotenone inhibits electron transport, the block being situated on the substrate side of cytochrome b, between diaphorase and cytochrome b (*117*).

II. General Remarks on Rotenone and the Rotenoids.

Rotenone appears to have been isolated crystalline for the first time from *Lonchocarpus (Robinia) nicou* by Geoffroy in 1895, who called it nicouline (*63*). Nagai, in 1912, isolated it from *Derris chinensis* and gave to it the name rotenone (*142*). The same compound was also isolated from *D. elliptica* under the names derrin and tubatoxin. Early chemical investigation was made by Kariyone (*100*) and considerably extended by Takei. Rapid development of its chemistry followed the work of Butenandt and of Haller and LaForge. Formal priority for the correct structure (I, p. 277) appears due to LaForge and Haller (*107*), but papers by Butenandt (*18*), Robertson (*161*) and Takei (*179*) reported the same conclusion within a very short space of time.

Within the last few years there has been a resurgence of interest in the chemistry, stereochemistry and synthesis of rotenone and rotenoids. The stereochemistry indicated in (I) was recently assigned by Büchi, Crombie and their collaborators (*9, 10*).

It is proposed that by definition a rotenoid must contain, as a minimum, the four-ring heterocyclic system indicated in (II), and that this supersedes

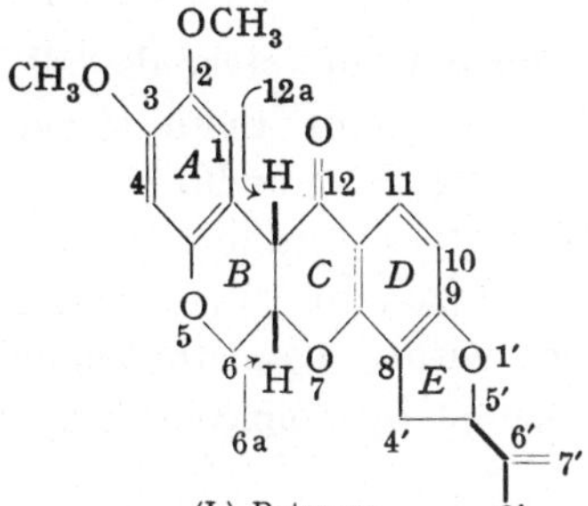

(I.) Rotenone.

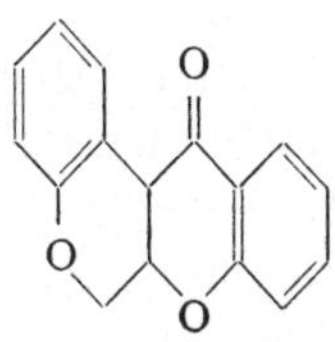

(II.) 6a,12a-Dihydro-6 *H*-rotoxen-12-one.

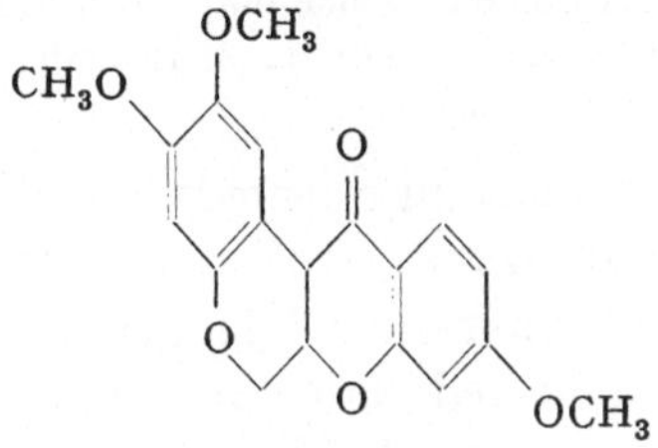

(III.) Munduserone.

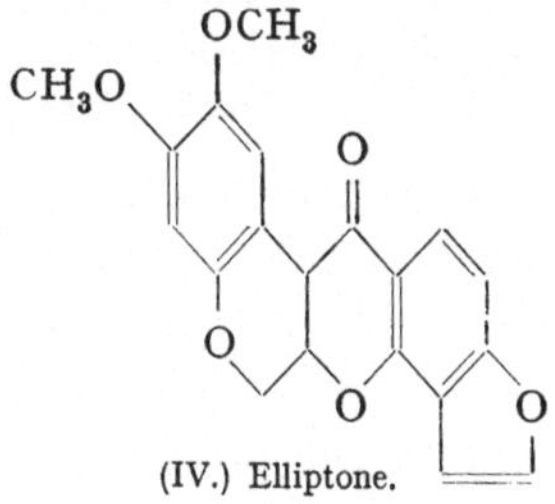

(IV.) Elliptone.

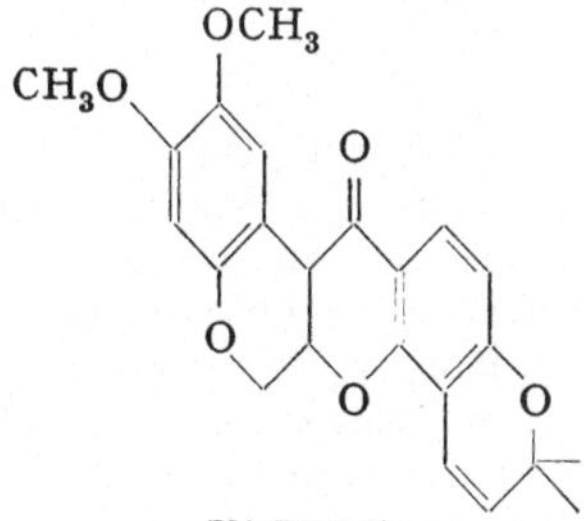

(V.) Deguelin.

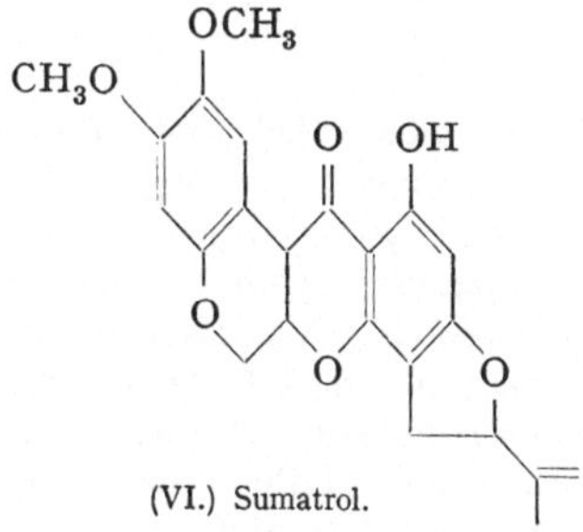

(VI.) Sumatrol.

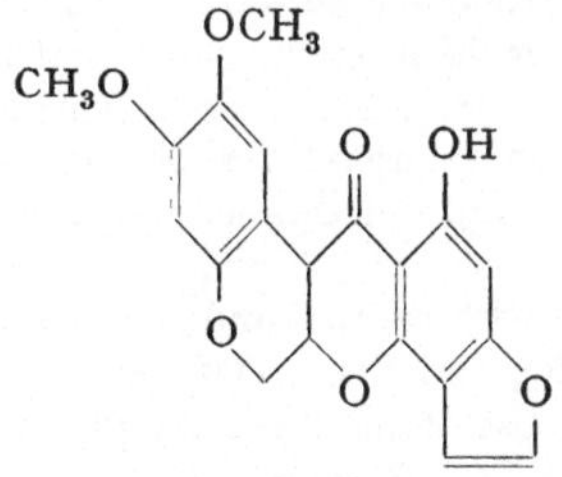

(VII.) Malaccol.

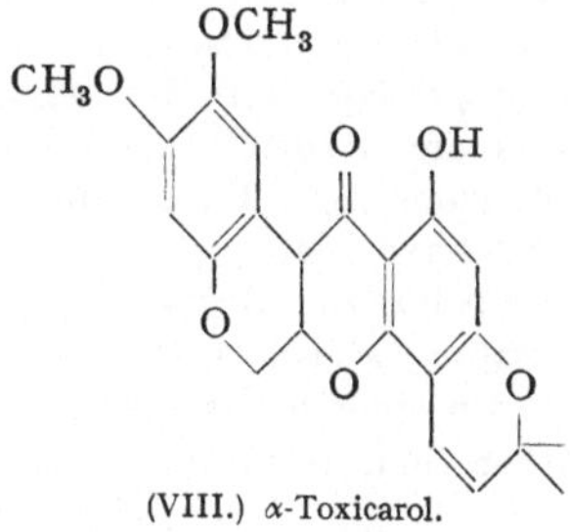

(VIII.) α-Toxicarol.

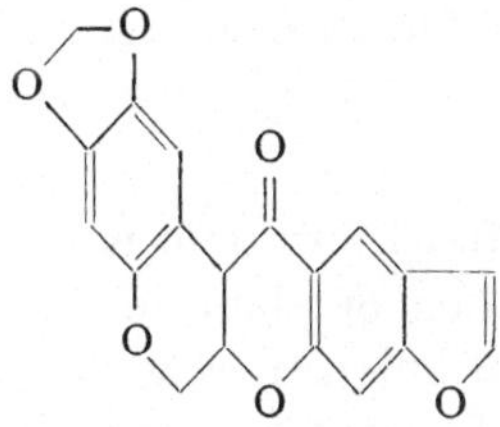

(IX.) Dolineone.

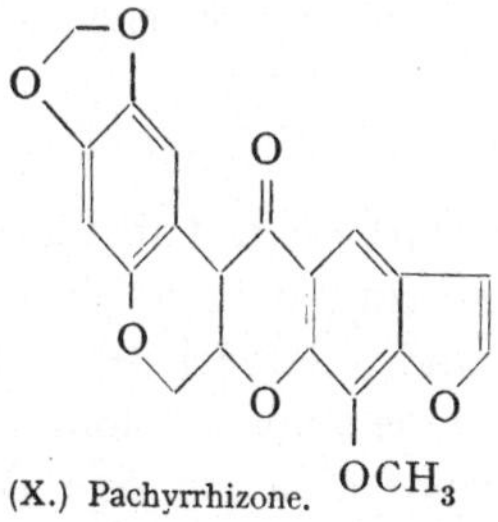

(X.) Pachyrrhizone.

Chart 1. Rotenone and Natural Rotenoids.

older criteria (*160*). On this basis, the better defined rotenoids fall into three groups (*Chart 1*, p. 277). The first contains those structures involving the absence or modification of ring *E*, munduserone (III), elliptone (IV) and deguelin (V). A further group contains the related 11-hydroxylated derivatives sumatrol (VI), malaccol (VII), and toxicarol (VIII). Finally, there are compounds with a 2,3-methylene-dioxy group and a linear *D/E* fusion, as represented by dolineone (IX) and pachyrrhizone (X).

A small group of less well defined compounds, some of which may be rotenoids, is known, and it seems certain that there will be many additions as investigation of extractives of the Papilionatae proceeds.

Earlier phases of the chemistry of rotenone and rotenoids have been well reviewed by LaForge, Haller and Smith (*110*) and Haller, Goodhue and Jones (*73*). The articles by Feinstein and Jacobson (*54*), and Venkataraman (*184*), which have appeared earlier in this Series are relevant to the present review. For insecticidal effects see (*54*).

1. Isolation.

The crushed dried root is extracted with ether or methylene chloride or fluorinated solvents (*98, 99*). A resin obtained by extracting South American *Lonchocarpus* species enters into commerce as "timbo resin" and contains about 30% of rotenone, together with other rotenoids. Rotenone is readily isolated from this by dissolving it in carbon tetrachloride and keeping at 0° overnight. The carbon tetrachloride solvate of rotenone crystallises out. This solvate is dissolved in trichloroethylene and filtered: on diluting the filtrate with methanol nearly pure rotenone crystallises and it is further crystallised from trichloroethylene (*9*). In the case of seeds, and some roots and barks, preliminary defatting with petrol is desirable before extraction. The resolution of rotenoid mixtures can be effected by chromatography, but alkaline alumina causes racemisation, and neutral or acid grades are needed. Counter-current extraction is useful for dealing with intractable gums (*47*). Paper chromatography (*118, 24*) and thin-layer chromatography (*47*) have been used.

Melting points as criteria of purity must be used with caution and spectroscopic characterisation is always desirable. Many rotenoids and their relatives (cf. *111*) exist in polymorphic forms and, apart from this, the use of alkaline soda-glass m. p. tubes rather than Pyrex ones may give inaccurately low m. ps. (*86, 97*). The alkalinity of the glass may induce enolisation and racemisation or, in the case of certain 11-hydroxy rotenoids, some linear-angular *D/E* equilibration.

2. Colour Tests.

The Durham test in its original and modified form (*96*) is frequently employed and it is applicable to investigation of plant sections. The Rogers and Calamari test (*169*) is also useful but neither test is specific for rotenone and rotenoids (*72*). Contrary to what was once believed, the absence of a positive Durham test does not imply the absence of a rotenoid. Another test often used is the Goodhue test (*65*) and it is applicable to quantitative work. The unspecific Meijer test is also available (*120*).

3. Nomenclature.

There are numerous trivial names for degradation products in the chemistry of rotenoids and it is convenient to use them for brevity. However, sometimes a name has been attached to as many as three different compounds in the field (e. g. rotenol) and specific naming is at times needed to remove ambiguity. A disadvantage of the "Ring Index" system is that the essential *A B C D* ring system may take on differing numbering depending on the absence of, or variations in, ring *E*. A procedure which is a compromise between utility and rigorous nomenclature procedure has therefore been suggested (*9*). In this, rotenoids are named from the heterocyclic system rotoxen (XI). Compound (II) then becomes 6a,12a-dihydro-6*H*-rotoxen-12-one, and natural rotenone with full configurational prefixes is 6aβ,12aβ,4′,5′-tetrahydro-2,3-dimethoxy-5′β-isopropenyl-furano-(3′,2′:8,9)-6*H*-rotoxen-12-one. (The 6a hydrogen atom is taken as the configurational reference point and called β.) The prefix *seco*-accomodates degradation products with broken *C* or *E* rings.

(XI.) Rotoxen.

III. Stereochemistry of Rotenone.

Rotenone, $[\alpha]_D^{20} - 228°$ (benzene) is dimorphic, m. p. 167–168° or 185–186°. Its chemistry is not easily discussed without taking stereochemical features into account, so this aspect, which has been the subject of recent work by Büchi, Crombie and their colleagues (*9, 10*), is considered first.

In brief, the problem has been approached by determining the absolute configuration at 5′, then again at 6a, and finally relating position 6a to 12a.

Rotenone is cleaved by alkali to tubaic acid (XII) (*74, 76, 178*) which on hydrogenation gives dihydrotubaic acid (XIII) in which the 5′ asymmetric centre of rotenone is retained. On exhaustive ozonolysis and oxidation with hydrogen peroxide, the acid (XIII) gave (+)-3-

(XII.) Tubaic acid.

(XIII.) Dihydrotubaic acid.

hydroxy-4-methylpentanoic acid (XIV). The $(\pm)$-form of this was synthesised and resolved, and degradation continued with the more

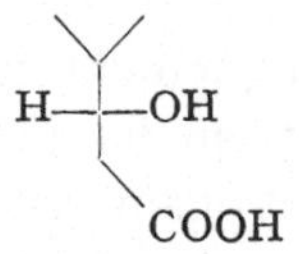

(XIV.) (+)-3-Hydroxy-4-methyl-   (XV.) (—)-form of (XIV). 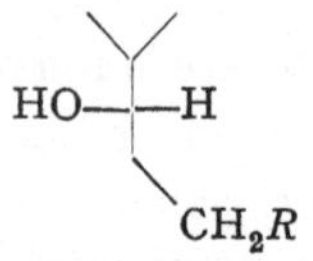   (XVI.) $R =$ H. (—)-2-Methylpentan-3-ol
pentanoic acid.

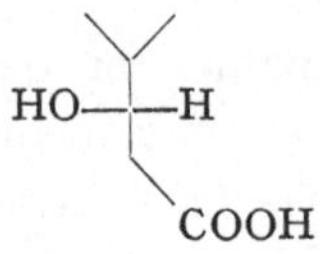

accessible (—)-form (XV). The latter was reduced to the (—)-diol (XVI, $R = $ OH) which was converted into the mono-toluene-p-sulphonate (XVI, $R = $ O $\cdot$ SO$_2$C$_6$H$_4$CH$_3$) and reduced with lithium aluminium hydride to (—)-2-methylpentan-3-ol (XVI, $R = $ H). Correlation of the latter with L-(—)-glyceraldehyde, in which the isopropyl group of the alcohol replaces the aldehyde and the ethyl group replaces the —CH$_2$OH, can be deduced from the literature. (+)-3-Hydroxy-4-methylpentanoic acid from (—)-dihydrotubaic acid is thus related to D-glyceraldehyde and hence rotenone has the (R) configuration at position 5′ (9, 10).

This configuration assignment has been confirmed by NAKAZAKI and ARAKAWA (143) who isolated the acid (XIV) by ozonolysis and peracetic acid oxidation of dihydrotubaic acid. The hydroxyacid was esterified and converted into the hydrazide and the latter, by treatment with nitrous acid, formed (—)-5-isopropyloxazolidone (XVII). (—)-Methyl-2-hydroxy-3-methylbutanoate (XIX, $R = $ OCH$_3$) was then made from D-valine (XVIII), converted into the amide (XIX, $R = $ NH$_2$) and reduced to the aminoalcohol (XX, $R = $ H). When treated with ethyl chlorocarbonate in pyridine the latter gave an ethyl carbamate derivative (XX, $R = $ COOC$_2$H$_5$) which, with sodium methoxide, yielded (+)-5-isopropyl-oxazolidone (XXI), enantiomeric with the specimen from rotenone. This confirms the (R) configuration assigned to position 5′.

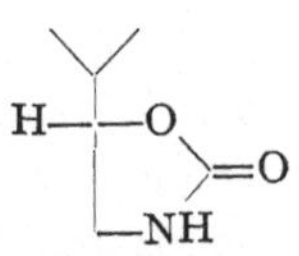

(XVII.) (—)-5-Isopropyl-   (XVIII.) D-Valine. 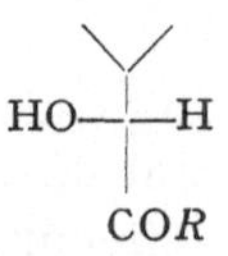   (XIX.) 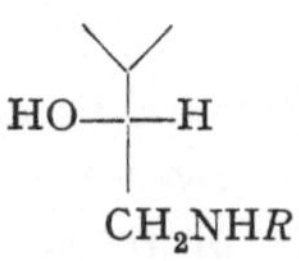   (XX.) 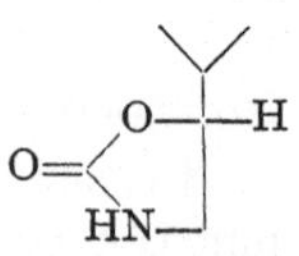   (XXI.) (+)-5-Isopropyl-
oxazolidone.   oxazolidone.

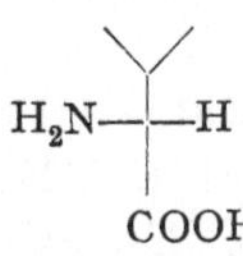

Since the configuration at $C_{(6a)}$ or $C_{(12a)}$ relative to $C_{(5')}$ would be difficult to establish, the absolute configuration at $C_{(6a)}$ was again determined (9, 10). The enol acetate of dihydrorotenone (XXII) was made by treating dihydrorotenone with isopropenyl acetate and acid under conditions which avoid racemisation at 6 a. It could also be prepared by treating rotenone with acetic anhydride and sodium acetate which racemises 6 a to give a pair of diastereoisomers, (XXIII) and (XXIV), only one of which, (XXIII), is crystalline. The latter is hydrolysed by acid to rotenone as expected and on partial hydrogenation it gives the enol acetate (XXII). Exhaustive ozonolysis of the latter gave (+)-3-hydroxy-4-methylpentanoic acid (XIV) containing $C_{(5')}$ and also (—)-gly-

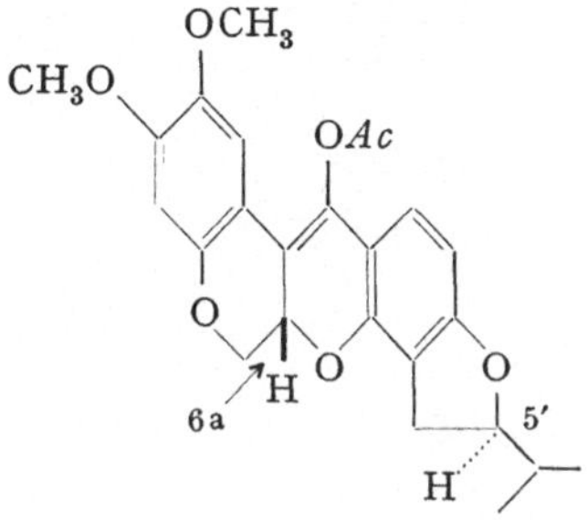

(XXII.) Dihydrorotenone enol acetate.

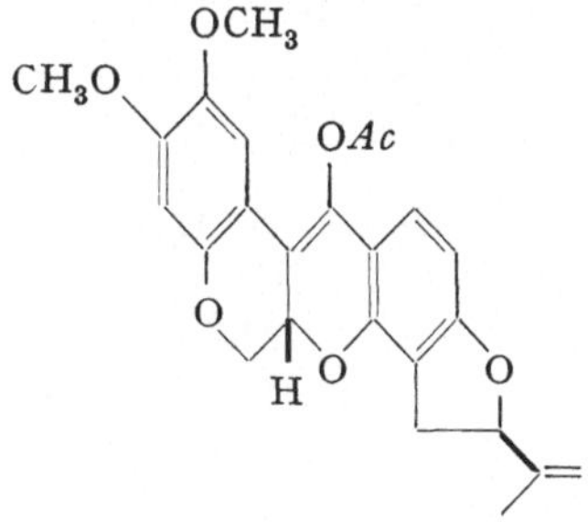

(XXIII.) 6a β-Rotenone enol acetate.

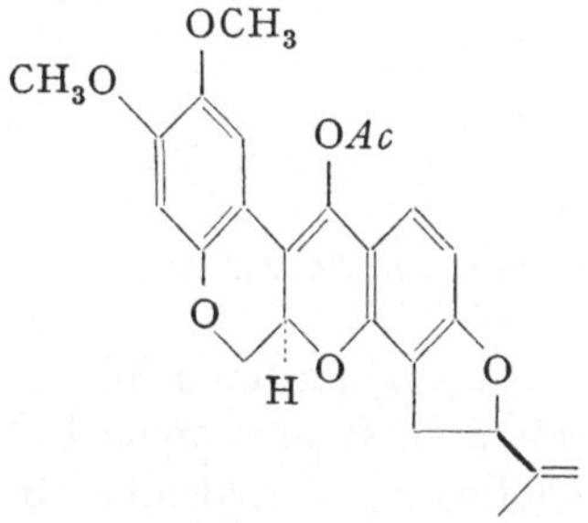

(XXIV.) 6a α-Rotenone enol acetate.

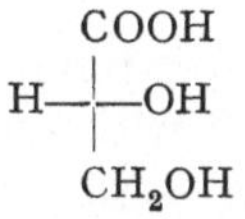

(XXV.)

ceric acid (XXV) incorporating the 6a centre of rotenone. Since (—)-glyceric acid has been related to D-glyceraldehyde, the 6a centre of rotenone is (S) $(9, 10)$ (cf. p. 277).

The third stereochemical point, the configuration at $C_{(12a)}$, was settled by deciding the *cis-* or *trans-* nature of the B/C fusion $(9, 10)$. For more than 20 years this has been known to be the thermodynamically stable one, and rotenone with the unstable fusion has not been isolated: this stable fusion has generally been considered *trans-* $(22, 135)$. The limiting conformers of the flexible *cis*-fusion (XXVI, $R = $ H) and (XXVII, $R = $ H) and the one rigid *trans*-fusion (XXVIII, $R = $ H) are represented below. Ring B is shown as a quasi-chair, though in each model it is readily converted into a quasi-boat (*Chart 2*, p. 282). A number of lines of evidence point to the stable fusion of rotenone being the *cis* and the conformer (XXVII) seems to be preferred in solution.

Rotenone is converted by sulphuric acid-acetic acid into iso-rotenone (XXIX) by carbonium ion shifts. In isorotenone the asymmetry

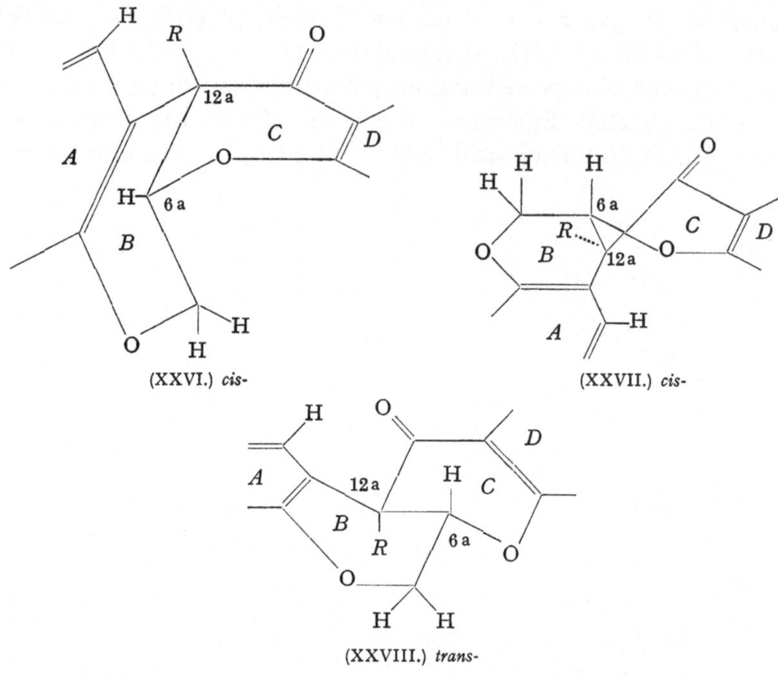

Chart 2. Conformers of *cis-* and *trans-B/C* Rotenoids.

at 5′ is destroyed but 6a,12a retain the same configuration as in rotenone. When reduced with potassium borohydride a crystalline alcohol (**XXX**) is obtained in > 80% yield and this alcohol, which can also be obtained by catalytic hydrogenation, is dehydrated by dilute acid to give the *A–D* conjugated compound (**XXXI**). The alcohol (**XXX**) shows intramolecular hydrogen bonding in dilute solution (3566 cm.$^{-1}$). Analysis using model compounds, and without making assumptions about the *B/C* fusion, shows that $O_{(5)}$ must be the acceptor seat of the hydrogen bond: bonding to aryl groups or other oxygen atoms in the molecule does not provide an adequate explanation. For hydrogen bonding to $O_{(5)}$ to be possible, the *B/C* fusion must be *cis-* as in (**XXXII**): with neither an α- nor a β-orientation at $C_{(12)}$ is such bonding possible with a *trans*-fusion *(Chart 3)*. Similar conclusions can be drawn for the borohydride reduction of products of (±)- and (—)-4′,5′-dihydrodeguelin (β-dihydrorotenone) and for rotenone itself, though in the latter case the alcohol crystallises only as a methanol solvate; so there is a band at 3641 cm.$^{-1}$ due to methanol as well as those due to the free and intramolecularly bonded secondary hydroxyl groups. Formation of an α-oriented hydroxyl in these cases, as in (**XXXII**), is also consistent with attack by borohydride from the less hindered side of the *cis*-molecule (**XXVII**).

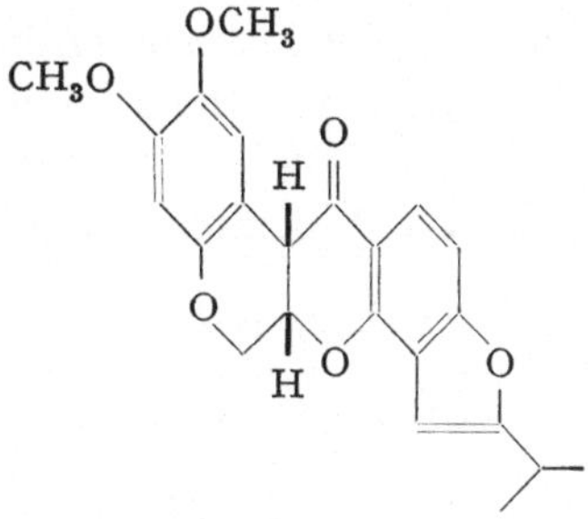

(XXIX.) Isorotenone.

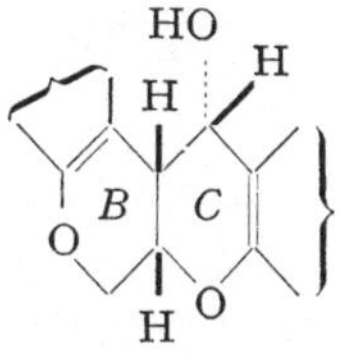

(XXX.)*

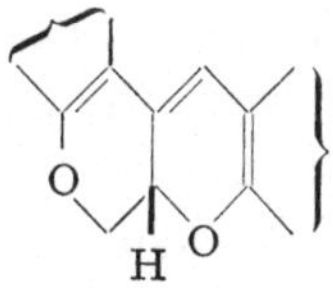

(XXXI.)*

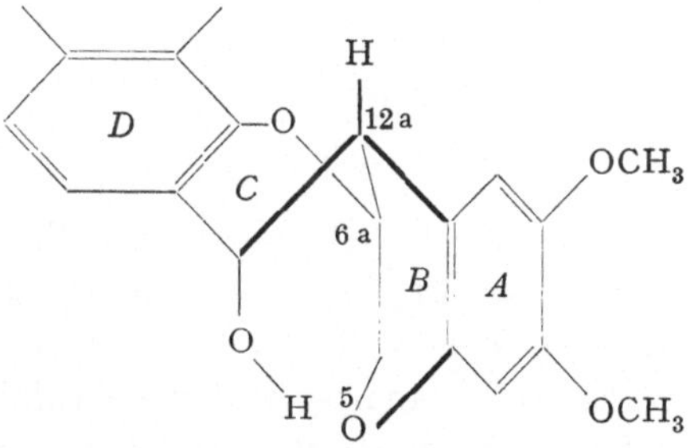

(XXXII.)

Chart 3. B/C Fusion of Isorotenone.

The nuclear magnetic resonance data of CROMBIE and LOWN (*42, 43*) also support a *cis*-fusion for rotenone. Although compounds with the unstable *B/C* fusion are not known for the readily enolisable rotenone, isorotenone and other rotenoids, in the case of their 12a-hydroxyl and methoxy compounds where enolisation is blocked, both *cis*- (XXXIII) and *trans*- (XXXIV) compounds are known and their configurational assignments are based on independent evidence (see below).

In the *cis*-compounds the τ value for the 1-hydrogen atom is at 3.2–3.6 whereas in the *trans*-compounds it is at 2.0–2.2. This variation in line position is caused by asymmetric shielding of the 1-hydrogen atom by

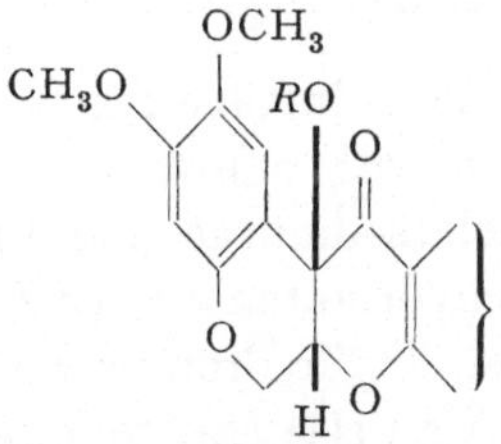

(XXXIII.)

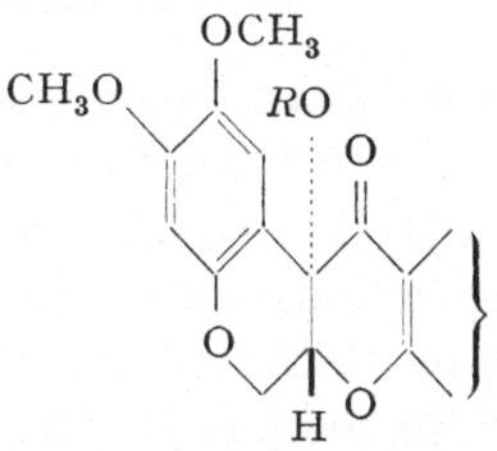

(XXXIV.)

---

* Reference back to such partial formulae is made when discussing other rotenoids later in this review. The remainder of the structure then corresponds to the rotenoid being discussed.

the 12-carbonyl, as shown qualitatively in (XXXV). Since 12-deoxy-rotenoids (XXXVI) have the 1-hydrogen line near 3.3, the shielding of the 1-hydrogen in the *cis*-fused ketones (XXXIII) is small, consistent with the hydrogen lying near the nodal surface (XXXV) [compare (XXVII)]

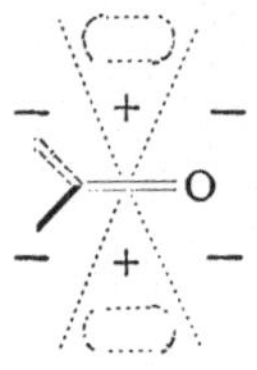

(XXXV.) 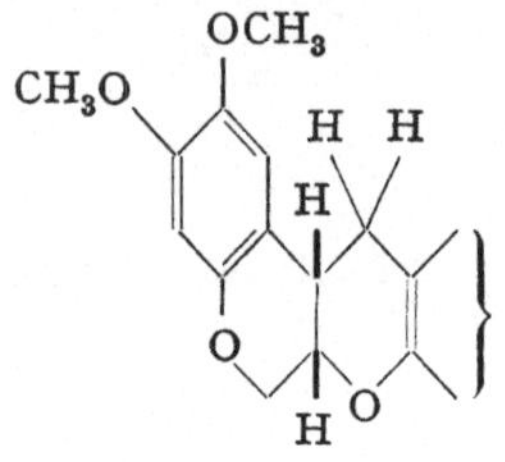   (XXXVI.)

of carbonyl magnetic shielding. In the rather flat *trans*-molecules, however, the 1-hydrogen is more nearly coplanar with the 12-carbonyl and is thus negatively shielded (cf. 6a,12a-dehydroisorotenone where it is still closer to coplanarity and the τ value for the 1-hydrogen is 1.67). The τ value for rotenone is 3.32, so it clearly belongs to the *cis*-series, and this comment applies to other known rotenoids (*42, 43*).

Rotenone thus has the (6aS,12aS,5'R) configuration (*9, 10*), and it will be seen later that the *cis*-fusion accommodates other points of its chemistry. Examination of models for non-bonded interactions which might destabilise one fusion relative to the other suggests that a likely cause is the interaction between the 1-hydrogen and the 12-ketonic oxygen atom. In the *trans*-fusion the distance between atomic centres is ~ 2.05 Å if ring B is a quasi-chair (~ 1.95 Å if a quasi-boat): this compression is much relieved in the *cis*-conformation (XXVII). According to FELDMAN's notation (*55*) natural rotenone is /6/-rotenone on the R/S symbolism but /7/-rotenone on the α/β system.

IV. Chemistry of Rotenone.

Some of the characteristic reactions of rotenone involve the oxidation levels of the B/C ring system. When shaken with manganese dioxide in acetone (*41*) or treated with iodine and ethanolic sodium acetate (*14, 112*), dehydrogenation occurs to give 6a,12a-dehydrorotenone (XXXVII) (*14, 114*) in which asymmetry at 5' is retained. On hydrolysis with base the dehydro compound gives derrisic acid (XXXVIII) (*14, 107*) and this behaviour distinguishes a dehydrorotenoid from an isoflavone which has similar spectroscopic characteristics. The latter gives a deoxybenzoin and formic acid. On oxidation, derrisic acid yields derric acid (XXXIX) (*Chart 4*) (*104, 115*), rissic acid (XL) (*104*), and tubaic acid (XII, p. 279).

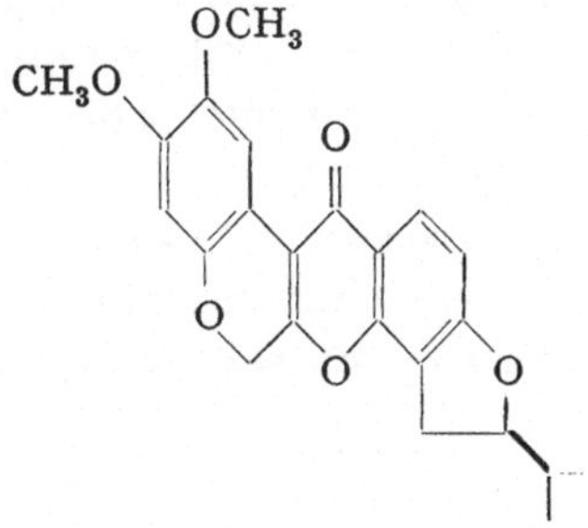

(XXXVII.) 6a,12a-Dehydrorotenone.

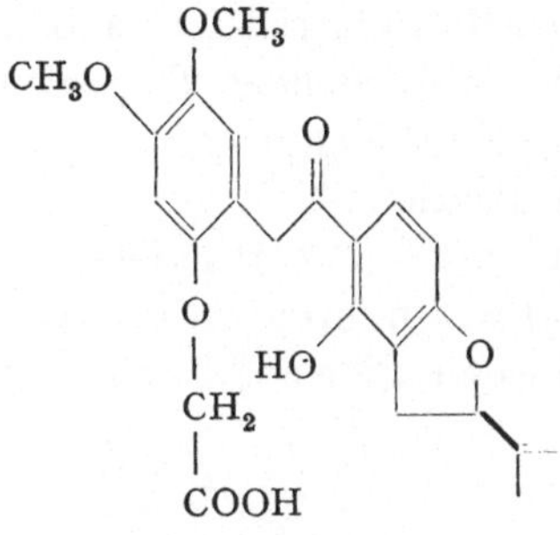

(XXXVIII.) Derrisic acid.

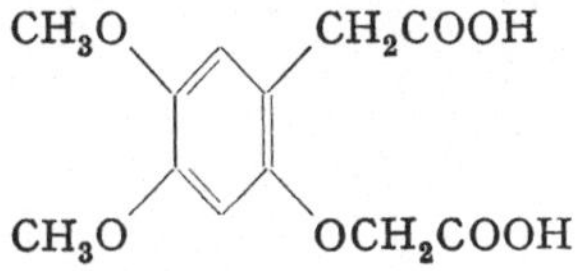

(XXXIX.) Derric acid.

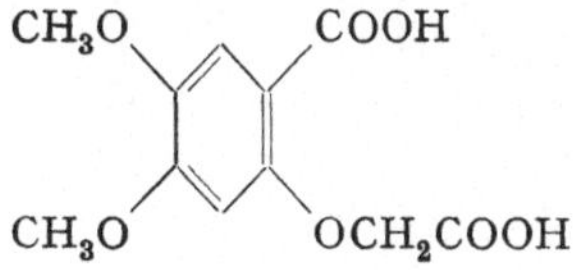

(XL.) Rissic acid.

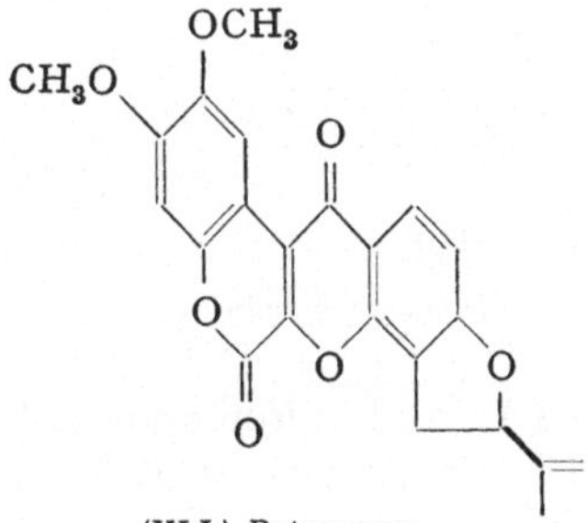

(XLI.) Rotenonone.

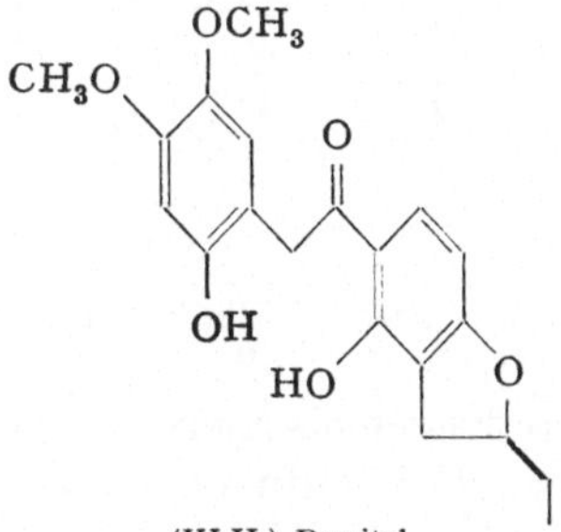

(XLII.) Derritol.

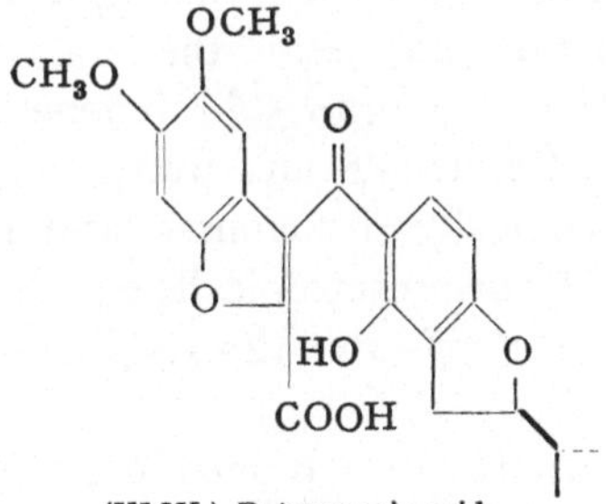

(XLIII.) Rotenononic acid.

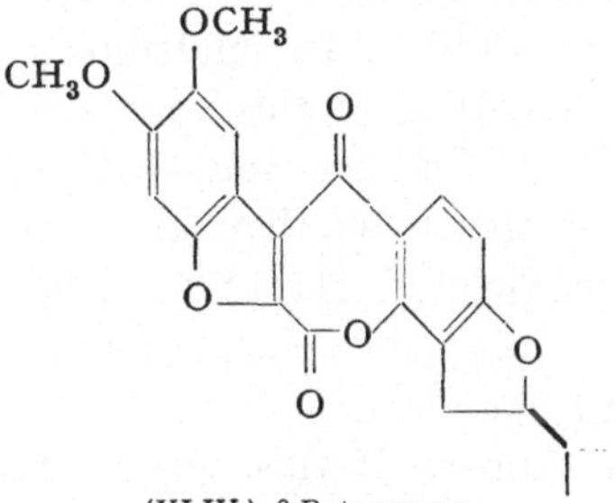

(XLIV.) β-Rotenonone.

Chart 4. Degradation Products of Rotenone.

Treatment of dehydrorotenone with nitrous acid, or oxidation with chromic acid, gives the yellow insoluble lactone rotenonone (XLI) (*18, 105*). This is hydrolysed by alkali under vigorous conditions to derritol (XLII) and oxalic acid (*105*). The hydrolysis of rotenonone under milder conditions has been studied by TAKEI, who considers that rotenononic acid (XLIII) is formed by isomerisation, and this can be lactonised by acid to yield an isomer of rotenonone, β-rotenonone (XLIV) (*179, 185*). β-Rotenonone is reported by Takei (*179*) to rearrange again to rotenonone in weak alkaline solution.

Many important reactions of rotenone occur in alkaline solution (*Chart 5*) which, even with as weak a base as sodium acetate in ethanol, causes racemisation at 6a and 12a. Rotenone epimerises to give a mixture

(XLV.) (XLVI.) (XLVII.) (XLVIII.)

(IL.) (L.) (LI.) (LII.)

Chart 5. Alkaline Treatment of Natural Rotenone.

of the two diastereoisomers, rotenone (XLV) and epirotenone (LII), the 5'-isopropenyl substituent remaining β in each case: this mixture is often referred to as mutarotenone (*22*). The process involves the removal of the 12a proton from rotenone ($6a\beta,12a\beta,5'\beta$) to give the $6a\beta$-enolate (XLVI) which can undergo β-elimination to yield the 6a,7-*seco*-ion (XLVII) (*17*): this can cyclise to give $6a\alpha,12a\alpha,5'\beta$-rotenone (epirotenone, LII). The *trans-B/C* forms ($6a\beta,12a\alpha,5'\beta$ and $6a\alpha,12a\beta,5'\beta$) are not shown as they are thermodynamically disfavoured and have not been isolated. The 6a,7-*seco*-ion (XLVII) can prototropically equilibrate under strong basic conditions with the $12a\beta,5'\beta$- and $12a\alpha,5'\beta$-ions (IL and LI) (*40, 41*).

In support of this scheme (*41*) it has recently proved possible to isolate the β,γ-unsaturated diastereoisomers (IL) and (LI), m. p. 150–151°, $[\alpha]_D^{20} + 203°$, and m. p. 122–123°, $[\alpha]_D^{20} - 235°$. As direct intermediates in the racemisation of rotenone, as was once proposed (*22*), they are not acceptable entities, since although boiling ethanolic sodium acetate racemises 6a it does not reconvert them into mutarotenone (potassium carbonate in wet acetone does, however, do this). The true intermediate (LIII, $R = H$) in the racemisation of the 6a and 12a centres of rotenone has however been isolated.

References, pp. 316—325.

When rotenone is treated in benzene with sodium or solid sodium hydroxide, a solution of a sodium derivative is obtained which with dilute acid under carefully defined conditions gives the yellow phenol (LIII, $R = H$) *(41)*; its acetate is obtained by pouring the benzene solution into acetylating mixture *(94)*. In a solvent of low dielectric constant, the sodium compound will exist as an ion-pair, inhibiting re-cyclisation.

JENNEN *(94)* suggests that the benzene solution contains a polymeric system based on 4-coordinated sodium. This explains why a derivative of (XLVII) rather than one of (XLVI) or (XLVIII), from which such a coordinated compound cannot be formed, is preferred. The structure of the intermediate (LIII, $R = H$) has been rigorously established by conversion into, and preparation from, rotenol (see below): it is converted into mutarotenone by alcoholic sodium acetate *(41)*.

A well-known reaction of rotenone is the formation of rotenol (LIX–LX) and derritol (LVIII) when it is treated with zinc and alkali *(14, 112, 113)*

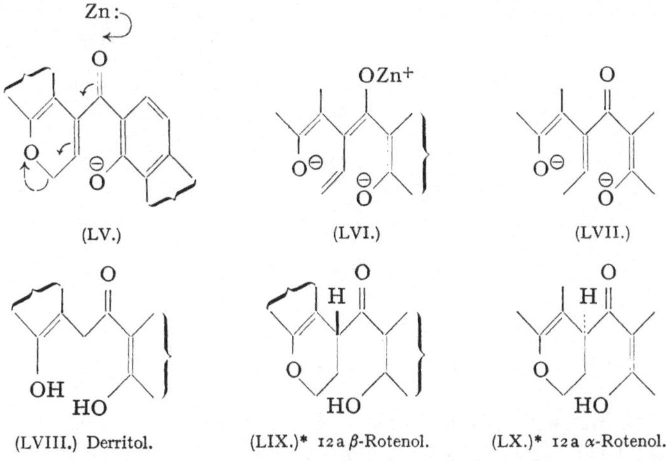

Chart 6. Rotenol and Derritol.

* These and similar structures are chelated.

(Chart 6): this is readily explained as involving the ion (XLVII). Rotenol arises by 1,4-reduction whilst the formation of derritol is rationalised as attack as in (LV) to give (LVI) which by prototropic shift gives (LVII): the reaction is completed by retroaldol condensation. Rotenol, long treated in the literature as an individual, has lately been separated into the expected two diastereoisomers (LIX and LX) (*41*). These have m. p. 141–143°, $[\alpha]_D^{21} - 245°$; and m. p. 90°, $[\alpha]_D^{21} + 33°$.

The oximation of rotenone also illuminates the significance of the ion (XLVII) in alkaline solution, as originally suggested by Butenandt (*14*). A normal oxime (LXI) can be obtained but in the presence of sodium hydroxide hydroxylamine hydrochloride yields an isoxazoline (isooxime) arising from attack on the ion (XLVII) with cyclisation as in (LXII). Spectral evidence suggests that the isoxazoline may be a tautomeric mixture (LXIII–LXIV) in solution (*41*). The formation of dehydronetoric (toxicaric) acid (LXV) when rotenone is oxidised by alkaline peroxide (Dakin reaction) is also readily explained on the basis of (XLVII) (*16, 18, 32*).

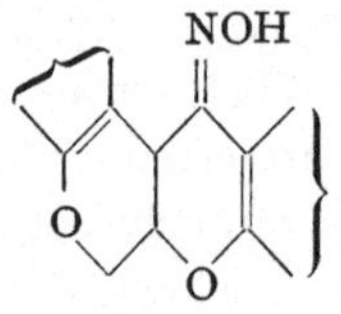

(LXI.) Oxime.

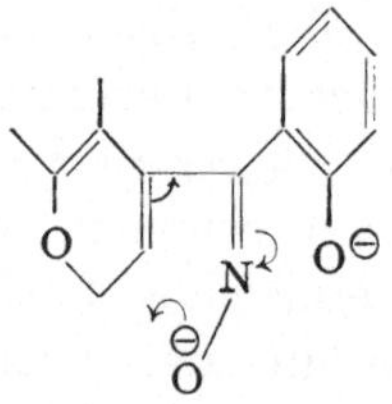

(LXII.)

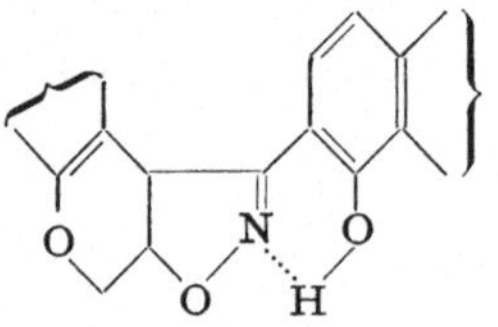

(LXIII.)

Isooxime.

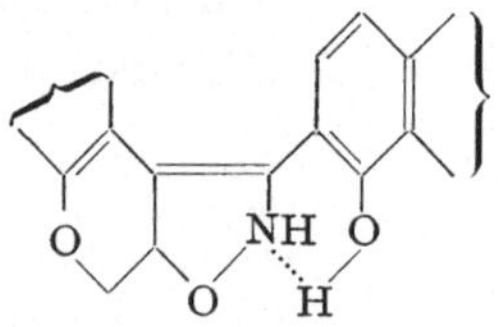

(LXIV.)

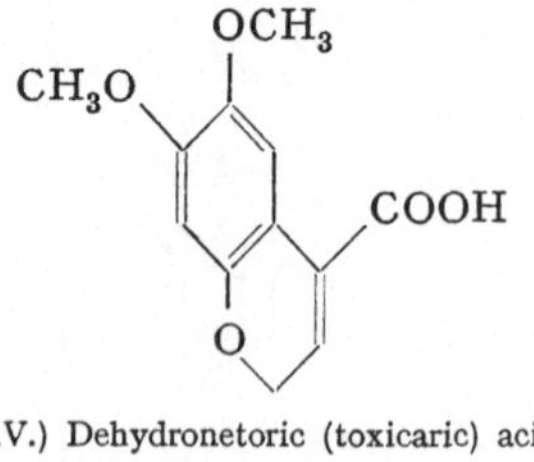

(LXV.) Dehydronetoric (toxicaric) acid.

It is of interest that the first claim to have isolated 6a,12a-dehydrorotenone (LIII, $R = $ H) was made by Haller and LaForge, but their claim was later withdrawn (*70, 74a*). They obtained a white compound having the correct analytical figures by treating rotenol with alkaline ferricyanide, and the same compound is obtained, together with 6a,12a-dehydrorotenone, by oxidising rotenol with manganese dioxide (*41*). It has lately been shown (*40, 41*) that the product (actually a mixture of two diastereoisomers) is in fact the spiro-compound (LXVI) formed via the mono- or di-radical (LXVII) *(Chart 7)*. Zinc and alkali re-converts it entirely into rotenol as in (LXVIII). The spiro-compound is also obtained via the carbonium ion (LXIX) when 6a,12a-dehydro-rotenol is treated with acid.

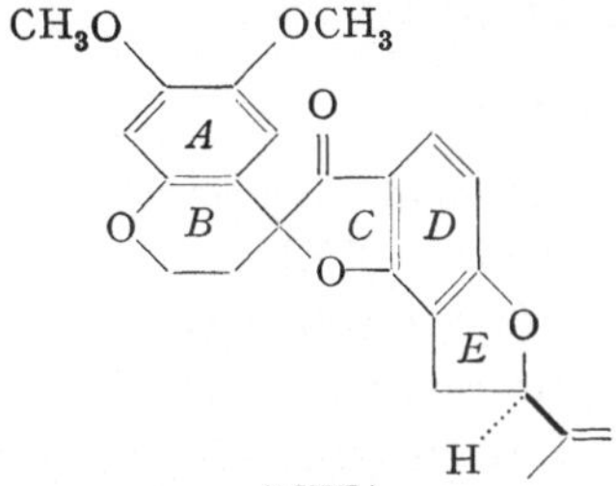

(LXVI.)

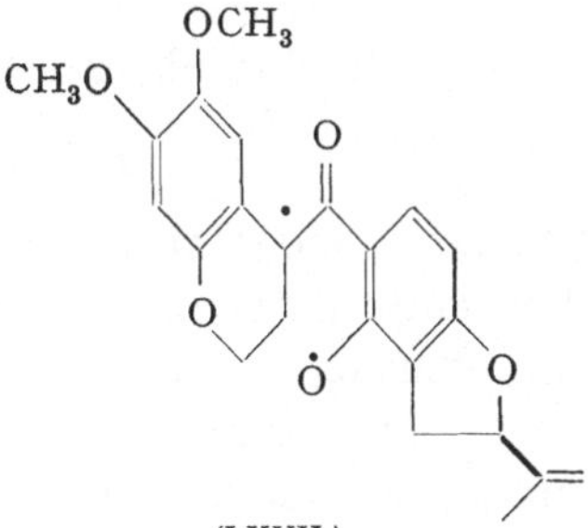

(LXVII.)

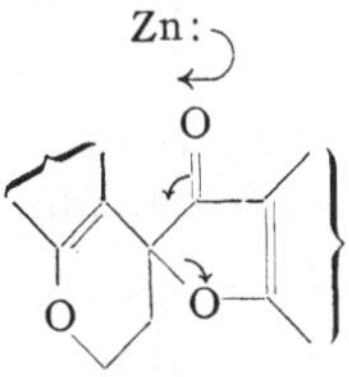

(LXVIII.)

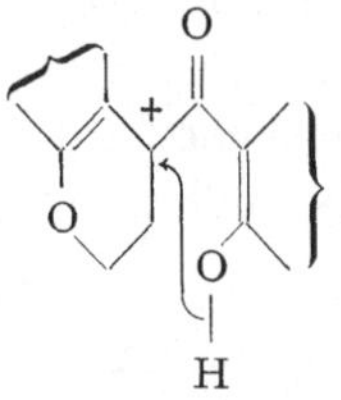

(LXIX.)

Chart 7. Spiro-isomerisation Product of Rotenone.

The catalytic hydrogenation of rotenone can lead to a number of products. Beside 6′,7′-dihydrorotenone (LXX) *(Chart 8)* which can be formed in high yield (77) and is of interest as a stabilised insecticide,

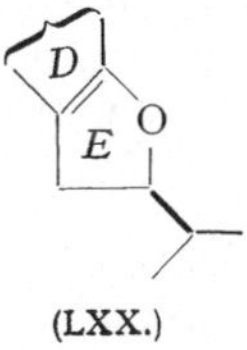

(LXX.)

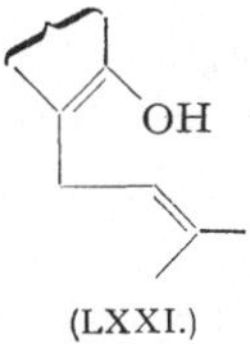

(LXXI.)

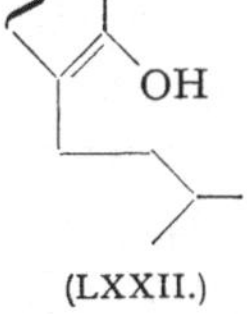

(LXXII.)

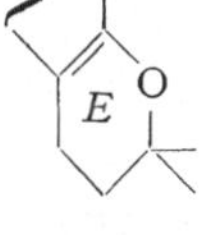

(LXXIII.)

Chart 8. Hydrogenation Products of Rotenone (Ring *E*).

the 1′,5′-*seco* product (LXXI) (*10, 77*), resulting from 1,4-addition, and its dihydro derivative (LXXII) can be obtained by varying the conditions (*107, 112*). The compound (LXXI) is of interest as it cyclises with acid to β-dihydrorotenone (LXXIII) (*70*), identical with 4′,5′-dihydrodeguelin. Other conditions affect ring *C*, giving the 6′,7′-dihydro-12-hydroxy compound (XXX, p. 283) or the 12-deoxy-derivative (XXXVI, p. 284). In catalytic hydrogenation of the alcohol (XXX), the olefin (XXXI), and the enol acetate (XXIII, p. 281) all give the same 6′,7′-dihydro-12-deoxy compound (XXXVI), which is obtained by direct hydrogenation of rotenone itself (*9*). This shows that attack of hydrogen from the catalyst surface occurs at the α-face of the molecule.

V. The Rotenolones and Isorotenolones.

1. The A and B Series.

LaForge and Haller (*108*) discovered that oxygenation of an alkaline suspension of (—)- or (±)-isorotenone (XXIX, p. 283) gave (±)-isorotenolone I, m. p. 203–104°, and (±)-isorotenolone II, m. p. 193–194°. During the preparation of dehydroisorotenone by the iodine-sodium acetate method an acetate is also formed (accessible too by the action of lead tetraacetate on isorotenone) which upon hydrolysis gives mainly isorotenolone I. Recently, these compounds have been re-investigated by Crombie and Godin (*38, 39*) who have developed methods for the stereochemical study of the optically active isorotenolones, the rotenolones, and the corresponding 12-hydroxy compounds.

Pure (±)-isorotenolone I has m. p. 208–209° and is re-named (±)-isorotenolone A. (±)-Isorotenolone II is a mixture of (±)-iso-rotenolone A and (±)-isorotenolone B, m. p. 243°. The 12a placing of the hydroxyl in both compounds follows from the methods of preparation, their hydrogenolysis to the deoxy-compound (XXXVI, p. 284), the absence of enolisation reactions, and their ready dehydration by dilute acid to 6a,12a-dehydroisorotenone. Zinc and alkali degradation gives rotenol and derritol from both iso-rotenolones A and B: this is explained in (LXXIV).

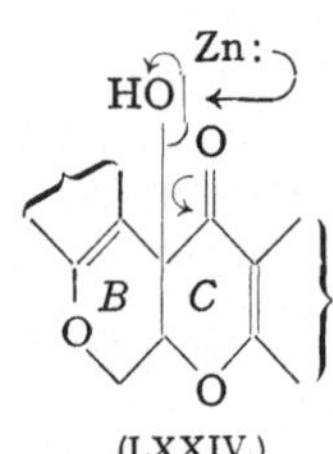
(LXXIV.)

The hydroxyl is eliminated and the reaction proceeds as discussed earlier. Decision as to which is the *cis*- and which the *trans*-compound rests on infrared evidence and the nuclear magnetic resonance data discussed on p. 284 (*42, 43*). In the infrared (LiF optics) iso-rotenolone A shows a hydroxyl group bonded to the ketone ν_{max} 3506 (hydroxyl, dilute solution) and 1676 cm.$^{-1}$ (ketone), whereas iso-rotenolone B shows a free or lightly bonded hydroxyl and ketone ν_{max} 3593 and 1697 cm.$^{-1}$. The *cis*-conformer (XXVII, R = OH, p. 282) is well oriented for hydrogen bonding whereas the *trans*-fusion (XXVIII, R = OH) has the hydroxyl poorly disposed towards and more distant from the ketone. (±)-Isorotenolone A is assigned the *cis*-fusion (LXXV

–LXXVI) and (±)-isorotenolone B the *trans-* (LXXVII–LXXVIII) *(Chart 9)*; infrared data for model compounds support these assignments.

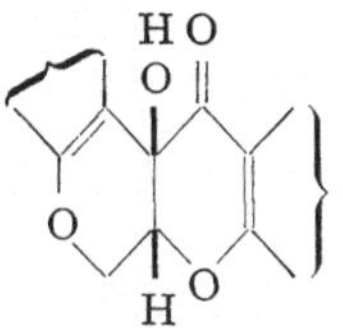

H O   H O   H O   H O

(LXXV.) 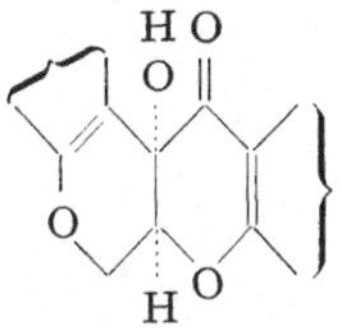   (LXXVI.) 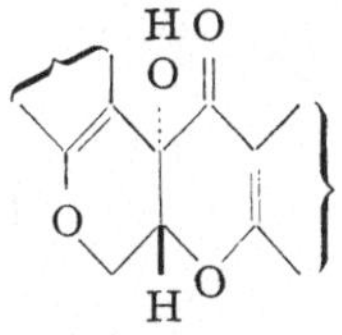   (LXXVII.) 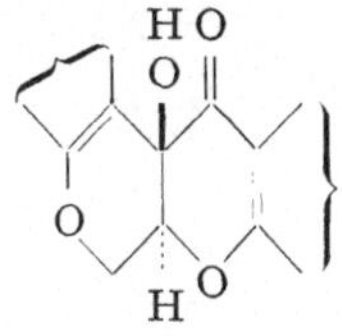   (LXXVIII.)

Chart 9. Rotenolones and Isorotenolones.

(±)-Isorotenolones A and B form different methyl ethers with methyl iodide-silver oxide. If the A ether is refluxed with methanolic sulphuric acid, it is recovered unchanged, except for the formation of 6a,12a-dehydro-compound. However, if the B ether is so treated, it gives the A ether. The corresponding rotenolone methyl ethers behave similarly and presumably the 12a carbonium ion is formed. Despite the apparent placing of the positive charge adjacent to the formal positive end of the carbonyl, the carbonium ion is resonance stabilised by the oxygenated ring A, and the carbonyl dipole is involved in resonance involving the oxygen substituents of ring D. This transformation of (±)-isorotenolone B methyl ether into that of the (±)-A stereoisomer again indicates the thermodynamic stability of the *cis-B/C* fusion. When acetylated with refluxing acetic acid-acetic anhydride containing sodium acetate, the isorotenolones A and B give the same acetate (and 6a,12a-dehydro-compound). This acetate is of the A series since it gives the racemic diol (LXXIX–LXXX) on treatment with lithium aluminium hydride (see below).

Because (—)-isorotenone (the sign refers to benzene solution), is readily racemised by base, alkaline oxygenation will not yield optically active isorotenolones. These can be obtained by oxidation with acid dichromate. (—)-6aβ,12aβ-Isorotenone gives the two diastereoisomers (LXXV and LXXVII) whilst its enantiomer, (+)-6aα,12aα-isorotenone, gives (LXXVI and LXXVIII). The pair (LXXV) and (LXXVI) are enantiomers as are (LXXVII) and (LXXVIII) and their identity as the *cis-* and *trans-B/C* compounds follows from the infrared hydrogen bonding criteria mentioned above (*38, 39*).

On reducing (±)-isorotenolone B (LXXVII–LXXVIII) with borohydride *(Chart 10)* two 12,12a-diols, (±)-B$_1$ and (±)-B$_2$ are obtained. In these the *trans-B/C* fusion is maintained and one diol must have a *trans-* and the other a *cis-*12,12a-diol arrangement. The (±)-B$_1$ diol must be *cis* (LXXXIII–LXXXIV) as it results when the racemate (LXXXVII–LXXXVIII) is *cis*-hydroxylated with osmium tetroxide,

and because it reacts more rapidly with lead tetraacetate than the
$(\pm)$-B$_2$ diol which is thus *trans*-12,12a (LXXXV–LXXXVI).

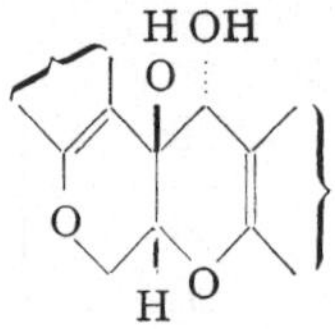

(LXXIX.) 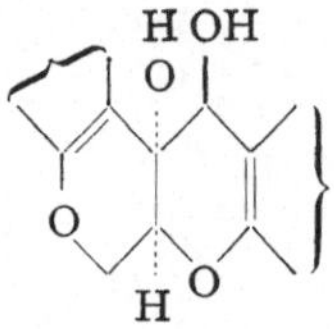   (LXXX.) 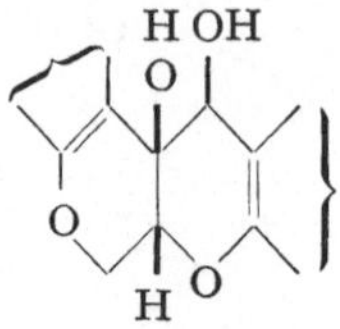   (LXXXI.) 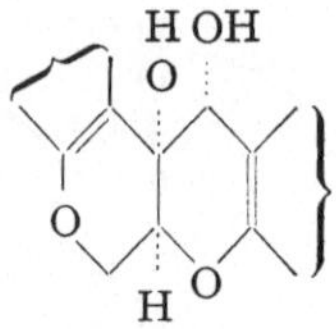   (LXXXII.)

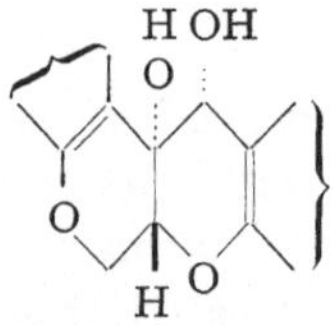

(LXXXIII.) 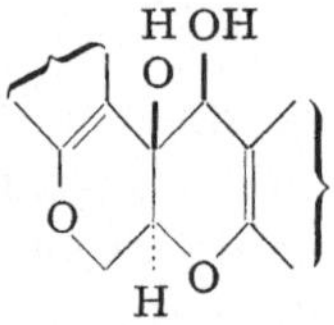   (LXXXIV.) 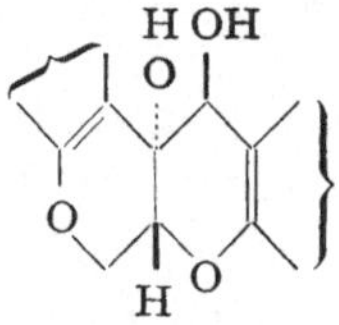   (LXXXV.) 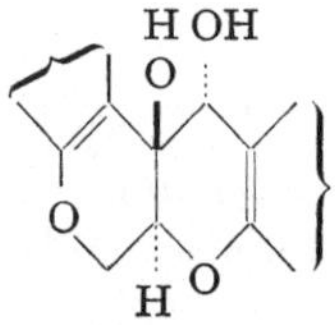   (LXXXVI.)

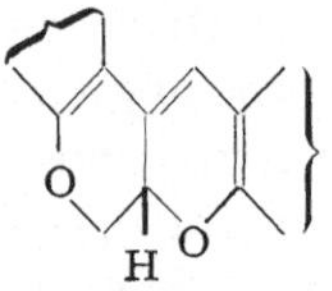

(LXXXVII.) 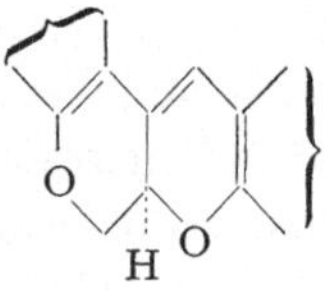   (LXXXVIII.) 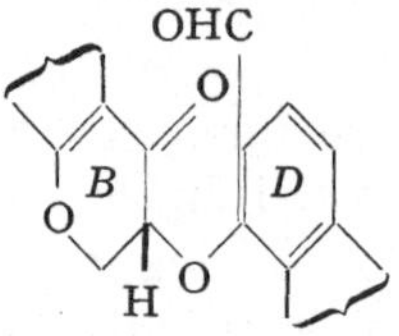   (LXXXIX.) 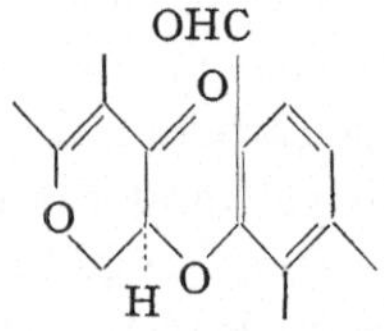   (XC.)

Chart 10. Borohydride Reduction Products of Rotenolones and Isorotenolones.

$(\pm)$-Isorotenolone B is re-formed by shaking either diol with manganese
dioxide suspension. When $(\pm)$-isorotenolone A is reduced with borohydride only one diol $(\pm)$-A is obtained. Its rate of reaction with lead
tetraacetate suggests it to be (LXXIX–LXXX). This is as expected,
for the α-face of the 6aβ,12aβ-isorotenolone molecule is hindered and
attack from the β-side would give an α-oriented 12-hydroxyl; from
6aα,12aα-isorotenolone the 12-hydroxyl would be β-oriented. When
treated with acid the diols $(\pm)$-A and $(\pm)$-B$_1$ are readily converted into
$(\pm)$-isorotenone.

On using optically active *cis-B/C* isorotenolones, the 6aβ,12aβ-ketol
(LXXV, p. 291) gave the 12α-diol (LXXIX) when treated with borohydride, whilst the 6aα,12aα-ketol (LXXVI) gave the 12β-diol (LXXX):
when crystallised together these two diols gave the $(\pm)$-A diol mentioned
above. The racemic diols $(\pm)$-A, $(\pm)$-B, and $(\pm)$-B$_2$ were all degraded
to the same racemic ketoaldehyde (LXXXIX–XC) by periodic acid:
the optically active diols (LXXIX) and (LXXX) gave the enantiomers
(LXXXIX) and (XC), respectively (*38, 39*).

The formation of isorotenolones by aerial oxidation in alkaline medium can be viewed as resulting from attack by molecular oxygen on the ion (XCI) to give the hydroperoxide ion. The hydroperoxide (XCII) may be hydrolysed to the ketol and hydrogen peroxide or react with (XCIII) to give in either case isorotenolones (*39*). Other modes of decomposition are possible for the hydroperoxide, and the ketoacid (XCIV) can be isolated as a by-product (*37*).

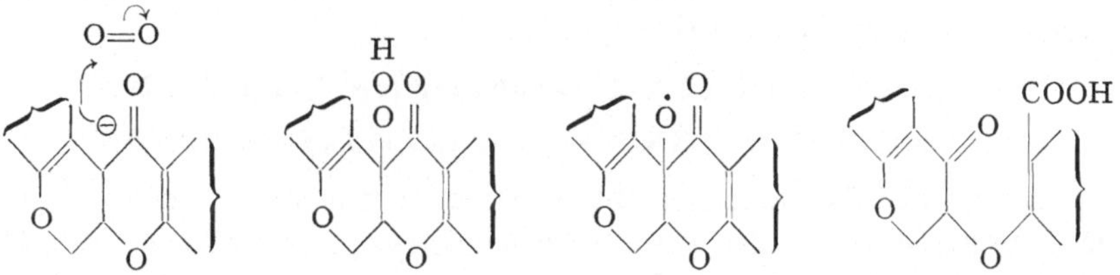

(XCI.) (XCII.) (XCIII.) (XCIV.)

The alkaline oxygenation of natural (—)-rotenone, originally described by TAKEI (*180*) has also been examined by CROMBIE and GODIN (*38, 39*) who have shown that the substances rotenolones I and II are not pure entities but mixtures of diastereoisomers. Diastereoisomers, and not racemates, as for the isorotenolones, are obtained since there is an additional 5'β-optical centre. Crystallisation of rotenolone II gives the two *trans B/C* diastereoisomers 6aβ,12aα,5'β (LXXVII) and 6aα,12aβ,5'β (LXXVIII). The two *cis-B/C* diastereoisomers contained in rotenolone I do not part cleanly on crystallisation (though their presence can be demonstrated by examination of the products of borohydride reduction), and they were made individually. The 6aβ,12aβ,5'β-compound (LXXV) was the major product from dichromate oxidation of rotenone, and the 6aα,12aα,5'β-compound (LXXVI) was made by treating epi-rotenone similarly. Assignments of ring fusion were again made on the basis of hydrogen bonding criteria. The configuration of the two *cis-B/C* diastereoisomers at 6a is shown by degradation to the 5'β-ketoaldehydes (LXXXIX) and (XC), the configurations of which are established (see below).

Each *trans*-rotenolone diastereoisomer gave a separable pair of 12,12a-diols when reduced with borohydride:

6aβ,12aα,5'β-ketol (LXXVII.) ↗ 6aβ,12aα,12α,5'β-diol (LXXXIII.)
6aβ,12aα,5'β-ketol (LXXVII.) ↘ 6aβ,12aα,12β,5'β-diol (LXXXV.)

6aα,12aβ,5'β-ketol (LXXVIII.) ↗ 6aα,12aβ,12β,5'β-diol (LXXXIV.)
6aα,12aβ,5'β-ketol (LXXVIII.) ↘ 6aα,12aβ,12α,5'β-diol (LXXXVI.)

The *cis*-12,12a structure is assigned to diols (LXXXIII) and (LXXXIV), because they are cleaved almost instantly with lead tetraacetate whilst diols (LXXXV) and (LXXXVI) are attacked more slowly. The diols (LXXXIII and (LXXXV) give the $6a\beta,5'\beta$-ketoaldehyde (LXXXIX) whilst (LXXXIV) and (LXXXVI) yield the $6a\alpha,5'\beta$-ketoaldehyde (XC). All the diols are oxidised back to their parent ketols by manganese dioxide.

Only one diol was isolated from each of the *cis-B/C* rotenolone diastereoisomers on borohydride reduction:

$6a\beta,12a\beta,5'\beta$-ketol (LXXV) $\rightarrow 6a\beta,12a\beta,12\alpha,5'\beta$-diol (LXXIX.)

$6a\alpha,12a\alpha,5'\beta$-ketol (LXXVI) $\rightarrow 6a\alpha,12a\alpha,12\beta,5'\beta$-diol (LXXX.)

The 12,12a-diol arrangement is based on the rate of cleavage with lead tetraacetate whilst the stereochemistry at 6a follows from that of their precursors. On oxidation with periodic acid the diol (LXXIX) gives ketoaldehyde (LXXXIX) whilst (LXXX) gives (XC). Thus, these ketoaldehydes are available to fix configurations in the *trans-B/C* series (*38, 39*).

The rotenolones have also been investigated by Matsui and Miyano (*133—135, 141*) who apparently treat rotenolones I and II as pure individuals: where comparisons are possible, they reach conclusions different from those stated above.

It is of interest that many of the *cis-B/C*-compounds above crystallise solvated from methanol (*39*) whereas in the *trans-B/C* series this rarely happens. Rotenone itself forms a number of solvates (*95*) and this may be a reflection of the bent *cis*-molecule (XXVI—XXVII) which can provide cavities in the crystal lattice.

2. The C and D Series.

$(\pm)$-Isorotenolone C, $C_{23}H_{22}O_7$, was discovered by LaForge and Haller (*108*) when isorotenone was heated with 30% hydrogen peroxide and 5% methanolic potassium hydroxide. It has been investigated by Crombie and Godin (*39*) who propose structure (XCV, $R = OH$), and enantiomer *(Chart 11)*. Analogous compounds are formed by dihydrorotenone and by deguelin (*108*), but from the reaction mixture containing isorotenolone C another compound, isorotenolone D, has been obtained (m. p. 189—190°): recent workers (*39*) have only succeeded in isolating a polymorph of C having this m. p.

Isorotenolone C contains one hydroxyl group (methyl ether, acetate, toluene-*p*-sulphonate) but gives no enol acetate suggesting that $C_{(12a)}$ carries no hydrogen. The hydroxyl is not benzylic (resistance to manganese dioxide oxidation and to hydrogenolysis) and not phenolic (failure to methylate with diazomethane, infrared criteria). The ketone stretching frequency (1709 cm.$^{-1}$) compared with that of isorotenone (1680 cm.$^{-1}$) suggests a 5-membered environment. Isorotenolone C has no hemiacetal properties and its chloro-replacement product (XCV, $R = $ Cl) has the stability of a β-halogenoether. Structure (XCV) is consistent with its

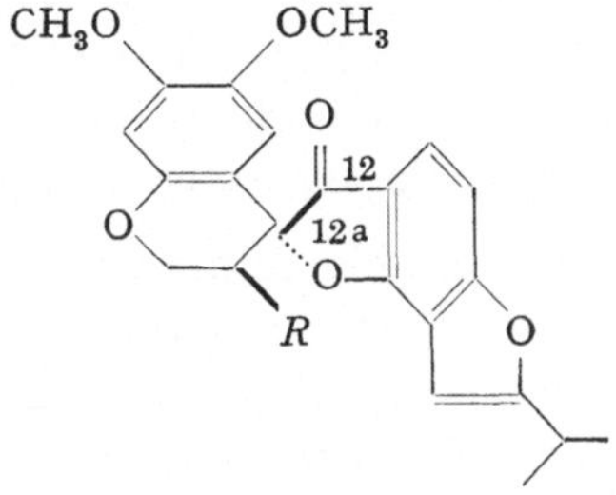

(XCV.) Isorotenolone C.

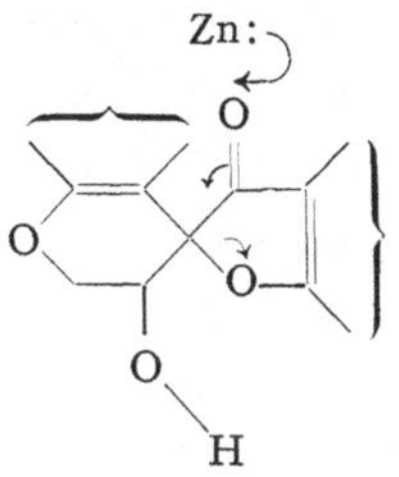

(XCVI.)

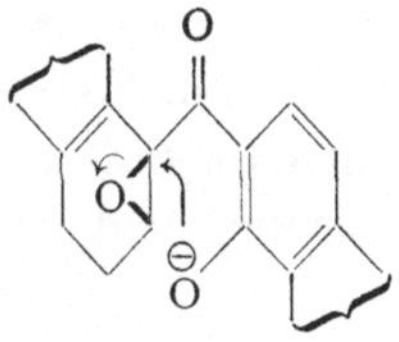

(XCVII.)

Chart 11. Formation and Degradation of Isorotenolone C.

reaction with zinc and alkali which gives rotenol and derritol, obviously derived from a ring *C* opening (XCVI) followed by dehydration and the steps described earlier. With borohydride, isorotenolone C yields a diol which is oxidised back to its precursor by manganese dioxide, confirming the location of the keto group. Nuclear magnetic resonance data (*43*) support the structure proposed.

The formation of isorotenolone C can be viewed as attack by $^-$OOH on the ion (XLVII) present in the alkaline solution to give the epoxide, followed by intramolecular opening as in (XCVII). The hydroxyl should therefore be *cis* to the spiro ketone (*39*).

VI. The Rotenoids.

1. **Stereochemistry.** The stereochemistry of rotenone (I, p. 277) has been determined in full and examination of its optical rotatory dispersion curve shows a strong positive Cotton effect.

The O. R. D. curves of five natural rotenoids have been examined by DJERASSI, OLLIS and RUSSELL (*51*) and all have a positive Cotton effect. It is thus very likely, as discussed by these authors, that all natural rotenoids have the same configuration as natural rotenone at 6a and 12a. The remaining known rotenoids measured since have all shown positive Cotton effects (*44*).

In the case of sumatrol with a 5' optical centre, derrisic acid (XXXVIII, p. 285) has been compared with the corresponding degradation product of sumatrol (*44*). In these the 6a and 12a centres have been destroyed and both components show the same negative plain curve suggesting that the configuration at 5' is probably the same in each case. In sumatrol and rotenone this negative plain curve is clearly superimposed on the normal Cotton effect curve (*51*). There thus appears to be remarkable stereochemical homogeneity in the rotenoid class.

2. Deguelin (V). Treatment of uncrystallisable resins from derris and cube roots, and from the roots of *Tephrosia (Cracca) toxicaria* and the leaves of *T. vogelii* yielded ($\pm$)-deguelin, m. p. 171° (or 165°) (*21, 26*). The optically pure rotenoid has never been isolated though concentrates containing it are reported; however, direct crystallisation of the hydrogenated resins does give optically pure dihydrodeguelin (*109*). Deguelin was discovered by Clark (*26*) and much structure work is due to him (*29, 30, 33*). On dehydrogenation a 6a,12a-dehydro compound is obtained which is hydrolysed to deguelinic acid, the analogue of derrisic acid

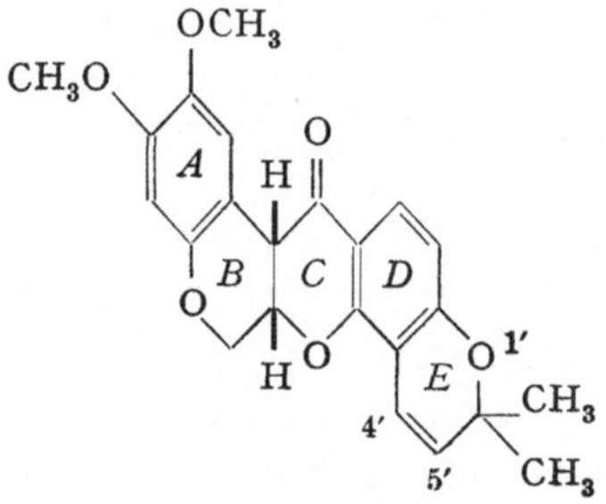

(V.) Deguelin. 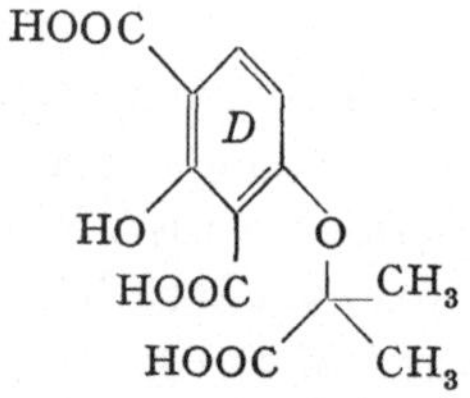   (XCVIII.) Nicouic acid.

(XXXVIII), or further oxidised to the ring *B* lactone level (cf. XLI). Oxidation of 6a,12a-dehydrodeguelin with permanganate gives rissic acid (XL, p. 285) and 2-hydroxy-4,5-dimethoxybenzoic acid representing rings *A* and *B* whilst nicouic acid (XCVIII) represents rings *C*, *D*, and *E*.

The identity of ($\pm$)-4',5'-dihydrodeguelin with ($\pm$)-β-dihydrorotenone (*70*), made by cyclising ($\pm$)-rotenonic acid (cf. LXXI, p. 289) in the presence of acid (*69, 75*), puts the structure of deguelin beyond doubt.

The alleged isodeguelin of Merz and Schmidt (*124*) from *T. vogelii* is spurious, as are allotephrosin and alloisotephrosin (*7, 39*).

($\pm$)-Tephrosin, m. p. 169–170°, was originally isolated by Hanriot from *T. vogelii* (*78*). It, and ($\pm$)-isotephrosin, m. p. 252°, were both isolated from Peruvian cube root by Clark (*27*). ($\pm$)-Tephrosin was investigated by Clark (*27, 34*) and by Butenandt (*16*), who showed it to be a hydroxydeguelin (cf. IC). On oxidation ($\pm$)-tephrosin gives tephrosin dicarboxylic acid (C) which, when heated in diphenyl ether gives tephrosin monocarboxylic acid (CI) and isobutyric acid *(Chart 12)*. ($\pm$)-Tephrosin is readily dehydrated to dehydrodeguelin, and on alkaline oxygenation deguelin gives ($\pm$)-tephrosin and ($\pm$)-isotephrosin *(36, 181)*. CROMBIE and GODIN *(39)* have shown that the former is the ($\pm$)-*cis-B/C* compound (IC and enantiomer) and the latter the ($\pm$)-*trans* form (LXXVII –LXXVIII). There is general agreement that tephrosin and isotephrosin from natural sources are artefacts derived from deguelin.

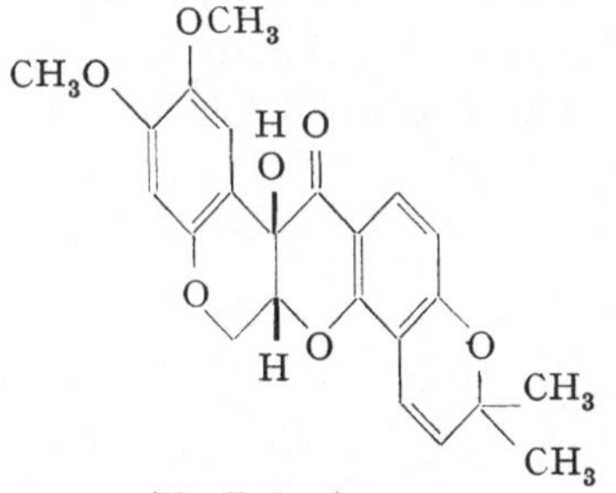

(IC.) Tephrosin.

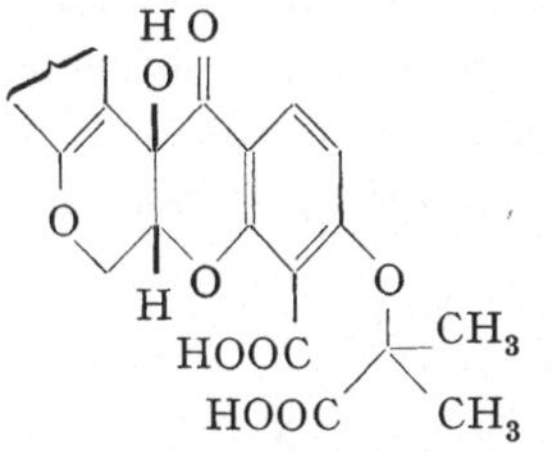

(C.) Tephrosin dicarboxylic acid.

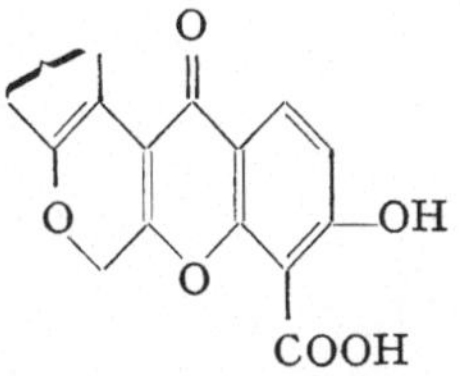

(CI.) Tephrosin monocarboxylic acid.

Chart 12. Tephrosin Derivatives.

3. Elliptone (IV). ($\pm$)-Elliptone, $C_{20}H_{16}O_6$, was first isolated by BUCKLEY *(11*, cf. *6)* from *Derris elliptica*, m. p. 180–181°. The optically active natural compound, m. p. 160° or 178.5–179.5°, $[\alpha]_D^{18} - 18°$ (benzene), was isolated by HARPER *(79, 80)* and by MEIJER and KOOLHAAS *(122)*. HARPER *(82, 83)* has extensively investigated its chemistry. It gives a 6a,12a-dehydro compound and the analogue of derrisic

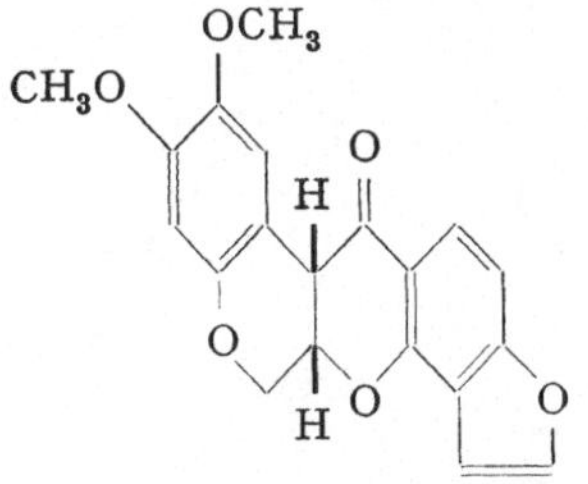

(IV.) Elliptone.

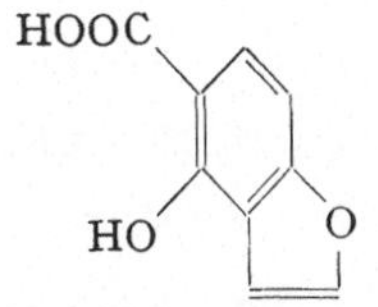

(CII.) 4-Hydroxycoumaran-5-carboxylic acid.

acid (XXXVIII, p. 285), elliptic acid. The latter is oxidised by peroxide to derric acid (XXXIX) and this, taken with the oxidation of ring *B* to the lactone level, established the nature of the *A/B/C* system. Rings *D* and *E* were identified by strong alkaline treatment of elliptone which gave 4-hydroxycoumaran-5-carboxylic acid (CII). The hydrogenation of elliptone has been extensively studied and partial synthetic work carried out.

4. **Munduserone** (III). This rotenoid, $C_{19}H_{18}O_6$, m. p. 162° $[\alpha_D^{16} + 103°$ (chloroform), was isolated from *Mundulea servica* bark by FINCH and OLLIS (56), to whom the structural work is due. With sodium acetate it gives ($\pm$)-munduserone, m. p. 171–172°. Munduserone yields a 6a,12a-dehydro compound, a rotenonone-type compound (cf. XLI, p. 285) and an analogue of derrisic acid (XXXVIII). On oxidation with alkaline peroxide *(Chart 13)* β-resorcyclic acid 4-methyl ether (CIII) accounting

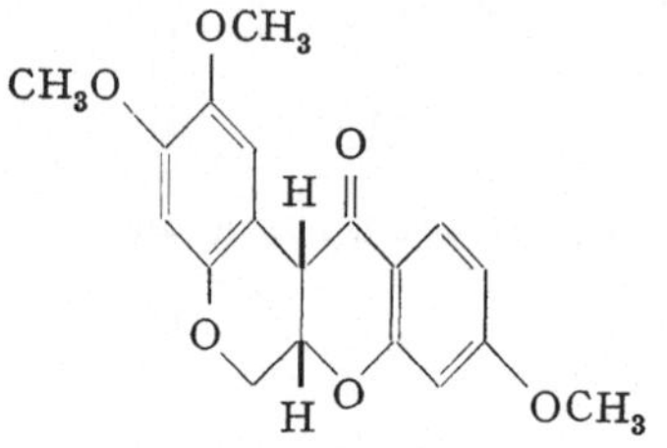

(III.) Munduserone.

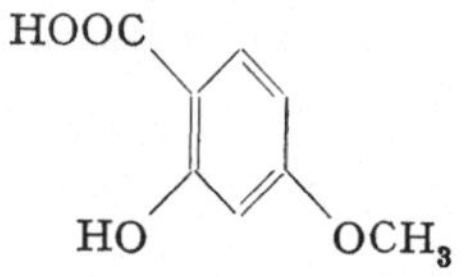

(CIII.) β-Resorcyclic acid 4-methyl ether.

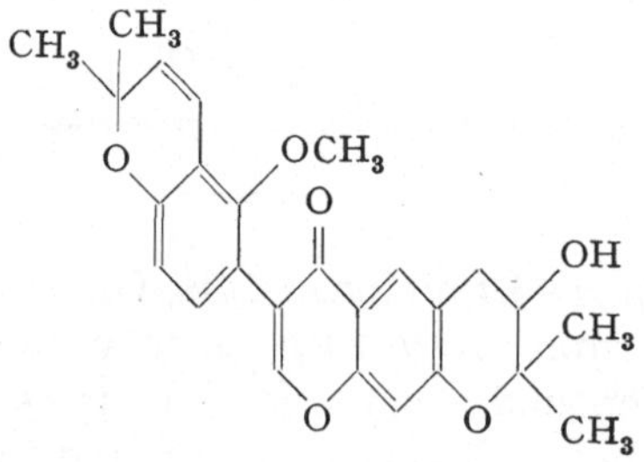

(CIV.) Mundulone.

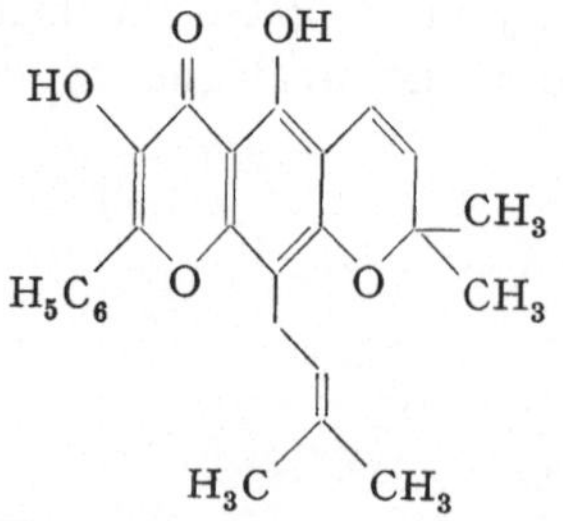

(CV.)

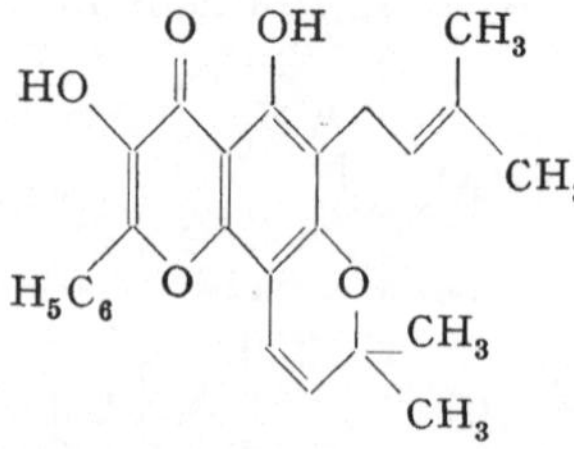

Sericetin.
(Alternative structures.)

(CVI.)

Chart 13. Munduserone and its Natural Relatives.

for rings C/D, and rissic acid (XL) accounting for rings A/B were obtained. Munduserone is the simplest natural rotenoid; the ($\pm$)-form has been synthesised. Occurring with it is mundulone (CIV) (*12*) and sericetin formulated as (CV) or (CVI) (*13*).

5. **α-Toxicarol (VIII).** ($\pm$)-α-Toxicarol, $C_{23}H_{22}O_7$, m. p. 231–232°, was isolated by CLARK (*25, 28, 31, 32, 35*) by alkaline treatment of *Tephrosia toxicaria* and *Derris* extracts. Natural (—)-α-toxicarol, m. p. 125–127°, $[\alpha]_D^{20}$ — 66° (benzene), was isolated by TATTERSFIELD and MARTIN (*182*) from Sumatra-type derris resin and was fully examined by CAHN and his colleagues (*22*): it is identical with "prototoxicarol".

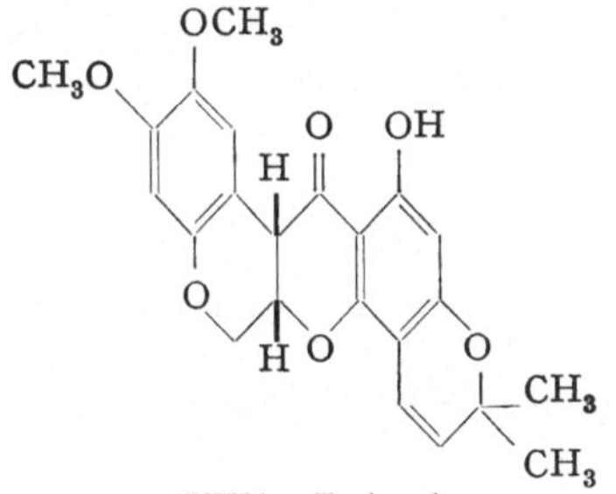

(VIII.) α-Toxicarol.

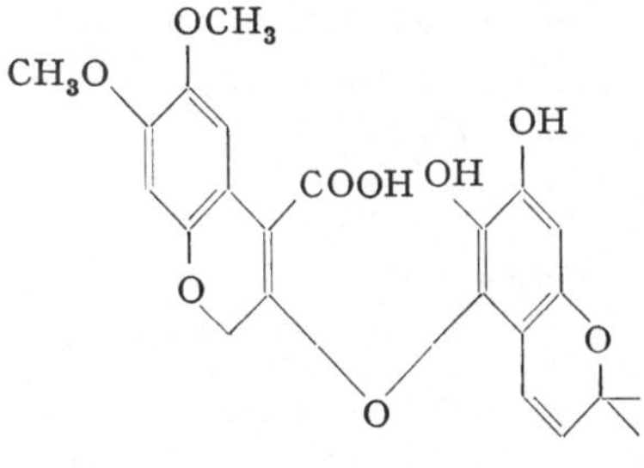

(CVII.)

α-Toxicarol contains an 11-hydroxyl group chelated to the 12-ketone and not methylated with diazomethane (ν_{max} 1638 cm.$^{-1}$; cf. rotenone

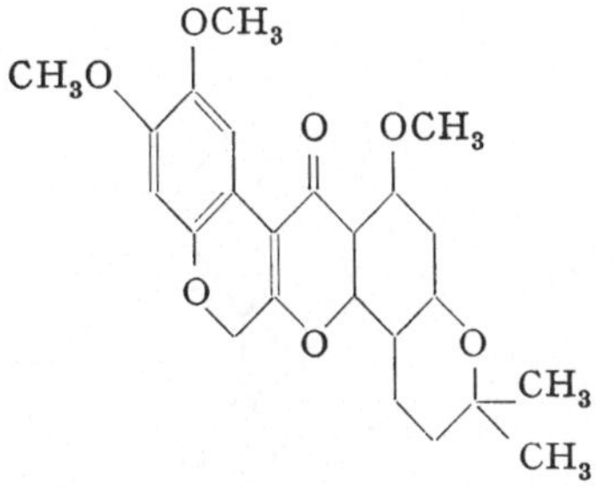

(CVIII.) 4′,5′-Dihydro-6 a,12 a-dehydrotoxicarol-11-methyl ether.

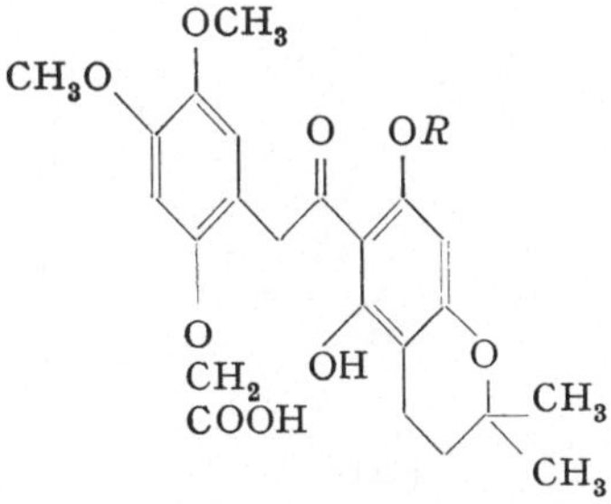

(CIX.) O-Methyl-dihydrotoxicarolic acid ($R = CH_3$).

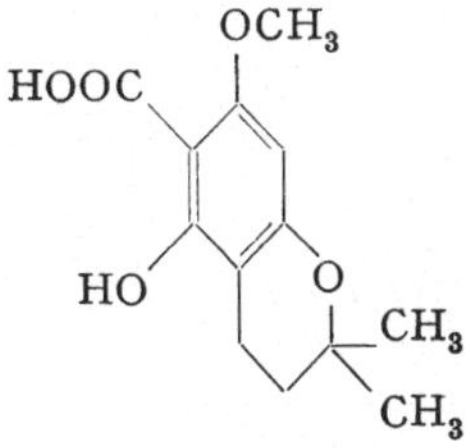

(CX.)

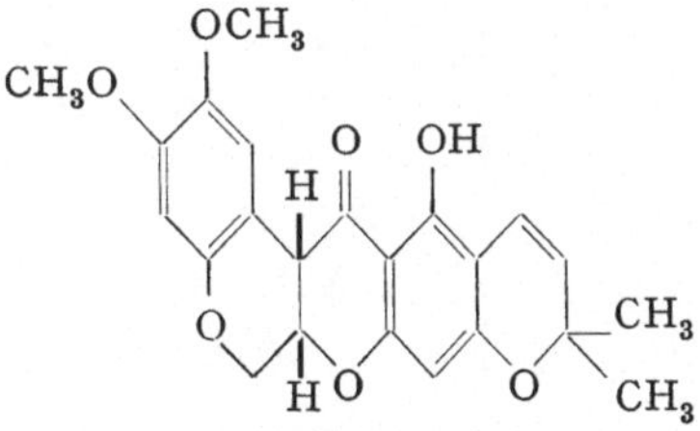

(CXI.) β-Toxicarol.

Chart 14. Linear D/E Fusion of β-Toxicarol.

1674 cm.$^{-1}$); its location has been elegantly shown by Clark (*32*), as the Dakin reaction on 6a,12a-dehydrotoxicarol gives (CVII), which is further oxidised to rissic acid. On alkaline hydrolysis dehydrotoxicarol gives the derrisic acid analogue (cf. XXXVIII, p. 285) which can be oxidised to derric acid (XXXIX). Degradation of 4′,5′-dihydro-6a,12a-dehydrotoxicarol-11-methyl ether (CVIII) (made by methylation with dimethyl sulphate-potassium carbonate) has been especially illuminating *(Chart 14)*. With 5% alkali it gives O-methyl-dihydrotoxicarolic acid (CIX, $R = CH_3$) which, when hydrolysed by 50% alkali yields derric acid and the chroman fragment (CX), known by synthesis. This observation reveals the fundamentals of the toxicarol structure and settles whether the D/E attachment in α-toxicarol is angular, as in (VIII), or linear, as in (CXI): in the latter case the methyl ether isomeric with (CX) would have been obtained (*64*).

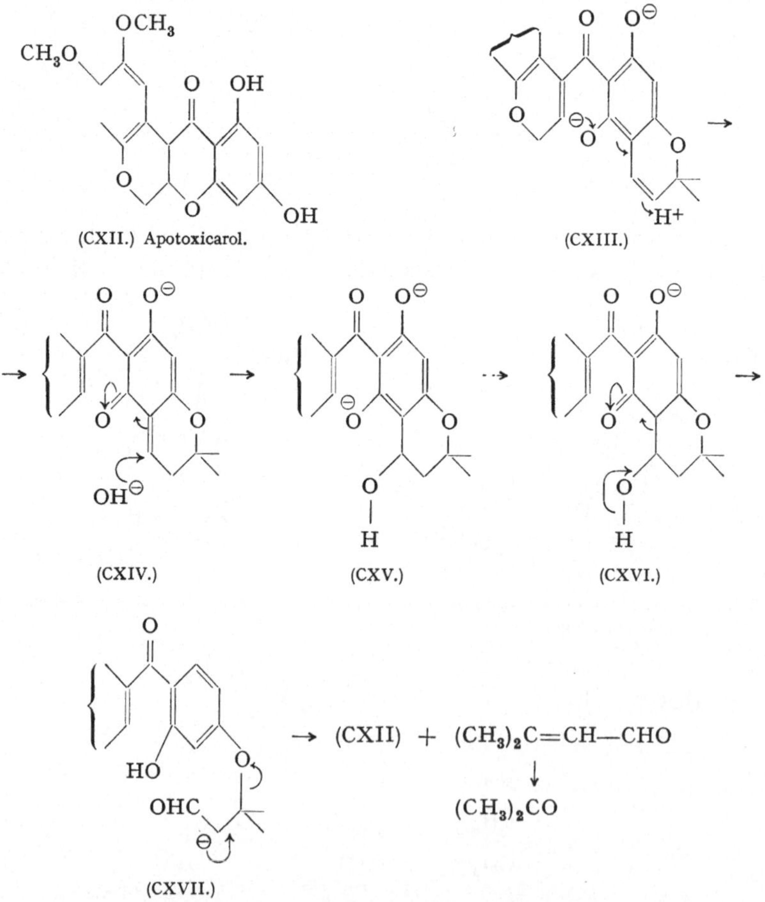

(CXII.) Apotoxicarol.

(CXIII.)

(CXIV.)

(CXV.)

(CXVI.)

→ (CXII) + $(CH_3)_2C{=}CH{-}CHO$

$(CH_3)_2CO$

(CXVII.)

Chart 15. Formation of Apotoxicarol.

When α-toxicarol is treated with alkali, an interesting reaction occurs forming apotoxicarol (CXII) and acetone (deguelin also gives acetone under rather severer conditions) (*8, 64, 91*). The removal of ring *E* and formation of acetone may be represented as (CXIII–CXVII) in *Chart 15* (cf. *151*). Like toxicarol itself, apotoxicarol gives dehydronetoric (toxicaric) acid (LXV) and hydroxynetoric acid when oxidised in alkaline solution.

The linear isomer of α-toxicarol, (±)-β-toxicarol, m. p. 169–170° (CXI), has been isolated and thoroughly characterised by CAHN (*22*). It forms in equilbrium when α-toxicarol is treated with potassium carbonate in wet acetone. When treated with alkali, both 4′,5′-dihydro-6a,12a-dehydro-α- and β-toxicarol give the same compound (CIX, $R = H$), so β-toxicarol must be the linear isomer of α-toxicarol. Application of the Gibbs reaction confirms this assignment (*44*).

The importance of the β-elimination intermediate (XLVII) (*17, 41, 94*) to the understanding of the chemistry of α- and β-toxicarol is apparent from the foregoing; it also explains the contrast between the methylation

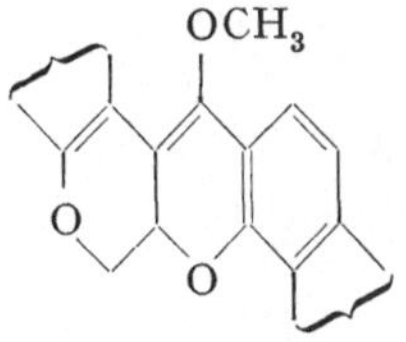

(CXVIII.) 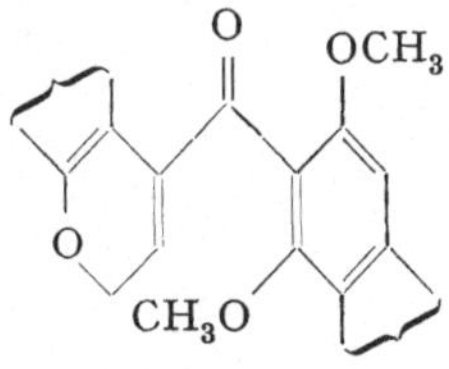   (CXIX.)

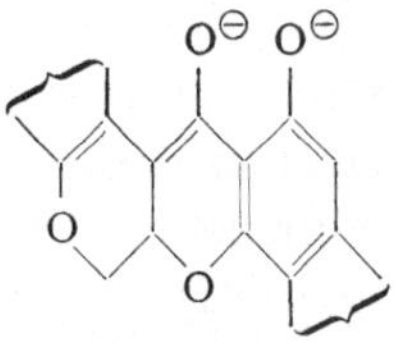

(CXX.) 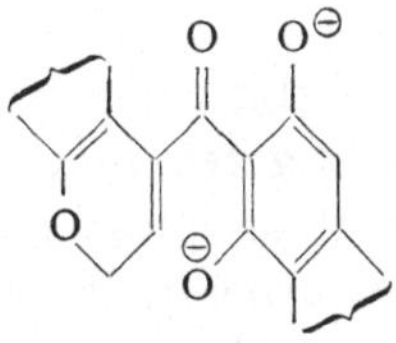   (CXXI.)

Chart 16. Dichotomous Methylation of Rotenone and Toxicarol.

of rotenone and toxicarol (*23, 41*). With dimethyl sulphate and alkali *(Chart 16)* rotenone gives the acid labile enol ether (CXVIII), whereas toxicarol gives the acid stable dimethyl ether (CXIX). In the case of toxicarol the ion (CXX), which would yield a dimethyl ether having one acid-labile methyl ether, is destabilised relative to (CXXI) because of the adjacent negative charges, and (CXXI) predominates in alkaline solution (*40, 41*).

6. **Sumatrol** (VI). This rotenoid, $C_{23}H_{22}O_7$, m. p. 194°, $[\alpha]_D - 184°$ (benzene), was isolated by CAHN and BOAM (*20*) from a Sumatra-type *Derris (D. malaccensis* var. *Sarawakensis).* It contains a chelated

11-hydroxyl group and in much of its general chemistry resembles rotenone: the 6a,12a-dehydro compound, the derrisic acid analogue and the rotenonone analogue are known. On hydrogenation ring E can be broken to give the dihydrorotenonic acid (LXXII) analogue. When degraded, this yields the derrisic acid type (CXXIII) via the 6a,12a-dehydro compound.

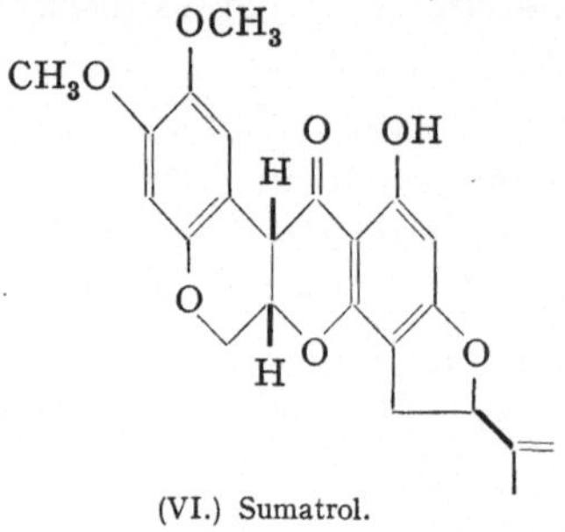

(VI.) Sumatrol.

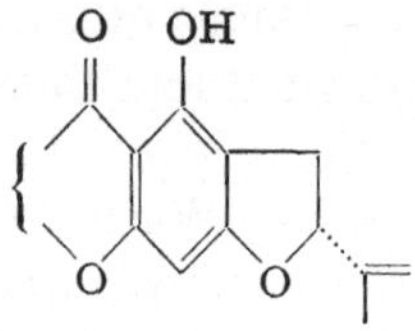

(CXXII.)

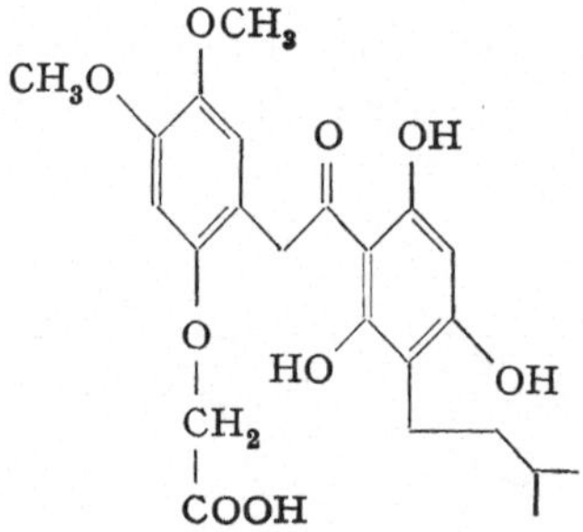

(CXXIII.)

The synthesis of this compound by Robertson and his colleagues (*102*), who have investigated much of the sumatrol chemistry (*166*), led to the angular (VI) or linear (CXXII) D/E formulae as the structure of sumatrol. The earlier preference for structure (VI) has been confirmed by a study of the Gibbs reaction of sumatrol (*44*), and its stereochemistry is as shown (*44, 51*). In the mixture obtained by sodium acetate treatment of sumatrol, there is much linear material (*44*).

7. Malaccol (VII). Kinta-type *Derris malaccensis* was examined by Meijer and Koolhaas (*122*) and by Harper (*84*), who isolated from it

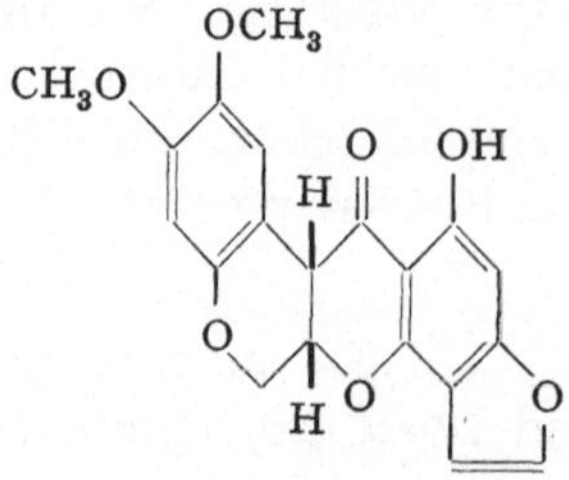

(VII.) Malaccol.

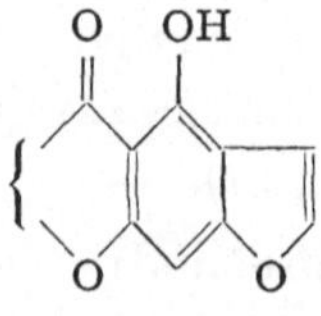

(CXXIV.)

an 11-hydroxyrotenoid malaccol, $C_{20}H_{16}O_7$, m. p. 225–227°, resolidifying and melting as ($\pm$)-malaccol, 249–250°, $[\alpha]_D^{20} + 67°$ (benzene). HARPER (*84, 85*) has investigated its chemistry.

The *B/C* ring system shows the expected reactions for a rotenoid, and on hydrogenation ring *E* can absorb four atoms of hydrogen to give (CXXV) (*88*). When degraded in the usual way this yields the derrisic acid type (CXXVI) which has been synthesised leaving (beyond the uncertainty of linear versus angular *D/E* fusion) little doubt as to

(CXXV.)   (CXXVI.)

the correctness of (VII). By the Gibbs reaction, natural malaccol has been shown to have angular fusion (VII) but unexpectedly ($\pm$)-malaccol of the literature, made by sodium acetate treatment, is the linear form (CXXIV) (*44*). Depending on the alkalinity and reaction conditions natural malaccol gives dehydro compounds derived from the angular or linear *D/E* fusion (*44*).

8. Pachyrrhizone (X). This compound, $C_{20}H_{14}O_7$, m. p. 272°, $[\alpha]_D$ 98° (chloroform), was isolated as "Compound II" by NORTON and HANSBERRY (*145*) from *Pachyrrhizus erosus* (yam beans) and later examined by MEIJER (*121*). The structure and detailed examination of its chemistry is due to BICKEL and SCHMID (*3*). Sodium acetate racemises it to ($\pm$)-pachyrrhizone, m. p. 265°, so no other optical centre is present. Pachyrrhizone contains one methoxy group, and infrared data and the Labat test suggest that a methylenedioxy group is present. The usual

(X.) Pachyrrhizone.   (CXXVII.)

reactions of the *B/C* ring system of a rotenoid are displayed by pachyrrhizone—the formation of an oxime, an enol acetate, a 6a,12a-dehydro compound, and compounds of the rotenone and derrisic acid type. The crucial degradation of pachyrrhizone with alkali gave the coumaran (CXXVII), the structure of which was settled by synthetic work.

9. **Erosone.** This compound was discovered by NORTON and HANSBERRY (*145*) in *P. erosus* and was considered by them to be an elliptone isomer. It was also referred to as "Compound V", $C_{20}H_{16}O_6$, m. p. 218°, $[\alpha]_D^{25} + 234°$ (benzene). Erosone is racemised by sodium acetate to give ($\pm$)-erosone, m. p. 232–234°, and it gives a dehydro compound. The expected derrisic acid type compound was apparently formed and is different from that of elliptone. In view of its occurrence with pachyrrhizone and dolineone, structure (CXXVIII) seems reasonable. But the isoflavone nepseudin (CXXIX) also occurs with dolineone (*45*, *47*), so an erosone structure with ring *A* oxygenation as in this, should also be considered.

(CXXVIII.)

(CXXIX.) Nepseudin.

10. **Dolineone** (Dolichone) (IX). This rotenoid has been isolated from *Neorautanenia pseudopachyrrhiza*, together with the closely related isoflavonone, neotenone (CXXX), by CROMBIE and WHITING (*46*, *47*). It also occurs in *Pachyrrhizus erosus* (*47*). Dolineone, $C_{19}H_{12}O_6$, m. p. 235° or 249°, $[\alpha]_D^{20} + 135°$ (chloroform), has been available in small amounts only (100 mg.). It has ν_{max} 1684 cm.$^{-1}$, and the characteristic ultra-

(IX.) Dolineone.

violet spectrum closely simulates that of neotenone. This, taken with the nuclear magnetic resonance assignments shown has led to structure (IX).

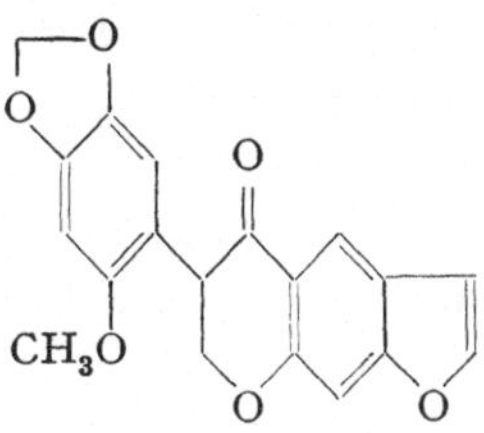

(CXXX.) Neotenone.

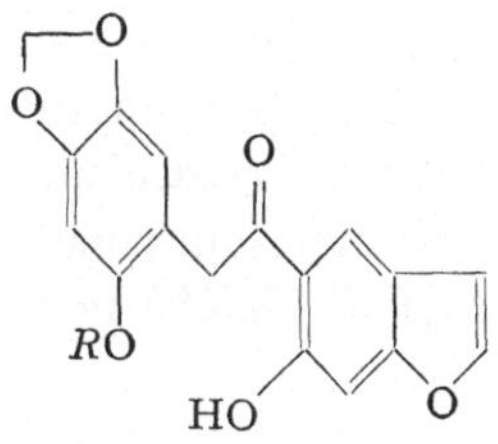

(CXXXI.)

In agreement with (IX) dolineone is dehydrogenated to the 6a,12a-compound (ν_{max} 1639 cm.$^{-1}$) and oxidised to the ring B-lactonic rotenonone type (ν_{max} 1748 and 1658 cm.$^{-1}$). This confirms the 5,6,6a,12a,12 sequence of atoms. Zinc dust and alkali treatment of dolineone gave a derritol analogue (CXXXI, $R = H$) which was methylated to (CXXXI, $R = CH_3$). The latter was purified and identified with the same degradative fragment obtained this time from neotenone (CXXX), by thin-layer chromatography. This result confirms the nature of the rest of the molecule (47).

VII. Biogenesis and Biogenetic Connections of the Rotenoids.

During investigation of *Derris malaccensis*, HARPER (85) isolated a compound related in structure to α-toxicarol and suggested that it might be the isoflavone (CXXXII, $R = CH_3$). Its structure has never been finally cleared up (89) but the isoflavone (CXXXIII) as well as the iso-flavanone (CXXX) occurs together with dolineone (IX) (47).

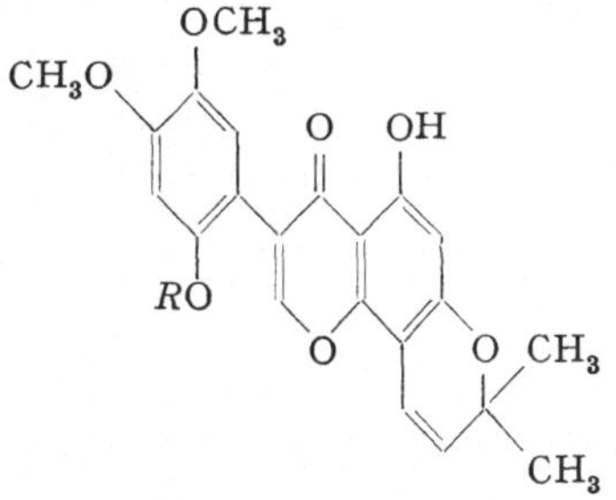

(CXXXII.)

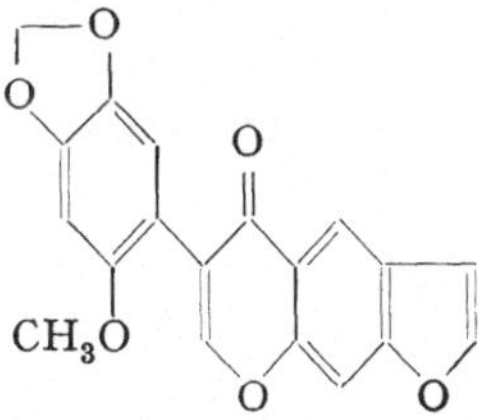

(CXXXIII.)

Although the view that rotenoids may be formed in Nature from coumarins and substituted resorcyclic acids has been put forward (168), a close connection with isoflavanoid biogenesis seems generally accepted.

The biogenesis of isoflavones *(Chart 17)* has recently been studied by Grisebach *(67)* using ^{14}C tracer techniques and has been reviewed *(66, 68)*. It has been demonstrated that ring A and the three carbon atoms a, b, and c in (CXXXIV) can be derived from phenylalanine (or cinnamic acid or p-coumaric acid as closer precursors), whilst ring D is derived from acetate (or more probably malonyl CoA). During formation of the isoflavone the aryl residue A originally attached to carbon (a) (cf. CXXXV) migrates to carbon (b). It has been suggested

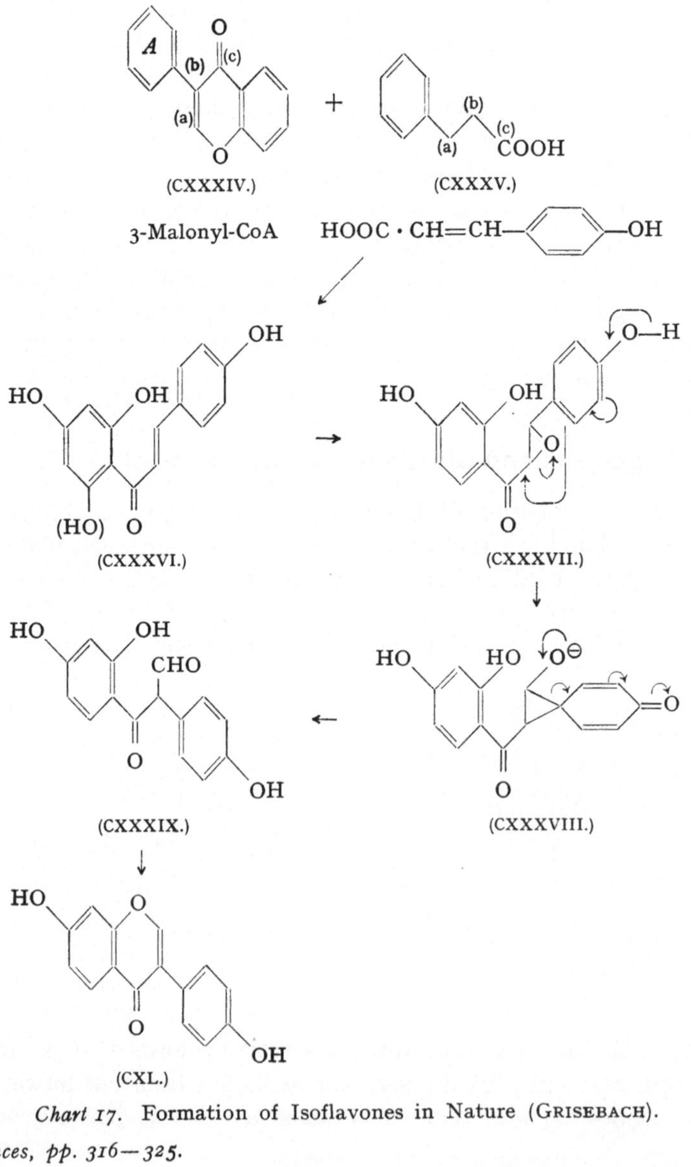

(CXXXIV.)   (CXXXV.)

3-Malonyl-CoA   HOOC · CH=CH—⟨ ⟩—OH

(CXXXVI.)   (CXXXVII.)

(CXXXIX.)   (CXXXVIII.)

(CXL.)

Chart 17. Formation of Isoflavones in Nature (Grisebach).

that biosynthesis and aryl migration in the case of formenonetin proceeds as in (CXXXVI–CXL) (*66*).

If it is assumed that rotenoid formation in Nature is closely linked with isoflavonoid formation, a supply of $C_{(5)}$ in the rotenoid has to be provided for. There are various views on this point. DUTTA (*52*) suggests

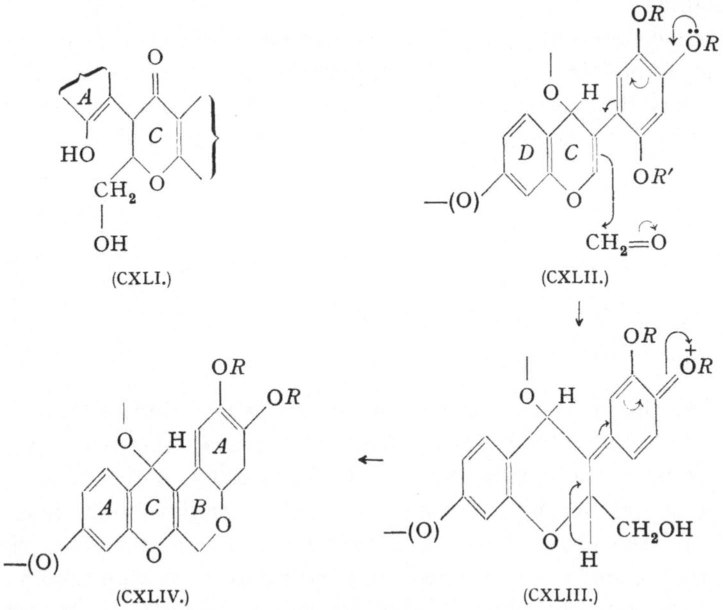

Chart *18*. Scheme for Rotenoid Formation in Nature (SCHMID et al.).

that the methyl group R in an isoflavone like (CXXXII) may supply this carbon. SESHADRI (*171*) suggests that it is supplied from a species (CXLI). SCHMID and his colleagues (*176*) put forward the more specific proposal (CXLII–CXLIV), that the rotenoid is initially formed at the 6a,12a-dehydro level via a Prins reaction with formaldehyde or an equivalent *(Chart 18)*.

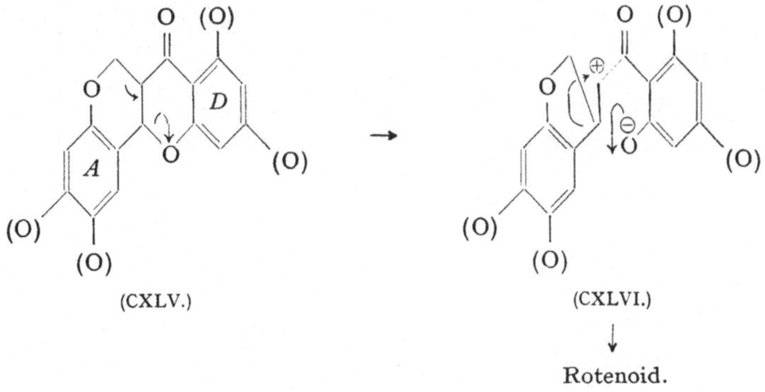

Grisebach and Ollis (*68*) suggest that a hydroxychalcone (cf. CXXXVI) might be an intermediate in rotenoid biogenesis. This, or the flavanone derived from it might, by unspecified processes, react with formaldehyde or its equivalent to give the tetracyclic nucleus (CXLV) which rearranges by a process such as indicated in (CXLVI) to give a rotenoid-type compound.

The development of ring E can be viewed as insertion of an isoprenoid residue, probably as isopentenyl-pyrophosphate or $\gamma\gamma$-dimethylallyl-pyrophosphate, via the mevalonic acid route. The furan ring E may be produced by degradation of a C_5 unit as suggested by Birch (*4*):

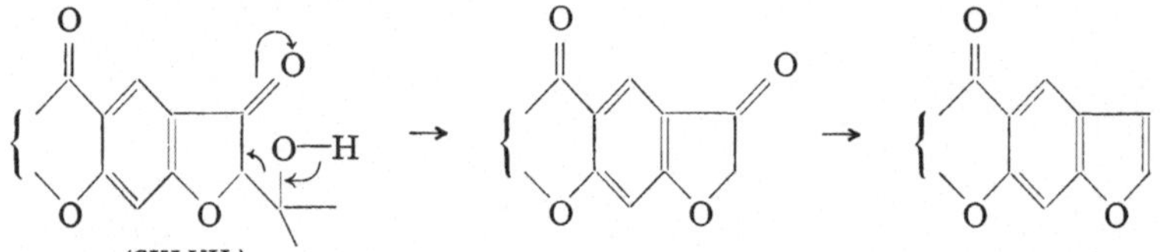

(CXLVII.)

Up to the present, no metabolic studies using ^{14}C-tracers have been reported in the rotenoid field, hence the present biogenetic views are speculative. In a broader context, understanding of the details of the biogenetic relationships between rotenoids, isoflavonoids and other closely related plant phenolics occurring together in the same plant or in the same group of plants will be of great interest. Within the Phaseolea tribe of the sub-family Papilionatae, for instance, a series of close relatives has been found which will undoubtedly be extended as investigation proceeds.

These include neotenone (CXXX) (*46, 47*), dehydroneotenone (CXXXIII) (*46, 47*), dolineone (IX) (*46, 47*), pachyrrhizone (X) (*3*), pachyrrhizin (CXLVIII) (*175*),

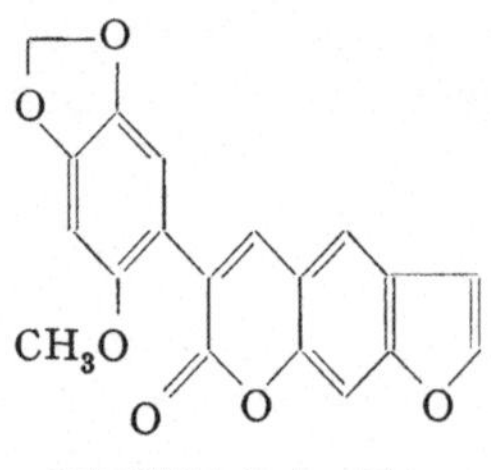

(CXLVIII.) Pachyrrhizin.

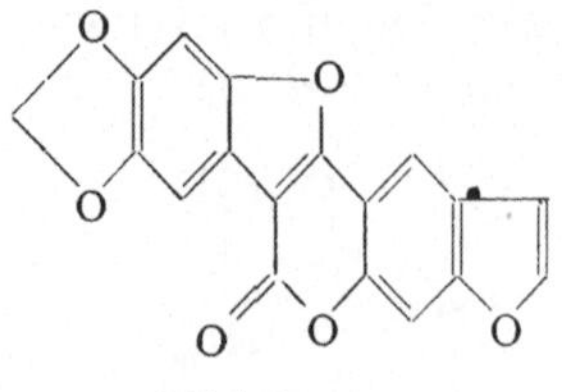

(CIL.) Erosnin.

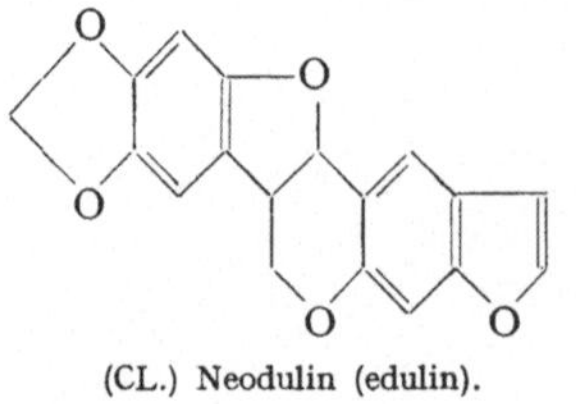

(CL.) Neodulin (edulin).

erosnin (CIL) (*53*), neodulin (edulin) (CL) (*47*, *183*), rotenone (I) (*145*), nepseudin (CXXIX) (*45*, *47*), and also erosone (*145*). A table of their occurrence in *Neorautanenia pseudopachyrrhiza*, *N. edulis* and *Pachyrrhizus erosus*, has been drawn up (*47*).

VIII. Synthesis in the Rotenoid Group.

Many of the degradation products of rotenone and rotenoids have been synthesised and are useful as intermediates for further synthesis. Among those containing ring *A* are rissic acid (XL, p. 285) (*161*) and derric acid (XXXIX) (*161*). Nitriles of the type (CLI) are important intermediates in the Hoesch syntheses (see below) and are often prepared from aldehydes (CLII) by the azlactone synthesis as indicated (*162*). The azlactone is hydrolysed to the α-keto-acid, which is converted into the oxime and then dehydrated to the nitrile.

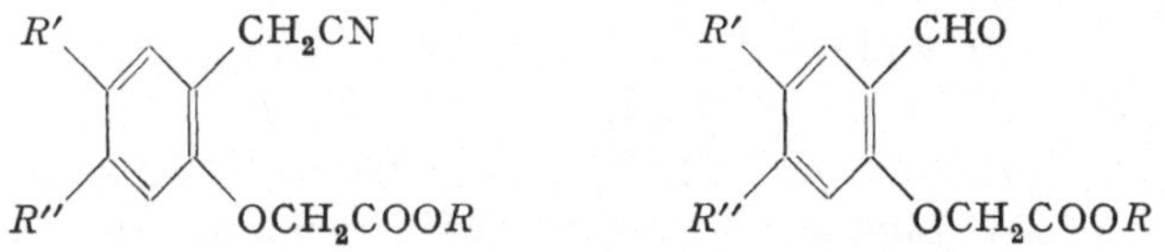
R'
CH_2CN
R''
OCH_2COOR
R'
CHO
R''
OCH_2COOR

(CLI.) (CLII.)

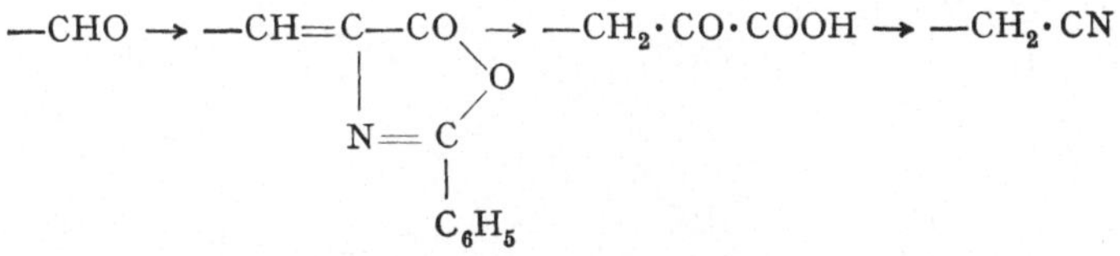
—CHO → —CH=C—CO → —CH_2·CO·COOH → —CH_2·CN
O
N=C
C_6H_5

Netoric acid and dehydronetoric (toxicaric) acids containing ring *A* can be prepared by cyclising derric ester (cf. XXXIX) with sodium to give the enol-ester (CLIII) (*5*, *146*, *165*). This, on catalytic reduction, yields hydroxynetoric acid (obtainable in *cis*- and *trans*-forms) (*129*) which is dehydrated to dehydronetoric acid. Hydrogenation and hydrogenolysis of the acetate from (CLIII) gives netoric acid. The isomer (CLIV) is also reported to be formed in the Dieckmann cyclisation (*128*, *129*).

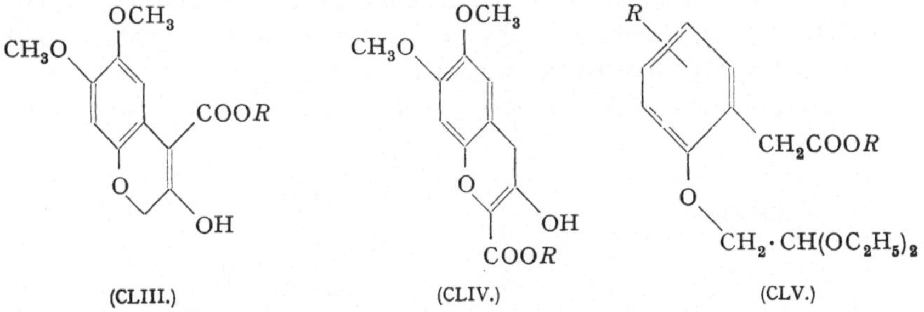
OCH_3
CH_3O
COOR
O
OH
OCH_3
CH_3O
O
OH
COOR
R
CH_2COOR
O
$CH_2·CH(OC_2H_5)_2$

(CLIII.) (CLIV.) (CLV.)

Compounds of the dehydronetoric acid type are alternatively available by treating suitable phenols with chloroacetaldehyde dimethyl acetal to give (CLV) and cyclising the aldehyde from this (*147*). Abutic acid (CLVI), the degradation product of rotenononic acid (XLIII) was prepared by Robertson (*93*). The 3-halogenocoumarin (CLVII), made by a modified Pechmann reaction on 2,4-dihydroxyanisole and methylation, was subjected to Perkin transformation.

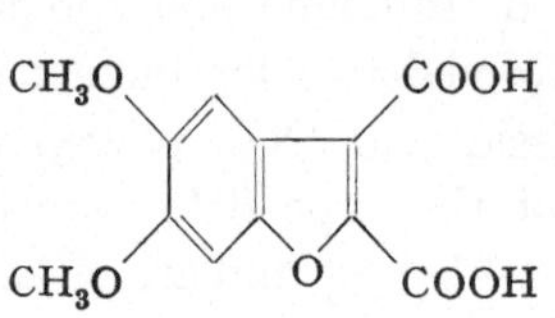

(CLVI.) Abutic acid. 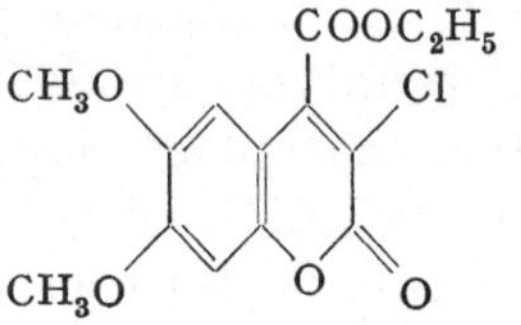   (CLVII.)

Among fragments containing ring D tetrahydrotubanol (CLVIII, $R = R' = H$) (*71, 167*), tetrahydrotubaic acid (CLVIII, $R = COOH$, $R' = H$) (*71, 164*), and the degradation products from sumatrol (CLVIII, $R = H$, $R' = OH$) and (CLVIII, $R = COOH$, $R' = OH$) (*102*) have been made. Isotubaic acid (CLIX) was prepared by Reichstein (*154*), who condensed sodium succinate with isopropylfurfural (CLX),

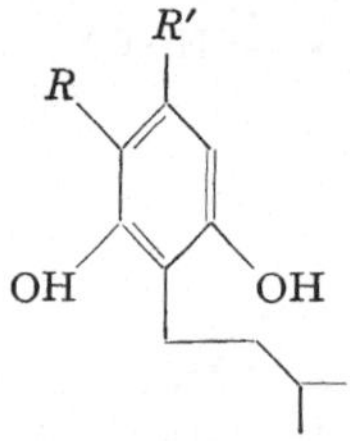

(CLVIII.) 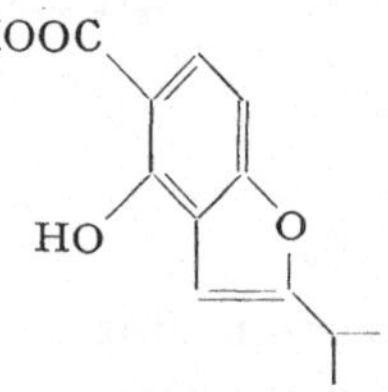   (CLIX.) 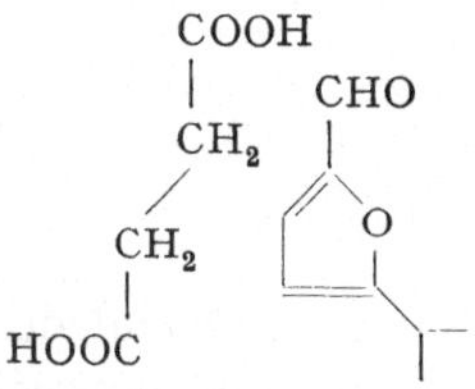   (CLX.)

decarboxylated the product, and then recarboxylated the isotubanol formed: the procedure was used to make the corresponding acid from elliptone (CII) (*83*). Synthesis of ($\pm$)-tubaic acid (CLXIV, $R = COOH$) was claimed by Shamshurin (*172–174*). The aldehyde (CLXI) was condensed with bromacetone and sodium methoxide and the product (CLXII) was hydrogenated, hydrolysed and decarboxylated to the ketone (CLXIII). Treatment with Grignard reagent, dehydration and carboxylation was said to give ($\pm$)-tubaic acid. Miyano and Matsui

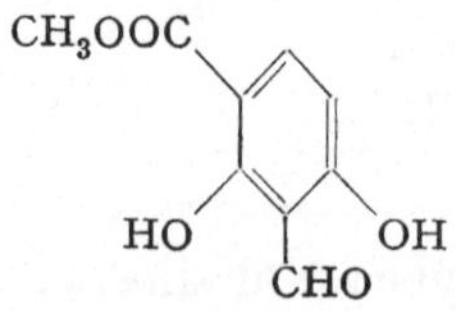

(CLXI.) 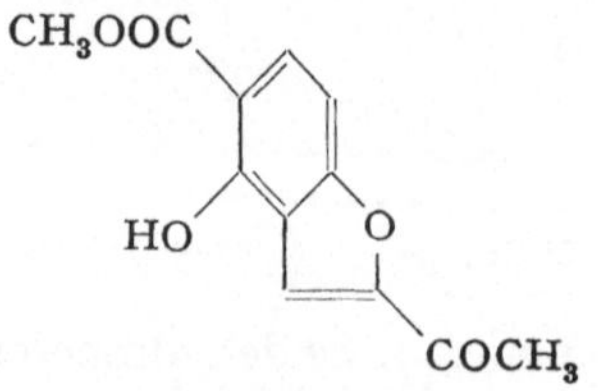   (CLXII.)

(*135, 136, 139, 140*) have modified this approach and have resolved the acid, but there are serious disagreements between the results of the Russian and Japanese authors.

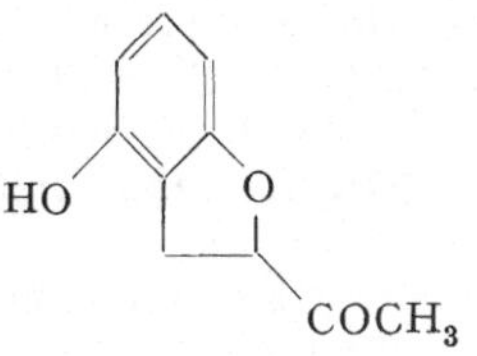

(CLXIII.) 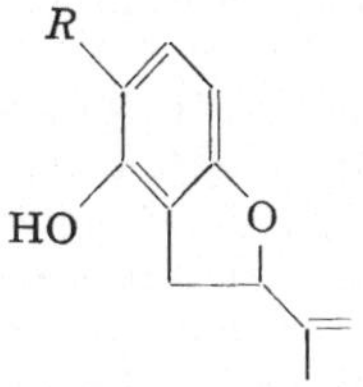   (CLXIV.)

NICKL (*144*) has developed a simple synthesis of the deguelin degradation product (CLXV) by condensing methyl 2,4-dihydroxybenzoic acid with 3-hydroxy-3-methylbut-1-yne and basic zinc chloride as in (CLXVI). Chromones of the type (CLXVII) are also synthetically useful and are

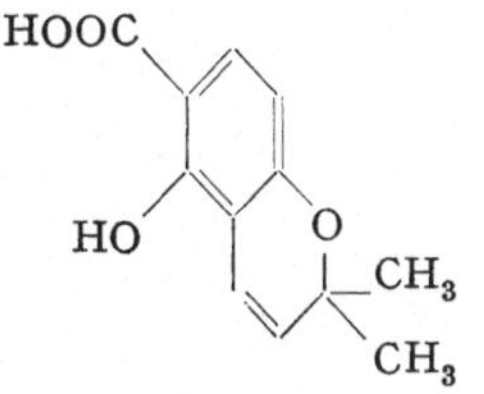

(CLXV.) 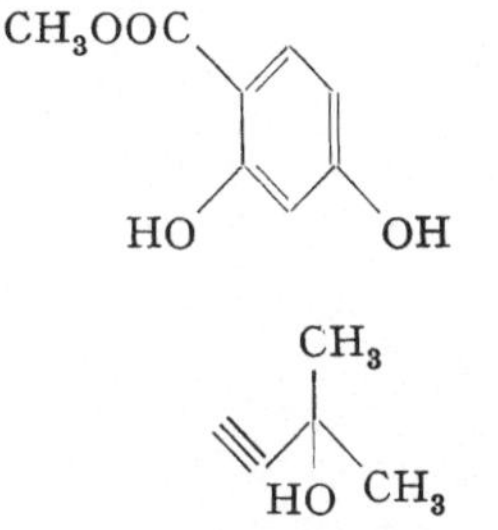   (CLXVI.) 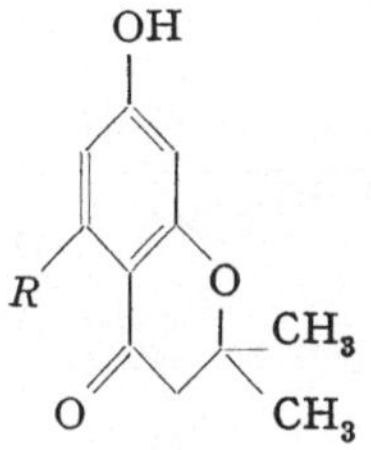   (CLXVII.)

made by condensing phloroglucinol with bromoisovaleryl chloride in the presence of aluminium chloride (*155*), or β-methylcrotonic acid and resorcinol in the presence of anhydrous hydrofluoric acid or antimony trichloride or zinc chloride (*130, 148, 149*).

Compounds of the derrisic acid type have been particularly important synthetic intermediates, and the use of the Hoesch synthesis was

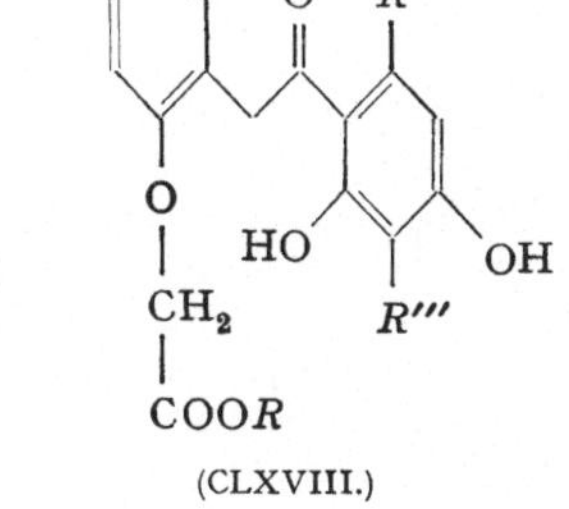

(CLXVIII.)

developed by Robertson (*162*). Nitriles (CLI) are condensed with phenols to give deoxybenzoin derivatives (CLXVIII).

In this way tephrosic acid (CLXVIII, $R' = R'' = OCH_3$, $R''' = R'''' = H$) (*163*) and the derrisic acid analogues from tetrahydrosumatrol ($R' = R'' = OCH_3$, $R''' = CH_2CH_2CH(CH_3)_2$, $R'''' = OH$) (*102*), tetrahydroelliptone ($R' = R'' = OCH_3$, $R''' = C_2H_5$, $R'''' = H$) (*88*), tetrahydromalaccol ($R' = R'' = OCH_3$, $R''' = C_2H_5$, $R'''' = OH$) (*88*), dihydrorotenonic acid ($R' = R'' = OCH_3$, $R''' = CH_2CH_2CH(CH_3)_2$, $R'''' = H$) (*163*), dihydrorotenone (CLXX, $R' = H$) (*127, 137*) and related compounds (*64*) were prepared.

Difficulties arise when there are not two free p-hydroxy positions present in the phenol undergoing Hoesch reaction (*155*). Thus, the nitrile (CLI, $R = R' = OCH_3$) and the tubanol derivative (CLXIX, $R' = OH$) are said to give both (CLXX, $R' = OH$ and Cl) and (CLXXI, $R' = OH$) (*125, 126*). The ester mixture (CLXX, $R' = OH$ and Cl, $R = CH_3$) is said to be dehydrated and hydrolysed

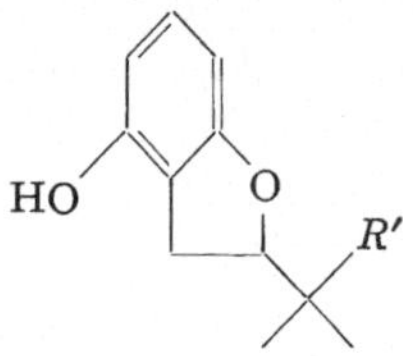

(CLXIX.)

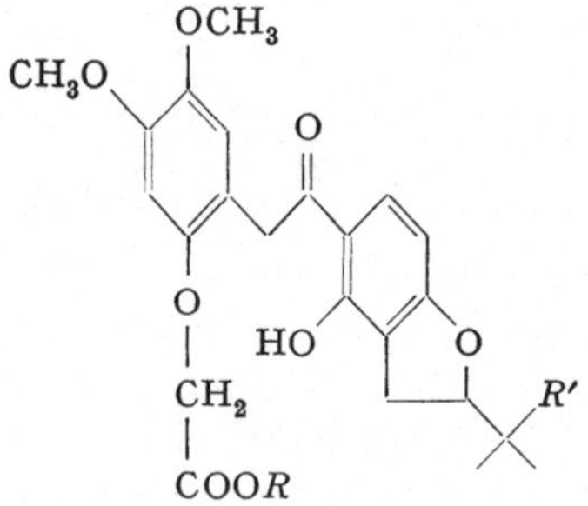

(CLXX.) 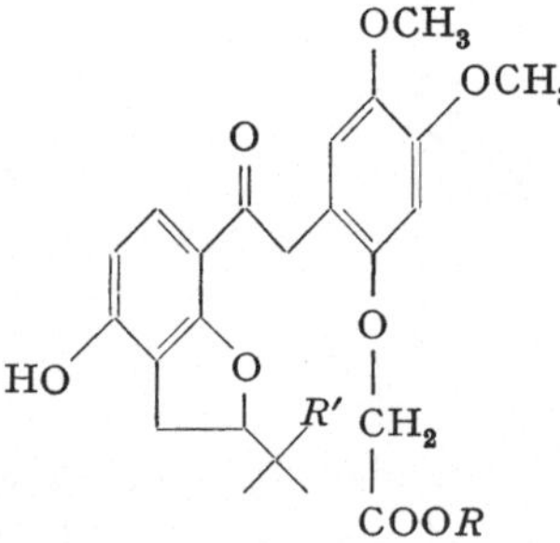   (CLXXI.)

to ($\pm$)-derrisic acid, but only 3 mg. of product having an unsatisfactory analysis was isolated. All attempts to carry out a Hoesch reaction between the nitrile (CLI, $R = R' = OCH_3$) and ($-$)-tubanol have failed (*126*).

In the case of linear D/E rotenoids the synthetic ambiguity reflected in the formation of (CLXX) and (CLXXI) does not arise as the p-position

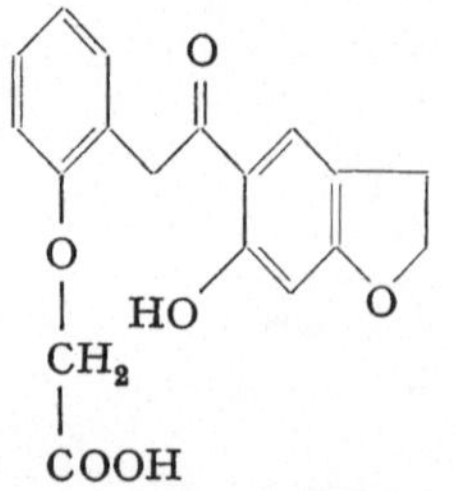

(CLXXII.) 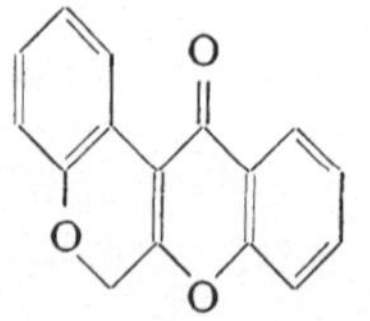   (CLXXIII.)

of the phenol is blocked. Thus the derrisic type compound (CLXXII) is satisfactorily available (*153*).

LaForge and Haller (*109a*) and Takei (*179*) have shown that derrisic acid compounds like (CLXX) cyclise to 6a,12a-dehydrorotenoids when treated with sodium acetate and acetic anhydride. Free hydroxyls are acetylated, but this normally causes no difficulty.

In this way dehydrorotenone (*125*, *126*), dehydrodihydrorotenone (*127*, *137*), dehydrotetrahydrorotenone (dehydrodihydrorotenonic acid) (*163*), dehydronordihydrorotenone (*138*), dehydrotetrahydroelliptone (*88*), dehydromunduserone (*90*), dehydrotetrahydromalaccol (*88*), dehydrotetrahydrosumatrol (*102*), dehydrodegeulin (59) and related compounds are available (*103*, *162*).

The parent dehydro-compound (CLXXIII) has also been obtained by LaForge (*106*). By benzoin condensation of salicylaldehyde methoxy methyl ether, followed by zinc and alkaline reduction, the deoxybenzoin (CLXXIV, $R = R' = CH_2OCH_3$) is prepared. This can be demethoxymethylated to the phenol (CLXXIV, $R = R' = OH$). When treated with ethyl bromoacetate and sodium ethoxide, only the unchelated hydroxyl reacts to give the derrisic type (CLXXIV, $R = CH_2COOC_2H_5$, $R' = H$), which can be cyclised to give (CLXXIII). The compound

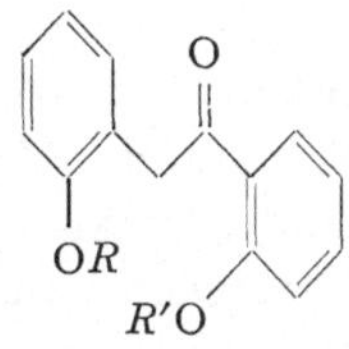

(CLXXIV.)

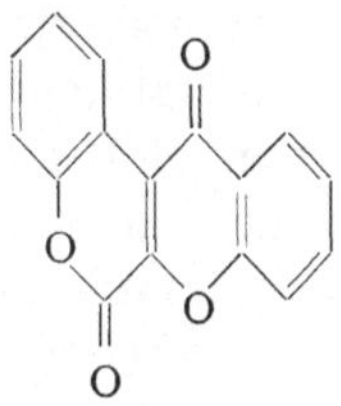

(CLXXV.)

(CLXXV), parent of the rotenonone series, is obtained by treating (CLXXIV, $R = R' = OH$) with $ClCOCOOC_2H_5$ and pyridine (*106*), a method later employed for isoflavone synthesis (*1*, *2*). Sodium acetate and dimethyl oxalate also react with derritol types (CLXXIV, $R = R' = OH$) to give rotenonones (*179*). During syntheses of 11-hydroxy-dehydrorotenoids or 11-hydroxy-rotenonone compounds via these kinds of reaction there is always ambiguity, for cyclisation of compounds of the type (CLXVIII, $R'''' = OH$) and its relatives can lead to (CLXXVI) or (CLXXVII), i. e. to linear D/E and angular D/E derivatives.

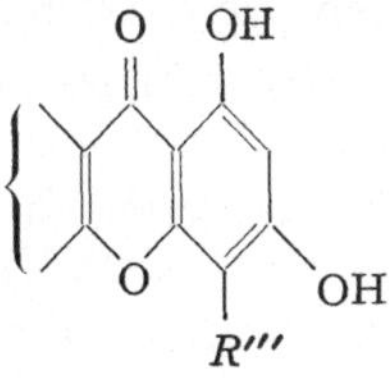

(CLXXVI.)

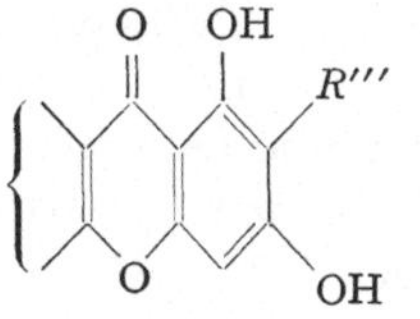

(CLXXVII.)

Two methods are available for the conversion of 6a,12a-dehydro-rotenoids into rotenoids. The first is catalytic hydrogenation. Conversion of dehydrotoxicarol into (±)-dihydrotoxicarol was originally reported by Clark, but this was overlooked for many years and a number of failures were reported. Clark's work is sound. Low pressure hydrogenation over a platinum catalyst (9, 61, 177) or high pressure hydrogenation over a palladium catalyst, followed by oxidation of the alcohol formed back to the ketone with chromic acid, may be used (48–50). In this way the parent of the rotenoid series, 6a,12a-dihydro-6H-rotoxen-12-one (II, p. 277) has been made (9, 49, 61). Dihydrorotenone is also accessible from dehydrorotenone. The procedure of Matsui and Miyano (131, 132) is superior in that hydrogenation of other olefinic linkages is not normally encountered. The 6a,12a-dehydrorotenoid is treated with sodium borohydride and undergoes 1,4-reduction and then reduction of the resulting ketone to the 12-hydroxy compound (XXX); the latter is oxidised back to the ketone level, without 6a,12a-dehydrogenation, by the Oppenauer reaction. Compound (II) has been obtained in this way as well as mutarotenone (131, 132), (±)-deguelin (59, 60), (±)-munduserone (90), and other relatives. Cahn (22) has shown that natural (—)-rotenone can be separated as the carbon tetrachloride solvate from the diastereoisomers which constitute mutarotenone.

By combination of the efforts of a number of workers who have made important contributions—LaForge and Haller, Takei, Robertson, Cahn, Shamshurin, Matsui and Miyano—it is possible to trace out a formal synthesis of natural (—)-rotenone which uses natural degradation products as relays. In doing so, the fact that (±)-derrisic acid has not been resolved is overlooked; but more serious is the poor characterisation of this synthetic compound and the unsatisfactory character of the Hoesch reaction (see above) (126).

The synthesis of (±)-deguelin (57–60, 141) involves Hoesch reaction between the nitrile (CLI, $R = CH_3$, $R' = R'' = OCH_3$, p. 309) and the compound (CLXXVIII). This again gives two products corresponding to (CLXX) and (CLXXI). The former is cyclised to 4',5'-dihydro-6a,12a-dehydrodeguelin. Its treatment with N-bromosuccinimide is said to give a little 6a,12a-dehydrodeguelin as well as compounds of the rotenonone type. Dehydrodeguelin is then converted into (±)-deguelin by the borohydride method (59).

(CLXXVIII.)

Other synthetic approaches to the rotenoid system have been tried. Seshadri (170, 171) prepared the isoflavone (CLXXIX, $R = CH_3$, $R' = Ac$, $R'' = H$) which was hydrolysed, demethylated and reacetylated to (CLXXIX, $R = R' = Ac$, $R'' = H$). When treated with

N-bromosuccinimide this gave (CLXXIX, $R = R' = $ Ac, $R'' = $ Br); treatment of the latter with alcoholic hydrogen chloride and then dry potassium carbonate yielded the 6a,12a-dehydrorotenoid (CLXXX). A modification uses the ethoxy compound (CLXXIX, $R = R' = OAc$, $R'' = OC_2H_5$): with hydrogen bromide this is converted into the bromo derivative (CLXXIX, $R = R' = $ OH, $R'' = $ Br) (*119*).

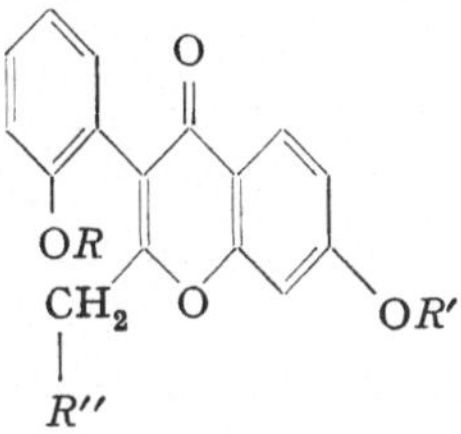

(CLXXIX.) 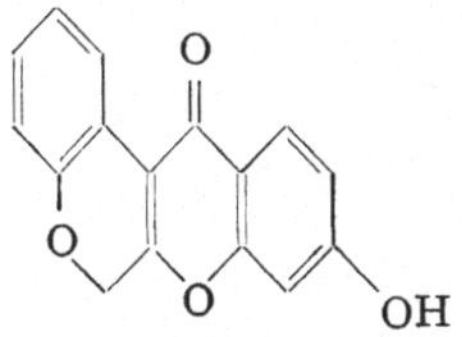 (CLXXX.) 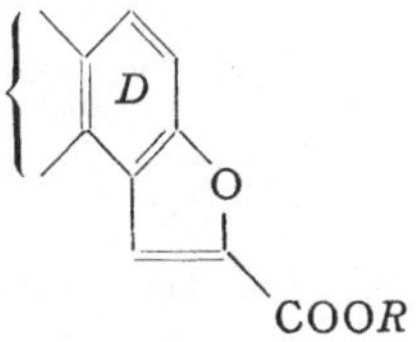 (CLXXXI.)

In recent work a furan ring has been built onto ring D by the introduction of an 8-formyl grouping via the hexamine method. This aldehyde, when condensed with bromomalonate and potassium carbonate gives the ester (CLXXXI), which is hydrolysed and decarboxylated to 6a,12a-dehydroelliptone (*100a, 101*). No comparison with the natural degradation product has been made (*100a*).

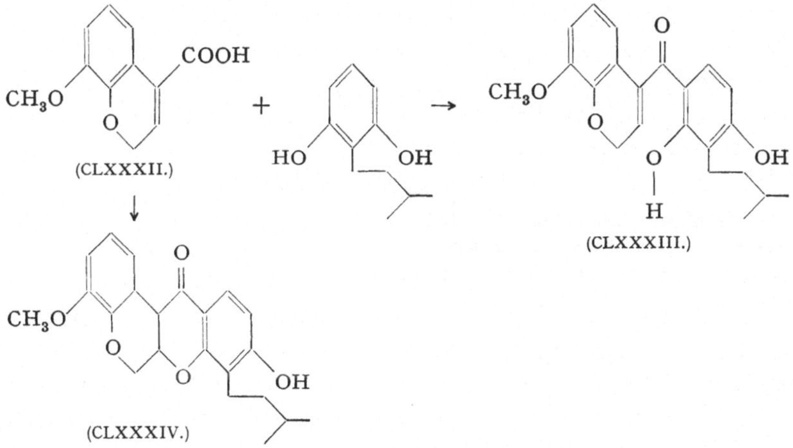

(CLXXXII.) (CLXXXIII.)

(CLXXXIV.)

Offe (*147, 148*) has attempted condensation of a dehydronetoric acid type (CLXXXII) with tetrahydrotubanol in the presence of anhydrous hydrogen fluoride at 70°. Three products were obtained, two being thought to be (CLXXXIII) and (CLXXXIV), but there is no positive evidence of their identities (*148*).

An interesting route to the 12-deoxyrotenone skeleton has been outlined by Herbert and Ollis (*90*). The aldehyde (CLXXXVI) and the enol acetate (CLXXXV) condense in the presence of piperidine to give the 12,12a-dehydro derivative of (CLXXXVII, $R = OH$), which can be hydrogenolysed and hydrogenated to (CLXXXVII, $R = H$).

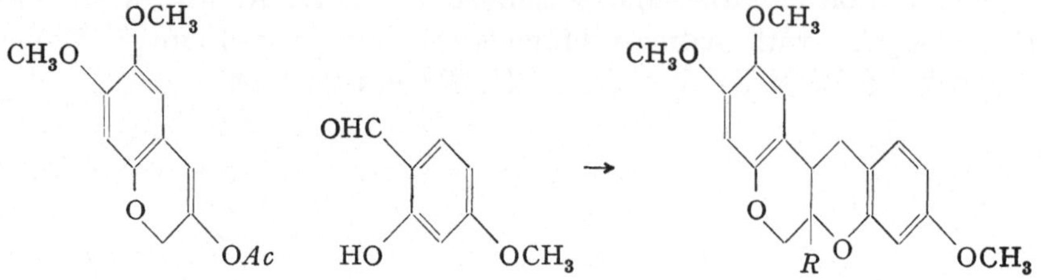

(CLXXXV.)   (CLXXXVI.)   (CLXXXVII.)

Finally, compounds of the rotenonone class may be prepared by condensing 3,7-dimethoxycoumarin-4-carboxylic acid chloride (CLXXXVIII) with phenols such as resorcinol under Friedel-Crafts conditions to give a compound typified by (CLXXXIX). The latter is cyclised by hydrogen bromide to the rotenonone type (CXC) (*152*).

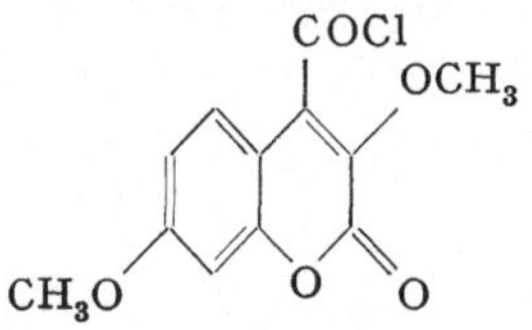

(CLXXXVIII.) 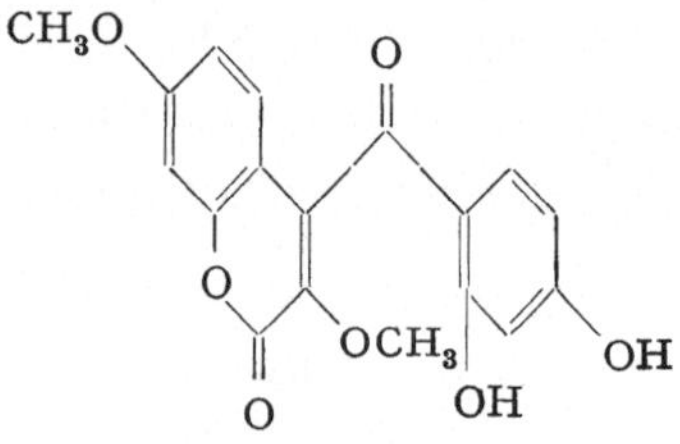   (CLXXXIX.)

(CXC.)

Addendum. Amorphin, isolated from the seeds of *Amorpha fructicosa*, is the glycoside of 8'-hydroxyrotenone (amorphigenin) and is the first rotenoid glycoside [CROMBIE, L. and R. PEACE, Proc. Chem. Soc. (London) **1963**, 246].

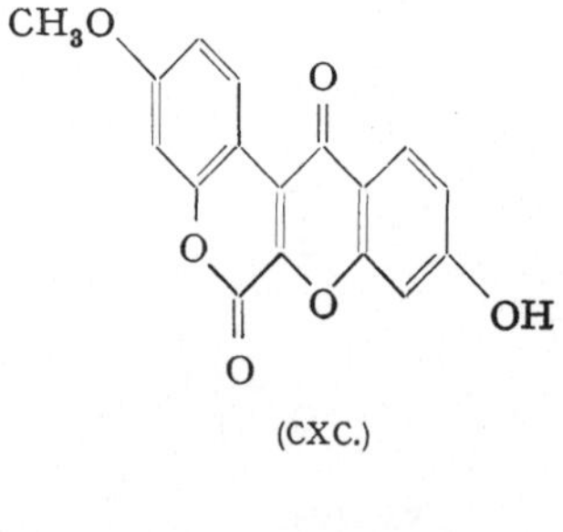

References.

1. BAKER, W., J. CHADDERTON, J. B. HARBORNE and W. D. OLLIS: A New Synthesis of *iso*Flavones. Part I. J. Chem. Soc. (London) **1953**, 1852.

2. BAKER, W., J. B. HARBORNE and W. D. OLLIS: A New Synthesis of *iso*Flavones. Part II. 5 : 7 : 2'-Trihydroxy*iso*flavone. J. Chem. Soc. (London) **1953**, 1860.

3. BICKEL, H. und H. SCHMID: Über die Konstitution des Pachyrrhizons. Helv. Chim. Acta **36**, 664 (1953).

4. BIRCH, A. J. and H. SMITH: Chem. Soc. Sympos., Bristol, 1958. Chem. Soc. (London), Special Publ. **1958**, No. 12, 1.

5. Birch, H. F., A. Robertson and T. S. Subramaniam: Experiments on the Synthesis of Rotenone and its Derivatives. Part. X. 6 : 7-Dimethoxychroman-4-one. J. Chem. Soc. (London) **1936**, 1832.

6. Boam, J. J. and R. S. Cahn: Buckley's Substance m. p. 183° from *Derris* Extract. J. Chem. Soc. (London) **1938**, 1818.

7. Boam, J. J., R. S. Cahn and A. Stuart: The Identification of Tephrosin and Deguelin from Different Sources. J. Soc. Chem. Ind. **56**, 91 T (1937).

8. Bridge, W., R. G. Heyes and A. Robertson: Experiments on the Synthesis of Rotenone and its Derivatives. Part XII. The 2 : 2-Dimethyl-Δ^3-chromen Residue of Toxicarol. J. Chem. Soc. (London) **1937**, 279.

9. Büchi, G., L. Crombie, P. J. Godin, J. S. Kaltenbronn, K. S. Siddalingaiah and D. A. Whiting: The Absolute Configuration of Rotenone. J. Chem. Soc. (London) **1961**, 2843.

10. Büchi, G., J. S. Kaltenbronn, L. Crombie, P. J. Godin and D. A. Whiting: The Stereochemistry of Rotenone. Proc. Chem. Soc. (London) **1960**, 274.

11. Buckley, T. A.: The Toxic Constituents of Derris Root. J. Soc. Chem. Ind. **55**, 285 T (1936).

12. Burrows, B. F., N. Finch, W. D. Ollis and I. O. Sutherland: Mundulone. Proc. Chem. Soc. (London) **1959**, 150.

13. Burrows, B. F., W. D. Ollis and L. M. Jackman: Sericetin. Proc. Chem. Soc. (London) **1960**, 177.

14. Butenandt, A.: Über das Rotenon, den physiologisch wirksamen Bestandteil der *Derris elliptica*. Liebigs Ann. Chem. **464**, 253 (1928).

15. Butenandt, A. und F. Hildebrandt: Untersuchungen über pflanzliche Fisch- und Insektengifte. II. 2. Mitt. über Rotenon, den physiologisch wirksamen Bestandteil der *Derris elliptica*. Liebigs Ann. Chem. **477**, 245 (1930).

16. Butenandt, A. und G. Hilgetag: Untersuchungen über pflanzliche Fisch- und Insektengifte. IV. Über die Inhaltstoffe der Derris- und Tephrosia-Arten. Liebigs Ann. Chem. **495**, 172 (1932).

17. — — Untersuchungen über pflanzliche Fisch- und Insektengifte. VI. Über die Beziehungen des Toxicarols zum Rotenon. Liebigs Ann. Chem. **506**, 158 (1933).

18. Butenandt, A. und W. McCartney: Untersuchungen über pflanzliche Fisch- und Insektengifte. III. 3. Mitt. über Rotenon, den physiologisch wirksamen Bestandteil des *Derris elliptica:* Die Konstitution des Rotenons. Liebigs Ann. Chem. **494**, 17 (1932).

19. Cahn, R. S.: The Correlation of Toxicity with Optical Activity of *Derris* Derivatives. J. Soc. Chem. Ind. **55**, 259 T (1936).

20. Cahn, R. S. and J. J. Boam: The Constituents of Derris Resin. J. Soc. Chem. Ind. **54**, 42 T (1935).

21. Cahn, R. S., R. F. Phipers and J. J. Boam: The Total Composition of Derris Extract. J. Soc. Chem. Ind. **57**, 200 T (1938).

22. — — — The Action of Alkali on Rotenone and Related Substances. J. Chem. Soc. (London) **1938**, 513.

23. — — — The Methylation and Ease of Ring-fission of Rotenone and Related Substances. J. Chem. Soc. (London) **1938**, 734.

24. Chen, Y.-L. and C.-S. Tsai: The Paper Chromatography of Rotenone. J. Taiwan Pharm. Assoc. **7**, 31 (1955).

25. Clark, E. P.: Toxicarol, A Constituent of the South American Fish Poison *Cracca (Tephrosia) toxicaria*. J. Amer. Chem. Soc. **52**, 2461 (1930).

26. — Deguelin. I. The Preparation, Purification and Properties of Deguelin, A Constituent of Certain Tropical Fish-Poisoning Plants. J. Amer. Chem. Soc. **53**, 313 (1931).

27. Clark, E. P.: Tephrosin. I. The Composition of Tephrosin and its Relation to Deguelin. J. Amer. Chem. Soc. **53**, 729 (1931).
28. — Toxicarol. II. Some Acetyl Derivatives of Toxicarol. J. Amer. Chem. Soc. **53**, 2264 (1931).
29. — Deguelin. II. Relationships Between Deguelin and Rotenone. J. Amer. Chem. Soc. **54**, 2369 (1931).
30. — Deguelin. III. The Orientation of the Methoxyl Groups in Deguelin, Tephrosin and Rotenone. J. Amer. Chem. Soc. **53**, 3431 (1931).
31. — Toxicarol. III. A Relation Between Toxicarol and the Rotenone Group of Fish Poisons. J. Amer. Chem. Soc. **54**, 1600 (1932).
32. — Toxicarol. IV. Concerning the Structure of Toxicarol. J. Amer. Chem. Soc. **54**, 2537 (1932).
33. — Deguelin. IV. The Structure of Deguelin and Tephrosin. J. Amer. Chem. Soc. **54**, 3000 (1932).
34. — Tephrosin. III. Some Acidic Derivatives of Tephrosin. J. Amer. Chem. Soc. **55**, 759 (1933).
35. — Toxicarol. V. 7-Hydroxytoxicarol and Related Compounds. J. Amer. Chem. Soc. **56**, 987 (1934).
36. Clark, E. P. and H. V. Claborn: Tephrosin. II. Isotephrosin. J. Amer. Chem. Soc. **54**, 4454 (1932).
37. Crombie, L.: Unpublished results.
38. Crombie, L. and P. J. Godin: Structure and Stereochemistry of the Rotenolones, Rotenolols, Isorotenolones and Isorotenolols. Proc. Chem. Soc. (London) **1960**, 276.
39. — — Structure and Stereochemistry of the Rotenolones, Rotenolols, Isorotenolones and Isorotenolols. J. Chem. Soc. (London) **1961**, 2861.
40. Crombie, L., P. J. Godin, K. S. Siddalingaiah and D. A. Whiting: Rotenoid Reactions Prefaced by Alkaline Attack. Proc. Chem. Soc. (London) **1961**, 19.
41. — — — — Some Chemistry of the B/C Ring-System of Rotenoids. J. Chem. Soc. (London) **1961**, 2876.
42. Crombie, L. and J. W. Lown: Determination of the Geometry of the B/C Fusion of Rotenoids by Means of Long-Range Asymmetric Magnetic Shielding of the Carbonyl Group. Proc. Chem. Soc. (London) **1961**, 299.
43. — — Proton Magnetic Studies of Rotenone and Related Compounds. J. Chem. Soc. (London) **1962**, 775.
44. Crombie, L. and R. Peace: Structure and Stereochemistry of Sumatrol and Malaccol. J. Chem. Soc. (London) **1961**, 5445.
45. Crombie, L. and D. A. Whiting: Structure of Nepseudin. A Trimethoxyfurano-isoflavanone. Chem. and Ind. **1962**, 1946.
46. — — The Constitution of Neotenone and Dolichone. Biogenetic Connexions in the Sub-Family Papilionatae. Tetrahedron Letters **1962**, No. 18, 801.
47. — — A Study of the Extractives of *Neorautanenia pseudopachyrrhiza*. The Isolation and Structure of a New Rotenoid and Two New Isoflavanones. J. Chem. Soc. (London) **1962**, 1569.
48. Dann, O. und G. Volz: Katalytische Hydrierung von Dehydrorotenon zu Dihydrorotenon. Liebigs Ann. Chem. **631**, 102 (1960).
49. — — 5 a,11 a-Dihydro-chromeno-[3,4 : b]-chromon, die Stammverbindung der Rotenoide. Liebigs Ann. Chem. **631**, 111 (1960).
50. — — Einfache synthetische Rotenoide. Naturwiss. **48**, 162 (1961).
51. Djerassi, C., W. D. Ollis and R. C. Russell: The Relative Stereochemistry of the Rotenoids. J. Chem. Soc. (London) **1961**, 1448.

52. DUTTA, N. L.: Chemical Investigation of *Mundulea suberosa*. II. Constitution of Munetone, the Principal Crystalline Product of the Root Bark. J. Indian Chem. Soc. **36**, 165 (1959).

53. EISENBEISS, J. und H. SCHMID: Struktur des Erosnin (Norton and Hansberry's "Compound I"). Helv. Chim. Acta **42**, 61 (1959).

54. FEINSTEIN, L. and M. JACOBSON: Insecticides Occurring in Higher Plants. Fortschr. Chem. organ. Naturstoffe **10**, 423 (1953).

55. FELDMAN, A.: Stereo Numbers: A Short Designation for Stereoisomers. J. Organ. Chem. (USA) **24**, 1556 (1959).

56. FINCH, N. and W. D. OLLIS: Munduserone. Proc. Chem. Soc. (London) **1960**, 176.

57. FUKAMI, H., M. NAKAYAMA and M. NAKAJIMA: Synthesis of Rotenoids. III. Synthesis of β-Tubanol Methyl Ether. Agric. Biol. Chem. (Tokyo) **25**, 243 (1961).

58. — — — Synthesis of Rotenoids. IV. Synthesis of β-Tubanol. Agric. Biol. Chem. (Tokyo) **25**, 247 (1961).

59. FUKAMI, H., J. ODA, G. SAKATA and M. NAKAJIMA: Total Synthesis of *dl*-Deguelin. Bull. Agric. Chem. Soc. Japan **24**, 327 (1960).

60. — — — — Synthesis of Rotenoids. V. Total Synthesis of *dl*-Deguelin. Agric. Biol. Chem. (Tokyo) **25**, 252 (1961).

61. FUKAMI, H., S. TAKAHASHI, K. KONISHI and M. NAKAJIMA: Synthesis of Rotenoids. I. Synthesis of Chromanochromanone and 2-Substituted Isoflavanones. Bull. Agric. Chem. Soc. Japan **24**, 119 (1960).

62. FUKAMI, J., T. NAKATSUGAWA and T. NARAHASHI: The Relation Between Chemical Structure and Toxicity in Rotenone Derivatives. Japan. J. Appl. Entomol. Zool. **3**, 259 (1959).

63. GEOFFROY, E.: Contribution à l'étude du *Robinia nicou* au point de vue botanique, chimique et physiologique. Ann. Inst. Colon. Marseilles **2**, 1 (1896).

64. GEORGE, S. W. and A. ROBERTSON: Experiments on the Synthesis of Rotenone and its Derivatives. Part XIV. The Structure of Toxicarol. J. Chem. Soc. (London) **1937**, 1535.

65. GOODHUE, L. D.: An Improvement on the Gross and Smith Colorimetric Method for the Determination of Rotenone and Deguelin. J. Assoc. Official Agric. Chem. **19**, 118 (1936).

66. GRISEBACH, H.: In: W. D. OLLIS, Recent Developments in The Chemistry of Natural Phenolic Compounds. London: Pergamon. 1961.

67. GRISEBACH, H. und G. BRANDNER: Einbau des 2′,4,4′,6′-Tetrahydroxychalcon-2′-glucosid-[β-^{14}C] in Isoflavone. Experientia **18**, 400 (1962) (and earlier papers).

68. GRISEBACH, H. and W. D. OLLIS: Biogenetic Relationships Between Coumarins, Flavonoids, Isoflavonoids and Rotenoids. Experientia **17**, 4 (1961).

69. HALLER, H. L.: Rotenone. XI. The Relation Between Isorotenone and Rotenone. J. Amer. Chem. Soc. **53**, 733 (1931).

70. — Rotenone. XXI. The Structure of Isorotenone, β-Dihydrorotenone and Dehydrorotenol. J. Amer. Chem. Soc. **54**, 2126 (1932).

71. — Rotenone. XXV. The Synthesis of Tetrahydrotubanol and Tetrahydrotubaic Acid. J. Amer. Chem. Soc. **55**, 3032 (1933).

72. — Some Color Tests for Rotenone Not Specific. Ind. Eng. Chem., Analyt. Ed. **16**, 277 (1944).

73. HALLER, H. L., L. D. GOODHUE and H. A. JONES: The Constituents of Derris and Other Rotenone-bearing Plants. Chem. Rev. **30**, 33 (1942).

320   L. CROMBIE:

74. HALLER, H. L. and F. B. LaForge: Rotenone. VII. The Structure of Tubanol and Tubaic Acid. J. Amer. Chem. Soc. **52**, 3207 (1930).

74a. — — Rotenone. XII. Some New Derivatives of Rotenol. J. Amer. Chem. Soc. **53**, 2271 (1931).

75. — — Rotenone. XIV. The Relation of the Optical Activity of Some Rotenone Derivatives to the Structure of Tubaic Acid. J. Amer. Chem. Soc. **53**, 3426 (1931).

76. — — Rotenone. XX. The Structure of Tubaic Acid. J. Amer. Chem. Soc. **54**, 1988 (1932).

77. HALLER, H. L. and P. S. SCHAFFER: Rotenone. XXVII. Note on the Hydrogenation of Rotenone. J. Amer. Chem. Soc. **55**, 3494 (1933).

78. HANRIOT, M.: Sur les substances actives du *Tephrosia vogelii*. C. R. hebd. Séances Acad. Sci. **144**, 150 (1907).

79. HARPER, S. H.: A New Compound from *Derris elliptica* Resin. Chem. and Ind. **1938**, 1059.

80. — The Precursor of Buckley's Compound. Chem. and Ind. **1939**. 292.

81. — The Active Principles of Leguminous Fish-poison Plants. Part I. The Properties of *l*-α-Toxicarol Isolated from *Derris malaccensis* (Kinta Type). J. Chem. Soc. (London) **1939**, 812.

82. — The Active Principles of Leguminous Fish-poison Plants. Part II. The Isolation of *l*-Elliptone from *Derris elliptica*. J. Chem. Soc. (London) **1939**, 1099.

83. — The Active Principles of Leguminous Fish-poison Plants. Part III. The Structure of Elliptone. J. Chem. Soc. (London) **1939**, 1424.

84. — The Active Principles of Leguminous Fish-poison Plants. Part IV. The Isolation of Malaccol from *Derris malaccensis*. J. Chem. Soc. (London) **1940**, 309.

85. — The Active Principles of Leguminous Fish-poison Plants. Part V. *Derris malaccensis* and *Tephrosia toxicaria*. J. Chem. Soc. (London) **1940**, 1178.

86. — The Melting Points of Toxicarol and Related Compounds. J. Chem. Soc. (London) **1941**, 878.

87. — The Active Principles of Leguminous Fish-poison Plants. Part VII. The Reduction of Elliptone. J. Chem. Soc. (London) **1942**, 587.

88. — The Active Principles of Leguminous Fish-poison Plants. Part VIII. The Synthesis of Dehydrotetrahydroelliptone and of Dehydrotetrahydromalaccol. J. Chem. Soc. (London) **1942**, 593.

89. — The Active Principles of Leguminous Fish-poison Plants. Part IX. The Synthesis of Furanoisoflavones Related to Rotenone. J. Chem. Soc. (London) **1942**, 595.

90. HERBERT, J. R., W. D. OLLIS and R. C. RUSSELL: Synthesis of $(\pm)$-Munduserone. Proc. Chem. Soc. (London) **1960**, 177.

91. HEYES, R. G. and A. ROBERTSON: Experiments on the Synthesis of Rotenone and its Derivatives. Part V. The Constitution of *apo*Toxicarol. J. Chem. Soc. (London) **1935**, 681.

92. HOLMAN, H. J.: A Survey of Insecticide Materials of Vegetable Origin. London: Imperial Institute. 1940.

93. HOLTON, G. W., G. PARKER and A. ROBERTSON: Experiments on the Synthesis of Rotenone and its Derivatives. Part XVI. The Synthesis of Abutic Acid and its Analogues. J. Chem. Soc. (London) **1949**, 2049.

94. JENNEN, A.: De Inwerking van Alkaliën en Alkali-metalen op Rotenon en Toxicarol. Bull. soc. chim. Belges **61**, 536 (1952).

95. JONES, H. A.: Crystalline Solvates of Rotenone. J. Amer. Chem. Soc. **53**, 2738 (1931).

96. JONES, H. A. and C. M. SMITH: A Colour Test for Rotenone. Ind. Eng. Chem., Analyt. Ed. 5, 75 (1933).

97. JONES, H. A. and J. W. WOOD: Depression of the Melting Point of α-Toxicarol and Related Compounds in Soft-glass Capillary Tubes. J. Amer. Chem. Soc. 63, 1760 (1941).

98. KAGEYAMA, I.: Extraction of Active Principles from Pyrethrum, Derris Root, Tobacco, etc. Japanese Patent 3649 (1952) [Chem. Abstr. 47, 8328 (1953)].

99. KAGEYAMA, I., J. YASAMURA and M. SATO: Rotenone from Cube or Derris Root. British Patent 708735 (1954) [Chem. Abstr. 48, 11715 (1954)].

100. KARIYONE, T. and S. KONDO: Tubaic Acid. J. pharmac. Soc. Japan 518, 376 (1925) [Chem. Abstr. 19, 2485 (1925)], and earlier papers.

100a. KAWASE, Y. and C. NUMATA: Synthesis of Furano [2″,3″ : 7,8] chromeno-[3′,4′ : 2,3] chromone. Chem. and Ind. 1961, 1361.

101. — — Synthetic Studies of Benzofuran Derivatives. Part 8. Synthesis of Furo[2,3-f]chromene[3,4-b]chromone. Bull. Chem. Soc. Japan 35, 1366 (1962).

102. KENNY, T. S., A. ROBERTSON and S. W. GEORGE: Sumatrol. Part II. The Synthesis of Dehydrotetrahydrosumatrol. J. Chem. Soc. (London) 1939, 1601.

103. KING, H. I. and A. ROBERTSON: Experiments on the Synthesis of Rotenone and its Derivatives. Part VI. Chromenochromones. J. Chem. Soc. (London) 1935, 993.

104. LaFORGE, F. B.: Rotenone. XV. The Structure of Derric Acid. J. Amer. Chem. Soc. 53, 3896 (1931).

105. — Rotenone. XXIII. The Structure of Rotenonone. J. Amer. Chem. Soc. 54, 3377 (1932).

106. — Rotenone. XXVI. Synthesis of the Parent Substances of Some Characteristic Rotenone Derivatives. J. Amer. Chem. Soc. 55, 3040 (1933).

107. LaFORGE, F. B. and H. L. HALLER: Rotenone. XIX. The Nature of the Alkali Soluble Hydrogenation Products of Rotenone and Derivatives and their Bearing on the Structure of Rotenone. J. Amer. Chem. Soc. 54, 810 (1932).

108. — — Rotenone. XXIX. The Isomerism of the Rotenolones. J. Amer. Chem. Soc. 56, 1620 (1934).

109. — — Rotenone. XXX. The Non-Crystalline Constituents of Derris Root. J. Amer. Chem. Soc. 56, 2415 (1934).

109a. LaFORGE, F. B., H. L. HALLER and L. E. SMITH: Rotenone. XVI. Interpretation of Some Characteristic Reactions of Rotenone. J. Amer. Chem. Soc. 53, 4400 (1931).

110. — — — The Determination of the Structure of Rotenone. Chem. Rev. 12, 181 (1933).

111. LaFORGE, F. B. and G. L. KEENAN: Rotenone. XVII. Note on the Dimorphic Forms of Dihydrorotenone. J. Amer. Chem. Soc. 53, 4450 (1931).

112. LaFORGE, F. B. and L. E. SMITH: Rotenone. I. Reduction Products of Rotenone. J. Amer. Chem. Soc. 51, 2574 (1929).

113. — — Rotenone. II. The Derivatives of Derritol. J. Amer. Chem. Soc. 52, 1088 (1930).

114. — — Rotenone. III. Dehydrorotenone. J. Amer. Chem. Soc. 52, 1091 (1930).

115. — — Rotenone. VI. Derric Acid. J. Amer. Chem. Soc. 52, 2878 (1930).

116. — — Rotenone. VIII. Isomeric Hydroxy Acids and their Relation to Dehydrorotenone. J. Amer. Chem. Soc. 52, 3603 (1930).

117. LINDAHL, P. E. and K. E. OBERG: Mechanism of the Physiological Action of Rotenone. Nature (London) 187, 784 (1960).

118. Matsumoto, H.: Qualitative Analysis of Rotenone (Derris Root) by Paper Partition Chromatography. Kagaku to Sosa **11**, No. 4, p. 18 (1958).

119. Mehta, A. C. and T. R. Seshadri: Synthetic Experiments in the Benzopyrone Series. Part LVI. A New Synthesis of 7-Hydroxy-chromano-(3′ : 4′ : 2 : 3)-chromone. Proc. Indian Acad. Sci. **42**, 192 (1955).

120. Meijer, T. M.: Approximate Colorimetric Determination of Derris Extract. Rec. trav. chim. Pays-Bas **55**, 954 (1936).

121. — The Insecticidal Constituents of *Pachyrrhizus erosus* Urban. I. Rec. trav. chim. Pays-Bas **65**, 835 (1946).

122. Meijer, T. M. and D. R. Koolhaas: New Constituents of Derris Root. I. Rec. trav. chim. Pays-Bas **58**, 207 (1939).

123. Merck Index. 7th Edition, p. 909. Rahway, N. J.: Merck and Co. Inc. 1960.

124. Merz, K. W. und G. Schmidt: Über die giftigen Inhaltsstoffe der Samen von *Tephrosia vogelii*. Arch. Pharmaz. **273**, 1 (1935).

125. Miyano, M., A. Kobayashi and M. Matsui: Synthesis and Configurational Elucidation of Rotenoids. XVIII. The Total Synthesis of Natural Rotenone. Bull. Agric. Chem. Soc. Japan **24**, 540 (1960).

126. — — — Synthesis and Configurational Analyses of Rotenoids. XIX. The Total Synthesis of Natural Rotenone. Agric. Biol. Chem. (Tokyo) **25**, 673 (1961).

127. Miyano, M. und M. Matsui: Synthesen und Konfigurationsermittlung in der Rotenoid-Reihe X. Totalsynthese des Dihydrorotenons. Proc. Japan Acad. **35**, 175 (1959).

128. — — Synthesen und Konfigurationsermittlung in der Rotenoid-Reihe. I. Ein Beitrag zur Synthese der Toxicarsäure. Bull. Chem. Soc. Japan **31**, 207 (1958).

129. — — Synthesen und Konfigurationsermittlung in der Rotenoid-Reihe. II. Zur Synthese und Konfiguration der *dl-cis-* und *dl-trans*-Oxynetorsäure. Bull. Chem. Soc. Japan **31**, 271 (1958).

130. — — Synthesen und Konfigurationsermittlung in der Rotenoid-Reihe. III. Über die Reaktion der α,β-ungesättigten Säuren mit Resorcin bzw. Phloroglucin: eine neue Synthese des Chromanonrings und ein Versuch zur Synthese des Rotenons. Bull. Chem. Soc. Japan **31**, 397 (1958).

131. — — Synthesen und Konfigurationsermittlung in der Rotenoid-Reihe. IV. Synthese des Rotenons. Bull. Agric. Chem. Soc. Japan **22**, 128 (1958).

132. — — Synthesen und Konfigurationsermittlung in der Rotenoid-Reihe. V. Partialsynthese des Rotenons und Dihydrorotenons. Chem. Ber. **91**, 2044 (1958).

133. — — Synthesen und Konfigurationsermittlung in der Rotenoid-Reihe. VI. Die Struktur und Konfiguration von Rotenolonen. Bull. Agric. Chem. Soc. Japan **22**, 335 (1958).

134. — — Synthesen und Konfigurationsermittlung in der Rotenoid-Reihe. VII. Eine Herstellung des natürlichen Rotenons aus Rotenolon II. Ein weiterer Beweis für die Struktur von Rotenolon II. Die Konfiguration des natürlichen Rotenons. Bull. Agric. Chem. Soc. Japan **22**, 337 (1958).

135. — — Synthesen und Konfigurationsermittlung in der Rotenoid-Reihe. VIII. Partialsynthesen des Rotenons aus Rotenolon I und II. Konstitution und Konfiguration der Rotenolone, Rotenolole und Deguelinole. Chem. Ber. **92**, 1438 (1959).

136. — — Synthesen und Konfigurationsermittlung in der Rotenoid-Reihe. IX. Synthese des Roteols und Hydroxyroteols. Bull. Agric. Chem. Soc. Japan **23**, 141 (1959).

137. MIYANO, M. und M. MATSUI: Synthesen und Konfigurationsermittlung in der Rotenoid-Reihe. XI. Synthese des Dihydrorotenons. Chem. Ber. **92**, 2487(1959).

138. — — Synthesen und Konfigurationsermittlung in der Rotenoid-Reihe. XII. Synthese des *d,l*-Nordihydrorotenons. Chem. Ber. **93**, 54 (1960).

139. — — Synthesen und Konfigurationsermittlung in der Rotenoid-Reihe. XIII. Synthese der Tubasäure. Bull. Agric. Chem. Soc. Japan **24**, 218 (1960).

140. — — Synthesen und Konfigurationsermittlung in der Rotenoid-Reihe. XIV. Synthese der Tubasäure und verwandter Cumarane. Chem. Ber. **93**, 1194 (1960).

141. MIYANO, M., T. NISHIKUBO und M. MATSUI: Synthesen und Konfigurationsermittlung in der Rotenoid-Reihe. XV. Partialsynthese des Deguelins. Untersuchung über Rotenonharz. Chem. Ber. **93**, 1746 (1960).

142. NAGAI, K.: J. Tokyo Chem. Soc. **23**, 744 (1902).

143. NAKAZAKI, M. and H. ARAKAWA: The Absolute Configuration of Rotenone. Bull. Chem. Soc. Japan **34**, 453 (1961).

144. NICKL, J.: Synthese von β-Tubasäure und von Lonchocarpin. Chem. Ber. **91**, 1372 (1958).

145. NORTON, L. B. and R. HANSBERRY: Constituents of the Insecticidal Resin of the Yam Bean *(Pachyrrhizus erosus)*. J. Amer. Chem. Soc. **67**, 1609 (1945).

146. O'DONNELL, R. W., F. P. REED and A. ROBERTSON: Experiments on the Synthesis of Rotenone and its Derivatives. Part IX. J. Chem. Soc. (London) **1936**, 419.

147. OFFE, H. A.: Derris- und Pyrethrum-Inhaltsstoffe. Übersicht über Versuche zur Synthese von Rotenoiden und Pyrethrinen. Angew. Chem. **60**, 9 (1948).

148. OFFE, H. A. und W. BARKOW: Über die Kondensation von Resorcin und Resorcin-Derivaten mit Carbonsäuren unter dem Einfluß wasserfreier Flußsäure. Chem. Ber. **80**, 458 (1947).

149. — — Über die Kondensation von Hydrochinon und Hydrochinon-Derivaten mit Carbonsäuren unter dem Einfluß von wasserfreier Flußsäure. Chem. Ber. **80**, 464 (1947).

150. OFFE, H. A. und H. JATZKEWITZ: Δ^3-Chromen-carbonsäuren-(4). Chem. Ber. **80**, 469 (1947).

151. OLLIS, W. D.: In: W. D. OLLIS, Recent Developments in the Chemistry of Natural Phenolic Compounds. London: Pergamon. 1961.

152. PARKER, G. and A. ROBERTSON: Experiments on the Synthesis of Rotenone and its Derivatives. Part XVII. The Rotenonone Nucleus. J. Chem. Soc. (London) **1950**, 1121.

153. PAVANARAM, S. K. and L. RAMACHANDRA Row: Synthetic Analogues of Rotenoids. Current Sci. (India) **26**, 145 (1957).

154. REICHSTEIN, T. und R. HIRT: Synthese von 4-Oxy-cumaronen und Synthese der Iso-tubasäure (Rotensäure). Helv. Chim. Acta **16**, 121 (1933).

155. RICHARDS, J. H., A. ROBERTSON and J. WARD: Experiments on the Synthesis of Rotenone and its Derivatives. Part XV. J. Chem. Soc. (London) **1948**, 1610.

156. ROARK, R. C.: A Digest of the Literature of *Derris (Deguelia)* Species used as Insecticides, 1747–1931. U. S. Dept. Agric., Misc. Publ. No. 120 (1932).

157. — *Lonchocarpus* species (barbasco, cube, nekoe and timbo). Used as Insecticides. Bur. Entomol. U. S. Dept. Agric., Mon. E-367 (1936).

158. — Tephrosia as an Insecticide — A Review of the Literature. Bur. Entomol. U. S. Dept. Agric., Mon. E-402(1937).

159. — *Lonchocarpus* (barbasco, cube and tumbo). A Review of Recent Literature. Bur. Entomol. U. S. Dept. Agric., Mon. E-453 (1938).

160. — Definition of the Word Rotenoid. J. Econ. Entomol. **33**, 416 (1940).

161. Robertson, A.: Experiments on the Synthesis of Rotenone and its Derivatives. Part II. The Synthesis of Rissic Acid and of Derric Acid, and the Constitution of Rotenone, Deguelin and Tephrosin. J. Chem. Soc. (London) **1932**, 1380.

162. — Experiments on the Synthesis of Rotenone and its Derivatives. Part III. The Dehydrorotenone Nucleus. J. Chem. Soc. (London) **1933**, 489.

163. — Experiments on the Synthesis of Rotenone and its Derivatives. Part IV. Dehydrodihydrorotenonic Acid and Tephrosic Acid. J. Chem. Soc. (London) **1933**, 1163.

164. Robertson, A. and G. L. Rusby: Experiments on the Synthesis of Rotenone and its Derivatives. Part VII. Tetrahydrotubaic Acid. J. Chem. Soc. (London) **1935**, 1371.

165. — — Experiments on the Synthesis of Rotenone and its Derivatives. Part VIII. Netoric Acid and Toxicaric Acid. J. Chem. Soc. (London) **1936**, 212.

166. — — Sumatrol. Part I. J. Chem. Soc. (London) **1937**, 497.

167. Robertson, A. and T. S. Subramaniam: Experiments on the Synthesis of Rotenone and its Derivatives. Part XI. Tetrahydrotubanol. J. Chem. Soc. (London) **1937**, 278.

168. Robinson, Sir R.: The Structural Relations of Natural Products. Oxford: Clarendon Press. 1955.

169. Rogers, H. D. and J. A. Calamari: Rotenone Determined by Colorimetric Means. Ind. Eng. Chem., Analyt. Ed. **8**, 135 (1936).

170. Sehgal, J. M. and T. R. Seshadri: Synthetic Experiments in the Benzo-pyrone Series. Part LV. A Synthesis of 7 : 7′-Dihydroxy-chromene-(3′:4′:2:3)-chromone. Proc. Indian Acad. Sci. **42 A**, 36 (1955).

171. Seshadri, T. R. and S. Varadarajan: Synthetic Experiments in the Benzo-pyran Series. Part XXXI. A Synthetis of 7-Hydroxy-chromeno-(3′:4′:2:3)-chromone. Proc. Indian Acad. Sci. **37 A**, 784 (1953).

172. Shamshurin, A. A.: Synthesis of Structure Fragments of Rotenone and its Satellites. I. Synthesis in the Tubaic Acid Series. J. Gen. Chem. (USSR) **16**, 1877 (1946).

173. — Synthesis of Tetrahydrotubanol and Tetrahydrotubaic Acid. J. Gen. Chem. (USSR) **19**, 1864 (1949).

174. — Synthesis of Structural Fragments of Rotenone and its Satellites. II. Synthesis in the Tubaic Acid Series. J. Gen. Chem. (USSR) **21**, 2068 (1951).

175. Simonitsch, E., H. Frei und H. Schmid: Die Konstitution des Pachyrrhizins. Monatsh. Chem. **88**, 541 (1957).

176. Stamm, O. A., H. Schmid und J. Büchi: Die Konstitution des Jamaicins. Helv. Chim. Acta **41**, 2006 (1958).

177. Takahashi, S., H. Fukami and M. Nakajima: Synthesis of Rotenoids. II. Synthesis of Four Diastereoisomers of Dihydrorotenone. Bull. Agric. Chem. Soc. Japan **24**, 123 (1960).

178. Takei, S. und M. Koide: Über Rotenon, den wirksamen Bestandteil der Derriswurzel, III: Über die Tubasäure. Ber. dtsch. chem. Ges. **62**, 3030 (1929).

179. Takei, S., S. Miyajima und M. Ono: Über Rotenon, den wirksamen Bestand-teil der Derriswurzel, IX. Mitt. Nachtrag zur Konstitution der Tetrahydro-tubasäure und des Rotenons. Synthesen einiger Abbauprodukte des Rotenons. Ber. dtsch. chem. Ges. **65**, 1041 (1932).

180. — — — Über Rotenon, den wirksamen Bestandteil der Derriswurzel, X. Mitt. Oxydation und Reduktion des Rotenons in schwach alkalischer Lösung. Ber. dtsch. chem. Ges. **66**, 479 (1933).

181. TAKEI, S., S. MIYAJIMA und M. ONO: Über Rotenon, den wirksamen Bestand-teil der Derriswurzel, XI. Mitt. Rotenonharz, Quantitative Bestimmung des Rotenons und des Deguelins in Rotenonharz. Ber. dtsch. chem. Ges. **66**, 1826 (1933).

182. TATTERSFIELD, F. and J. T. MARTIN: An Optically Active Constituent of Derris Resin Related to Toxicarol. J. Soc. Chem. Ind. **56**, 77 T (1937).

183. VAN DUUREN, B. L.: Chemistry of Edulin, Neorautone and Related Compounds. from *Neorautanenia edulis* C. A. SM. J. Organ. Chem. (USA) **26**, 5013 (1961).

184. VENKATARAMAN, K.: Flavones and Isoflavones. Fortschr. Chem. organ. Naturstoffe **17**, 1 (1959).

185. WHALLEY, W. B. and G. LLOYD: 3-Aroylcoumarones. J. Chem. Soc. (London) **1956**, 3213.

(Received, January 28, 1963.)

Namenverzeichnis. Index of Names. Index des Auteurs.

ABELES, R. H. 233, 264.
ACHARYA, S. P. 51, 70, 72.
ACHENBACH, H. 52, 55, 74.
ACKER, R. F. 56, 66, 75, 76.
AGRANOFF, B. W. 9, 10, 13, 15.
AHLERS, N. H. E. 56, 72.
AKANABE, S. 58, 64, 79.
ALBERT, A. 184, 203, 205, 207, 254.
ALEXANDER, N. 216, 234, 237, 254.
ALLEN, B. K. 241, 258.
ALLEN, W. 204, 211, 225, 254.
ALLFREY, V. (G.) 190, 225, 254.
ALLPORT, D. C. 53, 73.
ALTEN, B. 195, 196, 258.
AMBEKAR, G. R. 51, 57, 70, 72.
AMDUR, B. H. 7, 13.
AMES, B. N. 252, 255.
AMMANN, A. 24, 37, 50, 57, 64, 72, 74, 79.
ANDERSON, E. I. 233, 270.
ANDERSON, M. L. 199, 205, 206, 273, 274.
ANDERSON, R. C. 202, 209, 273.
ANDRÉ, T. 103, 116.
ANDRES, W. W. 111, 118.
ANGIER, R. B. 200—204, 221, 225, 255, 256, 261, 262, 265, 270—272.
ANSLOW, W. K. 32, 40, 72.
ARAI, T. 19, 62, 72.
ARAKAWA, H. 280, 323.
ARAVINDAKSHAN, I. 195, 256.
ARCAMONE, F. 56, 58, 72.
ARCHER, B. L. 8, 10, 14.
ARIMA, N. 58, 72.
ARISHIMA, N. 64, 72.
ARNOLD, N. (H.) 114, 115, 119.
ARNSTEIN, H. R. V. 231—233, 237, 255.
ARREGUIN, B. 4, 6, 14.
ARTH, G. E. 202, 209, 273.
ASAHI, Y. 226, 255.
ASAHINA, T. 58, 64, 79.
ASAHINA, Y. 6, 14.
ASCHNER, M. 68, 74.
ASHESHOV, I. N. 150, 179, 181.
ASHMORE, J. 231, 258.

AWAPARA, J. 243, 258.
AYREY, G. 10, 14.

BACHHAWAT, B. K. 6, 7, 14.
BACKUS, E. J. 60, 72.
BAKER, H. 187, 227, 255, 262.
BAKER, W. 313, 316.
BAKERMAN, H. A. 208, 255.
BALL, S. 19, 32, 64, 72.
BALLARD, S. A. 212, 270.
BANDURSKI, R. 6, 14.
BANGERT, R. 54, 80, 110, 111, 113, 119.
BARDOS, T. J. 207, 211—213, 265.
BARGER, F. L. 207, 211—213, 265.
BARKEMEYER, H. 192, 228, 264.
BARKER, H. A. 248, 255.
BARKOW, W. 311, 315, 323.
BARNARD, D. 8, 14.
BARNER, H. D. 240, 255.
BARTNER, E. 26, 46, 57, 68, 72.
BATT, R. D. 239, 255.
BAUER, H. 247, 271.
BAUER, K. 176, 180.
BAVLEY, A. 83, 119.
BEACHEM, M. 204, 270.
BECK, J. V. 234, 248, 249, 255, 268.
BEDFORD, C. T. 108, 120.
BEEREBOOM, J. J. 102, 103, 105, 108, 116, 117, 120.
BEERSTECHER, E., Jr. 220, 227, 236, 273.
BEINERT, H. 6, 14.
BELEN'KII, B. G. 19, 57, 70, 78.
BELT, M. 221, 259.
BENDICH, A. 257.
BEN-EFRAIM, D. A. 57, 77.
BERG, P. 233, 243, 255.
BERGMANN, F. 197, 255.
BERKOZ, B. 28, 30, 37, 38, 64, 72.
BERTAZZOLI, C. 58, 72.
BERTINO, J. R. 199, 251, 255.
BESSELL, C. J. 19, 32, 40, 64, 72.
BESSEY, O. A. 202, 264.
BEZBORODOV, A. M. 70, 78.
BHATE, D. S. 51, 57, 70, 72.

BHATNAGAR, K. K. 51, 57, 70, 72.
BICKEL, H. 303, 308, 316.
BINKLEY, S. B. 188, 189, 199—201, 255, 267.
BIRCH, A. J. 27, 43—45, 52—54, 58, 72, 73, 113, 120, 163, 179, 308, 316.
BIRCH, H. F. 309, 317.
BIRD, O. D. 188—190, 199, 200, 255, 267.
BIRNIE, G. D. 240, 256.
BITLER, B. A. 82, 96, 100, 101, 114, 117, 118.
BLACK, S. 6, 15.
BLACKWOOD, R. K. 102, 105, 106, 108, 110—112, 117, 120.
BLAIR, D. 245, 268.
BLAKLEY, R. L. 197, 198, 204—209, 216, 221, 228, 234, 237—241, 256, 265, 268, 273.
BLINOV, N. O. 179, 182.
BLOCH, K. 5, 7, 9, 13—15.
BLOOM, E. S. 188, 189, 199—201, 204, 207, 255, 266, 267.
BLOOMER, J. L. 18, 56, 60, 73.
BLY, R. K. 18, 31, 32, 34, 64, 73.
BOAM, J. J. 281, 286, 296, 297, 299, 301, 314, 317.
BOHONOS, N. 188, 189, 200, 261.
BOLDT, P. 126, 132, 133, 135, 137, 139, 141, 143, 145, 147, 179, 180.
BOLHOFER, W. A. 19, 24, 35, 40, 62, 64, 78.
BOL'SHAKOVA, L. O. 70, 78.
BOND, T. J. 240, 272.
BONNER, J. 1, 2, 4—11, 14—16, 54.
BONVICINO, G. E. 101, 117.
BOON, W. R. 202, 205, 207, 256.
BOOTHE, J. H. 95, 96, 101, 103—105, 109—111, 115, 117—120, 200—204, 221, 225, 255, 256, 261, 262, 265, 270—272.
BORCHERS, I. 176, 177, 180.
BOROWSKI, E. 28—30, 50, 51, 57, 66, 70, 72, 77.
BOSKA, J. A. 114, 119.
BOYACK, G. 58, 74.
BOYER, P. D. 214, 268.
BRAGANCA, B. M. 195, 199, 256, 263.
BRANDNER, G. 306, 319.
BRATTON, A. C. 202, 225, 226.
BREGOFF, H. M. 233, 256.
BREMER, J. 233, 257.
BRENNER-HOLZACH, O. 193, 257.
BRETT, R. A. 56, 72.

BRIDGE, W. 301, 317.
BROCK, T. D. 24, 37, 57, 64, 79.
BROCKMAN, J. A., Jr. 211, 268.
BROCKMANN, H. 82, 117, 121—127, 131—141, 143, 145—147, 150—152, 168, 171, 173—178, 180, 181.
BROCKMANN, H., Jr. 122, 124, 126, 127, 130, 132, 134—137, 140, 141, 143, 145, 146, 150—152, 155, 160, 161, 163, 168, 179—181.
BRODE, E. 186, 207, 209, 213, 218, 222, 238, 240, 250, 257, 262.
BROQUIST, H. P. 211, 220, 221, 224, 258, 262, 268.
BROSCHARD, R. W. 81, 93—96, 101, 117, 118, 120.
BROWN, G. B. 233, 264.
BROWN, G. M. 193—195, 257, 268.
BROWN, P. K. 22, 75.
BROWN, R. 19, 58, 62, 72, 74.
BROWN, R. A. 188, 200, 255, 267.
BRUNINGS, K. J. 81—83, 91, 92, 97, 99, 102, 111, 118—120.
BUCHANAN, J. M. 187, 213, 215, 219, 229, 234, 236, 242—245, 249—252, 257, 260, 264, 268, 271.
BÜCHI, G. 276, 278—281, 284, 289, 290, 314, 316, 317.
BÜCHI, J. 307, 324.
BUCKLEY, T. A. 297, 317.
BUCKWALTER, F. H. 68, 75.
BUDDE, G. 131, 133, 180.
BUDZIKIEWICZ, H. 28, 32, 37, 39, 64, 73.
BUERGER, M. J. 81, 120.
BUHR, G. 110, 111, 113, 119.
BU'LOCK, J. D. 53, 73.
BURNS, J. 60, 73.
BURROWS, B. F. 299, 317.
BURROWS, E. P. 18, 31, 32, 34, 64, 73.
BUTENANDT, A. 276, 284—288, 296, 301, 317.
BUTLER, K. 112, 116, 117.
BUYSKE, D. A. 105, 118.

CAHILL, G. F., Jr. 231, 258.
CAHN, R. S. 281, 286, 296, 297, 299, 301, 314, 317.
CALAMARI, J. A. 278, 324.
CALKINS, D. G. 188, 200, 255, 267.
CAMPBELL, C. J. 188, 200, 255, 267.
CANEVAZZI, G. 58, 72.
CAPELL, L. T. 200, 267.
CARPENTER, K. J. 225, 257.

CARTER, H. E. 24, 37, 57, 64, 74, 79.
CATHOU, R. E. 219, 236, 242—244, 257, 260, 264.
CAVALIERI, L. F. 257.
CEDER, O. J. 18, 31, 32, 34, 64, 73.
CELMER, W. D. 56, 57, 60, 62, 68, 75.
CHADDERTON, J. 313, 316.
CHAIET, L. 225, 268.
CHALLENGER, F. 243, 257.
CHAN, P. C. 194, 195, 262.
CHARNEY, J. 19, 24, 35, 40, 62, 64, 78.
CHAYKIN, S. 9, 14.
CHEN, Y. L. 278, 317.
CHENEY, L. C. 99, 117.
CHOU, T. C. 6, 14.
CIGANEK, E. 18, 31, 32, 34, 64, 73.
CLABORN, H. V. 297, 318.
CLARK, E. P. 288, 296, 297, 299, 300, 314, 317, 318.
CLEMENS, D. H. 215, 257.
COCKBAIN, E. G. 10, 14.
CODNER, R. C. 155, 164, 181.
COHEN, I. R. 57, 75.
COHEN, S. S. 234, 239—241, 255, 259, 267.
COLLIE, J. N. 163.
COLLINS, K. H. 202, 257.
CONOVER, L. H. 81—83, 91, 92, 97, 100, 102, 111, 112, 116—120.
COON, M. J. 6, 7, 14.
COPE, A. C. 18, 30—34, 36, 38, 40, 64, 73.
CORKE, C. T. 66, 75.
CORNFORTH, J. W. 10, 14.
CORUM, C. J. 57, 66, 75.
COSTA PLÁ, L. 146, 180.
COSULICH, D. B. 96, 120, 201—204, 209, 211—213, 220, 221, 224, 255—258, 261, 265, 268, 269, 272.
COYLE, T. 6, 14.
CRAVERI, R. 19, 26, 50, 57, 68, 70, 73.
CRAWFORD, E. J. 199, 202, 241, 259, 264, 265.
CRESSON, E. L. 7, 15.
CROMBIE, L. 275, 276, 278—281, 283, 284, 286—297, 301—305, 308, 309, 314, 317, 318.
CROSBIE, G. W. 231, 232, 240, 256, 258, 259, 267.

DABROWSKA, W. 190, 200, 258.
DAFT, F. S. 247, 271.
DAHMS, G. 193, 231, 272.
DANN, M. 60, 72.
DANN, O. 314, 318.

DAVISSON, J. W. 19, 56, 58, 73.
DAY, P. L. 188, 195, 196, 241, 258.
DE FERRARI, G. A. 82, 114, 120.
DE HAAN, P. C. 231, 273.
DELLUVA, A. M. 229, 257.
DELWICHE, C. C. 233, 256.
DEODHAR, S. 237, 258.
DeRENZO, E. C. 197, 258.
DESPOIS, R. 19, 58, 73.
DEWEY, V. C. 227, 228, 258, 261.
DHAR, M. L. 27, 28, 30—34, 37, 38, 64, 73.
DICKENS, F. 239, 255.
DI MARCO, A. 58, 72.
DINNING, J. S. 195, 196, 241, 258, 261.
DISRAELY, H. N. 195, 269.
DITURI, F. 7, 14.
DIVEKAR, P. V. 18, 56, 60, 73.
DJERASSI, C. 27, 28, 30—32, 37—39, 43—45, 52, 54, 58, 64, 72, 73, 295, 296, 302, 318.
DOCTOR, V. M. 197, 199, 243, 258.
DOERSCHUK, A. P. 82, 96, 100—103, 108, 114, 115, 117—119.
DOLEŽILOVÁ, L. 19, 78.
DONALDSON, K. O. 206, 207, 219, 220, 236, 242, 243, 258, 263.
DONOHUE, D. M. 251, 255.
DONOVAN, E. 241, 259.
DONOVICK, R. 58, 68, 73, 74.
DORNBUSH, A. C. 117.
DOUDOROFF, M. 6, 15. .
DOWLING, H. F. 80, 117.
DROST, G. 40, 56, 60, 77.
DUGGAR, B. M. 80, 91, 117.
DURR, I. F. 7, 14.
DURY, K. 91, 118.
DUTCHER, J. D. 26—30, 37, 43—45, 50, 52, 54, 56, 58, 62, 66, 68, 72—74, 78, 79.
DUTTA, N. L. 307, 319.
DUUREN, B. L. VAN 309, 325.
DU VIGNEAUD, V. 243, 271.
DWORSCHACK, R. G. 19, 26, 50, 57, 68, 70, 73.

EAKIN, R. E. 209, 220, 227, 236, 259, 273.
EASTHAM, J. F. 18, 56, 60, 73.
EBAUGH, F. G., Jr. 191, 211, 227, 228, 269.
EEK, T. VAN 40, 56, 60, 77.
EGGERER, H. 9, 10, 13, 15.
EIGEN, E. 229, 273.
EISENBEISS, J. 309, 319.

ELION, G. B. 205, 258.
ELWYN, D. 229, 231, 258.
EMMETT, A. D. 189, 199, 200, 255, 267.
EMMONS, W. D. 215, 257.
ENTSCHEL, R. 130, 151, 181.
EPSTEIN, L. F. 26, 76.
ESPOSITO, R. G. 193, 258.
ETTLINGER, L. 123, 124, 131, 137, 147, 151, 175, 181.
EUGSTER, C. H. 130, 151, 181.

FAHRENBACH, M. J. 201—204, 211, 255—257, 261, 268, 272.
FAIVRE-AMIOT, A. 70, 77.
FEINSTEIN, L. 278, 319.
FELDMAN, A. 284, 319.
FERGUSON, J. J., Jr. 7, 14.
FERRARI, G. A. de 82, 114, 120.
FIELDS, T. L. 115, 117, 118, 120.
FIGARD, P. H. 233, 257.
FILIPPOVA, A. I. 19, 57, 70, 78.
FINCH, N. 298, 299, 317, 319.
FINLAY, A. C. 19, 56, 58, 73, 81, 82, 117.
FISHER, W. P. 19, 24, 35, 40, 62, 64, 78.
FLAKS, J. G. 229, 234, 239, 241, 252, 257, 259.
FLEMING, L. W. 231, 259.
FLETCHER, A. M. 193, 258.
FLETCHER, D. L. 32, 40, 72.
FLON, H. 179, 181.
FLYNN, E. H. 211, 267.
FLYNN, R. M. 6, 15.
FOLKERS, K. 7, 14, 15, 202, 209, 273.
FORREST, H. S. 193, 204, 259.
FOSTER, M. A. 190, 236, 244, 259, 260.
FOX, S. 58, 74.
FOX, S. M. 100, 101, 118.
FRANCK, B. 124, 126, 131, 133, 135, 137, 140, 141, 181.
FRANKLIN, A. L. 221, 259.
FRAZIER, W. R. 114, 119.
FREDIANI, H. A. 225, 264.
FREI, H. 308, 324.
FRIEDKIN, M. 199, 206, 222, 234, 240, 241, 259, 265, 266, 271.
FRIEDMAN, S. 190, 236, 244, 260.
FRISELL, W. R. 233, 264.
FROMMER, W. 125, 181.
FRYTH, P. W. 95, 118.
FUJII, S. 77.
FUJIMORI, E. 221, 259.
FUJITA, H. 56, 57, 60, 70, 74, 77.
FUKAMI, H. 313, 314, 319, 324.

FUKAMI, J. 276, 319.
FUTTERMAN, S. 195, 197, 204, 209, 234, 259.

GABRIO, B. W. 197—199, 240, 255, 269.
GALLO, G. G. 82, 114, 120.
GALSTON, A. W. 1, 14.
GARDINER, R. C. 191, 196, 211, 227, 228, 236, 248, 269, 270.
GARDNER, J. N. 53, 74.
GARNY, B. J. 252, 255.
GARRETSON, A. L. 19, 76.
GATENBECK, S. 54, 74, 113, 117.
GÄUMANN, E. 123, 124, 131, 137, 147, 151, 175, 181.
GAZZOLA, A. L. 204, 256.
GEFFEN, C. 225, 254.
GEOFFROY, E. 276, 319.
GEORGE, S. W. 300—302, 310, 312, 313, 319, 321.
GHANEKAR, D. S. 195, 256.
GHIONE, M. 58, 72.
GIBBS, M. H. 7, 15.
GILLIS, B. T. 18, 31, 32, 34, 64, 73.
GILNER, D. 74.
GIOLITTI, G. 19, 68, 73.
GIRDWOOD, R. H. 220, 259.
GLADNER, J. A. 229, 257.
GLAZKO, A. J. 225, 259.
GODIN, P. J. 276, 278—281, 284, 286—297, 301, 314, 317, 318.
GOLD, W. 26, 45, 56, 58, 62, 66, 68, 74.
GOLDIN, A. 199, 265.
GOLDMAN, A. A. 81, 93—95, 118, 120, 202, 271.
GOLDTHWAIT, D. A. 229, 234, 251, 259.
GOLYAKOV, P. N. 19, 57, 70, 78.
GOODHUE, L. D. 278, 319.
GOODMAN, J. J. 82, 86, 114, 117.
GORDON, J. J. 131, 132, 134, 137, 150, 151, 154, 155, 157, 164, 176, 179—181.
GORDON, M. 209, 236, 259.
GORDON, P. N. 81, 82, 92, 97, 100, 111, 118, 120.
GORDON, S. 95, 117, 118.
GOTTLIEB, D. 24, 37, 50, 55—57, 60, 64, 72, 74, 79.
GOTTSTEIN, W. J. 99, 117.
GREEN, A. 96, 109—111, 117, 118.
GREEN, D. E. 6, 14.
GREEN, M. 240, 259.
GREENBERG, D. M. 197—199, 206, 216, 221, 232—234, 236, 237, 240, 241, 243, 248, 254, 257, 260—262, 265—267.

GREENBERG, G. R. 190, 204, 207—210, 214, 216, 217, 228, 229, 234, 236, 239, 248—251, 259, 260.
GREGORY, F. J. 18, 26, 50, 66, 78.
GREIN, A. 58, 72.
GRIER, J. 204, 270.
GRIFFITH, I. 202, 269.
GRISEBACH, H. 52, 55, 74, 194, 271, 306—308, 319.
GRISEBACH, U. C. 52, 74.
GROSJEAN, M. 160, 181.
GROSSOWICZ, N. 68, 74.
GROWICH, J. A. 82, 96, 108, 114, 117, 119.
GRUBER, W. 234, 249, 268.
GUEST, J. R. 190, 233, 236, 244, 260, 262.
GUNSALUS, I. C. 232, 234, 249, 268.
GURIN, S. 6, 7, 14, 15.

HAAGEN-SMIT, A. J. 9, 10, 11, 15.
HAAN, P. C. DE 231, 273.
HAENSELER, C. M. 66, 75.
HAGEMANN, G. 19, 70, 74.
HAKALA, M. T. 190, 199, 222, 260, 274.
HALLER, H. L. 276, 278, 279, 284, 288—290, 294, 296, 310, 313, 314, 319.
HAMADA, K. 58, 64, 79.
HAMAMURA, N. 19, 46, 75.
HANRIOT, M. 296, 320.
HANSBERRY, R. 303, 304, 309, 323.
HARBORNE, J. B. 313, 316.
HARPER, S. H. 278, 297, 302, 303, 305, 310, 312, 313, 320.
HARRIS, S. A. 202, 209, 273.
HARRISS, E. 191, 246, 265.
HARTMAN, S. C. 187, 213, 215, 229, 234, 250—252, 257, 260.
HASTINGS, A. B. 231, 258.
HATCH, F. T. 236, 242, 244, 260, 271.
HATEFI, Y. 206—208, 212, 214, 229, 234, 238, 239, 248, 260, 261, 263, 266.
HATTORI, K. 25, 46, 50, 57, 66, 74, 76.
HAUSSER, K. W. 22, 74.
HAYAISHI, O. 231, 269.
HAZEN, E. L. 19, 58, 62, 72, 74.
HEINEMANN, B. 68, 75.
HEISLER, C. R. 197, 199, 260.
HELE, P. 6, 14.
HELLEINER, C. W. 233, 244, 260.
HENDERSON, R. F. 241, 261.
HENDRICKS, S. B. 2, 14.
HENIS, Y. 68, 74.
HENNING, U. 9, 10, 13—15.
HERBERT, J. R. 313—315, 320.

HERBERT, V. 220, 261.
HERMAN, E. C. 199, 205, 274.
HERMICH, M. R. 228, 261.
HESS, G. B. 83, 119.
HEUSER, L. J. 114, 119.
HEYES, R. G. 301, 317, 320.
HEYROVSKÝ, J. 226, 261.
HIBBITS, W. 27, 28, 30, 43, 50, 68, 74.
HICKEY, R. J. 57, 66, 74, 75.
HIDY, P. H. 57, 75.
HIFT, H. 6, 14.
HIGGINS, G. M. C. 8, 14.
HILDEBRANDT, F. 317.
HILGETAG, G. 286, 288, 296, 301, 317.
HILL, R. J. 232, 263.
HIMES, R. H. 189, 228, 250, 261, 267.
HIRATA, Y. 25, 57, 66, 74, 76.
HIROKAWA, S. 82, 118.
HIRSCH, U. 82, 96, 103, 115, 118, 119.
HIRT, R. 310, 323.
HITCHINGS, G. H. 199, 205, 226, 258, 273.
HLAVKA, J. J. 103—105, 117—119.
HO, P. P. K. 217, 261.
HOBBY, G. L. 81, 82, 117.
HOCHSTEIN, F. A. 81—83, 91, 92, 97, 98, 111, 118—120.
HOETTE, I. 40, 56, 60, 77.
HOFFMAN, C. H. 7, 14, 15.
HOGAN, A. G. 189, 199, 267.
HOLLAND, J. F. 220, 261.
HOLLY, F. W. 7, 14.
HOLMAN, H. J. 276, 320.
HOLMLUND, C. E. 111, 118.
HOLTMAN, D. F. 18, 56, 60, 73.
HOLTON, G. W. 310, 320.
HOLZAPFEL, C. W. 27, 43—45, 54, 58, 73.
HONJO, M. 57, 68, 77.
HOOGERHEIDE, J. C. 40, 56, 60, 77.
HOOPER, I. R. 68, 75.
HOSOYA, S. 19, 46, 66, 75.
HOTCHKISS, R. D. 194, 273.
HSU, P. T. 233, 270.
HUENNEKENS, F. M. 187, 197—199, 206—208, 212, 214, 217, 222, 224, 227—229, 234, 237—240, 248, 251, 255, 260, 261, 263, 265, 266, 269, 272, 273.
HUFF, J. W. 7, 15.
HULLIN, R. P. 7, 14.
HULTQUIST, M. E. 201—204, 209, 211—213, 220, 221, 255—257, 261, 268, 269, 272.

HULYALKAR, R. K. 51, 57, 70, 72.
HUMPHREYS, G. K. 234, 240, 241, 260, 261.
HUMPHREYS, S. R. 199, 265.
HUTCHINGS, B. L. 81, 93—95, 117, 118, 120, 188, 189, 200—204, 221, 225, 228, 229, 255, 256, 261, 262, 265, 270—273.
HUTNER, S. H. 187, 227, 255, 262.
HÜTTER, R. 123, 124, 131, 137, 147, 151, 175, 181.

ICHIHARA, A. 232, 262.
IGARASHI, M. 62, 68, 75.
IGARASHI, S. 57, 62, 64, 76.
IKAWA, M. 237, 265.
INOUE, S. 18, 40, 64, 76.
ISELIN, B. 173, 181.
ISHIKAWA, M. 28, 32, 37, 39, 64, 73.
ITO, T. 18, 40, 64, 76.

JACKMAN, L. M. 131, 132, 134, 154, 157, 179, 181, 299, 317.
JACKSON, R. W. 19, 26, 50, 57, 68, 70, 73.
JACOBSON, M. 278, 319.
JAENICKE, L. 183, 186, 187, 190, 193—195, 207—210, 214, 216—219, 228, 229, 234, 236—243, 245, 246, 249—251, 257, 260, 262, 268, 273.
JATZKEWITZ, H. 323.
JENNEN, A. 287, 301, 320.
JENNY, E. 231, 262.
JENSEN, E. R. 82, 96, 102, 105, 114, 117—119.
JOHNSON, B. C. 200, 262.
JOHNSON, H. E. 18, 30—34, 36, 38, 40, 64, 73.
JOHNSON, L. F. 28, 32, 37, 39, 64, 73, 155, 157.
JOHNSON, S. 114, 115, 119.
JOHNSTON, J. A. 6, 14.
JOHNSTON, J. D. 112, 116, 117.
JONES, E. J. 2, 14.
JONES, E. R. H. 22, 75.
JONES, H. A. 278, 294, 319—321.
JONES, K. M. 190, 233, 244, 259, 260, 262.
JONES, M. E. 6, 15.
JONES, R. G. 211, 267.
JOSSE, S. 239, 264.
JUKES, T. H. 211, 220, 221, 224, 226, 227, 259, 262, 268.

KACZKA, E. A. 202, 209, 273.
KAGEYAMA, I. 278, 321.

KALCKAR, H. M. 197, 262.
KALTENBRONN, J. S. 276, 278—281, 284, 289, 290, 314, 317.
KANE, J. H. 56, 58, 73, 81, 82, 117.
KANEKO, J. 77.
KAPLAN, M. A. 68, 75.
KAPLAN, N. O. 5, 14.
KARIYONE, T. 276, 321.
KARRER, P. 130, 151, 181, 203, 262.
KATSUNUMA, N. 193, 194, 262, 263.
KATZ, E. 54, 75.
KAUFMAN, B. 211, 226—228, 270.
KAUFMAN, S. 206—208, 263.
KAUFMANN, H. P. 56, 75.
KAWASE, Y. 315, 321.
KAY, L. D. 206, 208, 212, 214, 234, 238, 239, 248, 260, 261, 263, 266.
KAZENKO, A. 190, 200, 258.
KEENAN, G. L. 278, 321.
KEGLEVIĆ, D. 231, 232, 237, 255.
KEKWICK, R. G. O. 8, 14.
KELLER-SCHIERLEIN, W. 123, 124, 131, 137, 146, 147, 150, 151, 175, 180, 181.
KENDE, A. S. 115, 117, 118.
KENKARE, U. W. 199, 256, 263.
KENNY, T. S. 302, 310, 312, 313, 321.
KERESZTESY, J. C. 195, 196, 206, 207, 209, 211, 219, 220, 225, 227, 236, 242, 243, 248, 258, 263, 268—270.
KESSEL, I. 9, 15.
KHOKHLOV, J. M. 178, 182.
KHOPKO, G. V. 70, 78.
KIDDER, G. W. 227, 228, 258, 261, 263.
KIHARA, R. 68, 78.
KIMBEL, K. H. 209, 227, 268.
KING, C. G. 190, 225, 254.
KING, F. E. 203, 263.
KING, H. I. 313, 321.
KISLIUK, R. L. 216, 222, 234, 237, 239, 242—244, 263.
KJELDGAARD, N. D. 197, 262.
KLEIN, L. 245, 266.
KLENOW, H. 197, 262.
KLOPOTOWSKI, T. 253, 263.
KLOSTERMAN, H. J. 6, 15.
KNOWLES, R. 204, 227, 269.
KNOX, W. E. 230, 265.
KOBAYASHI, A. 312—314, 322.
KOCH, J. 219.
KODICEK, E. 225, 257.
KOE, B. K. 56, 57, 60, 62, 68, 75.
KOEPPE, R. E. 232, 263.

KOHLER, A. R. 117.
KOHN, J. 247, 263.
KOIDE, M. 279, 324.
KOLMAN, Z. 222, 228, 270.
KOMATSU, N. 66, 75.
KOMENDH, J. 229, 264.
KONDO, S. 276, 321.
KONISHI, K. 314, 319.
KOOLHAAS, D. R. 297, 302, 322.
KORENIAKO, A. J. 179, 181.
KORFF, R. W. v. 6, 14.
KORNBERG, A. 234, 239—241, 259, 264.
KORNBERG, S. R. 239, 264.
KORST, J. J. 112, 116, 117.
KORTE, F. 192, 228, 264.
KOSAKA, H. 64, 75.
KOVAL, G. J. 196, 209, 227, 236, 248, 269, 270.
KRADOLFER, F. 123, 124, 131, 137, 147, 151, 175, 181.
KRASSILNIKOV, N. A. 179, 181.
KRAZINSKI, H. 103—105, 118.
KREUTZER, A. 113, 115, 119.
KROPP, E. 57, 75.
KRUPKA, G. 117.
KUH, E. 201, 203, 204, 221, 255, 256, 261, 272.
KUHN, H. 26, 77.
KUHN, R. 22, 74, 91, 118.
KÜHN, K. 19, 64, 75.
KULESZA, J. S. 26, 46, 57, 68, 72.
KUNG, P. Y. 60, 76.
KUROYA, M. 66, 75.
KUSHNER, S. 95, 115, 117, 118.
KUTZBACH, C. 183, 213, 214.
KWIETNY, H. 197, 255.

LaFORGE, F. B. 276, 278, 279, 284, 285, 287—290, 294, 296, 313, 314, 320, 321.
LANSFORD, M. 207, 211—213, 265.
LARRABEE, A. R. 219, 236, 242, 244, 260, 264.
LARSEN, A. 195, 270.
LASCELLES, J. 192, 232, 264, 266.
LASKOWSKI, M. 190, 200, 258, 265.
LAW, J. 9, 14.
LAW, L. W. 211, 226—228, 270.
LAZIER, W. A. 82, 119.
LECHEVALIER, H. 19, 25, 29, 30, 51, 54, 57, 66, 70, 72, 75—77.
LEE, A. 202, 269.
LEFIMINE, D. V. 117.
LEGRAND, M. 160, 161, 181, 182.

LEIGH, T. 202, 205, 207, 256.
LEIN, J. 68, 75.
LENK, W. 123—125, 131, 135—138, 146, 147, 150—152, 168, 174, 180, 181.
LEPPER, M. H. 80, 118.
LEUTHARDT, F. 193, 231, 257, 262.
LEVENBERG, B. 197, 229, 252, 257, 264.
LEVY, C. C. 197, 264.
LINDAHL, P. E. 276, 321.
LINDENBEIN, W. 123, 181.
LINDNER, F. 19, 56, 64, 75, 99, 120.
LIN WANG, E. 64, 75.
LIPMANN, F. 5, 6, 14, 15.
LLOYD, G. 286, 325.
LÖCHEL, W. 217, 272.
LOEB, J. N. 22, 75.
LOVELL, F. M. 82, 118.
LOWE, G. 53, 74.
LOWN, J. W. 283, 284, 290, 295, 318.
LOWRY, O. H. 202, 264.
LOWY, D. A. 233, 264.
LUGLI, A. M. 19, 73.
LUKENS, L. 229, 252, 257.
LUZZATI, M. 253, 263.
LYNEN, F. 5, 6, 9, 10, 13—15.

McCANN, M. P. 195, 269.
McCARTHY, F. J. 19, 24, 35, 40, 62, 64, 78.
McCARTNEY, W. 285, 288, 317.
McCORMICK, J. R. D. 82, 96, 100—103, 105, 108, 114, 115, 117—119.
McDOUGALL, B. M. 197, 198, 206, 221, 234, 240, 241, 256, 265.
MACKENZIE, C. G. 233, 264.
McNUTT, W. S. 197, 264.
MacRAE, G. D. E. 7, 15.
McSWEENEY, G. P. 8, 10, 14.
McTAGGART, N. G. 56, 72.
MADER, W. J. 225, 264.
MAEDA, K. 64, 75.
MAGASANIK, B. 252, 253, 265.
MALEY, F. 241, 264.
MALYSHKINA, M. A. 70, 78.
MANGUM, J. H. 245, 264.
MARCO, A. DI 58, 72.
MARGALITH, P. 190, 271.
MARSH, W. S. 19, 76.
MARSHALL, E. K., Jr. 202, 225, 256.
MARTIN, G. J. 220, 221, 264.
MARTIN, J. T. 299, 325.
MARTIN, R. G. 252, 255.
MASON, S. F. 224, 264.

MATHEWS, C. K. 197, 198, 206, 208, 224, 264, 265.
MATRISHIN, M. 82, 96, 114, 117.
MATSUI, M. 281, 294, 309—314, 322, 323.
MATSUMOTO, H. 278, 322.
MATSUOKA, M. 56, 60, 75.
MATSUURA, S. 205, 207, 254.
MAY, M. 207, 211—213, 265.
MEBANE, A. D. 17, 18, 26, 56, 57, 66, 75, 76.
MEHLER, A. H. 230, 247, 265, 269, 271.
MEHTA, A. C. 315, 322.
MEIJER, T. M. 279, 297, 302, 303, 322.
MEISTER, A. 231, 265.
MELERA, A. 157.
MERZ, K. W. 296, 322.
METZLER, D. E. 237, 265.
MILLER, A. 204, 246, 265.
MILLER, G. A. 155, 164, 181.
MILLER, M. W. 82, 98, 119.
MILLER, P. A. 82, 96, 102, 108, 114, 115, 117—119.
MILLERD, A. 6, 15.
MIMS, V. 190, 200, 265.
MINOR, W. F. 99, 117.
MINTHORN, M. L., Jr. 232, 263.
MISRA, D. K. 199, 265.
MITCHELL, H. K. 188, 189, 200, 265, 270.
MITOMA, C. 229, 236, 265.
MIYAJIMA, S. 276, 286, 293, 297, 313, 324, 325.
MIYAKE, A. 62, 68, 75, 77.
MIYANO, M. 281, 294, 309—314, 322, 323.
MOE, R. A. 26, 46, 57, 68, 72.
MÖLLER, E. F. 192, 272.
MOLLIN, D. L. 191, 246, 247, 263, 265, 272.
MOLNAR, D. A. 195, 257.
MONTERMOSO, J. C. 5, 16.
MOORE, C. G. 8, 14.
MORELAND, W. T. 81, 91, 120.
MORITA, Y. 19, 62, 72.
MORTIMER, A. (M.) 19, 32, 40, 64, 72.
MORTON, J., II 95, 118.
MÖSLEIN, E. M. 9, 10, 14, 15.
MOSS, J. 221, 264.
MOWAT, J. H. 200—204, 221, 225, 255, 256, 261, 262, 265, 270—272.
MOYED, A. S. 252, 253, 265.
MOZINGO, R. 202, 209, 273.
MULLER, W. H. 100, 101, 118.
MÜLLER, W. 133, 134, 138, 180.
MURAI, K. 82, 83, 97, 99, 102, 118, 120.

MUSSELMAN, M. M. 80, 119.
MUXFELDT, H. 54, 80, 102, 110, 111, 113, 115, 116, 119.
MYDLINSKI, I. 68, 75.

NAGAI, K. 276, 323.
NAGASE, M. 60, 78.
NAGER, U. F. B. 57, 75.
NAKADA, H. I. 231, 266.
NAKADA, S. 40, 64, 78.
NAKAJIMA, M. 313, 314, 319, 324.
NAKAMURA, S. 23, 56, 57, 62, 66, 76, 78.
NAKANO, H. 25, 26, 28—31, 46—50, 57, 66, 74, 76.
NAKAO, A. 233, 243, 266.
NAKAO, Y. 57, 62, 64, 76.
NAKATSUGAWA, T. 276, 319.
NAKAYAMA, M. 319.
NAKAZAKI, M. 280, 323.
NAKAZAWA, K. 57, 62, 68, 76, 77.
NALESNYK, S. 115, 119.
NAMESTNIKOVA, U. P. 70, 78.
NARAHASHI, T. 276, 319.
NATH, R. 198, 206, 241, 260, 266.
NATHAN, H. A. 187, 262.
NAYLER, P. 56, 57, 76.
NEFELOVA, M. V. 70, 76.
NEIDLE, A. 252, 266.
NEIPP, L. 123, 124, 131, 137, 147, 151, 175, 181.
NESEMANN, G. 19, 64, 75.
NEUBERGER, A. 233, 255.
NICHOL, C. A. 197—199, 205, 207—209, 216, 222, 227, 237, 241, 260, 266, 268, 272, 274.
NICKL, J. 311, 323.
NIEMEYER, J. 133, 141, 143, 145, 180.
NIMMO, C. C. 9, 10, 11, 15.
NIMMO-SMITH, R. H. 192, 266.
NINET, L. 19, 58, 73.
NISHIKUBO, T. 294, 314, 323.
NISHIO, M. 18, 40, 64, 76.
NODA, H. 193, 263.
NOLLSTADT, K. 245, 266.
NOMINE, G. 19, 70, 74.
NORONHA, J. M. 189, 191, 199, 207, 211, 227, 228, 246, 266, 270, 273.
NORTHEY, E. H. 201, 203, 204, 255, 256, 261, 272.
NORTON, L. B. 303, 304, 309, 323.
NÖTHER, H. 26, 77.
NUMATA, C. 315, 321.

OBERG, K. E. 276, 321.
ODA, J. 313, 314, 319.
O'DELL, B. L. 188, 189, 199—202, 204, 205, 207, 266, 267, 273.
O'DONNELL, R. W. 309, 323.
OFFE, H. A. 310, 311, 315, 323.
OGATA, K. 57, 62, 64, 68, 75, 76.
OGAWA, H. 18, 40, 64, 76.
OI, K. 64, 75.
OKAMI, Y. 23, 56, 57, 62, 66, 68, 76, 78.
OKAMOTO, E. 58, 72.
OKAYA, Y. O. 82, 118.
OKI, K. 77.
OLESON, J. J. 221, 261.
OLIVERIO, V. T. 228, 266.
OLLIS, W. D. 123, 130—132, 134, 137, 150, 151, 154, 155, 157, 164, 167, 176, 179—181, 295, 296, 298, 299, 301, 302, 306, 308, 313—320, 323.
ONO, M. 276, 286, 293, 297, 313, 324, 325.
ONUMA, H. 40, 64, 78.
OOKA, M. 19, 39, 57, 64, 78.
ORIGONI, V. E. 100, 101, 118.
OROSHNIK, W. 17, 18, 26, 57, 66, 76.
OSBORN, M. J. 187, 197, 206—208, 212, 214, 217, 227—229, 234, 237, 239, 240, 248, 251, 260, 261, 263, 266, 273.
OSWALD, E. J. 70, 76.
OTTKE, R. C. 5, 15.

PAGANO, J. F. 26, 45, 56, 58, 62, 66, 68, 74.
PAO, C. C. 60, 76.
PARK, R. B. 5, 7, 8, 15.
PARKER, G. 310, 316, 320, 323.
PARKER, R. P. 209, 211—213, 257, 268.
PARKS, R. E., Jr. 227, 258.
PASTERNACK, R. 81—83, 91, 92, 97, 111, 118—120.
PASTERNAK, R. L. 204, 211, 225, 254.
PASTORE, E. J. 241, 259, 266.
PATRICK, J. B. 27—29, 31, 40, 42, 48, 60, 76.
PATT, P. 126, 127, 177—180.
PATTERSON, A. M. 200, 267.
PATTON, T. 243, 258.
PAVANARAM, S. K. 313, 323.
PEABODY, R. A. 229, 251, 259.
PEACE, R. 295, 296, 301—303, 316, 318.
PEEL, J. L. 246, 267.
PÉNASSE, L. 19, 70, 74.
PENNINGTON, F. C. 100, 120.
PEPINSKY, R. 82, 118.
PERAULT, A. M. 206, 214, 221, 267.

PEREGO, M. 56, 58, 72.
PERLMAN, D. 52, 72, 114, 119.
PETERS, J. M. 197—199, 221, 236, 240, 248, 267.
PETERSON, E. A. 228, 270.
PETERSON, W. H. 188, 189, 270.
PETISI, J. (P.) 95, 96, 101, 103, 105, 117—119.
PETTY, M. A. 82, 96, 114, 117.
PEZER, L. I. 239, 267.
PFIFFNER, J. J. 188—190, 199—202, 204, 207, 255, 266, 267, 273.
PFLEIDERER, W. 206, 267.
PHELPS, A. S. 82, 96, 114, 117.
PHILLIPS, A. H. 9, 14.
PHIPERS, R. F. 281, 286, 296, 299, 301, 314, 317.
PIDACKS, C. 117.
PILGRIM, F. J. 81, 82, 91, 92, 97, 100, 111, 118, 120.
PINNERT-SINDICO, S. 19, 58, 73.
PITTS, J. D. 231, 232, 267.
PLAN, S. Y. 81, 82, 117.
PLEDGER, R. A. 19, 25, 76.
POHLAND, A. 211, 267.
POLHEMUS, L. 5, 16.
POLLARD, C. 9—11, 15.
POPJÁK, G. 10, 14, 15.
PORATH, J. 228, 271.
PORTER, J. W. 10, 16.
PORTER, R. F. 18, 31, 32, 34, 64, 73.
POTE, H. L. 55, 56, 60, 74.
POZMOGOVA, I. N. 70, 76.
PRELOG, V. 123, 124, 131, 137, 146, 147, 150, 151, 175, 176, 180, 181.
PREUD'HOMME, J. 19, 58, 73.
PRICER, W. E., Jr. 207, 214—216, 228, 234, 247, 249, 250, 267, 268.
PRIDE, E. 52, 72.
PRIDHAM, T. G. 19, 26, 50, 57, 68, 70, 73.
PULLMAN, B. 206, 214, 221, 267.
PURCELL, A. E. 7, 11, 15, 16.

QUAYLE, J. R. 231, 267.

RABINOWITCH, E. 26, 76.
RABINOWITZ, J. C. 187, 189, 207, 213—216, 228, 234, 247—250, 261, 267, 268, 271.
RABINOWITZ, J. L. 6, 7, 14, 15.
RACHELE, J. A. 233, 264.
RACUSEN, D. W. 6, 14.
RAMACHANDRA ROW, L. 313, 323.
RAMAKRISHNAN, C. V. 6, 14.

RAMASASTRI, B. V. 207, 234, 239, 268.
RANDALL, W. A. 70, 76.
RAO, K. V. 19, 56, 57, 60, 62, 68, 75, 76.
RAPOPORT, S. 231, 271.
RASMUSSEN, R. S. 212, 270.
RAUBITSCHECK, F. 56, 76.
RAUEN, H. M. 187, 197, 202, 209, 210, 227, 229, 236, 268.
RAVEL, J. M. 207, 209, 211—213, 236, 259, 265.
READ, G. 53, 74.
REED, F. P. 309, 323.
REED 130.
REEDY, R. J. 70, 76.
REGNA, P. P. 81, 82, 91, 92, 97, 111, 117—120.
ŘEHÁČEK, Z. 19, 78.
REICHENTHAL, J. 100, 101, 115, 118, 119.
REICHERT, E. 5, 15.
REICHSTEIN, T. 173, 181, 310, 323.
REID, J. 28, 78.
REINHOLD, D. 204, 270.
RENNHARD, H. H. 102, 103, 105, 106, 110—112, 116, 117, 120.
RENZO, E. C. DE 197, 258.
REUSSER, P. 123, 124, 131, 137, 147, 151, 175, 181.
REYNOLDS, J. J. 193, 268.
RHODES, A. 32, 40, 72.
RICHARDS, J. H. 311, 312, 323.
RICHARDSON, A. C. 28, 76.
RICKARDS, R. W. 27, 43—45, 52, 54, 58, 72, 73.
RICKES, E. L. 209, 225, 268.
RILLING, H. 7, 13.
ROARK, R. C. 276, 278, 323.
ROBBINS, E. A. 214, 268.
ROBBINS, M. 190, 200, 255.
ROBERTS, DeW. 195, 241, 268.
ROBERTSON, A. 276, 300—302, 309—314, 316, 317, 319—321, 323, 324.
ROBINSON, R. 305, 324.
ROBINSON, W. G. 6, 7, 14.
RODE 140.
ROGALSKI, W. 102, 116, 119.
ROGERS, H. D. 278, 324.
ROHRBAUGH, T. 243, 245, 268.
ROLLAND, G. 82, 114, 120.
ROOZE, V. 245, 268.
ROSENBACH, L. M. 247, 263.
ROSENTHAL, S. 219, 245, 264, 268.
ROSS, W. C. J. 49, 76.

ROTH, B. 202, 209, 211—213, 257, 268.
ROUTIEN, J. B. 81, 82, 117.
ROW, L. R. 313, 323.
RÜCKER, B. 242, 243, 245, 273.
RUDNEY, H. 6, 7, 14, 15.
RUEFF, L. 5, 6, 15.
RÜHL, E. 197.
RUPP, W. 204, 272.
RUSBY, G. L. 302, 309, 310, 324.
RUSSELL, R. C. 295, 296, 302, 313—315, 318, 320.
RYHAGE, R. 30, 31, 33, 64, 73.

SAFFERMAN, R. S. 54, 77.
SAGERS, R. D. 232, 234, 249, 268.
SAKAMI, W. 204, 216, 219, 227, 229—231, 234, 236, 237, 239, 243, 245, 258, 263, 268—270.
SAKAMOTO, J. 58, 64, 72.
SAKAMOTO, J. M. J. 40, 46, 56—58, 64, 66, 77.
SAKATA, G. 313, 314, 319.
SALTZA, M. H. v. 28, 78.
SATO, M. 278, 321.
SAUBERLICH, H. E. 188, 211, 269.
SCHACH VON WITTENAU, M. 82, 97, 102, 105, 108, 117, 118, 120.
SCHÄFER, G. 203, 272.
SCHAFFNER, C. P. 28—30, 50, 51, 54, 57, 66, 67, 70, 72, 77.
SCHAFFNER, P. S. 289, 320.
SCHEINDLIN, S. 202, 269.
SCHLIEP, H. J. 193, 231, 272.
SCHMID, H. 303, 307—309, 316, 319, 324.
SCHMIDT, G. 296, 322.
SCHMIDT-THOMÉ, J. 19, 56, 64, 75.
SCHMIED-KOWARZIK, V. 202—204, 224, 272.
SCHNABEL, E. 26, 77.
SCHNELLER, A. 103, 105, 118.
SCHWARTZMAN, L. H. 56, 79.
SCHWEIGERT, B. S. 197, 199, 260.
SCHWYZER, R. 203, 262.
SCOTT, A. I. 108, 120.
SCRIMGEOUR, K. G. 217, 239, 245, 261, 264, 269.
SEAMAN, W. 204, 211, 225, 254.
SEEGER, D. R. 201—204, 220, 221, 255—257, 261, 269, 272.
SEEGMILLER, J. E. 247, 269.
SEELEY, D. B. 81, 82, 117.
SEHGAL, J. M. 314, 324.
SEIDEL, P. C. 27, 43—45, 54, 58, 73.

SEITZ, G. 22, 74.

SEKI, M. 25, 57, 66, 74, 76.

SEMAR, J. B. 114, 119.

SEMB, J. 200—204, 221, 225, 255, 256, 261, 262, 265, 270—272.

SENSI, P. 82, 114, 120.

SESHADRI, T. R. 307, 314, 315, 322, 324.

SGARZI, B. 19, 73.

SHAMSHURIN, A. A. 310, 314, 324.

SHAW, E. 193, 271.

SHAY, A. J. 111, 118.

SHEMIN, D. 231, 232, 236, 239, 269, 272.

SHERMAN, J. H. 27, 28, 30, 43, 50, 68, 74.

SHIBATA, M. 57, 68, 77.

SHIBATA, Y. 77.

SHIMAZONO, H. 231, 269.

SHIOTA, T. 194, 195, 269.

SHIO(T)ZU, S. 19, 39, 57, 64, 78.

SHIRLEY, D. A. 18, 56, 60, 73.

SHIVE, W. 192, 207, 209, 211—213, 220, 227, 236, 259, 265, 267, 269, 273.

SHOCKMAN, G. H. 229, 271.

SHODA, A. 193, 194, 262.

SHODA, T. 193, 263.

SHOOLERY, J. N. 28, 32, 37, 39, 64, 73, 157.

SHOTWELL, O. L. 19, 26, 50, 57, 68, 70, 73.

SHULL, G. M. 81, 82, 117.

SHUNK, C. H. 7, 14.

SICKELS, J. P. 201, 203, 204, 255, 256, 261, 272.

SIDDALINGAIAH, K. S. 276, 278—281, 284, 286—290, 301, 314, 317, 318.

SIEDEL, W. 99, 120.

SILBER, R. 197, 198, 240, 269.

SILVERMAN, M. 189, 191, 195, 196, 207, 209, 211, 214, 226—228, 236, 246—248, 250, 259, 260, 263, 266, 269—271, 273.

SIME, J. T. 195, 196, 258.

SIMMONS, B. 251, 255.

SIMON, H. 193, 203, 231, 272.

SIMONITSCH, E. 308, 324.

SJOLANDER, N. O. 82, 96, 103, 108, 114, 115, 118, 119.

SKEGGS, H. R. 7, 15, 195, 274.

SLAVÍK, K. 198, 221, 222, 228, 270.

SLAVÍKOVÁ, V. 198, 221, 222, 228, 270.

SLETZINGER, M. 204, 270.

SLOANE, N. H. 188, 189, 200, 261.

SLONIMSKI, P. P. 253, 263.

SMAKULA, A. 22, 77.

SMITH, C. M. 278, 321.

SMITH, C. W. 212, 270.

SMITH, E. L. 244, 260.

SMITH, F. 6, 15.

SMITH, H. 308, 316.

SMITH, J. M., Jr. 201—204, 209, 211—213, 220, 221, 224, 255—258, 261, 268, 269, 272.

SMITH, L. E. 278, 284, 287, 289, 313, 321.

SMITH, L. L. 100, 101, 118.

SNELL, E. E. 188, 189, 200, 227, 229, 237, 238, 265, 270, 273.

SNELL, J. F. 113, 120.

SOBER, H. A. 228, 270.

SOBIN, B. A. 56, 57, 60, 62, 68, 75, 81, 82, 117.

SOBOTKA, H. 227, 255.

SOEDA, M. 57, 60, 66, 70, 75, 77.

SOEDER, F. 99, 120.

SOLOMONS, I. A. 19, 56, 58, 73, 81, 82, 117, 119.

SOLOV'EV, S. N. 19, 57, 70, 78.

SONDHEIMER, F. 57, 77.

SONNE, J. C. 229, 257.

SONODA, Y. 66, 75.

SOUTHWICK, P. L. 202, 209, 273.

SPENCER, J. L. 103, 105, 117, 119.

SPENSLEY, P. C. 203, 263.

SPOHLER, E. 171, 172, 177, 180, 181.

SPRINSON, D. B. 229, 236, 237, 258, 270.

SREENIVASAN, A. 199, 266.

SRINIVASAN, P. R. 74.

STADTMAN, E. R. 6, 15.

STADTMAN, T. C. 238, 274.

STÄLLBERG-STENHAGEN, S. 30, 31, 33, 64, 73.

STAMM, O. A. 307, 324.

STAMM, W. 209, 227, 268.

STANLEY, R. G. 7, 15.

STARON, T. 70, 77.

STEIGLER, A. 19, 64, 75.

STEIN, W. J. 81, 93, 96, 120.

STEINBERG, B. A. 68, 73.

STEINMAN, I. D. 54, 77.

STEKOL, J. A. 233, 270.

STENHAGEN, E. 30, 31, 33, 64, 73.

STEPHENS, C. R. 81—83, 91, 92, 97, 99, 100, 102, 103, 105, 106, 108, 110—112, 116—118, 120.

STEVENS, A. L. 237, 243, 245, 258, 270.

STEWART, J. A. 231, 267.

STILLER, E. T. 19, 45, 56, 58, 68, 78.

STOKES, J. L. 195, 270.

STOKSTAD, E. L. R. 187—189, 199—204, 211, 220, 221, 225, 227, 255, 256, 259, 261, 262, 265, 268, 270, 272.
STOUT, H. A. 58, 68, 74.
STRECKER, H. J. 270.
STRELITZ, F. 179, 181.
STRIEGLER, K. 102, 116, 119.
STRUYK, A. P. 40, 56, 60, 77.
STUART, A. 296, 317.
SUBBAROW, Y. 188, 189, 200—204, 221, 225, 255, 256, 261, 262, 265, 270, 272.
SUBRAMANIAM, T. S. 309, 310, 317, 324.
SUD, R. K. 56, 75.
SUND, I. P. 231, 266.
SUTHERLAND, G. L. 207, 211—213, 265.
SUTHERLAND, I. O. 123, 131, 132, 134, 137, 150, 151, 154, 155, 157, 164, 167, 176, 179—181, 299, 317.
SWOBODA, O. P. 203, 272.
SYNNATSCHKE, G. 192, 228, 264.
SZUMSKI, S. A. 82, 96, 114, 117.

TABER, W. A. 18, 26, 46, 50, 57—59, 66, 76—78.
TABOR, H. 213, 215, 216, 222, 228, 234, 246, 247, 249, 269, 271.
TADOKORO, I. 77.
TAKAHASHI, I. 66, 77.
TAKAHASHI, S. 314, 319, 324.
TAKAMIZAWA, Y. 19, 62, 72.
TAKEI, S. 276, 279, 285, 286, 293, 297, 313, 314, 324, 325.
TAKEUCHI, Y. 81, 120.
TAKEYAMA, S. 244, 271.
TALBERT, P. T. 207, 217, 227—229, 237, 260, 266.
TAMELEN, E. E. VAN 82, 120.
TAMURA, G. 7, 15.
TANABE, K. 77.
TANAKA, Y. 39, 57, 64, 78.
TANNER, F. W., Jr. 19, 56—58, 60, 62, 68, 73, 75, 82, 97, 118.
TATTERSFIELD, F. 299, 325.
TATUM, E. L. 5, 15.
TAVORMINA, P. A. 7, 15.
TAYLOR, E. C. 206, 267.
TCHEN, T. T. 9, 14, 16.
TEAS, H. 5, 16.
TEPLEY, L. J. 225, 254.
THALLER, V. 27, 28, 30—34, 37, 38, 64, 73.
THOMAS, R. 27, 43—45, 54, 58, 73.
THOMPSON, G. A., Jr. 7, 11, 15, 16.
THOMSON, P. J. 52, 72.

THOMSON, P. L. 113, 120.
THRUM, H. 19, 46, 62, 70, 77.
TIMRECK, A. E. 82, 119.
TISHLER, M. 204, 270.
TOENNIES, G. 229, 271.
TOKUI, Y. 57, 68, 77.
TOLMAN, L. 221, 264.
TORII, M. 190, 271.
TREBST, A. 194, 271.
TSAI, C. S. 278, 317.
TS'AI, J. S. 60, 76.
TSCHESCHE, R. 193, 271.
TSYGANOV, V. A. 19, 57, 70, 78.
TURNER, W. B. 53, 73.
TYTELL, A. A. 19, 24, 35, 40, 62, 64, 78.

UKSTINS, I. 219, 236, 245, 266, 269.
ULLBERG, S. 103, 116.
UMEZAWA, H. 23, 56, 57, 60, 62, 64, 66, 68, 75, 76, 78.
UMEZAWA, S. 19, 39, 40, 57, 64, 78.
URBAHN, H. 231, 271.
URSPRUNG, J. J. 103, 105, 116.
USDIN, E. 228, 229, 271.
UTAHARA, R. 23, 56, 57, 62, 66, 76, 78.

VANDENBELT, J. M. 190, 200, 202, 204, 207, 255, 266, 273.
VANDEPUTTE, J. 19, 26, 45, 56, 58, 62, 66, 68, 73, 74, 78.
VAN DUUREN, B. L. 309, 325.
VAN EEK, T. 40, 56, 60, 77.
VANĚK, Z. 19, 78.
VARADARAJAN, S. 307, 314, 324.
VEAL, P. L. 151, 155, 167, 181.
VELLUZ, L. 182.
VENKATARAMAN, K. 278, 325.
VERCAMER, F. N. 227, 237, 266.
VERONESI, U. 57, 68, 73.
VIEIRA, E. 193, 271.
VIGNEAUD, V. DU 243, 271.
VINING, L. C. 18, 26, 46, 50, 57—59, 66, 76—78.
VINSON, J. W. 81, 82, 117.
VOLCANI, B. E. 190, 271.
VOLZ, G. 314, 318.
VONDERBANK, H. 80, 120.
VON SALTZA, M. H. 28, 78.

WACHTEL, J. L. 19, 45, 56, 58, 68, 78.
WACKER, A. 192, 194, 202, 204, 224, 271, 272.

WACKER, H. 193, 231, 272.
WAEHNELDT, TH. 124, 127, 171—173, 175, 178, 181, 182.
WAELSCH, H. 204, 246, 252, 265, 266.
WAGNER, A. F. 7, 14.
WAGNER, K. J. 100, 120.
WAHBA, A. J. 206, 222, 241, 271.
WAISVISZ, J. M. 40, 56, 60, 77.
WAKAKI, S. 58, 64, 79.
WAKSMAN, S. A. 56, 66, 75—77.
WALD, G. 22, 75.
WALDMANN, H. 202, 227, 268.
WALDSCHMIDT, M. 193, 231, 272.
WALKER, J. 204, 259.
WALLER, C. W. 81, 93—96, 101, 117, 118, 120, 200—204, 221, 225, 255, 256, 261, 262, 265, 270—272.
WALLHÄUSSER, K. H. 19, 56, 64, 75.
WALTERS, D. R. 27, 28, 30, 43, 44, 50, 58, 68, 74, 79.
WALTON, E. 7, 14.
WANG, E. L. 64, 75.
WANZLICK, H. W. 217, 272.
WARD, J. 311, 312, 323.
WARREN, L. 229, 252, 257.
WATERS, A. H. 191, 246, 265, 272.
WATJEN, A. 233, 270.
WEBB, J. S. 27—29, 31, 40, 42, 48, 60, 76, 96, 120.
WEBB, M. 231, 272.
WEIDENMÜLLER, H. 56, 75.
WEINHOUSE, S. 231, 266, 272.
WEISLOGEL, O. 240, 272.
WEISMAN, R. A. 195, 257.
WEISS, S. 233, 270.
WEISS, U. 179, 181.
WELCH, A. D. 190, 195, 198, 216, 237, 260, 272, 274.
WELCH, W. 231, 258.
WELLIKY, I. 231, 272.
WERKHEISER, W. C. 197, 198, 221, 272.
WESSELY, L. 6, 15.
WESTLEY, J. 27, 43—45, 54, 58, 73.
WEYGAND, F. 192—194, 202—204, 224, 231, 271, 272.
WHALLEY, W. B. 286, 325.
WHITFIELD, G. B. 24, 37, 57, 64, 79.
WHITING, D. A. 276, 278—281, 284, 286—290, 301, 304, 305, 308, 309, 314, 317, 318.
WHITING, M. C. 27, 28, 30—34, 37, 38, 56, 57, 64, 73, 76.

WHITLEY, H. R. 187, 214, 234, 251, 261, 272, 273.
WHITTAKER, V. K. 241, 273.
WIELAND, O. 6, 15.
WIELAND, O. P. 228, 229, 273.
WILDMAN, S. G. 2, 14.
WILKINSON, R. G. 96, 109, 110, 115, 117, 118, 120.
WILKINSON, R. W. 101, 117.
WILLIAMS, J. H. 81, 93—95, 118, 120, 228, 229, 273.
WILLIAMS, R. J. 188, 189, 200, 220, 227, 236, 265, 273.
WILLIAMS, R. P. 27—29, 31, 40, 42, 48, 60, 76.
WILLIAMSON, D. H. 239, 255.
WILLS, L. 187.
WILMANNS, W. 199, 242, 243, 245, 251, 273.
WILSON, A. N. 7, 14.
WILSON, E. M. 238, 273.
WIMMER, E. 134, 138, 140, 143, 147, 181.
WINKLER, K. C. 231, 273.
WINSTEIN, W. A. 229, 273.
WINTERBOTTOM, R. 100, 101, 118.
WINTERSTEINER, O. (P.) 28, 43, 44, 50, 58, 74, 78, 79.
WITTENAU, M. S. v. 82, 97, 102, 105, 108, 117, 118, 120.
WITTENBERG, J. B. 189, 211, 227, 228, 273.
WITTING, L. A. 10, 16.
WITTLE, E. L. 202, 273.
WOERNLEY, D. L. 201, 273.
WOLF, B. 194, 273.
WOLF, C. F. 27—29, 31, 40, 42, 48, 60, 76, 81, 93—96, 118, 120.
WOLF, D. E. 7, 14, 15, 202, 209, 273.
WOLF, L. M. 225, 259.
WOLOVSKY, R. 57, 77.
WOOD, B. J. 4, 6, 14.
WOOD, J. W. 278, 321.
WOOD, R. C. 199, 226, 273.
WOODS, D. D. 190, 192, 232, 233, 244, 259, 260, 262—264, 266, 273.
WOODS, G. F. 56, 79.
WOODWARD, R. B. 27, 79, 81—83, 91, 92, 97, 99, 111, 112, 116—118, 120.
WOOLDRIDGE, W. E. 70, 79.
WORK, P. S. 195, 196, 258.
WRIGHT, B. E. 187, 189, 199, 205, 206, 211, 238, 273, 274.
WRIGHT, L. D. 7, 15, 195, 274.

WYNGAARDEN, C. 213, 215, 216, 222, 228, 246, 247, 271.

YACUBOV, G. Z. 179, 182.
YAJIMA, T. 79.
YASAMURA, J. 278, 321.
YOSHISHIRO, S. 77.
YOUNG, M. B. 27, 28, 30, 43, 50, 68, 73.
YOUNG, S. R. 241, 258.
YUNAGUCHI, T. 66, 75.

ZABIN, I. 5, 15.
ZÄHNER, H. 123, 124, 131, 137, 147, 151, 175, 181.
ZAKRZEWSKI, S. F. 197—199, 205, 207, 209, 222, 234, 260, 274.
ZECHMEISTER, L. 25, 79.
ZIEGLER, I. 193, 274.
ZIMMERMAN, S. B. 239, 264.
ZINNES, H. 26, 46, 57, 68, 72.
ZOTTU, S. 231, 258.

Sachverzeichnis. Index of Subjects. Index des Matières.

Abutic acid, synthesis 310.

Acetat-Einheiten, Kopf-Schwanz-Verknüpfung 163.

Acetat-Hypothese (Anthracyclinone) 164.

—, (rubber) 6.

Acetoacetyl CoA, in rubber biosynthesis 6.

Acetyl CoA, in rubber biosynthesis 6.

ACF (Abkürzung für Anhydro-citrovorumfactor) 212.

Adenin (Einkohlenstoff-Reservoir) 230.

Adenosyl-methionin, Beteiligung an der Methionin-Bildung 246.

—, als Katalysator (biol.) 245.

Adenylphosphoryl-CO_2 (rubber biosynthesis) 7.

Adipinsäure, β-äthyl- 139.

—, pteroyl- 221.

ADP (Abkürzung für Adenosin-diphosphat) 184.

Aerobacter aerogenes, Amethopterin-gehemmte Kulturen 231.

AICAR (Abkürzung für 4-Amino-5-imidazol-carboxamid-ribotid) 234, 251.

Akitamycin, spectrum 56, 60.

Aklavin 123.

—, → Aklavinon 179.

—, Hemmwirkung 179.

—, aus *Streptomyces* 179.

—, Wirkung gegen Phagen 154, 179.

Aklavinon 123.

—, aus Aklavin 154, 179.

—, Aromatisierung, Ring A 134.

—, → Benzol-tetracarbonsäure-(1,2,3,4) 154.

—, Biogenese 166.

—, α-Hydroxygruppen 128.

—, KMR-Spektrum 154, 155, 156.

—, Konstitution 129, 154.

—, lineare Anellierung des alicyclischen Ringes 133.

—, UV- und IR-Spektrum 132, 154.

—, bisanhydro-, Struktur 154.

—, 7-desoxy- 123, 154.

—, —, aus *Streptomyces* spp. 154, 155.

—, triacetat, Spektrum 154.

Alcaligenes metalcaligenes, Desaminierung der Folsäure 197.

Aliomycin 62.

—, S-content 27.

Allo-iso-tephrosin 296.

Allo-tephrosin 296.

Ameisensäure, „aktivierte" 211.

—, —, im Purin-Stoffwechsel 248.

Ameisensäure-Stoffwechsel, Teilnahme der Methylen-tetrahydrofolsäure 236, 237.

Amethopterin, als Folat-Antagonist 220, 221.

—, Hemmung der Folat-Reduktase 198.

p-Aminobenzoesäure, Antagonisten 194.

—, (Biosynthese von Folat-aktiven Substanzen) 194.

p-Aminobenzoyl-glutaminsäure-diäthylester, und Redukton 204.

Aminoimidazol-carboxamid-ribotid-Transformylase 234.

Aminopterin, als Folat-Antagonist 220, 221.

—, Hemmung der Folat-Reduktase 198.

—, di- und tetrahydro- 222.

Amorpha fruticosa, amorphin 316.

Amorphigenin 316.

Amorphin = 8′-hydroxyrotenone glycoside 316.

Amphotericin A, B 45, 50.

—, → ammonia 45.

—, IR spectrum 26, 45.

Amphotericin A, properties, spectrum 56, 58.

Amphotericin B, all-*trans* configuration 26.

—, carbohydrate moiety 28.

—, medical use 18.

—, properties, spectrum 57, 68.

—, spectral "degradation" by association 26.

—, perhydro- 29, 30.

Anämie, perniciöse, und Folsäure 191.

Anämie-Faktor (Leber) 189.

Anhydro-citrovorumfactor 212.

Anhydro-leucovorin 212.

Anthrachinon, 1,8-dihydroxy- 130.

—, hydroxy-, Biogenese 167.

—, 1,4,5,8-tetrahydroxy- 130.

—, —, Spektrum 132.

—, —, Deriv. 138, 140.

—, 1,4,5-triacetoxy-, Spektrum 148.

—, 1,4,5-trihydroxy- 130, 145, 151.

—, —, Spektrum 133.

Anthracycline 121, 122.

—, von Aklavinon 179.

—, basische Glykoside 124.

—, Glykoside von Anthracyclinonen mit Rhodosamin und N-freien Zuckern 170.

—, Isolierung 123.

—, von ε-Pyrromycinon 174.

—, von Rhodomycinonen 176.

—, tabellarische Übersicht 123.

—, Trennung von Anthracyclinonen 124.

—, Zucker 171.

Anthracyclinone 121, 122.

—, Aromatisierung von Ring A, und Spektrum 133, 134.

—, Asymmetriezentren und Biogenese 168.

—, Äthylgruppe 135.

—, Bezifferung, Schreibweise der Formeln 137.

—, Biogenese 163.

—, —, Acetathypothese 164, 168.

—, —, Einbau der Äthylgruppe durch Propionat-Einheit 164.

—, —, Schema, noch fehlende Typen 169.

—, —, Hydroxy-, Carbomethoxy- und Carboxy-Gruppen 167.

—, Carbomethoxy-Gruppe 131, 135.

—, mit Carbomethoxy an $C_{(10)}$, Konfiguration 162.

—, Chromophor 131.

—, funktionelle Gruppen 130.

—, → gelbe Acetate 131.

—, mit Hydroxygruppe an $C_{(7)}$, Konfiguration 162.

—, IR-Spektrum und H-Brücken 131.

—, Isolierung 123.

—, Konstitution 127, 130, 137.

—, KMR-Spektrum 131, 155, 156, 157, 159.

—, —, und Konstellation 159.

—, lineare Anellierung des alicyclischen Ringes 133.

—, Massenspektroskopie 130.

Anthracyclinone, Mol.-Gew. 130.

—, reduzierende Acetylierung, Spektrum 131.

—, Stammverbindung, Bezifferung 128.

—, Stereochemie 160.

—, aus *Streptomyces* spp. 122.

—, Substituenten, Stellung in Ring A 134.

—, —, Kombinationen 135.

—, tabellarische Übersicht 123.

—, = 1,4,5,8-Tetrahydroxy-, 1,4,5-Trihydroxy- oder 1,8-Dihydroxy-anthrachinon-Deriv. 130.

—, Trennung von Anthracyclinen 124.

—, Unterklassen, bedingt durch Anzahl von OH 128.

—, Zirkulardichroismus-Kurven 161.

—, —, und Asymmetriezentren 161.

(+)-*trans*-Anthranilsäure, 2,3-dihydro-3-hydroxy-, in *Str. aureofaciens* 115.

Antibiotics, polyene, antifungal 17.

Antifungal antibiotics 17.

Antifungin 4915 70.

Antimycoin A, biosynthesis 54.

—, spectrum 56.

Apo-toxicarol, → dehydronetoric and hydroxynetoric acids 301.

—, formation, structure 300.

—, from α-toxicarol 301.

Ascosin 66.

—, identity (?) with trichomycin 50.

—, spectrum 21, 25, 57.

Asparaginsäure, 4-amino-pteroyl-, als Hemmstoff 221.

—, pteroyl- 220, 221.

Atebrin, Hemmung von Folat-Reduktase 199.

Äthylendiamin, diphenyl-, → N,N′-Diphenyl-imidazolin 216.

—, N-formyl-diaryl-, Cyclisierung 213.

—, N-formyl-N,N′-diphenyl-, → N,N′-Diphenylimidazolinium-Verbindung 212, 213.

ATP (Abkürzung für Adenosin-triphosphat) 184.

Aureofacin 68.

Aureomycin 91.

—, → Anhydro-aureomycin 93.

—, → Anhydro-tetracyclin 93.

—, → 5-Chlor-salicylsäure 91.

—, = 7-Chlor-tetracyclin 81.

—, → Desdimethylamino-aureomycinsäure 93.

Aureomycin, → Iso-aureomycin 93.
—, Konfiguration 82.
—, Röntgenstruktur-Analyse 82.
—, Spektrum 91.
—, aus *Str. aureofaciens* 80.
—, Struktur 81, 92.
—, → Tetracyclin 91.
—, Unterschied von Terramycin 91.
—, anhydro- 93.
—, —, → Aureomycin (biol.) 114.
—, —, → 5 a, 11 a - Dehydro - tetracyclin
  (biol.) 114.
—, —, licht-katalysierte Oxydation 108.
—, anhydro-2-acetyl-descarboxamido-, →
  2-Acetyl-descarboxamido-tetracyclin
  (biol.) 114.
—, desdimethylamino- 91, 101, 112.
—, —, → Anhydrid 92, 93.
—, —, Struktur 92.
—, desdimethylamino-anhydro- 112, 113.
—, —, Struktur 116.
—, —, Synthese 115.
—, desdimethylamino-12 a-desoxy- 91.
—, —, → Anhydrid 92, 93.
—, —, Struktur 92.
—, desdimethylamino-12 a-desoxy-
  anhydro-, → Tetracen 93.
—, hydrochlorid, pK-Werte 98.
—, methyläther, → 3-Methoxy-6-chlor-
  phthalsäure und Phthalid-Deriv. 95.
—, monomethyläther, desdimethylamino-
  12 a-desoxy-anhydro- 112.
α- und β-Aureomycinsäure 95.
—, → Desdimethylamino-aureomycin-
  säure 95.
Aureomycinsäure, desdimethylamino- 93,
  94.
—, —, → Aureonamid (Cyclisierung) 93.
—, —, → Cyclopentan-1,3-dion Deriv. 93.
—, —, Struktur 94.
—, nitril, IR-Spektrum 93.
Aureon, Struktur 94.
Aureonamid 93.
—, → Aureon 93.
—, → Aureonchinon-amid 93, 94.
Aureonchinon-amid, Struktur 94.
Ayfactin (AYF), properties 68.

Bacillus mycoides, Hemmung durch
  Aklavin 179.
B. subtilis, 5-Formyl-pteroyltriglutamat
  190.
—, Hemmung durch Aklavin 179.

B. subtilis, Hemmung durch Rhodo-
  mycin A 177.
B. megatherium, Hemmung durch Cine-
  rubine 176.
B. pumilis, pumilin 51.
BAL (Abkürzung für Dimercaptopropa-
  nol) 208.
Balata 2.
Barbasco (insecticide) 276.
B_c-Konjugat = Pteroyl-heptaglutamin-
  säure 200.
Benzoic acid, 2-hydroxy-4,5-dimethoxy-
  296.
Biosynthesis (Biogenese), Aklavinon 166.
—, Anthrachinone, hydroxy- 167.
—, Anthracyclinone 163, 167.
—, antimycoin A 54.
—, carbomycin 52, 55.
—, Cynodontin 167.
—, erythromycin 54.
—, filipin 53, 54.
—, Folat-aktive Substanzen 194.
—, Folsäure 192.
—, formenonetin 307.
—, Helminthosporin 167.
—, Islandicin 167.
—, isoflavanoids 305.
—, isoflavones 306.
—, isoflavonoids, linked to rotenoids 305,
  306, 307.
—, isoprenoids 5.
—, Isorhodomycinone 167.
—, macrolides 53.
—, magnamycin 52.
—, Methionin 185, 219, 244, 245.
—, nystatin 54.
—, polyacetylenes 53.
—, polyene antifungal antibiotics 51,
  52.
—, Pteroyl-Konjugate 192, 195.
—, Purine 229.
—, Pyrromycinone 167.
—, Rhodomycinone 167.
—, rotenoids 305.
—, rubber 1.
—, Serin 232.
—, *Streptomyces* polyene antibiotics 54.
—, Tetracycline 54, 113.
—, Thymin 185.
—, Valin 185.
L-Boivinose, aus Anthracyclinen 171, 172.
—, Struktur 171.

Brenztraubensäure, phosphoroklastische Spaltung 230.

2-L : 3-D *(erythro)*-1,2,3-Butanetriol, from fungichromin 36.

(—)-Butanoate, methyl, 2-hydroxy-3-methyl-, → (+)-5-isopropyl-oxazolidone 280.

Cabicidin 64.

—, spectrum 57.

Candicidin 66.

—, identity (?) with trichomycin 50.

—, spectral "degradation" by association 26.

—, spectrum 25, 57.

Candida albicans, and nystatin, trichomycin 18.

C. vulgaris, Hemmung durch Cinerubine 176.

Candidin 45, 50, 66.

—, all-*trans* configuration 26.

—, IR spectrum 26.

—, spectral "degradation" by association 26.

—, spectrum 21, 57.

Candimycin 68.

—, spectrum 57.

Capacidin 46, 62.

Carbomycin, biosynthesis 52, 55.

—, elimination of sugar by alkali 29.

Carbomycin B 52.

Carcinom und Cinerubin 176.

CF (Abkürzung für Citrovorum-Faktor) 184.

Chalcone, hydroxy-, intermediate in rotenoid biogenesis 308.

Cholesterol, acetate precursor 5.

Chromin, properties 58.

Chrysazin 166.

Chrysazin-anthranol-Derivat (Biogenese der Anthracyclinone) 166.

Chrysophansäure (Biogenese von Hydroxy-anthrachinonen) 167.

Cinerubin A, B 123.

—, antibiotische Wirkung 176.

—, Gewinnung aus *Str. antibioticus* 124.

—, Giftigkeit 176.

—, → ε-Pyrromycinon 175.

—, → Rhodosamin und 2-Desoxy-L-fucose 175.

—, Trennung 124.

—, UV- und IR-Spektrum 175.

Citrovorum-Faktor 188, 211.

—, = Folinic acid 211.

—, Wuchsstoff-Aktivität 226.

—, anhydro- 212.

Cladinose 52.

Clostridium acidi-urici, Purin-Vergärung 248.

Cl. cylindrosporum, Folsäure-Konjugate 189.

—, Formiminoglycin-tetrahydrofolat-Transferase 249.

—, Purin-Vergärung 248.

Cl. sticklandii, Pterin-Reduktase 199.

—, Serin-Aldolase 238.

CMP (Abkürzung für Cytosin-monophosphat) 184.

CoA (abbreviation for coenzyme A) 5.

Cobamid, methyl-, als Methyldonator 244.

CoC (Cofaktoren) 238.

Coenzyme A, acetoacetyl-, in rubber biosynthesis 6.

—, acetyl-, in rubber biosynthesis 6.

—, β-hydroxy-β-methylglutaryl-, → isoprenoids and β-methylcrotonyl CoA 6.

—, —, in rubber biosynthesis 6, 7.

—, β-methylcrotonyl-, in rubber biosynthesis 7.

—, β-methylglutaryl-, → mevalonic acid 7.

Coenzym F = 5,6,7,8-Tetrahydro-folsäure 185.

Cofaktoren, Einteilung 219.

Corynebacterium diphtheriae, Hemmung durch Cinerubine 176.

—, Hemmung durch Rhodomycin A 177.

C. sp., Folat-Synthese 193.

Coumaran-5-carboxylic acid, 4-hydroxy-, from elliptone 297.

Coumarin-4-carboxylic acid, 3,7-dimethoxy, → rotenonones 316.

Cracca toxicaria, deguelin 296.

Crotonic acid, β-methyl- (rubber biosynthesis) 7.

Cryptocidin 66.

—, IR spectrum 46.

—, spectrum 21, 57.

Cryptococcal meningitis and amphotericin B 18.

Cube (insecticide) 276.

Cube root, deguelin, tephrosin, isotephrosin 296.

Cyclodesaminase 234, 247, 249.

Cycloheximide, in *Str. noursei* 19.

Cyclohydrolase 234, 236, 249, 252.

Cyclopentan, 1,3-dion 94.

—, —, aus Aureomycin 93.

—, 1,2,4-trion 94.

—, —, aus Aureomycin 93.

Cynodontin, Biogenese 167.

Cytosin, 5-hydroxymethyl-, Biosynthese, und Folsäure-Coenzyme 185.

—, —, (Einkohlenstoff-Reservoir) 230.

Dandelion, Russian, rubber 2.

Deacylase, Wirkung auf 10-Formyl-tetrahydrofolsäure 247.

1,3,5,7,9-Decapentaene, spectrum 56.

2,4,6,8-Decatetraene, spectrum 56.

Deguelin, degradation product, synthesis 311.

—, from dehydro compound 314.

—, structure 277, 278, 296.

—, synthesis 314.

—, → ($\pm$)-tephrosin and ($\pm$)-isotephrosin 297.

—, 6a,12a-dehydro- 296, 313, 314.

—, —, → rissic acid, 2-hydroxy-4,5-dimethoxy-benzoic acid, and nicouic acid 296.

($\pm$)- and (—)-Deguelin, 4',5'-dihydro-282, 289, 290.

($\pm$)-Deguelin, 4',5'-dihydro-, = ($\pm$)-β-dihydrorotenone 296.

—, —, 6a,12a-dehydro- 314.

Deguelinic acid 296.

Derric acid, from elliptic acid 298.

—, structure 285.

—, synthesis 309.

Derrin 276.

Derris, α-toxicarol 299.

Derris spp., fish poison, insecticides 275, 276.

D. *chinensis*, rotenone 276.

D. *elliptica*, elliptone 297.

—, rotenone 276.

D. *malaccensis*, malaccol 302.

—, sumatrol 301.

Derris root, deguelin 296.

Derrisic acid 296, 297, 298.

—, → derric acid, rissic acid, tubaic acid 284.

—, rotatory dispersion 296.

—, structure 285.

($\pm$)-Derrisic acid, synthesis 312.

Derrisic acid, analogues, synthesis 311, 312.

Derrisic acid type compounds, cyclization 313.

Derritol 285.

—, structure 285, 287.

Desosamine 52.

Desoxy-cytosinmonophosphat-Hydroxy-methylase 239.

Desoxycytidylat-Transhydroxymethylase 234.

Diaryl-imidazolinium/Diaryl-imidazolidin, Redoxpotential 218.

Dihydrofolat-Reduktase 197, 198, 234.

Dihydrofolsäure: NAD(P)-Oxydoreduktase 234.

C_{10}-Di-isoprenoid alcohol pyrophosphate 10.

Dimedon, 2-acetyl-, Spektrum 98.

γ,γ-Dimethylallyl-pyrophosphate 9.

—, in rotenoid biosynthesis 308.

Diplococcus glycinophilus, Glycin → Formiat 232.

Distamycin B 62.

—, spectrum 56.

DMF (abbreviation for dimethyl-formamide) 58.

1,3,5,7,9,11-Dodecahexaene, spectrum 57.

Dodecanedioic acid, α-methyl- 38.

2,4,6,8,10-Dodecapentaene, spectrum 56.

Dolineone 304.

—, connection with isoflavanoid biogenesis 305.

—, dehydrogenation 305.

—, NMR data 304.

—, occurrence, structure 304.

—, in *Phaseolea* 308.

—, reactions 305.

—, structure 277, 278.

DPN (Abkürzung für Diphospho-pyridin-nucleotid) 184.

Durham test (rotenoids) 278.

EDTA (Abkürzung für Äthylendiamin-tetraacetat) 208.

Edulin, in *Phaseolea* 309.

—, structure 308.

Elliptic acid, → derric acid 298.

Elliptone, from *Derris elliptica* 297.

—, → 4-hydroxycoumaran-5-carboxylic acid 297.

—, structure 277, 297.

—, 6a,12a-dehydro-, synthesis 315.

Elliptone, dehydro-tetrahydro-, synthesis 313.
—, tetrahydro-, synthesis 312.
Endomyces albicans, Hemmung durch Cinerubine 176.
Endomycin 46.
Endomycin A 58.
Endomycin B 46, 66.
—, spectrum 21, 57.
Entamoeba histolytica und Cinerubine 175.
Enterococcus, Hemmung durch Rhodomycin A 177.
Enzym „205-2" 242.
Enzyme, Folsäure-Konjugat-spaltende 190.
Enzym-Reaktionen, Folat-katalysierte, im Stoffwechsel 236.
4-Epi-aureomycin 100.
Epi-rotenone 286.
—, dichromate cxidation 293.
6-Epi-terramycin, 6-desoxy- 102.
4-Epi-tetracyclin 100.
—, Struktur 111.
—, anhydro- 100.
—, 12 a-desoxy-, Struktur 110.
—, —, → 4-Epi-tetracyclin 111.
—, —, → Desdimethylamino-4 a,12 a-dehydro-12 a-desoxy-tetracyclin 111.
6-Epi-tetracyclin, 6-desoxy- 102, 106.
Erosnin, in *Phaseolea* 309.
—, structure 308.
Erosone 304.
—, from *Pachyrrhizus erosus* 304.
—, in *Phaseolea* 309.
—, dehydro- 304.
Erythromycin 52.
—, biosynthesis 54.
—, elimination of sugar by alkali 29.
—, dihydro-, lactone peak (IR) 27.
Escherichia coli, und Cinerubine 175.
—, Folat-aktive Substanzen 194.
—, Folat-synthetisierendes System 195.
—, 5-Formyl-pteroyltriglutamat 190.
—, Hemmung durch Aklavin 179.
—, Hemmung durch Rhodomycin A 177.
—, Methionin-Synthese 244, 245.
—, Mutanten, und Folsäure-Konjugate 190.
—, —, und T_2- oder T_5-Bakteriophagen 241.
Etruscomycin 58.
—, spectrum 56.

Eurocidin 62.
—, spectrum 56.
Eurocidin-group methylpentaenes (antibiotics) 64.
Eurotin A 70.
—, spectrum 57.

F (Abkürzung für Folsäure) 184.
FAD (Abkürzung für Flavin-adenin-dinucleotid) 184.
Farnesol pyrophosphate 10.
Fermentation, *L. casei*-factor 188, 189.
—, ein Pteroyltriglutamat aus *Corynebacterium* 200.
FH_2 und FH_4 (Abkürzungen für Dihydro- und Tetrahydro-folsäure) 184.
FIASP (Abkürzung für Formimino-asparaginsäure) 247.
Ficus elastica, rubber 1.
FIGLU (Abkürzung für α-Formimino-*L*-glutaminsäure) 246.
FIGLY (Abkürzung für Formimino-glycin) 247, 249.
Filipin 37, 64.
—, biosynthesis 53, 54.
—, chromophore 32.
—, → hexanaldehyde 30, 37.
—, NMR spectrum 28, 37.
—, and periodic acid 37.
—, structure 39.
—, UV and IR spectrum 24, 57.
—, perhydro-, → α-methyl-dodecanedioic acid 37, 38.
—, —, and nitric acid 37.
Flavacid 46, 66.
—, spectrum 21, 57.
Flavobacterium polyglutamicum, Konjugase 190.
FMN (Abkürzung für Flavin-mono-nucleotid) 184.
Folat-Antagonisten, Blockierung der Folatreduktase-Reaktion 221.
—, Dihydro- und Tetrahydro-Stufen 222.
—, Wirkungsmechanismus 221.
Folat-Coenzyme, Reaktions-Cyclen 253.
Folat-Cofaktor, mit gebundenem Formaldehyd 187.
Folat-Cofaktor-abhängige Enzym-Reaktionen 234, 235.
Folat-Enzyme 183.
Folat-Formylase-Reaktion, Push-pull-Mechanismus 250.

Folat-gebundene Kohlenstoff-Einheiten 186.

Folat-katalysierte Bildung der Methylgruppe von Methionin 242.

Folat-katalysierte Enzym-Reaktionen, im Stoffwechsel 236.

Folat-katalysierte Vorgänge und Oestrogene 199.

Folat-Reduktase 197, 198, 234.

—, Hemmung durch Folsäure-Antagonisten, Atebrin, Riboflavin 198, 199.

Folat-Verbindungen, Spektren 222, 223.

—, tetrahydro-, Asymmetrie und Wuchsstoffaktivität 227.

Folate, $C_{(9)}$- oder $N_{(10)}$-substituiert, und Reduktase 198.

Folate, tetrahydro-, formyl-, Bildung, Umwandlung 210.

—, —, —, Hydrolysenenergie, LCAO-Berechnung 214, 215.

—, —, —, → 5,10-Methinyl-tetrahydrofolsäure (Cyclisierungs-Geschwindigkeit) 213.

Folinic acid = Citrovorumfactor 211.

Folinic acid-SF 211.

Folinsäure-Isomerase 236.

Folsäure 183, 188, 210.

—, → p-Amino-benzoylglutaminsäure und Dihydropterin-6-aldehyd → Dihydropterin-6-carbonsäure und -6-carbinol 202.

—, = [(2-Amino-4-oxy-6-pteridyl)-methyl]-p-aminobenzoyl-L-glutaminsäure 201.

—, Bildung in Mikroorganismen 191.

—, biol. Abbau 195.

—, Biosynthese 192.

—, —, Dihydropteroinsäure als Primärprodukt 195.

—, —, Einfügen der Ribose 193.

—, → Citrovorumfaktor 199.

—, Desaminierung zu 2,4-Dihydroxypteroylglutaminsäure 197.

—, enzymat. Reduktion zum Cofaktor 197.

—, Formylierung 189, 209, 247.

—, $N_{(10)}$-Formylierung, aerobe 197.

—, → 5-Formyl-tetrahydrofolsäure (Leber-Enzym) 199.

—, gemischte 189.

—, Geschichte 187.

—, mit zusätzlichen Glutamyl-Resten = Konjugate 189.

Folsäure, IR- und NMR-Spektrum 201.

—, Isolierung aus Leber, Hefe, Spinat 199, 200.

—, künstliche Analoge 220.

—, Mangel-Anämien 191.

—, (dl), mikrobiol. Aktivität 200.

—, Nomenklatur, Numerierung 200, 201.

—, photochem. Spaltung 202.

—, → Pterin-6-aldehyd und p-Aminobenzoylglutaminsäure (biol.) 195.

—, → Pterin-6-carbonsäure 202.

—, = Pteroylglutaminsäure 201.

—, Reduktion zu Di- und Tetrahydrofolsäure 197, 204.

—, Reinigung 204.

—, Spektrum 222, 223.

—, Stoffwechsel, Herkunft der Ameisensäure 229.

—, Struktur 200, 201.

—, substituiert mit C_1-Körpern 209.

—, Synthese 203.

—, als Vitamin 187.

—, und Vitamin B_{12}, Zusammenhänge 244.

—, und Vitamin B_{12}-Mangel 191.

—, Vorkommen, Bedarf, Ausscheidung 190, 191.

—, Wuchsstoff-Aktivität 226.

—, dihydro- 204, 207.

—, —, enzymat. Reduktion 206.

—, —, ⇄ Folsäure 204, 205.

—, —, Problem von Tautomeren 207.

—, —, Spektrum 222, 223.

—, —, → Tetrahydrofolsäure 197.

—, —, /tetrahydro-, Redoxpotential 208.

—, —, Wuchsstoff-Aktivität 226.

—, 5,6-dihydro- 205.

—, —, 5-methyl- 207, 220.

—, —, —, als Methionin-Methyl-Donator 243.

—, 5,8-dihydro- 205.

—, 7,8-dihydro- 205.

—, —, Biogenese 192.

—, —, 10-formyl-, in Leberhomogenaten 247.

—, 10-methyl- 220, 221.

—, 10-nitroso- 202.

—, tetrahydro- → p-Aminobenzoylglutaminsäure und Formaldehyd (Luft) 208.

—, —, Bildung, Umwandlung 210.

—, —, Cyclodeaminase-Reaktion 215, 216.

Folsäure, tetrahydro-, → 5,6-dihydro- → 7,8-dihydro- 206.

—, —, und Formaldehyd, Transhydroxy-Methylierungs-Reaktion 218.

—, —, Kondensation mit Glycxylsäure 217.

—, —, Luftoxydation 208.

—, —, → Methylen-tetrahydro-folsäure, spektrale Änderung 216, 217.

—, —, Schlüsselsubstanz des Stoffwechsels 254.

—, —, Spektrum 208, 222, 223.

—, —, Stabilisierung 208.

—, —, Wuchsstoff-Aktivität 226.

—, 5,6,7,8-tetrahydro, Biogenese 192.

—, —, aus Folsäure 207.

l,L-Folsäure, tetrahydro- (natürliche) 208.

dl-L-(+)-Folsäure, tetrahydro-, Diastereomeren-Gemisch 208.

Folsäure, tetrahydro-, 5-formimino- 215, 247, 249.

—, —, —, → Methinyl-tetrahydrofolsäure und NH_3 247.

—, —, formylanhydro-, → Methylen-tetrahydrofolsäure 217.

—, —, 5-formyl- 210, 211.

—, —, —, biol. Abbau (Deformylierung, Transformylierung) 195, 196.

—, —, —, Diastereoisomeren, Trennung 224.

—, —, —, und Isomerase 248.

—, —, —, Polarographie 204.

—, —, —, Spektrum 223.

—, —, —, Stabilität 205.

—, —, —, Wuchsstoff-Aktivität 226.

—, —, 5-formyl-10-nitroso- 212.

—, —, 10-formyl- 191, 209, 210.

—, —, —, ⇄ 5,10-anhydroformyl- (biol.) 252.

—, —, —, → 5-formyl- 211.

—, —, —, → 10-Formyl-folsäure 212.

—, —, —, Spektrum 223.

—, —, —, Wuchsstoff-Aktivität 226.

—, —, 5-hydroxymethyl- 216.

—, —, 5,10-methinyl- 210, 212.

—, —, —, chlorid, Spektrum 212.

—, —, —, Fluoreszenz 213.

—, —, —, → 10-Formyltetrahydrofol-säure → 5-Formyltetrahydrofolsäure 212, 249.

—, —, —, isomere Formen 213.

Folsäure, tetrahydro-, 5,10-methinyl-, Orthoamide als Zwischenstufe 215.

—, —, —, Spektrum 223.

—, —, —, Wuchsstoff-Aktivität 226.

—, —, 5-methyl- 191, 219.

—, —, —, (Bildung der Methylgruppe von Methionin) 242, 243.

—, —, —, → 5-Methyl-dihydrofolsäure 220.

—, —, —, = Prefolic A 220.

—, —, —, Spektrum 223, 224.

—, —, —, Wuchsstoff-Aktivität 220, 226.

—, —, 5,10-methylen- 216, 217, 242.

—, —, —, = „aktiver" Formaldehyd 218.

—, —, —, Dehydrogenase-Reaktion 218.

—, —, —, enzymat. Reaktionen 240.

—, —, —, Spektrum 223, 224.

—, —, —, Wuchsstoff-Aktivität 226.

Folsäure-aktive Substanzen, natürliche 188.

Folsäure-Analoge, künstliche, Struktur 220.

—, —, mit 2,4-Diaminopyrimidin- oder 2,4-Diaminotriazin-Ring 221.

—, —, Hemmung von *St. faecalis* 221.

Folsäure-Coenzyme in Stoffwechsel-Systemen 185.

Folsäure-Cofaktoren, Auf- und Abbau 192.

—, Funktion 187.

—, Reaktionen 253.

Folsäure-Konjugate 189.

—, Geschichte 187.

—, Synthese aus *p*-Aminobenzoyl-*γ*-glut-amyl-glutaminsäureester 204.

—, = Pteroyl-*γ*-glutamyl-glutamate 200.

Folsäure-Konjugat-spaltende Enzyme 190.

Folsäure-Verbindungen, Analyse, Trennung, Bestimmung 225.

—, im Blut, Chromatographie 228.

—, Chromatographie, Elektrophorese 227, 228, 229.

—, Differenzierungs-Analyse mit *L. casei*, *St. faecalis*, *P. cerevisiae* 226.

—, Einkohlenstoff-Reservoir 229, 230.

—, Fluoreszenz 224.

—, mikrobiol. Wachstumstest 226.

—, optisch aktive 224.

—, Polarographie 225.

—, Wachstumstest mit *B. coagulans* und *T. geleii* 226, 227.

Formaldehyd (Einkohlenstoff-Reservoir) 230.
—, gebunden (Folat-Cofaktoren) 187.
—, gebunden an Folat 216, 217.
—, gebunden an Tetrahydrofolsäure (Transportmetabolit) 233.
Formaldehyd-Oxydation (biol.), Energie-Koppelung 251.
Formamino-Transferase 247.
Formenonetin, biosynthesis 307.
Formiat-aktivierendes Enzym 250.
Formimino-asparaginsäure, enzymat. Umsatz 247.
N-Formimino-L-glutamat: Tetrahydrofolsäure-5-Iminotransferase 234.
Formimino-glutaminsäure 215.
Formimino-glycin 215.
—, enzymat. Umsatz 247.
Formiminoglycin-tetrahydrofolat-Transferase aus *Clostridium* 249.
N-Formiminoglycin: Tetrahydrofolsäure-5-Iminotransferase 234.
5-Formimino-tetrahydrofolat-Ammoniak-lyase (cyclisierend) 234.
Formimino-tetrahydrofolat-Cyclo-desaminase 234, 249.
Formimino-Übertragung auf Tetrahydrofolsäure 249.
Formylase 249.
—, aus Erythrozyten, aus Leukozyten 251.
—, aus *Micrococcus aerogenes* 250, 251.
Formyl-folat-Reduktase 239.
Formyl-glutamat, enzymat. Umsatz 247.
Formyl-glycinamidin-ribotid 252.
10-Formyl-tetrahydrofolat-Amidohydrolase 234.
Formyl-tetrahydrofolat-Deacylase 234.
Formyl-tetrahydrofolat-Deformylase 234.
5-Formyl-tetrahydrofolat-Glutaminsäure-Transferase 248.
Formyl-tetrahydrofolat-Ligase 234.
Formyl-tetrahydrofolat-Reduktase 234.
Formyl-tetrahydrofolat-Synthetase 234.
Formyl-Transferase 236.
L-Fucose, 2-desoxy-, aus Cinerubin A, B 175.
—, —, in γ-Rhodomycinen 178.
—, —, Struktur 171.
Fungichromatin 62.
Fungichromin 64.
—, → 2-L:3-D *(erythro)*-1,2,3-butanetriol 36.

Fungichromin, chromophore 32.
—, → 7,21-dimethyl-tritriacontane 35.
—, esterified hydroxyl 36.
—, → hexanaldehyde 30.
—, → 2-n-hexyl-tridecanoic acid 34.
—, IR spectrum 40.
—, = Lagosin 18, 32.
—, macrocyclic ring 33.
—, → 2-methyl-dodecapentaenedial 32.
—, periodic acid fission 32.
—, a polyol 35.
—, spectrum 20, 24.
—, structure 34.
—, monohydrate 40.
—, perhydro-, deoxygenation 35.
Fungicidin A-94, properties 60.

GAR (Abkürzung für Glycinamid-ribotid) 234, 252.
Geraniol pyrophosphate, → farnesol pyrophosphate 10.
—, *(cis)*, (rubber biosynthesis) 10.
Glu (Abkürzung für Glutaminsäure) 184.
Glutamat-Formimino-Transferase 234.
Glutaminsäure, formimino- 215.
Glutamyl-Transferase 247.
Glutarsäure, β-(4-chlor-7-hydroxy-phthalid-3) 96.
Glycin, Abbau (biol.) 232.
—, Desaminierung 231.
—, → Formaldehyd und CO_2 (biol.) 232.
—, → Glyoxylat → Formiat (biol.) 231, 232.
—, Quelle von Einkohlenstoff-Körpern (Folat-Stoffwechsel) 231.
—, ⇄ Serin 231.
—, formimino- 215.
Glycinamid-ribotid → Formyl-glycinamid-ribotid 252.
Glycinamid-ribotid-Transformylase 234, 252.
Glycin-Formimino-Transferase 234.
Goodhue test (rotenoids) 278.
Guanin (Einkohlenstoff-Reservoir) 230.
Guanosin → Dihydrofolsäure 193.
Guayule, rubber 2, 5.
Gutta, *trans* configuration 2.
Guttapercha 2.

Haiari (insecticide) 276.
Hamycin 70.
—, polypeptide moiety 51.
—, from soil *Streptomyces* 51.

Hamycin, spectrum 57.
Helixin A, B 58, 66.
Helminthosporin, Biogenese 167.
Heptadecanedioic acid, 2-methyl- 44.
Heptaenes (polyene antibiotics) 46, 66.
—, antifungal potencies 46.
—, spectrum 21, 24, 57.
cis-Heptaenes (polyene antibiotics), spectrum, *cis*-peak 25.
—, stereoisomerization 25.
Heptaene (polyene antibiotics), from *Str. abikoensis* 68.
—, No. 757 57, 68.
—, 26/1 57, 70.
—, AE-56 70.
—, F-17-C 26, 57, 70.
—, 2814-H 70.
—, PA-150 57, 68.
Heptaglutamat (Folsäure) 189.
Heptamycin 68.
Hevea brasiliensis, rubber 2.
—, section through phloem in bark 4.
—, tapping 3.
Hexaenes (polyene antibiotics) 45, 66.
—, spectrum 21, 24, 57.
Hexose, tridesoxy-dimethylamino- 124.
Histidin, Abbau und Folsäure-Coenzyme 185.
—, biol. Abbau 246.
—, (Einkohlenstoff-Reservoir) 230.
Hoesch reaction (rotenoid synthesis) 312, 314.
Homocystein, S-hydroxymethyl- 243.
—, → Methionin (Beteiligung von Tetrahydrofolsäure, Vitamin B_{12}) 233.
—, als Methyl-Acceptor 245.
—, N-acetyl-, als Methyl-Acceptor 245.
Huhn, Wuchsstoff 188.
Hühnerleber, formylierte Folsäure-Konjugate 189.
Hundeleber, Konjugate 190.
β-Hydroxy-β-methyl-glutarate in plants 6.
Hydroxymethyl-tetrahydrofolat-Dehydrogenase 239.
Hydroxypyruvat, im Serin-Stoffwechsel 232.
—, 3-phospho-, im Serin-Stoffwechsel 232.

Imidazol-carbonsäureamid-ribotid, 4-amino- 253.
5-Imidazol-carboxamid-ribotid, 4-amino-, → Inosinsäure 251.

Imidazolidin, N,N'-diphenyl- 217.
Imidazolyl-glycerylphosphat 253.
Inosinicase 251.
IpPP (abbreviation for Δ^3-isopentenyl-pyrophosphate) 9.
Islandicin, Biogenese 167.
Iso-aureomycin 93.
—, Struktur 94.
—, desdimethylamino-12a-desoxy-, IR-Spektrum 91.
—, —, Struktur 92.
Iso-deguelin 296.
Iso-flavanoids, biogenesis, connection with rotenoid biogenesis 305.
Iso-flavones, biosynthesis 306.
—, synthesis 313.
Δ^3-Isopentenyl-pyrophosphate 9.
—, and dimethylallyl-pyrophosphate 10.
—, in latex rubber formation 9, 10, 12.
—, polymerization 10, 11.
—, in rotenoid biosynthesis 308.
Isopentenyl-pyrophosphate isomerase 10.
Isoprenoids, acetate precursors 5.
—, distribution in plants 1.
η-Iso-pyrromycinon = Bisanhydro-ε-iso-rhodomycinon 147, 149.
—, aus η-Pyrromycinon 147.
—, descarbomethoxy- 147.
—, —, Struktur 149.
Iso-rhodomycine, Gewinnung aus *Streptomyces purpurascens* 123, 126.
—, Isolierung 127.
Iso-rhodomycin A 123, 127, 178.
—, Hemmung von *Staphylococcus aureus* 179.
—, → β-Iso-rhodomycinon 178.
Iso-rhodomycin B aus *Streptomyces purpurascens* 179.
Iso-rhodomycinone 122.
—, Gemisch mit Rhodomycinon, Auftrennung 126.
—, und Glykoside, Gewinnung 125.
—, α-Hydroxy-Gruppen 128.
—, Konstitution 129, 137.
—, lineare Anellierung des alicyclischen Ringes 133.
—, Ring *A*, Aromatisierung 134.
—, Spektrum 131.
—, Stellung der Äthyl- und Carbomethoxy-Gruppen 136.
—, aus *Streptomyces* spp. 123, 126.

β-Iso-rhodomycinon 123, 127, 140.
—, Biogenese 167.
—, Konstitution 140.
—, bisanhydro-, Konstitution 140.
γ-Iso-rhodomycinon 123.
—, Biogenese 167.
—, Konstitution 140.
ε-Iso-rhodomycinon 123, 151.
—, → β-Äthyl-adipinsäure 138.
—, Biogenese 167.
—, Konstitution 137, 138, 139.
—, KMR-Spektrum 138, 155, 156.
—, Reduktion zu 1,4,5,8-Tetrahydroxy-
    anthrachinon-Deriv. 137, 138.
—, Spektrum 132.
—, Verseifung 152.
—, Zirkulardichroismus-Kurven 161, 162.
—, bisanhydro- 138.
—, —, = η-Iso-pyrromycinon 139, 147,
    149.
—, —, Spektrum, bathochrome Wirkung
    der Carbomethoxy-Gruppe 138.
—, —, Struktur 138, 139.
—, tetramethyläther, und 2,2-Dimeth-
    oxypropan 163.
ζ-Iso-rhodomycinon 123, 139.
—, Biogenese 167.
—, Konstitution 139.
—, Zirkulardichroismus-Kurven 161, 162.
Iso-rotenolones, borohydride reduction
    292.
—, formation by aerial oxidation 293.
—, fusion of B/C rings 291.
—, opt. activity 291, 292.
—, opt. active cis-B/C, and borohydride
    292.
—, stereochemistry 290.
($\pm$)-Iso-rotenolone I = ($\pm$)-Iso-roteno-
    lone A 290.
($\pm$)-Iso-rotenolone II, mixture 290.
($\pm$)-Iso-rotenolone A 290.
—, cis-fusion 290.
—, IR spectrum 290.
—, → rotenol and derritol 290.
—, methyl ether 291.
($\pm$)-Iso-rotenolone B 290.
—, and borohydride, → diols 291.
—, IR spectrum 290.
—, → rotenol and derritol 290.
—, $trans$-fusion 291.
—, methyl ether 291.
($\pm$)-Iso-rotenolone C 294, 295.

($\pm$)-Iso-rotenolone C, degradation 295.
—, NMR data 295.
—, → rotenol and derritol 295.
Iso-rotenolone D 294.
$6\,a\,\alpha,12\,a\,\alpha$-Iso-rotenolone 292.
$6\,a\,\beta,12\,a\,\beta$-Iso-rotenolone 292.
Iso-rotenone, B/C fusion of rings 283.
—, from rotenone 281.
—, structure 283.
—, 6a,12a-dehydro- 290.
(—)-Iso-rotenone 291.
($\pm$)- and (—)-Isorotenone → ($\pm$)-Iso-
    rotenolone I and II 290.
($+$)-$6\,a\,\alpha,12\,a\,\alpha$-Iso-rotenone 291.
(—)-$6\,a\,\beta,12\,a\,\beta$-Iso-rotenone 291.
Iso-tephrosin 296, 297.
Iso-terramycin, desdimethylamino-12 a-
    desoxy- 92.
Iso-tetracyclin, chlor- 105, 106.
Iso-tubaic acid, synthesis 310.

α-Ketosäuren (Einkohlenstoff-Reservoir)
    230.
KMR-Spektrum (NMR) (Anthracycli-
    none) 130, 155, 156.
Konjugase (Folsäure-Konjugate) 190.
Kupferoxinat, Entgiftung durch Pterine
    193.

$Lactobacillus$ $casei$, Wuchsstoff 188, 189.
$L.$ $casei$-Faktor = 5-Methyl-tetrahydro-
    folsäure 246.
Lagosin 64.
—, chromophore 32.
—, = fungichromin 32.
—, → hexanaldehyde 30.
—, IR spectrum 40.
—, a macrocyclic lactone 27.
—, → 2-methyl-dodecapentaenedial 32.
—, periodic acid fission 32.
—, structure 34.
Latex (rubber tree), composition 2.
—, enzymes, rubber synthesizing 13.
—, —, —, formation in ribosomes 13.
—, incubated with 2-C^{14}-mevalonic acid 8.
—, Δ^3-isopentenyl-pyrophosphate 12.
—, regulation of rubber concentration 12.
Leber-$L.$ $casei$-Faktor 188.
$Leuconostoc$ $citrovorum$, Wuchsstoff 188.
Leucovorin 211.
—, anhydro- 212.
—, chlorid, anhydro- 213.
Leucovorin A, anhydro- 213.

Levulinic acid from rubber 8.
Lewatit KSN 124.
Liver, polymerization of Δ^3-isopentenyl-pyrophosphate 10, 11.
Lonchocarpus spp., fish poison, insecticides 275, 276.
L. nicou, cryst. rotenone 276.
Lucensomycin 58.
L-Lyxohexose, 2,3,6-tridesoxy-3-dimethylamino- 173.

Macrolides, biosynthesis 53.
—, elimination of sugar by alkali 29.
Macrolide antibiotics 27.
Magnamycin, biosynthesis 52.
Malacccl 303.
—, from *Derris malaccensis* 302.
—, structure 277, 278, 302, 303.
—, dehydro-tetrahydro-, synthesis 313.
—, tetrahydro-, synthesis 312.
Mannich-Reaktion (Folat-Deriv.) 218.
Mediocidin 66.
—, spectrum 57.
Meijer test (rotenoids) 279.
5,10-Methinyl-tetrahydrofolat-Cyclohydrolase 247, 249.
Methionin, Biosynthese 219, 240, 243.
—, —, in *E. coli* 244, 245.
—, —, und Folsäure-Coenzyme 185.
—, —, im Säugetier 244, 245.
—, not utilized in biosynthesis of nystatin 54.
—, im Stoffwechsel, Herkunft der Methyl-Gruppe 240, 243.
Methionin-Synthetase 236.
β-Methylcrotonic acid (rubber biosynthesis) 7.
Methyl-Gruppe, Acceptoren (biol.) 245.
—, Bildung durch Hydridverschiebung 242.
—, labile (biol.) 233.
Methylenfolat-Reduktase 242.
Methylen-tetrahydrofolat-Dehydrogenase 234, 239.
5,10-Methylen-tetrahydrofolat: NADP-Oxydoreduktase 234.
Methylen-tetrahydrofolat-Reduktase 236, 242.
Methylpentaenes (polyene antibiotics), spectrum 24, 64.
Methylpentaene, Glaxo A-246 64.
—, from *Str. sanguineus* 64.

Methymycin 52.
—, elimination of sugar by alkali 29.
Mevalonic acid, incorporation into cholesterol 7.
—, $\rightarrow$ isopentenyl and dimethylallyl pyrophosphates 9.
—, metabolized to simple terpenes, carotenes, squalene 7.
—, $\rightarrow$ monomer for isoprenoid polymerization (yeast) 9.
—, not utilized in biosynthesis of nystatin 54.
—, $\rightarrow$ 5-phosphate 9.
—, in rubber biosynthesis 7.
—, pyrophosphate 9.
Mevalonic acid kinase 9.
Micrococcus pyogenes aureus, Hemmung durch Cinerubine 176.
Mikroorganismen, Bildung von Folsäure 191.
Millettia spp., fish poison 275.
Mimusops balata, gutta 2.
Moldcidin A 40, 64.
—, spectrum 57.
Moldcidin B 64.
—, = pentamycin 18, 40.
Mundulea spp., fish poison 275.
M. servica, Munduserone 298.
Mundulone, structure 298, 299.
Munduserone 298.
—, from dehydro compound 314.
—, dehydrogenation, structure 298.
—, and natural relatives 298.
—, $\rightarrow$ β-resorcyclic acid 4-methyl ether and rissic acid 298, 299.
—, structure 277, 278.
—, dehydro-, synthesis 313.
Mutarotenone 286, 287.
—, from dehydro compound 314.
Mycaminose, epimeric with mycosamine 28.
Mycetin-Violarin-Gruppe, Antibiotica 179.
Mycobacterium avium, Folsäure-Synthese 194.
—, Katalyse der Bildung von p-Aminobenzoyl-glutaminsäure 194.
M. Tbc, Hemmung durch Cinerubine 176.
Mycosamine, configuration 28.
—, from nystatin, pimaricin, candidin, trichomycin, etc. 28.
—, structure 28.

Mycosamine-containing antibiotics → NH_3 29.
Mycoses, systemic, and amphotericin B 18.

NAD und NADP (Abkürzungen für Di- und Triphospho-pyridin-nucleotid) 184.
Neodulin, in *Phaseolea* 309.
—, structure 308.
Neorautanenia spp., fish poison 275.
N. edulis, phenolics 309.
N. pseudopachyrrhiza, dolineone, neotenone, phenolics 304, 309.
Neotenone 308.
—, occurrence, structure 304, 305.
—, dehydro- 308.
Nepseudin, an iso-flavone 304.
—, in *Phaseolea* 309.
Nerol pyrophosphate *(trans)*, (rubber biosynthesis) 10.
Netoric acid, synthesis 309.
—, dehydro- 301.
—, —, structure 288.
—, —, synthesis 309.
—, hydroxy- 301.
Nicouic acid, structure 296.
Nicouline 276.
NMR spectrum, see KMR-Spektrum.
Nocardia acidophilus, acetate → mycomycin 54.
Norit-Eluat-Faktor 188, 189.
Nor-rotenone, dehydro-dihydro, synthesis 313.
Nystatin 45.
—, biogenetic tracer experiments 54.
—, chromophore 43, 44.
—, cleavage with alkali 45.
—, medical use 18.
—, → 2-methyl-heptadecanedioic acid 43, 44.
—, oxidative fission 44.
—, partial formula 45.
—, properties 58.
—, spectrum 20, 43.
—, in *Str. noursei* 19.
—, → succinic acid 43.
—, → tiglic aldehyde 44.
—, perhydro- 43.

1,3,5,7-Octatetraene, spectrum 56.
Oestrogene, und Folat-katalysierte Vorgänge 199.

(+)- and (—)-Oxazolidone, 5-isopropyl- (stereochemistry of rotenone) 280.
Oxytetracycline in *Str. rimosus* 19.

P_a (Abkürzung für Orthophosphat) 184.
pAB (Abkürzung für p-Aminobenzoesäure) 184.
Pachyrrhizone 303.
—, degradation to coumaran deriv. 304.
—, from *Pachyrrhizus erosus* 303.
—, in *Phaseolea* 308.
—, reactions 304.
—, structure 277, 278, 303, 308.
Pachyrrhizus erosus, dolineone, erosone, pachyrrhizone, phenolics 303, 304, 309.
Palaquium gutta, gutta 2.
Pankreas, Konjugase 190.
β-Parinaric acid, spectrum 23, 56.
Parthenium argentatum (rubber biosynthesis) 1, 5.
Pasteurella pestis, und Cinerubine 175.
Pediococcus cerevisiae, Permease 199.
—, Wuchsstoff 188, 189, 190.
Pentaenes (polyene antibiotics) 45, 62.
—, spectrum 20, 24, 56.
Pentaene from *Str. effluvius* 62.
—, spectrum 56.
Pentaene (polyene antibiotics), 2814-P 46, 62.
—, No. 83 and 90 62.
—, PA-153 56, 62.
—, —, spectrum 20.
Pentaenes, methyl- (polyene antibiotics), spectrum 24.
—, —, properties 64.
Pentamycin 64.
—, identity (?) with fungichromin 39, 40.
—, UV and IR spectrum 40, 57.
(+)-Pentanoic acid, 3-hydroxy-4-methyl- 280, 281.
—, —, from (—)-dihydrotubaic acid, relation to D-glyceraldehyde 280.
—, —, (stereochemistry of rotenone) 281.
(—)-Pentanoic acid, 3-hydroxy-4-methyl- 280.
(—)-Pentan-3-ol, 2-methyl- 280.
γ-Peptidase 190.
Perimycin 70.
—, absence of carboxyl 27, 51.
—, → p-amino-phenylacetone 30, 51.
—, → perosamine 29, 51.
—, spectrum 57.

Perosamine, from perimycin 29.
Phosphomevalonic acid kinase 9.
5'-Phosphoribosyl-aminoimidazol-carbox-amid-Transformylase 234.
5'-Phosphoribosyl-5-formamido-4-imid-azol-carboxamid: Tetrahydrofolsäure-10-Formyltransferase 234.
5-Phosphoribosyl-N-formylglycinamid: Tetrahydrofolsäure-Formyltransferase 234.
5-Phosphoribosyl-glycinamid-Formyl-transferase 234.
Phthalid, 3-methyl-3-chlor-7-hydroxy- 91.
—, —, Spektrum 92.
Physalia physalis, formylierte Folsäure-Konjugate 189.
Pimaricin 40, 60.
—, alkaline cleavage 31.
—, → ammonia 41.
—, decarboxylation 40.
—, epoxide ring 40.
—, → 13-hydroxy-tetradecapentaenal 31, 41.
—, → mycosamine 41.
—, structure 42.
—, = tennecetin 18.
—, UV and IR spectrum 27, 40, 42, 43, 56.
—, N-acetyl-, and periodic acid 41.
—, —, perhydro- 41.
—, perhydro-, → pimelic acid 41.
—, —, decarboxylation, dehydration 40, 42.
Pimelinsäure, pteroyl- 221.
Polyacetate chains (biosynthesis of natural products) 52.
Polyacetylenes, biosynthesis 53.
Polyamid-Pulver (Rhodomycinon/Iso-rhodomycinon-Chromatographie) 127.
Polyene antifungal antibiotics 17, 18.
—, basic N, zwitterion nature 27.
—, biogenetic relationships 51.
—, classification (tetraenes, pentaenes, hexaenes, heptaenes) 18.
—, formation from acetaldehyde (via acetyl-coenzyme A) 52.
—, glycoside link, scission 29.
—, heptaenes 46, 66.
—, *cis*-heptaenes, stereoisomerization 25.
—, hexaenes 45, 66.
—, lactone group (IR) 27.
—, methylpentaenes 39, 64.
—, —, IR spectrum 24.

Polyene antifungal antibiotics, pentaenes 45, 62.
—, *poly*hydroxy compounds 27, 28.
—, purification, structure 19, 26.
—, retro-aldol cleavage 30.
—, spectra 19, 20, 21, 22, 23, 24, 25, 56, 57.
—, tetraenes 43, 58.
Polyenynes, fungal, antibiotic properties 53.
Präfolsäure A 242.
Prefolic A = 5-Methyl-tetrahydrofolsäure 220.
—, Wuchsstoff-Aktivität 226.
Propion- und Essigsäure in Biogenese (Anthracyclinone) 164.
Protocidin 58.
—, spectrum 56.
Proto-toxicarol = (—)-α-toxicarol 299.
Pteridin, 6-aldehyd, 2-acetamido-4-oxy- 204.
—, 2-amino-4-hydroxy-, Baustein der Folsäure 200.
—, —, 6-carboxy- 193.
—, —, 6-formyl- 192.
—, 6-carbinol, 7,8-dihydro-, 2-amino-4-hydroxy- 192.
—, 6-carbinol-diphosphat, 7,8-dihydro-, 2-amino-4-hydroxy-, in Folsäure-Bio-synthese 192, 195.
—, dihydro- 193.
—, 4,7-dihydroxy- → 4,7-Dihydroxy-5,6-dihydropterin 205.
—, tetrahydro-, 2-amino-4-hydroxy-5-formyl-6-methyl- 211.
Pterine, Bildung in Schmetterlingsflügeln 193.
—, biol. Abbau 196.
—, 6-aldehyd, und Folat-Reduktase 198.
—, —, und Pteroylglutaminsäure-Bio-synthese 192.
—, —, dihydro- 202.
—, 2-amino-4-hydroxy-6-hydroxymethyl- 194.
—, 6-carbinol → Folat-aktive Substanzen 194.
—, —, dihydro-, → 6-Methyl-pterin 202.
—, —, 7,8-dihydro-, 2-amino-4-hydroxy-, Pterin-Donator in Folsäure-Biosyn-these 194.
—, 6-carbonsäure, Folatbildung 193.
—, —, aus Folsäure 202.

Pterine, 2,4-diamino-, als Hemmstoff 221.
—, 6-formyl-, Hemmstoff der Purin- und Pterin-Oxydation 197.
—, 6-phenyl-2,4,7-triamino-, als Hemmstoff 221.
—, tetrahydro-, 6,7-dimethyl- → 5,6-Dihydro-Verbindung 206.
Pteroinsäure 201.
—, und Folat-Reduktase 198.
—, → Rhizopterin 209.
—, Spektrum 223.
—, Wuchsstoff-Aktivität 226.
—, dihydro-, Primär-Produkt in Folsäure-Biosynthese 195.
—, 7,8-dihydro- 192.
—, 10-formyl- 209.
Pteroyl-adipinsäure 221.
Pteroyl-asparaginsäure 220, 221.
Pteroyl-glutaminsäure, Biogenese 192.
—, = Folsäure 188, 201.
—, 2,4-dihydroxy-, aus Folsäure 197.
—, tetrahydro-5-formyl- 188.
Pteroyl-heptaglutaminsäure 188.
—, Wuchsstoff-Aktivität 226.
Pteroyl-Konjugate, Biosynthese 195.
Pteroyl-pimelinsäure 221.
Pteroyl-triglutamat 188.
—, Wuchsstoff-Aktivität 226.
—, 5-formyl-, aus *E. coli* und *B. subtilis* 190.
—, tetrahydro-, Wuchsstoff-Aktivität 226.
Pumilin, from *B. pumilis* 51.
Purine, Biogenese, Rolle der Ameisensäure 229, 249.
—, biol. Endabbau 246.
—, Biosynthese, und Folsäure-Coenzyme 185.
—, Einbau von Formiat 252.
—, enzymat. Aufbau 251.
—, Gärung, und Folsäure-Coenzyme 185, 248.
—, → Pterine (in vitro) 193.
Purin-Stoffwechsel, aktivierte Ameisensäure 248.
Push-pull-Mechanismus 238, 250.
Pyrimethamin, als Hemmstoff 221.
Pyrimidin, 2,4-diamino-, Deriv., als Hemmstoff 221.
—, 2,4-diamino-5-*p*-chlorphenyl-6-äthyl-, als Hemmstoff 221.
—, 2,4-diamino-5-formamido- 211.

Pyrimidin, 2,4,5-triamino-6-hydroxy- (TAOP) 203.
—, —, → N-[(2-Amino-4-hydroxy-6-pteridyl)-methyl]-pyridiniumjodid 203.
—, —, Kondensation mit *p*-Aminobenzoyl-*L*-glutaminsäure und C₃ zur Folsäure 203.
—, —, Kondensation mit α,β-Dibrompropionaldehyd und *p*-Aminobenzoylglutaminsäure 203.
Pyrromycin 123.
—, aus Cinerubin 175.
—, Hemmwirkung 174.
—, Isolierung aus *Streptomyces* 124, 125, 174.
—, → ε-Pyrromycinon und Rhodosamin 174.
—, Struktur 174.
—, tetraacetat 174.
Pyrromycinone 121, 122.
—, Gewinnung 124.
—, und Glykoside 121.
—, α-Hydroxygruppen 128.
—, KMR-Spektrum 157.
—, Konstitution 129, 146.
—, Spektrum 132.
—, Stellung der Äthyl- und Carbomethoxy-Gruppen 136.
—, aus *Streptomyces* spp. 123, 124, 125.
—, Übersicht 123.
ε-Pyrromycinon 123, 125.
—, Aglykon von Cinerubin A, B, Pyrromycin 151.
—, Anthracycline 174.
—, Biogenese 164.
—, KMR-Spektrum 151, 155, 157.
—, Konstitution 149, 151.
—, und Pyroboracetat 151.
—, → ζ-Pyrromycinonsäure 152.
—, → η-Pyrromycinon 146, 151.
—, aus *Streptomyces*-Kulturen 151.
—, Verseifung 152.
—, Zirkulardichroismus-Kurven 161, 162.
ε-Pyrromycinon-Glykoside, Gewinnung 124.
ζ-Pyrromycinon 123, 125.
—, Biogenese 167.
—, KMR-Spektrum 152, 155, 157.
—, Konstitution 149, 151.
—, → ζ-Pyrromycinon-HBr-Produkt 152.

ζ-**Pyrromycinon,** → η-Pyrromycinon 151.
—, reduzierend acetyliertes, Spektrum 147.
—, Spektrum 152.
—, aus *Streptomyces*-Kulturen 151.
—, Wasser-Eliminierung und Konfiguration 163.
—, Zirkulardichroismus-Kurven 161, 162.
—, acetat, Spektrum 148.
—, anhydro- 152, 153.
—, HBr-Produkt, → Descarbomethoxy-η-pyrromycinon 152.
—, —, Struktur 149, 152.
—, P_2O_5-Produkt, Struktur 145.
—, —, KMR-Spektrum 153, 157.
—, —, UV- und IR-Spektrum 153.
—, —, Zirkulardichroismus-Kurven 161.
—, PTS-Produkt 149, 153.
—, —, Zirkulardichroismus-Kurven 161.
ζ-**Pyrromycinonsäure** → ζ-Pyrromycinon 152.
η-**Pyrromycinon** 123, 125.
—, = Bisanhydro-rutilantinon 150.
—, → η-Iso-pyrromycinon 138, 147.
—, Konstitution 146, 149.
—, und Pyroboracetat, Fluoreszenz 147.
—, aus ε-Pyrromycinon 146, 151.
—, aus ζ-Pyrromycinon 151.
—, reduzierend acetyliertes, Spektrum 147.
— Spektrum 147, 148.
—, aus *Streptomyces* spp. 146.
—, acetat, Spektrum 148.
—, descarbomethoxy-, Synthese, Spektrum 147, 148, 149, 151.
—, —, → Benzol-tricarbonsäure-(1,2,4) 150.
η-**Pyrromycinonsäure** 147.
Pyruvat, phosphoroklastische Spaltung 230.
—, hydroxy-, im Serin-Stoffwechsel 232.
—, —, 3-phospho-, im Serin-Stoffwechsel 232.

Quatrimycine 100.

β-**Resorcyclic acid** 4-methyl ether, from munduserone 298.
Retro-aldol cleavage, polyene antifungal antibiotics 30, 31.
Rhesusaffe, Wuchsstoff 188.
Rhizopterin 209.
—, Wuchsstoff-Aktivität 226.

Rhizopus nigricans, Rhizopterin 209.
Rhodinose, aus γ-Rhodomycinen 173, 178.
—, KMR-Spektrum 174.
—, Struktur 173.
Rhodomycine 123.
—, Gewinnung aus *Streptomyces purpurascens* 126.
—, Isolierung 127.
Rhodomycin A, B 123, 127.
Rhodomycin A, antibiotische Wirkung 177.
—, → Rhodomycin B 177.
—, aus *Streptomyces purpurascens* 176.
—, hydrochlorid, → β-Rhodomycinon und Rhodosamin 176, 177.
Rhodomycin B, Hemmung von *Staphylococcus aureus* 177.
—, → β-Rhodomycinon und Rhodosamin 177.
γ-**Rhodomycin I—IV** 123.
—, Baustein-Analyse (γ-Rhodomycinon, Rhodosamin, 2-Desoxy-*L*-fucose, Rhodinose) 178.
Rhodomycinone, Anthracycline 176.
—, Gemisch mit Iso-rhodomycinon, Auftrennung 126.
—, und Glykoside 121, 125.
—, α-Hydroxy-Gruppen 128.
—, KMR-Spektrum 155.
—, Konstitution 129, 140.
—, Spektrum 132.
—, Stellung der Äthyl- und Carbomethoxy-Gruppen 136.
—, aus *Streptomyces* spp. 122, 123, 126.
β-**Rhodomycinon** 123, 127.
—, Biogenese 167.
—, → Desoxyverbindung 143.
—, Entfernung von Hydroxy-Gruppen 143.
—, KMR-Spektrum 155, 156.
—, Konstitution 141, 142.
—, relative Konfiguration 162.
—, Stellung der Hydroxyle 143.
—, Zirkulardichroismus-Kurven 161.
—, bisanhydro-, Konstitution 142.
—, —, Spektrum 143.
—, Desoxyverbindung, KMR-Spektrum 145.
—, HJ-Produkt = Descarbomethoxy-bis-anhydro-ε-rhodomycinon 142, 143, 144.
γ-**Rhodomycinon** 123, 178.
—, Biogenese 167.

γ-Rhodomycinon, KMR-Spektrum 155, 156.
—, relative Konfiguration 162.
—, aus *Streptomyces purpurascens* 145.
—, Struktur 142, 145.
—, Zirkulardichroismus-Kurven 161.
—, bisanhydro-, Struktur 142.
δ-Rhodomycinon 123.
—, KMR-Spektrum 155, 156, 157.
—, Konstitution 145, 146.
—, aus *Streptomyces purpurascens* 145.
—, Zirkulardichroismus-Kurven 161.
—, bisanhydro- 145, 146.
—, —, Spektrum 146.
ε-Rhodomycinon 123.
—, Biogenese 167.
—, KMR-Spektrum 155, 156, 158.
—, Konstitution 140, 141, 142, 159.
—, Spektrum 133.
—, aus *Streptomyces purpurascens* 140.
—, Verseifung 152.
—, Zirkulardichroismus-Kurven 161, 162.
—, bisanhydro-, → Bisanhydro-ε-iso-rho-domycinon 141.
—, —, Konstitution 142.
—, —, Spektrum 140.
—, descarbomethoxy-bisanhydro- 142, 143.
ζ-Rhodomycinon 123.
—, Biogenese 167.
—, = hydriertes ε-Rhodomycinon 141.
—, Konstitution 141, 142.
—, aus *Streptomyces purpurascens* 141.
—, Zirkulardichroismus-Kurven 161, 162.
—, HBr-Produkt, Struktur 142.
—, —, → Descarbomethoxy-bisanhydro-ε-rhodomycinon 142.
Rhodosamin 124.
—, in Anthracyclinen 170, 171, 177, 178.
—, aus Cinerubin A, B 175.
—, → Dimethylamin, 2-Desoxy-*L*-fucose, *L*-Boivinose 171.
—, in γ-Rhodomycinen 178.
—, Struktur 171.
—, = 2,3,6-Tridesoxy-3-dimethylamino-*L*-lyxohexose 173.
—, acetate, KMR-Spektrum 172.
—, —, anomere 172.
Ribinia nicou, cryst. rotenone 276.
Riboflavin, Hemmung von Folat-Reduk-tase 199.
Rimocidin 45, 58.
Rimocidin, spectrum 56.
—, in *Str. rimosus* 19.
Rissic acid 296, 300.
—, structure 285.
—, synthesis 309.
Rogers and Calamari test (rotenoids) 278.
Rotenoids, biogenesis 305, 307.
—, biogenetic relationships with iso-flavonoids and related phenolics 308.
—, conformers of cis- and trans-B/C rings 282.
—, Cotton effect 295.
—, definition, classification 276, 278.
—, Durham test 278.
—, fish poisons, insecticides 275, 276.
—, Goodhue test 278.
—, isolation 278.
—, linear D/E rings, synthesis of derrisic type 312, 313.
—, Meijer test 279.
—, natural 275, 277.
—, nomenclature 279.
—, Rogers and Calamari test 278.
—, stereochemistry 295.
—, synthesis 309.
—, —, parent compound 313.
—, 6a,12a-dehydro-, → rotenoids 314.
—, —, synthesis 315.
—, 12-deoxy-, NMR data 284.
—, 11-hydroxy-dehydro-, synthesis 313.
Rotenoid glycoside 316.
Rotenol, diastereoisomers 288.
—, 12aα- and 12aβ-, structure 287.
—, 6a,12a-dehydro-, → spiro-compounds 289.
Rotenolones, A, B, C and D series 290, 294.
Rotenolones I and II, diastereoisomers 293.
cis-B/C-Rotenolone diastereoisomers → diols 294.
trans-Rotenolone diastereoisomers → separable pairs of 12,12a-diols 293, 294.
Rotenone, abs. configuration 279, 280, 281.
—, and alkali 286.
—, biol. action, toxicity 276.
—, 6a center, correlation with (—)-glyceric acid 281.
—, Cotton effect 295.
—, degradation products 285.

Rotenone, → dehydro-netoric acid 288.
—, → 6 a,12 a-dehydrorotenone 284.
—, dichotomous methylation 301.
—, dichromate oxidation 293.
—, epimerization 286.
—, fusion of *B/C* rings 281.
—, hydrogenation 289.
—, → (+)-3-hydroxy-4-methylpentanoic acid 279, 280.
—, → isorotenone 281.
—, → (—)-2-methyl-pentan-3-ol 280.
—, numbering 277.
—, occurrence 276.
—, in *Phaseolea* 309.
—, racemization 286.
—, → rotenol and derritol 287.
—, and sodium 287.
—, spiro-isomerization product 289.
—, stereochemistry 279, 295.
—, → tubaic acid → dihydrotubaic acid 279.
6 a α,12 a α,5′β-Rotenone 286.
(—)-Rotenone, natural 314.
—, —, = /6/-rotenone on the *R/S* symbolism = /7/-rotenone on the α/β-system 284.
—, —, = 6 a β,12 a β,4′,5′-tetrahydro-2,3-dimethoxy-5′β-isopropenylfurano-(3′,2′:8,9)-6 H-rotoxen-12-one 279.
—, —, oxygenation to diastereoisomeric rotenolones I and II 293.
Rotenone, 6 a,12 a-dehydro- 285, 288.
—, —, → derrisic acid 284.
—, dehydro-, → dihydro-rotenone 314.
—, —, synthesis 313.
—, dehydro-dihydro-, synthesis 313.
—, dehydro-tetrahydro-, synthesis 313.
—, 12-deoxy-skeleton, synthesis 315.
—, 6′,7′-dihydro- 289.
—, β-dihydro- 282.
—, —, = 4′,5′-dihydrodeguelin 289, 296.
—, —, NMR data 283, 284.
—, —, synthesis 312.
—, 6 a α- and 6 a β-enol acetate 281.
—, —, dihydro- 281.
—, 8′-hydroxy-glycoside 316.
—, oxime and iso-oxime 288.
Rotenone series, parent compound, synthesis 313.
Rotenonic acid, cyclization 296.
—, dehydro-dihydro-, synthesis 313.
—, dihydro-, synthesis 312.

Rotenonone, structure 285.
—, synthesis 313.
—, 11-hydroxy-, synthesis 313.
β-Rotenonone → rotenonone 286.
—, structure 285.
Rotenonone class, synthesis from 3,7-dimethoxy-coumarin-4-carboxylic acid 316.
Rotenononic acid, structure 285.
Rotoxen, numbering 279.
6 H-Rotoxen-12-one, 6 a,12 a-dihydro- 277, 314.
—, 6 a β,12 a β,4′,5′-tetrahydro-2,3-dimethoxy-5′β-isopropenylfurano-(3′,2′:8,9)- 279.
Rubber, acetate as precursor 4, 5.
—, all-*cis*-polyterpene 2.
—, biosynthesis 1, 4, 11.
—, —, enzymes 11.
—, —, inhibition by high rubber concentration 12.
—, chemical degradation 8.
—, distribution 1.
—, enzymatic synthesis 8.
—, from labeled mevalonic acid, by latex enzymes 8.
—, → levulinic acid 8.
—, precursors 5.
—, radioactive 4.
—, structure 2.
Russian dandelion, rubber 2.
Rutilantine 123, 176.
—, → ε-Pyrromycinon (= Rutilantinon) 176.
—, Rutilantinon 150, 176.
—, = ε-Pyrromycinon 150.
—, bisanhydro-, → Benzol-tetracarbonsäure-(1,2,3,4) 150.
—, —, = η-Pyrromycinon 150.
—, —, Spektrum 150.
—, —, Struktur 150.

Salmonella typhosa, und Cinerubine 175.
Sarkom und Cinerubine 176.
Sarkosin, Abbau 233.
—, → Serin (biol.) 233.
Schweineniere, Konjugase-ähnliches Enzym 190.
Sepiapterin → Folsäure 193.
Sericetin, structure 298.
Serin, Biosynthese, und Folsäure-Coenzyme 185.
—, als Einkohlenstoff-Donator 232.

Serin, → Glycin und Formaldehyd (biol.) 232, 236.
—, aus Sarkosin 233.
Serin-Aldolase 234, 236.
—, aus *Cl. sticklandii* 238.
—, = Serin-Desmolase = Serin-Trans-hydroxymethylase 237.
Serin-Desmolase 234.
Serin-Hydroxymethylase, Reaktions-mechanismus 234, 238.
L-Serin: Tetrahydrofolsäure-5,10-Hydr-oxymethyl-Transferase 234.
Sistomycosin 58.
—, spectrum 56.
Spinatblätter, Folsäure 189.
Staphylococcus aureus, Hemmung durch Aklavin 179.
—, Hemmung durch Rhodomycin A, B, Iso-rhodomycin A 177, 179.
Stoß-Zug-Reaktion 238.
Streptococcus, Hemmung durch Rhodo-mycin A 177.
Str. faecalis, Hemmung durch Cinerubine 176.
—, Hemmung durch Folat-Analogen 221.
—, Wuchsstoff 188, 189, 190.
Str. mitis, Hemmung durch Cinerubine 176.
Str. pyogenes, Hemmung durch Cinerubine 176.
Streptomyces, polyene antifungal anti-biotics 18, 19, 54.
—, —, absence of acetylenic bonds 54.
—, —, biosynthesis, interpolation of pro-pionate units 54.
—, polyene macrolides 54.
Streptomyces spp., anthracycline, anthra-cyclinone 122.
—, 7-Desoxy-aklavinon 154, 155.
—, ε- und ζ-Pyrromycinon 151.
Str. abikoensis, heptaene 68.
Str. antibioticus, Pyrromycinone und Glykoside 153.
Str. aureofaciens, Aureomycin 80.
—, und Br⁻ → 7-Brom-tetracyclin 114.
—, 7-Brom-tetracyclin 95.
—, und Cl⁻ → Aureomycin 114.
—, (+)-*trans*-2,3-Dihydro-3-hydroxy-anthranilsäure 115.
—, Mutanten, Umwandlung von Tetra-cyclin-Abbauprodukten in Tetra-cycline 114.

Str. aureofaciens, und Sulfaguanidin → 6-Desmethyl-7-chlor-tetracyclin 114.
—, Verwandlung von Anhydro-tetra-cyclinen in Tetracycline 114.
Str. DOA 1205, Pyrromycinone und Glykoside 123.
Str. galileus, Pyrromycinone und Glyko-side 123.
Str. gilvosporeus, tetraene, properties 60.
Str. noursei, nystatin, cycloheximide 19.
Str. niveoruber, Pyrromycinone und Glykoside 123.
Str. purpurascens, Rhodomycin A, Iso-rhodomycin B 176, 179.
—, Rhodomycinone, Rhodomycine, Iso-rhodomycinone, Iso-rhodomycine 123.
—, ε-, ζ-, γ- und δ-Rhodomycinon 140, 141, 145.
Str. rimosus, Mutante, Umwandlung von Tetracyclin-Abbauprodukten in Tetra-cycline 114.
—, radioaktives Terramycin 113.
—, rimocidin, oxytetracycline 19.
—, Terramycin 82.
Sumatrol, angular *D/E* rings 302.
—, from *Derris malaccensis* 301.
—, stereochemistry 302.
—, structure 277, 278.
—, 6a,12a-dehydro- 302.
—, dehydro-tetrahydro-, synthesis 313.
—, tetrahydro-, synthesis 312.

TAOP (Abkürzung für 2,4,5-Triamino-6-hydroxy-pyrimidin) 203.
Taraxicum kok saghyz, rubber 2.
Tennecetin 60.
—, = pimaricin 18.
Tephrosia spp., fish poison 275, 276.
T. toxicaria, α-toxicarol, deguelin 296, 299.
T. vogelii, deguelin, tephrosin 296.
Tephrosic acid, synthesis 312.
Tephrosin, artefact 297.
—, → dehydrodeguelin 297.
—, occurrence 296.
—, structure 297.
—, deriv. 297.
—, mono- and dicarboxylic acid 297.
Terracinsäure 83, 84.
—, Entstehung aus Terramycin 85.
—, iso-descarboxy- 86.
Terramycin 81, 82.
—, Abbau mit Methyljodid 101.

Terramycin, alkalischer Abbau 83.
—, Chromophor 87.
—, → 6-Desoxy-6-epi-terramycin 102.
—, = 5-Hydroxy-tetracyclin 81.
—, IR-Spektrum 83.
—, Konfiguration 82.
—, reduktiver Abbau 90.
—, → Salicylsäure und 3-Hydroxy-benzoesäure 83.
—, saurer Abbau 87.
—, aus *Str. rimosus* 82, 113.
—, Struktur 81, 90.
—, Teilformel 83, 85, 86.
—, Unterschied von Aureomycin 91.
—, anhydro-, → α- und β-Apo-terramycin 87.
—, —, Teilformel 87, 89.
—, —, → Terramycin (biol.) 114.
—, —, Spektrum 87.
—, α- und β-apo-, → 2,5-Dihydroxy-benzochinon 89.
—, —, pK$_a$-Werte 89.
—, —, Spektrum 87.
—, —, Teilformel 87, 89.
—, —, → Terranaphthoesäure 88.
—, —, → Terrinolid 89.
—, Benzolsulfonyl-nitril (IR-Spektrum) 83.
—, desdimethylamino-, Spektrum 90.
—, —, Teilformel 90, 91.
—, desdimethylamino - 12 a - desoxy- 90, 92.
—, —, → desdimethylamino-terrarubein 91.
—, —, Spektrum 90.
—, 6-desoxy-, Struktur 107, 108.
—, hydrochlorid, pK-Werte 98.
—, jodmethylat 102.
—, 6-methylen-, Addition von Benzyl-mercaptan 107.
Terranaphthoesäure, Struktur 84.
Terranaphthol, = 1,8-Dihydroxy-3-hydroxymethyl-4-methyl-naphthalin 84.
Terrarubein, Struktur 111.
—, desdimethylamino-, aus Desdimethyl-amino-desoxy-terramycin 91.
—, —, Struktur 90.
Terrinolid 88.
—, pK$_a$-Werte 89.
—, Teilformel 88, 89.
—, descarboxamido- 88, 89, 97.
—, —, pK$_a$-Werte 89.

Terrinolid, descarboxamido-, penta-methyläther, und Salpetersäure 88.
Tetracenchinon-(5,12), 6,11-diacetoxy-, reduzierend acetyliertes, Spektrum 147.
—, 1,11-dihydroxy- 154.
—, hydroxy-, Deriv. 137.
—, 9-hydroxy-9-äthyl- 128.
—, 8-hydroxy-8-äthyl-7,8,9,10-tetra-hydro- 127, 128.
—, hydroxy-7,8,9,10-tetrahydro- 133, 134.
—, 1,4,6,10,11-pentahydroxy- 140.
—, 7,8,9,10-tetrahydro- 122, 127.
—, 1,4,6,11-tetrahydroxy- 138.
—, 1,4,6-triacetoxy-, Spektrum 148.
—, 1,4,6-trihydroxy-, Spektrum 148.
—, 1,4,6-trihydroxy-7,8,9,10-tetrahydro-, Carbomethoxyäthyl-deriv. = ζ-Pyrromycinon 151.
—, 1,6,11-trihydroxy-, Spektrum 148.
—, trihydroxy-, Deriv. 134.
Tetracycline 80.
—, Abbauprodukte → Tetracycline (biol.) 114.
—, → Anhydro-tetracyclin 102.
—, aus Aureomycin 91.
—, Bezifferung 81.
—, Biogenese 54, 113.
—, biol. Vorstufen 115.
—, → 6-Desoxy-6-epi-tetracyclin 102.
—, → 12 a-Desoxy-4-epi-tetracyclin 110.
—, → 11a-Fluor-tetracyclin-6,12-hemi-ketal 106.
—, Hydrierung 102.
—, Isomerisierung an C$_{(4)}$ 101.
—, Konfiguration 81.
—, Reaktionen an C$_{(2)}$ 99.
—, Reaktionen an C$_{(4)}$ 100.
—, Reaktionen an C$_{(6)}$ 102.
—, Reaktionen an C$_{(11a)}$ und C$_{(12a)}$ 108.
—, Röntgenstruktur-Analyse 81.
—, und sec. Amine und Formaldehyd 99.
—, Struktur 81, 92.
—, synthetische Versuche 115.
—, und 9-Xanthenol 99.
Tetracyclin, 7-acetoxy-6-desmethyl-6-desoxy- 104.
—, 2-acetyl-2-descarboxamido-5-hydroxy- 82.
—, —, → Descarboxamido-terrinolid 97.
—, —, → Terracinsäure 97.
—, —, UV- und IR-Spektrum 97, 98.

Tetracyclin, anhydro- 93, 94.
—, —, → Terramycin (biol.) 114.
—, —, → Tetracyclin (biol.) 114.
—, —, totalsynthetische Versuche 112.
—, —, Vorstufe von Tetracyclinen 115.
—, anhydro-desmethyl- 96.
—, —, → Desmethyl-tetracyclin (biol.) 114.
—, 7-brom- 82, 95.
—, 7-brom-6-desmethyl- 105.
—, 7-chlor-2-descarboxamido-2-acetyl- 82, 98.
—, 7-chlor-6-desmethyl-6-desoxy- 103, 104, 105.
—, 9-chlor-desdimethylamino-anhydro- 109.
—, 11a-chlor-6-desoxy-6-desmethyl- 105, 106.
—, 4a,12a-dehydro-, aus 12a-O-Formyl-tetracyclin 111.
—, 4a,12a-dehydro-desdimethylamino-12a-desoxy- 109, 110.
—, 4a,12a-dehydro-6-desmethyl-6-desoxy- 112.
—, 5a,11a-dehydro-, biol. Vorstufe von Tetracyclinen 115.
—, 5a,11a-dehydro-7-chlor- 82, 96, 108.
—, —, → Tetracyclin 96.
—, descarboxamido-2-acetyl- 82, 97, 98.
—, desdimethylamino- 101.
—, —, → Desdimethylamino-11a-brom-tetracyclin 108.
—, desdimethylamino-anhydro- 109, 113.
—, desdimethylamino-11a-brom- 108, 109.
—, desdimethylamino-9-brom-anhydro- 108.
—, desdimethylamino-6-desmethyl-12a-desoxy-7-chlor- 96.
—, desdimethylamino-desmethyl-12a-desoxy-7-chlor-anhydro-, Synthese 115, 116.
—, desdimethylamino-12a-desoxy- 109.
—, —, → Desdimethylamino-12a-desoxy-11a,12a-dibrom-tetracyclin 109, 110.
—, desdimethylamino-12a-desoxy-11a,12a-dibrom-, Struktur 109.
—, desdimethylamino-6,12a-didesoxy-6-desmethyl-7-chlor-, Synthese 115, 116.
—, 6-desmethyl- 82, 95, 96.
—, —, → 6-Desmethyl-6-desoxy-tetracyclin 102.

Tetracyclin, 6-desmethyl- (7-H[3]) 103.
—, 6-desmethyl-7-chlor- 82, 96, 105.
—, —, → β-(4-Chlor-7-hydroxy-phthalid-3)-glutarsäure 96.
—, 6-desmethyl-7-chlor-anhydro- 96.
—, —, → 7-Chlor-desmethyl-tetracyclin (biol.) 114.
—, —, → 12a-O-Formyl-tetracyclin 111.
—, 6-desmethyl-6,12a-didesoxy- 111, 112.
—, 6-desoxy- 106, 107.
—, —, partial- synthetische Deriv. 103, 104.
—, 12a-desoxy-anhydro- 110.
—, —, → 12a-Desoxy-tetracyclin (biol.) 114.
—, —, keine biol. Vorstufe für Tetracyclin 115.
—, 6-desoxy-6-desmethyl- 103, 104, 106.
—, —, → 7-Nitro-6-desoxy-6-desmethyl-tetracyclin 103.
—, —, synthetisch, biol. Aktivität 116.
—, 7-fluor-6-desmethyl-6-desoxy- 104, 105.
—, 11a-fluor- 106.
—, 11a-fluor-6-methylen-, → 6-Desoxy-tetracyclin 108.
—, 12a-O-formyl-, → 2a,12a-Dehydro-tetracyclin 111.
—, —, → 6-Desmethyl-6,12a-didesoxy-tetracyclin 111.
—, —, → 12a-Desoxy-4-epi-tetracyclin 110.
—, —, Struktur 110.
—, 7-formyloxy-6-desmethyl-6-desoxy- 104.
—, 6,12-hemiketal, 11a-chlor- 105.
—, —, 11a-fluor- 106, 107.
—, hydrochlorid, pK-Werte 98.
—, 7-hydroxy-6-desmethyl-6-desoxy- 104.
—, 5-hydroxy-6-methylen-1a-chlor- 105.
—, 5-hydroxy-6-methylen-7-chlor- 106.
—, 5-hydroxy-6-methylen-7,11a-dichlor- 106.
—, 7-jod-6-desmethyl-6-desoxy- 105.
—, 7-jod[131]-6-desmethyl-6-desoxy- 105.
—, 6-methylen- 105, 107.
—, —, Addition von Benzylmercaptan 107.
—, —, 11a-chlor- 105.
—, —, 11a-fluor- 106, 107.
—, monomethyläther, desdimethylamino-anhydro- 112.

Tetracyclin, nitril, anhydro-, keine biol. Vorstufe für Tetracyclin 115.
—, —, —, → Tetracyclin-nitril (biol.) 114.
—, 7- und 9-nitro-6-desoxy-6-desmethyl- 103.
—, oxy- (from *Streptomyces*) 19.
—, N-pyrrolidinomethyl- 99.
1,3,5,7,9,11,13-Tetradecaheptaene, spectrum 57.
2,4,6,8,10,12-Tetradecahexaene, spectrum 57.
Tetradecapentaenal, 13-hydroxy- 31.
Tetraenes (polyene antibiotics) 43, 58.
—, spectrum 20, 23, 56.
—, 7071-RP 58.
—, PA-166 20, 29, 60.
—, —, spectrum 56.
—, from *Str. gilvosporeus* 60.
Tetrahydrofolat-Formylase 234, 249, 250.
Tetrahydrofolsäure: NAD(P)-Oxydo-reduktase 234.
Tetralon, 2-acetyl-8-hydroxy- 86, 87.
—, —, Spektrum 98.
Tetrin 60.
—, spectrum 56.
Thiamin-pyrophosphat, hydroxymethyl-, Kondensations-Mechanismus 219.
Thymidylat-Synthese (biol.) 240, 242.
Thymidylat-Synthetase 206, 207, 222, 234, 241.
—, aus *Str. faecalis* 251.
Thymidylsäure-Bildung und Vitamin B_{12} 241.
Thymin, Biosynthese, und Folsäure-Coenzyme 185.
—, Herkunft der Methylgruppe 240.
Tiglic aldehyde, from nystatin 44.
Timbo (insecticide) 276.
TMP (Abkürzung für Thymidin-mono-phosphat) 184.
Toxicaric acid 301.
—, structure 288.
—, synthesis 309.
Toxicarol, methylation 301.
—, dehydro-, → (±)-dihydrotoxicarol 314.
—, —, → derric acid 300.
—, 6a,12a-dehydro-, → rissic acid 300.
—, 11-methyl ether, 4′,5′-dihydro-6a,12a-dehydro-, cleavage with alkali 300.
α-Toxicarol, → apo-toxicarol 301.
—, structure 277, 278, 299.
—, 4′,5′-dihydro-6a,12a-dehydro- 301.

(—)-α-Toxicarol, from derris resin 299.
—, = proto-toxicarol 299.
(±)-α-Toxicarol, from *Tephrosia toxicaria* and *Derris* 299.
β-Toxicarol, linear D/E fusion 299.
—, linear isomer of α-toxicarol 301.
—, structure 299.
—, 4′,5′-dihydro-6a,12a-dehydro- 301.
(±)-β-Toxicarol, from α-Toxicarol 301.
Toxicarolic acid, O-methyl-dihydro- 300.
TPN (Abkürzung für Triphosphopyridin-nucleotid) 184.
TPNH (abbreviation for reduced tri-phosphopyridine-nucleotide) 7.
Transformylase 251, 252.
Transformylierung, in Biosynthese hetero-cyclischer Körper 251.
Transformylierungs-Cofaktor (Tauben-leber) 250.
Transformylierungs-Cyclen 246, 252.
Transhydroxy-methylierungs-Cyclus 236, 239, 240.
—, (Tetrahydro-folsäure und Formalde-hyd) 218.
Triazine, 2,4-diamino-, als Hemmstoffe 221.
Trichomycin 19, 46, 50.
—, alkaline fission 47.
—, → p-amino-acetophenone 48.
—, → β-diketone 48.
—, → N-dinitrophenyl ethylmycos-aminide 47.
—, medical use 18.
—, oxidative fission 49.
—, partial structures 49.
—, → polyene aldehyde (alkaline fission) 47.
—, retro-aldol cleavage 48.
—, spectrum 25, 26.
—, perhydro-, and periodic acid 48.
all-*trans*-Trichomycin, formation 25.
Trichomycin A 46, 66.
—, → p-amino-acetophenone 30.
—, → octaenal 31.
—, spectrum 25, 57.
—, perhydro- 29.
Triphosphopyridin-nucleotid, dihydro-, und enzymat. Reduktion von Folsäure 198.
Tritriacontane, 7,21-dimethyl- 35.
Tuba (insecticide) 276.

Tubaic acid, structure 279.
—, synthesis 310.
—, dihydro-, structure 279.
—, —, → (—)-5-isopropyloxazolidone 279.
—, tetrahydro-, synthesis 310.
Tubatoxin 276.

UMP (Abkürzung für Uridin-mono-
    phosphat) 184.
Unamycin 46, 60.
—, spectrum 56.
Undecanedioic acid, α-methyl- 37.
Uridin-nucleotid, desoxy-, → Thymidyl-
    säure 240.

Valeriansäure, α-keto-, phosphoroklasti-
    sche Spaltung 230.
Valin, Biosynthese, und Folsäure-Co-
    enzyme 185.
Violarin 179.
Vitamin B_{12}, Beteiligung an der Methio-
    nin-Biosynthese 244, 246.
—, und Folsäure, Zusammenhänge 244.

Vitamin B_{12}-Mangel und Folsäure 191.
Vitamin B_c 188, 189, 199.
—, Konjugat 188.
Vitamin M, Heilwirkung 187, 188.
Vitamine R, S, U, aus Hefe 187.

Xanthin (biol. Abbau des Pterins aus
    Folsäure) 196, 197.
—, (Einkohlenstoff-Reservoir) 230.
—, 8-carbonsäure (biol. Abbau von Fol-
    säure) 196, 197.
Xanthinoxydase 196, 197.
Xanthopterin → 7,8-Dihydroxanthopterin
    205.

Yam bean, pachyrrhizone 303.
Yeast, polymerization of Δ^3-isopentenyl
    pyrophosphate 10, 11.

Zirkulardichroismus-Kurven, Anthra-
    cyclinone 160, 161.
—, und Konfiguration, Asymmetrie-
    zentren 160.

Manzsche Buchdruckerei, Wien IX.